Biofuels and Biorefineries

Volume 12

Series Editor

Zhen Fang, Biomass Group, Nanjing Agricultural University, Nanjing, Jiangsu, China

Annual global biomass production is about 220 billion dry tons or 4,500 EJ, equivalent to 8.1 times the world's energy consumption in 2020 (556 EJ). On the other hand, world-proven oil reserves at the end of 2020 reached 1732.4 billion barrels, which can only meet just over 51.5 years of global production. Therefore, alternative resources are needed to both supplement and replace fossil oils as the raw material for transportation fuels, chemicals and materials in petroleum-based industries. Renewable biomass is a likely candidate, because it is prevalent over the Earth and is readily converted to other products. Compared with coal, some of the advantages of biomass are: (i) its carbon-neutral and sustainable nature when properly managed; (ii) its reactivity in biological conversion processes; (iii) its potential to produce bio-oil (ca. yields of 75%) by fast pyrolysis because of its high oxygen content; (iv) its low sulfur and lack of undesirable contaminants (e.g. metals, nitrogen content); (v) its wide geographical distribution and (vi) its potential for creating jobs and industries in energy crop productions and conversion plants. Many researchers, governments, research institutions and industries are developing projects for converting biomass including forest woody and herbaceous biomass into chemicals, biofuels and materials and the race is on for creating new "biorefinery" processes needed for future economies. The development of biorefineries will create remarkable opportunities for the forestry sector, biotechnology, materials, chemical processing industry, and stimulate advances in agriculture. The Biofuels and Biorefineries Series strives to highlight methods and technologies that will help industries to shift to renewable and carbon-neutral resources which will lead to the creation of sustainable society.

Zhen Fang • Richard Lee Smith Jr • Lujiang Xu
Editors

Production of N-containing Chemicals and Materials from Biomass

Springer

Editors
Zhen Fang
College of Engineering
Nanjing Agricultural University
Nanjing, Jiangsu, China

Richard Lee Smith Jr
Graduate School of Environmental Studies
Tohoku University
Sendai, Miyagi, Japan

Lujiang Xu
Biomass Group
Nanjing Agricultural University
Nanjing, China

ISSN 2214-1537 ISSN 2214-1545 (electronic)
Biofuels and Biorefineries
ISBN 978-981-99-4579-5 ISBN 978-981-99-4580-1 (eBook)
https://doi.org/10.1007/978-981-99-4580-1

This Springer imprint is published by the registered company Springer Nature Singapore Pte Ltd.
The registered company address is: 152 Beach Road, #21-01/04 Gateway East, Singapore 189721, Singapore

Preface

N-containing products, especially N-containing chemicals, and N-doped functional carbon materials have been used in a wide range of applications, including life, medicine, agriculture, environment, and energy industries. Biomass obtained from waste material or waste agriculture that is not used for food or feed consists of lignocelluloses, alga, and lipids. Due to the unique structure and composition of biomass, these bioresources have unique features in the synthesis of nitrogen-containing chemicals and nitrogen-containing functional carbon materials when compared with petrochemical resources. Therefore, the substitution of traditional fossil resources with waste biomass resources as sustainable feedstocks is being intensively investigated for the manufacture of high value-added N-containing chemicals and functional N-doped carbon materials that could be directly applied for promoting the pharmaceutical and specialty chemical manufacturing processes. This book provides a compilation of state-of-the-art techniques related to sustainable synthesis of nitrogen-containing compounds from biomass, including hydrolysis of nitrile compounds, reductive amination of bio-based molecules, depolymerization of lignocellulosic biomass, production of organic nitrogen compounds from chitin, preparation of nitriles from biomass, and synthesis of nitrogen-doped carbon materials. The book also explores the applications of microbial host metabolic engineering, biochar engineering for CO_2 capture, and nanocrystalline cellulose-derived thermosets. Throughout the book, there is a focus on green chemical processes and the use of renewable resources for sustainable production. Aimed at improving conversion effectiveness and developing innovative techniques for value-added N-containing chemicals and materials. This book was conceived as a collection of studies on state-of-the-art techniques and developed specifically for producing N-containing chemicals and materials from biomass via sustainable recycling methods. Discussion on related topics in terms of recent advances and their assessment and the promise and prospects of new methods or new technological strategies are provided to readers in a concise and informative format. Each individual chapter was contributed by globally selected experts or professionals and was peer-reviewed and edited for content and consistency in terminology.

This book is the twelfth book of the series entitled, *Biofuels and Biorefineries*, and contains 12 chapters contributed by leading experts in the field. The text is arranged into five key parts:

1. Part I. Production of N-Containing Compounds by Chemical Catalytic Processes (Chaps. 1, 2, 3, 4, 5, and 6)
2. Part II. Production of N-Containing Compounds via Biological Processes (Chap. 7)
3. Part III. Application of N-Containing Biomass to Manufacture of Chemicals and Materials (Chaps. 8, 9, 10, and 11)
4. Part IV. N Transformations During Thermal Processes (Chap. 12)

Chapter 1 introduces the properties of near-critical water and its important applications in green chemical processes, reviews progress in the hydrolysis of nitrile compounds in NCW and discusses their hydrolysis path, kinetics, mechanisms, and compares the reactivity of functional groups on their hydrolysis rate. Chapter 2 focuses on the reductive amination of the main platform molecules namely furfural, 5-hydroxymethyl furfural, and levulinic acid to bio-based amines and pyrrolidones, and discusses different heterogeneous catalytic systems using different noble and non-noble metals for this process with essential findings and fundamentals of operation of these catalysts. Chapter 3 first introduces the key intermediates of lignocellulosic biomass depolymerization, focuses on the synthesis of five-membered N-heterocyclic compounds and hexa-membered N-heterocyclic compounds by bio-based platform molecules, and reviews the synthesis of other biomass-based compounds such as quinoline and quinoxaline. Chapter 4 focuses on the sustainable production of organic nitrogen compounds from the concept of an end-shell biorefinery, including the depolymerization/deacetylation of chitin in aqueous media or organic solvents/co-solvents to acetylglucosamine or glucosamine, as well as the conversion to important platform chemicals such as amino acids and furans through oxidation and dehydration, and introduces and validates the conversion strategies and reaction pathways of chitin (such as hydrothermal method, hydrogenation). Chapter 5 focuses on the preparation of nitriles from biomass and biomass-derived compounds through sustainable synthesis methods, provides a systematic summary of substrates (including nitrogen-containing components, non-nitrogenous components, and raw biomass) and reaction mechanisms and typical reaction pathways that can be used to prepare bio-based nitriles, and provides key challenges and perspectives to stimulate discussion among researchers and industry. Chapter 6 describes the advantages of the one-pot method of reducing amination using bio-based ketoacids and hydrogen supply sources, discusses the mechanisms and reaction parameters involved (e.g., reaction simulation temperature, pressure, and reactor type), emphasizes the functional requirements of catalysts, discusses in detail the effects of catalyst composition and properties on its activity and reaction pathway, and looks forward to the development of promising biomass conversion and reduction systems. Chapter 7 summarizes the research progress and relevant applications of using microbial host metabolic engineering to produce nitrogen-containing chemical substances, discusses the significant impact of genetic engineering technology on metabolic engineering and the production of valuable nitrogen-containing chemical substances, and details the discussion on these chemical substances applied as feed additives, biological plastics precursors, or molecules with pharmacological activities. Chapter 8 provides an overview of biochar engineering for CO_2 capture, with a focus on the impact of various modification processes on the properties of biochar materials (such as specific surface area, pore volume, pore size, multi-level pore structure, and surface chemistry) and their relationship with CO_2 absorption. The chapter also highlights the key factors affecting the texture, porosity, aromatization, and hydrophobicity of biochar, including raw material type, pyrolysis chemical conditions, and surface chemical modification of functional groups. Chapter 9 provides a comprehensive summary of the recent advances in sustainable high-performance thermosets derived from nanocrystalline cellulose (NCC). It discusses the sources and acquisition of bio-based raw materials for NCCs used in thermosets, reviews the synthesis and structural properties of N-containing chemicals in typical bio-based thermosets such as epoxy resins, benzoxazines, and polyurethanes. Chapter 10 aims at covering the main aspects related to the synthesis of nitrogen-doped carbon materials from lignocellulosic biomass residues together with some representative examples of their use in energy-related applications. Chapter 11 comprehensively reviews preparation methods for green N-methyl biochar materials and one-step "pyrolysis-activation-doping" via pyrolysis, and discusses the formation mechanism of N-doped biochars and their corresponding catalytic properties, as well as the need for advanced online characterization methods and detection of the structural evolution of intermediates and radicals during reaction processes. Chapter 12 investigates the impact of different stages of meat and bone meal (MBM) gasification on the final fuel nitrogen distribution and NH_3 generation. It reveals the correlations between temperature, residence time, and steam use with NH_3 production in the pyrolysis and char conversion stages.

The text should be of interest to professionals in academia and industry who are working in the fields of natural renewable materials, biorefinery of lignocellulose, biofuels and environmental engineering. It can also be used as comprehensive references for university students with backgrounds in chemical engineering, material science, and environmental engineering.

Nanjing, China Zhen Fang
Sendai, Japan Richard L. Smith Jr
Nanjing, China Lujiang Xu

Acknowledgments

First and foremost, we would like to cordially thank all the contributing authors for their great efforts in writing and revising the chapters and insuring the reliability of the information given in their chapters. Their contributions have really made this project realizable.

Apart from the efforts of authors, we would also like to acknowledge the individuals listed below for carefully reading the book chapters and giving many constructive comments that significantly improved the quality of the chapters:

Prof. Ken-Ichi Fujita, Kyoto University, Japan

Dr. Atsushi Gabe, National Institute of Technology, Kurume College, Japan

Dr. Lucía García, Universidad de Zaragoza, Spain

Dr. Haixin Guo, Chinese Academy of Agricultural Sciences (Tianjin), China

Dr. Jian He, Jishou University, China

Prof. Liang-Nian He, Nankai University, China

Dr. Yaobing Huang, North China Electric Power University, China

Prof. Sara Iborra, Universitat Politècnica de Valencia-Consejo Superior de Investigaciones Científicas (UPV-CSIC), Spain

Prof. Mingjie Jin, Nanjing University of Science and Technology, China

Dr. Peter Kornatz, DBFZ (the German Centre for Biomass Research), Germany

Dr. Sha Li, Nanjing University of Technology, China

Dr. Yuxuan Liu, Chinese Academy of Sciences (Dalian), China

Prof. Qiang Lu, North China Electric Power University, China

Prof. Songqi Ma, Chinese Academy of Sciences (Ningbo), China

Prof. Tiancheng Mu, Renmin University of China, China

Dr Yoshinao Nakagawa, Tohoku University, Japan

Dr. Sonil Nanda, University of Saskatchewan, Canada

Prof. Dr. mont. Michael Nelles, University of Rostock, Germany

Dr Mitsumasa Osada, Shinshu University, Japan

Dr. Hu Pan, Jiaxing University, China

Prof. Likun Pan, East China Normal University, China

Prof. Zhiyan Pan, Zhejiang University of Technology, China

Prof. Alirio Rodrigues, University of Porto, Portugal
Prof. Takehiko Sasaki, The University of Tokyo, Japan
Prof. Takafumi Sato, Utsunomiya University, Japan
Prof. Feng Shen, Chinese Academy of Agricultural Sciences (Tianjin), China
Prof. Zhuohua Sun, Beijing Forestry University, China
Prof. Shurong Wang, Zhejiang University, China
Dr. Lujiang Xu, Nanjing Agricultural University, China

We are also grateful to Dr. Mei Hann Lee (Senior Editor, SpringerNature) for her encouragement, assistance, and guidance during preparation of the book.

We would like to express our deepest gratitude towards our families for their love, understanding, and encouragement, which helped us to complete this project. Especially, ZF would like to dedicate this book to his father (Mr. Hong Fang), who passed away on January 2, 2023 in Taining, Fujian, for all of his love and support even during his final illness.

Zhen FANG, April 19, 2023 in Nanjing
Richard L. Smith, Jr., April 19, 2023 in Sendai
Lujiang Xu, April 19, 2023 in Nanjing

About the Book

N-containing products, especially N-containing chemicals and N-doped functional carbon materials, have been used in a wide range of applications including life, medicine, agriculture, environment, and energy industries. Biomass obtained from waste material or waste agriculture that is not used for food or feed consists of lignocelluloses, alga, and lipids. Due to the unique structure and composition of biomass, these bioresources have unique features in the synthesis of nitrogen-containing chemicals and nitrogen-containing functional carbon materials when compared with petrochemical resources. This book provides a compilation of state-of-the-art techniques related to sustainable synthesis of nitrogen-containing compounds from biomass, including hydrolysis of nitrile compounds, reductive amination of bio-based molecules, depolymerization of lignocellulosic biomass, production of organic nitrogen compounds from chitin, preparation of nitriles from biomass, and synthesis of nitrogen-doped carbon materials. This text provides a compilation of state-of-the-art techniques for producing N-containing chemicals and materials from biomass via sustainable recycling methods. Chapter topics include manufacture of value-added products such as N-containing chemicals and N-containing functional materials from biomass feedstocks. All chapters are contributed by respected global experts in their field to provide readers with a broad range of perspectives on cutting-edge applications. The text is an ideal reference guide for academic researchers and industrial engineers in the fields of sustainability management, environmental chemistry, environmental engineering, natural renewable materials, biorefineries, and biofuel production, and it can also be used as a comprehensive reference source for university students in chemical engineering, material science, and environmental engineering.

Dr. Zhen Fang is Professor of Energy Engineering and Leader and Founder of the Biomass Group, Nanjing Agricultural University, China.

Dr. Richard L Smith, Jr. is Professor of Chemical Engineering in the Graduate School of Environmental Studies, Tohoku University, Japan.

Dr. Lujiang Xu is Associate Professor in Biofuel and Catalyst Synthesis, Biomass Group, Nanjing Agricultural University, China.

Contents

Editors and Contributors

About the Editors

Zhen Fang is professor and leader of biomass group, College of Engineering, Nanjing Agricultural University. He is the inventor of the "fast hydrolysis" process, the elected fellow of the Canadian Academy of Engineering. Professor Fang specializes in thermal/biochemical conversion of biomass, nanocatalysts synthesis and their applications, pretreatment of biomass for biorefineries, and supercritical fluid processes. He obtained his PhDs from China Agricultural University (Beijing) and McGill University (Montreal). Professor Fang is Associate Editor of the international journals, *Biotechnology for Biofuels* and *Journal of Supercritical Fluids (2018–2020)*. He has more than 20-year international research experience at top universities and institutes around the world, including 1 year in Spain (University of Zaragoza), 3 years in Japan (Biomass Technology Research Center, AIST; Tohoku University), and more than 8 years in Canada (McGill) in renewable energy and green technologies.

Richard L. Smith Jr is professor of Chemical Engineering, Graduate School of Environmental Studies, Research Center of Supercritical Fluid Technology, Tohoku University, Japan. Professor Smith has a strong background in physical properties and separations and obtained his PhD in chemical engineering from the Georgia Institute of Technology (USA). His research focuses on developing green chemical processes especially those that use water and carbon dioxide as the solvents in their supercritical state. He has expertise in physical property measurements, separation techniques with ionic liquids, hydrothermal and solvothermal carbonization, catalysis in reaction systems and has published more than 290 scientific papers, patents, and reports in the field of chemical engineering. Professor Smith is the Asia Regional Editor for the *Journal of Supercritical Fluids* and has served on editorial boards of major international journals associated with properties and energy.

Lujiang Xu is Associate Professor in the College of Engineering, Nanjing Agricultural University. Dr. Xu obtained his PhD in Chemistry from the University of Science and Technology of China (USTC, China). His research focuses on thermochemical conversion of biomass into biofuels and biochemicals. Dr. Xu has published more than 30 research papers, reviews, and patents in thermochemical conversion-related topics and has served as peer-reviewer for major scientific journals.

Contributors

Zainab Afailal Thermochemical Processes Group, Aragon Institute for Engineering Research (I3A), Chemical and Environmental Engineering Department, University of Zaragoza, Zaragoza, Spain

Ananda S. Amarasekara Department of Chemistry, Prairie View A&M University, Prairie View, TX, USA

Rubenildo Andrade Center for Excellence in Thermal and Distributed Generation (NEST), Institute of Mechanical Engineering, Federal University of Itajubá, Itajubá, Brazil

Bhalchandra M. Bhanage Department of Chemistry, Institute of Chemical Technology, Mumbai, India

Luciana F. Brito Department of Biotechnology and Food Science, Norwegian University of Science and Technology, Trondheim, Norway

Qi Cao State Key Laboratory of Fine Chemicals, Frontiers Science Center for Smart Materials Oriented Chemical Engineering, Dalian University of Technology, Dalian, China

Diego Cazorla-Amorós Department of Inorganic Chemistry and Materials Institute, University of Alicante, Alicante, Spain

Jessica Chaparro-Garnica Department of Inorganic Chemistry and Materials Institute, University of Alicante, Alicante, Spain

Wei Chen Biomass Group, College of Engineering, Nanjing Agricultural University, Nanjing, China
State Key Laboratory of Coal Combustion, School of Power and Energy Engineering, Huazhong University of Science and Technology, Wuhan, China

Xi Chen China-UK Low-Carbon College, Shanghai Jiao Tong University, Pudong, Shanghai, China

Yingquan Chen State Key Laboratory of Coal Combustion, School of Power and Energy Engineering, Huazhong University of Science and Technology, Wuhan, China

Chengyu Dong Biomass Group, College of Engineering, Nanjing Agricultural University, Nanjing, China

Yuhan Du Krzysztof Kapusta-Główny Instytut Górnictwa (Central Mining Institute), Katowice, Poland

Peigao Duan Shaanxi Key Laboratory of Energy Chemical Process Intensification, School of Chemical Engineering and Technology, Xi'an Jiaotong University, Xi'an, Shaanxi, China

Zhen Fang Biomass Group, College of Engineering, Nanjing Agricultural University, Nanjing, China

Isabel Fonts Thermochemical Processes Group, Aragon Institute for Engineering Research (I3A), Chemical and Environmental Engineering Department, University of Zaragoza, Zaragoza, Spain

Shuai Gao Biomass Group, College of Engineering, Nanjing Agricultural University, Nanjing, China

Noemí Gil-Lalaguna Thermochemical Processes Group, Aragon Institute for Engineering Research (I3A), Chemical and Environmental Engineering Department, University of Zaragoza, Zaragoza, Spain

Tejas A. Gokhale Department of Chemistry, Institute of Chemical Technology, Mumbai, India

Jinshu Huang National Key Laboratory of Green Pesticide, Key Laboratory of Green Pesticide & Agricultural Bioengineering, Ministry of Education, State-Local Joint Laboratory for Comprehensive Utilization of Biomass, Center for R&D of Fine Chemicals, Guizhou University, Guiyang, Guizhou, China

Yiyuan Jiang National Key Laboratory of Green Pesticide, Key Laboratory of Green Pesticide & Agricultural Bioengineering, Ministry of Education, State-Local Joint Laboratory for Comprehensive Utilization of Biomass, Center for R&D of Fine Chemicals, Guizhou University, Guiyang, Guizhou, China

Fernando Léo Thermochemical Processes Group, Aragon Institute for Engineering Research (I3A), Chemical and Environmental Engineering Department, University of Zaragoza, Zaragoza, Spain
Center for Excellence in Thermal and Distributed Generation (NEST), Institute of Mechanical Engineering, Federal University of Itajubá, Itajubá, Brazil

Hu Li National Key Laboratory of Green Pesticide, Key Laboratory of Green Pesticide & Agricultural Bioengineering, Ministry of Education, State-Local Joint Laboratory for Comprehensive Utilization of Biomass, Center for R&D of Fine Chemicals, Guizhou University, Guiyang, Guizhou, China

Jiahui Li State Key Laboratory of Fine Chemicals, Frontiers Science Center for Smart Materials Oriented Chemical Engineering, Dalian University of Technology, Dalian, China

Yixuan Liu National Key Laboratory of Green Pesticide, Key Laboratory of Green Pesticide & Agricultural Bioengineering, Ministry of Education, State-Local Joint Laboratory for Comprehensive Utilization of Biomass, Center for R&D of Fine Chemicals, Guizhou University, Guiyang, Guizhou, China

Electo Lora Center for Excellence in Thermal and Distributed Generation (NEST), Institute of Mechanical Engineering, Federal University of Itajubá, Itajubá, Brazil

Emilia Morallón Department of Physical Chemistry and Materials Institute, University of Alicante, Alicante, Spain

Miriam Navlani-García Department of Inorganic Chemistry and Materials Institute, University of Alicante, Alicante, Spain

Fernando Pérez-García Department of Biotechnology and Food Science, Norwegian University of Science and Technology, Trondheim, Norway

David Salinas-Torres Department of Chemical and Environmental Engineering, Polytechnical University of Cartagena, Cartagena, Spain

Jack Shee School of Biomedical Engineering, University of Melbourne, Parkville, VIC, Australia

Richard L. Smith Jr Graduate School of Environmental Studies, Tohoku University, Sendai, Japan

Bing Song Scion, Titokorangi Drive, Rotorua, New Zealand

Yudi Wang China-UK Low-Carbon College, Shanghai Jiao Tong University, Pudong, Shanghai, China

Volker F. Wendisch Genetics of Prokaryotes, Faculty of Biology & Center for Biotechnology (CeBiTec), Bielefeld University, Bielefeld, Germany

Zhihuan Weng State Key Laboratory of Fine Chemicals, Frontiers Science Center for Smart Materials Oriented Chemical Engineering, Dalian University of Technology, Dalian, China

Geliang Xie Biomass Group, College of Engineering, Nanjing Agricultural University, Nanjing, China

Lujiang Xu Biomass Group, College of Engineering, Nanjing Agricultural University, Nanjing, China

Haiping Yang State Key Laboratory of Coal Combustion, School of Power and Energy Engineering, Huazhong University of Science and Technology, Wuhan, China

Linxin Yin Shaanxi Key Laboratory of Energy Chemical Process Intensification, School of Chemical Engineering and Technology, Xi'an Jiaotong University, Xi'an, Shaanxi, China

Huan Zhang College of Engineering, Nanjing Agricultural University, Nanjing, China

Yue Zheng China-UK Low-Carbon College, Shanghai Jiao Tong University, Pudong, Shanghai, China

Guoqiang Zhu Biomass Group, College of Engineering, Nanjing Agricultural University, Nanjing, China

Part I
Production of N-Containing Compounds by Chemical Catalytic Processes

Chapter 1
Hydrolysis of Nitrile Compounds in Near-Critical Water

Linxin Yin, Yuhan Du, Peigao Duan, and Krzysztof Kapusta

Abstract With its unique properties, near-critical water (NCW) is expected to effectively replace traditional solvents and catalysts and to lead in the development of pollution-free, selective, efficient, and fast environment-friendly chemical processes. Nitrile compounds are a commercially vital class of reactive intermediates and solvents, finding applications in petrochemical, polymers and plastics, pharmaceutical, and pesticides industries. In the synthesis of nitrile compounds, nitrile byproducts are inevitably transferred into aqueous waste streams, which increases their potential for environmental impact. Hence, it is very important to understand the hydrolysis path and mechanism of nitrile compounds in NCW for efficient and clean chemical production. This chapter reviews progress in the hydrolysis of nitrile compounds in NCW and discusses their hydrolysis path, kinetics, mechanisms, and compares the reactivity of functional groups on their hydrolysis rate.

Keywords Near-critical water · Environmental chemistry · Nitrile compounds · Hydrolysis

1.1 Introduction

To alleviate the environmental pressure and to address a series of environmental issues brought out by the chemical industry, researchers began to advocate the concept of "green chemistry" in the 1990s, with the goal to transform chemical production processes with green chemistry philosophy [1]. Currently, green chemistry research mainly focuses on the atom economy of reaction systems and the use

L. Yin · Y. Du · P. Duan (✉)
Shaanxi Key Laboratory of Energy Chemical Process Intensification, School of Chemical Engineering and Technology, Xi'an Jiaotong University, Xi'an, Shaanxi, People's Republic of China
e-mail: pgduan@xjtu.edu.cn

K. Kapusta
Główny Instytut Górnictwa, Central Mining Institute, Katowice, Poland

Z. Fang et al. (eds.), *Production of N-containing Chemicals and Materials from Biomass*, Biofuels and Biorefineries 12,
https://doi.org/10.1007/978-981-99-4580-1_1

of environmentally friendly solvents, raw materials, and catalysts, as well as the application of renewable resources and the recycling of products [2–4]. Among them, the exploration of environmentally friendly reaction media mainly includes water, carbon dioxide, room temperature ionic liquids, and other reaction media [5–7]. Water is the most commonly used chemical reaction media, which has attracted wide attention due to its characteristics such as being non-toxic, cheap, and highly abundant.

The near-critical water (NCW) state refers to the condition of compressed liquid water at temperatures generally between 150 °C and 370 °C. As an environmentally friendly reaction solvent, NCW is widely used in the fields of organic chemical reactions and waste recycling due to its unique physical and chemical properties, such as autocatalysis and temperature-adjustable properties [8–10].

Nitrile compounds are important chemical products. The hydrolysis of nitrile compounds is widely used in the synthesis of amino acids, amides, carboxylic acids, and a series of derivatives, which plays an important role in organic synthesis reactions [11]. Traditional industries use large amounts of acids or bases to catalyze the hydrolysis of nitrile compounds [12]. Although these methods are simple and efficient, they also lead to complicated processes such as neutralization and desalting which cause serious environmental pollution and other problems. As such, a series of studies have been carried out on the catalysis of nitrile compounds in NCW without additives. During the synthesis of nitrile compounds, nitrile byproducts are inevitably transferred into aqueous waste streams, which increases their potential for environmental impact. Hence, it is very important to understand the hydrolysis path and mechanism of nitrile compounds in NCW for efficient and clean chemical production.

This chapter briefly introduces the properties of near-critical water and its important applications in green chemical processes. The hydrolysis of typical nitrile compounds in NCW is reviewed. The hydrolysis path, kinetics, and mechanism of these compounds in NCW are discussed. The hydrolysis of different nitrile compounds is compared and the influencing factors of the hydrolysis of nitrile compounds are introduced.

1.2 Near-Critical Water

Changes in temperature and pressure can significantly affect the physical and chemical properties of water. The critical temperature (T_C) of water is 374.2 °C and the critical pressure (P_C) is 22.1 MPa. The density, viscosity, solubility, heat capacity, dielectric constant, and other physical properties of the fluid change dramatically near the critical point [6, 10]. According to conditions of temperature and pressure, water as reaction media can be divided into three categories: room temperature water, near-critical water (NCW, the temperature in 150–370 °C, saturation pressure ranges from 0.48 MPa to 22.0 MPa), and supercritical water (SCW, temperature above 374.2 °C, pressure above 22.1 MPa).

NCW has unique advantages over organic reaction medium: (1) NCW has a high ionization constant meaning that it can be dissociated to produce abundant H^+ and OH^-, which are conducive to promoting acid and base catalysis reactions without the addition of traditional acid or base catalysts; (2) The dielectric constant of water decreases gradually with increasing temperature, which contributes to water's ability to dissolve less polar or even nonpolar organic matter. The increase of pressure increases the density of water, which enables the NCW system to dissolve some parts of inorganic matter. This allows many organics with low solubility in ambient liquid water to undergo homogeneous reactions in NCW, which greatly improves the reaction efficiency. In particular, the insoluble organic phase can be easily separated from the water phase after cooling and depressurization; (3) Conditions of NCW can be varied over a wide range of temperatures and pressure, which results in adjustable physical and chemical properties. NCW has different solvent properties and reaction properties in different states, which can meet the needs of different experiments [6, 13]. Hence, organic chemical reactions in NCW (such as hydrolysis reaction, dehydration/hydration reaction, alkylation, and acylation reaction) have been extensively studied due to these properties [14–16].

1.3 Nitrile Compounds

Nitrile compounds, which are organic compounds containing the cyano group (-CN), are important starting materials for the synthesis of amines, acids, and polymers. The simplest nitrile is acetonitrile, which is soluble in any proportion with water, while other nitriles are generally only slightly soluble in water. Acrylonitrile is commonly used as a monomer in synthetic polymer materials, such as polyacrylonitrile fiber, nitrile rubber, and polyacrylamide [17, 18]. Malonitrile is a mediator for the preparation of many important drugs such as vitamin B1, aminobenzene, and adenine [19, 20]. Adiponitrile is an important intermediate for the synthesis of nylon. Moreover, nitrile compounds are widely used in the synthesis of agrochemicals, liquid crystal materials, dye, and incense materials [21–23]. Nitrile compounds are not only numerous in species but also widely distributed in nature. Natural nitrile is widely distributed in plants, animals, and microorganisms, but its concentration is relatively low. For example, microorganisms and plants metabolize to produce aromatic nitrile or aliphatic nitrile. Lepidium apetalum contains phenyl acetonitrile and vinyl acetonitrile exists in some plants, which serves as a nitrogen source and can protect the organism from predators. Nitrile compounds have a certain toxicity. Long-chain nitriles in general, have low toxicity or are non-toxic, the toxicity of some lower nitrile and unsaturated nitriles is larger [24]. For example, the toxicity of butyronitrile, propionitrile, and other short-chain fatty nitriles is close to that of hydrogen cyanide.

Nitrile compounds have very strong molecular polarity, which leads to their boiling point being higher than ketones, aldehydes, hydrocarbons, and other compounds that have similar molecular masses, and lower than those of corresponding carboxylic acids. Due to the unique chemical activity of the cyano group, nitrile compounds mainly undergo two types of reactions: (1) reactions on the cyano group: acid/base catalytic hydrolysis of nitrile compounds to prepare amides; (2) reactions on the α-H: condensation with carbonyl compounds.

1.4 Hydrolysis Reaction of Typical Nitriles in NCW

1.4.1 Hydrolysis Reaction in NCW

Hydrolysis is a typical chemical unit process, which mainly uses water as the reaction medium to decompose a compound into two or more simple compounds. Hydrolysis of organic compounds is a reaction in which an atom or atomic group in an organic molecule is replaced by a hydrogen atom or hydrogen and oxygen group in a molecule of water. However, many organic compounds need to add strong acids or strong bases as catalysts into NCW hydrolysis systems. In addition, the hydrolysis rate is relatively slow, and the conversion rate and yield are significantly lower at room temperature. Water plays multiple roles in a series of organic reactions in an NCW system [6]. On one hand, water acts as both reactant and reaction medium due to the increase of the ionization constant and acidic enhancement, which can reduce or even avoid the use of strong acid and base catalysts and further avoid neutralization and separation after the reaction. For example, ester and polyester can be completely hydrolyzed in the NCW without external catalysts. In addition, pH value, reaction conversion, and product composition and distribution are controlled by adjusting the reaction temperature and other conditions. The object of study of hydrolysis reactions in NCW includes nitriles, esters, ethers, glucose, fructose, xylose, and other small molecule compounds, and starch, protein, cellulose, and other macromolecular compounds [25–30]. Moreover, hydrolysis of industrial wastewater containing complex polymer materials in NCW is a means to recover valuable chemical resources. Hydrolysis of biomass in NCW can provide an alternative source of chemical feedstock.

1.4.2 Hydrolysis of Nitrile Compounds

1.4.2.1 Hydrolysis of Nitrile Compounds Catalyzed by Acid/Base Catalysts

Hydrolysis of nitriles is widely performed to prepare chemical raw materials such as carboxylic acids, amides, and amino acids, which play an important role in the organic synthesis industry. The hydrolysis of nitrile compounds at room temperature and pressure requires the addition of acids or bases catalyst [31, 32]. In general, the products of simple alkyl nitriles hydrolysis (RCN, R = CH_3-, CH_3CH_2-, and

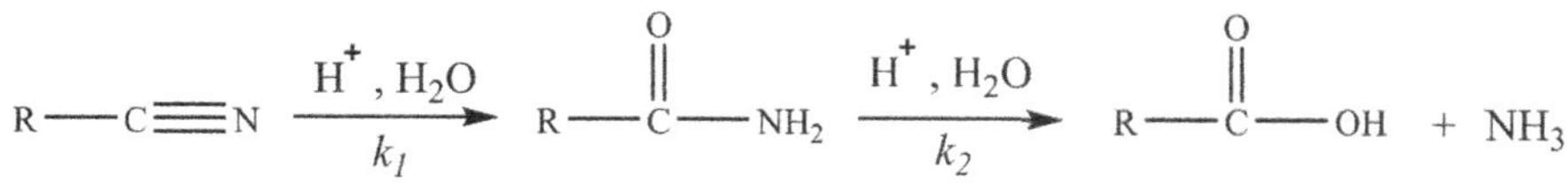

Scheme 1.1 General reaction pathway of simple alkyl nitriles (RCN, R = CH_3–, CH_3CH_2–, and $(CH_3)_2CH$–) hydrolysis

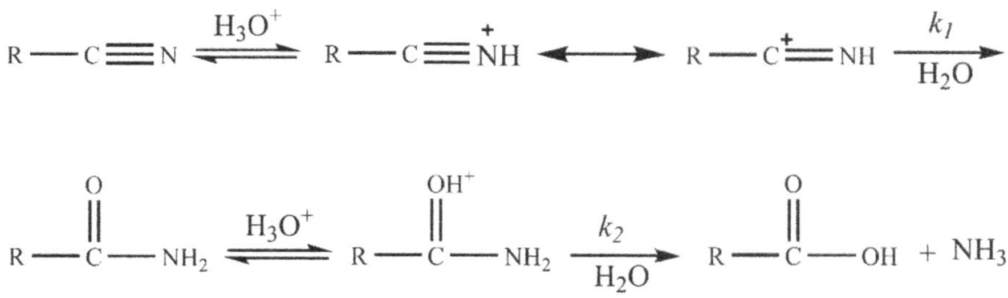

Scheme 1.2 Reaction pathway of simple alkyl nitriles hydrolysis catalyzed by HCl

$(CH_3)_2CH$–) include carboxylic acids substances and ammonia which can act as a base catalyst an acid or base catalyst, as shown in Scheme 1.1. Therefore, it is necessary to understand the hydrolysis pathways of simple nitrile compounds catalyzed by acid or base to better explore the hydrolysis pathway of nitrile-containing complex functional groups in NCW.

The hydrolysis rates of aliphatic nitriles catalyzed by different acids concentration are different due to the consumption of catalyst by byproduct NH_3 in the later stages. In addition, the type of acids also affects the hydrolysis rate and selectivity of aliphatic nitriles. HNO_3 oxidizes and hydrolyzes aliphatic nitrile to CO_2 and other products due to its strong oxidizing. Compared to HCl, H_2SO_4 is a poor catalyst to catalyze the nitriles, especially at higher concentrations [30]. Scheme 1.2 shows the reaction pathway of simple alkyl nitriles catalyzed by HCl. A large quantity of H^+ protonates nitrile under acidic conditions, which facilitates nucleophilic attack by H_2O molecules and further produces the corresponding amide. Similarly, protonated amide undergoes nucleophilic attack by H_2O molecule to form corresponding carboxylic acids and ammonia. As the reaction progresses, the accumulated NH_3 combines with H^+ in the system, which gradually consumes the catalyst [30, 33, 34].

The product of base-catalyzed hydrolysis of simple nitriles is also the corresponding amide which continues to be hydrolyzed to the corresponding carboxylic acid and ammonia, as shown in Scheme 1.3. Take the alkaline hydrolysis of propionitrile as an example, the rate of alkaline hydrolysis of propionitrile is given by the rate of formation of total ammonia and intermediate amide rather than by that of total ammonia alone. In addition, the rate of propionamide hydrolysis is higher than that of propionitrile over the whole alkali concentration range (relative rates approximately 1:10) [35]. Different from acid-catalyzed hydrolysis, only a small amount of OH^- in the system is consumed by the generated carboxylic acids substance, and the accumulation of NH_3 is conducive to the progress of the reaction.

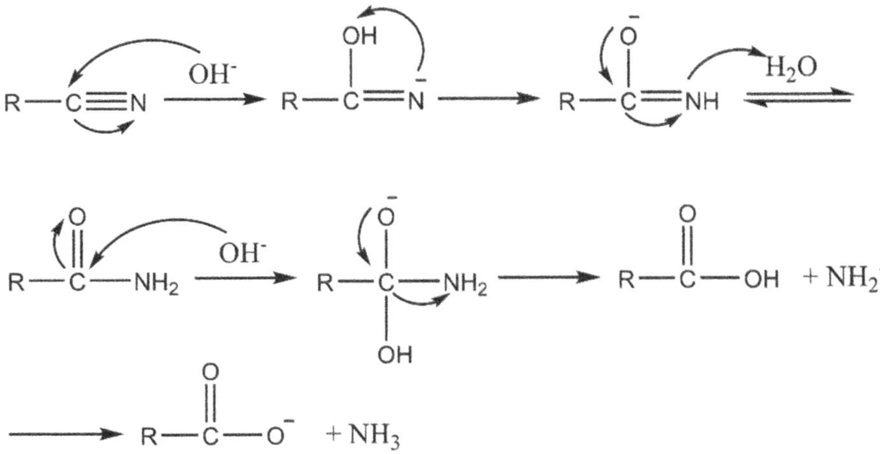

Scheme 1.3 Reaction pathway of simple alkyl nitrile hydrolysis catalyzed by bases

Although acid-base catalytic hydrolysis methods are simple and efficient, they have inherent problems. For example, the reaction process needs to add a large amount of strong acid and base catalysts that are followed by neutralization, salt removal, and other processes that cause serious environmental pollution.

1.4.2.2 Hydrolysis of Saturated and Unsaturated Nitriles in NCW

Aliphatic nitrile compounds, which are the most important type of cyano-containing compounds, are widely used in industrial production, such as acetonitrile, propyl nitrile, and nitrile. Aliphatic nitrile compounds have a simple structure and good stability, so their hydrolysis products are relatively simple, mainly the corresponding amides and carboxylic acids [32, 36]. Hydrolysis reactions of several typical aliphatic nitrile compounds in NCW are described below.

Propionitrile (CH_3CH_2CN) is a typical saturated nitrile compound and acrylonitrile ($CH_2 = CHCN$) is a typical unsaturated nitrile compound. Propionitrile is used as raw material, solvent, and resin additive for organic synthesis, and acrylonitrile is the raw material for the polymerization of fibers, rubber, and plastics [37, 38]. According to Izzo et al., the conversion of propionitrile is much lower than that of acrylonitrile [39]. The overall conversion of acrylonitrile at 250 °C reaches 70% in 1 h of reaction time. It almost achieves complete conversion within 4 h, whereas the reactivity of its saturated analog, propionitrile, is much lower, achieving a conversion of 32% in 5 h. The hydrolysis of propionitrile is relatively simple and mainly produces propionamide and propionic acid. Acrylonitrile has a more complex reaction pathway and produces a large number of by-products (e.g., acrylamide, 3-hydroxypropionitrile, acrylic acid, acetonitrile, and so on). The hydrolysis reactions of corresponding amides at the same conditions further reveal the hydrolysis pathways of the nitriles. The reactivity of propionamide is significantly

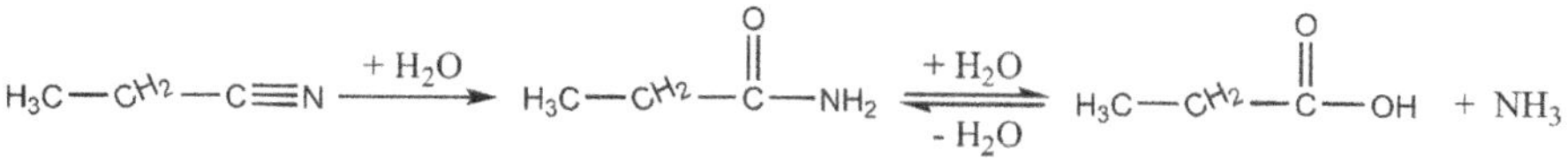

Scheme 1.4 Hydrothermal pathways of propionitrile

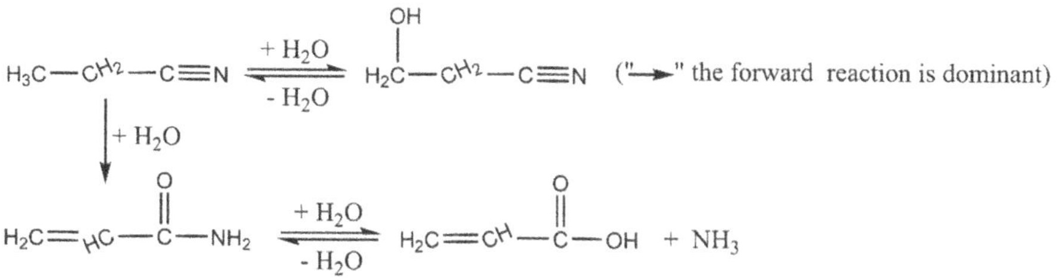

Scheme 1.5 Hydrothermal pathways of acrylonitrile

higher than that of the corresponding nitriles. In addition, the major reaction pathway for both nitriles is the nucleophilic addition of water. Both acrylamide and 3-hydroxypropanitrile are first-order products from acrylonitrile hydrolysis, the former is irreversible, while the latter is reversible (the forward reaction is dominant). The reaction of amide hydrolysis to carboxylic acid and ammonia is reversible, and the aliphatic acid produced from its parent nitrile is the stable end product. Scheme 1.4 and Scheme 1.5 summarize the straightforward sequential network of propionitrile and similarly saturated nitriles, respectively.

1.4.2.3 Hydrolysis of Aliphatic and Aromatic Nitriles in NCW

Phenyl-nitriles are nitrile compounds that contain a benzene ring and are raw materials for preparing pharmaceuticals. Benzonitrile is the simplest phenyl nitrile compound (aromatic nitrile). Izzo et al. compared the hydrolysis rate of butyronitrile (aliphatic nitriles), benzonitrile (aromatic nitriles), and their intermediates in NCW. The overall reactivities of acetonitrile and benzonitrile at 300 °C and 8.7 MPa were comparable. Although both substrates display characteristics of autocatalytic kinetics, the aromatic amide reactivity is significantly lower than the aliphatic reactivity for both catalyzed and uncatalyzed cases [40].

Phenylacetonitrile is another important representative of phenyl nitrile compound, which is the main hydrolytic product from benzyl cyanide, which is widely used in pesticides, medicines, spices, and as raw material in other chemical industries. The carboxyl group, methylene hydrogen, and benzene ring of phenylacetic acid can undergo a series of typical chemical reactions, so it has very important application value in the industry [41–43]. The compound, 2,6-difluorobenzonitrile (DFBN) and its derivatives are typical halogenated phenylnitrile compounds, which are important intermediates in pesticide synthesis. Both the hydrolysis reactions of phenylacetonitrile and DFBN are typical chain reactions, as shown in Scheme 1.6

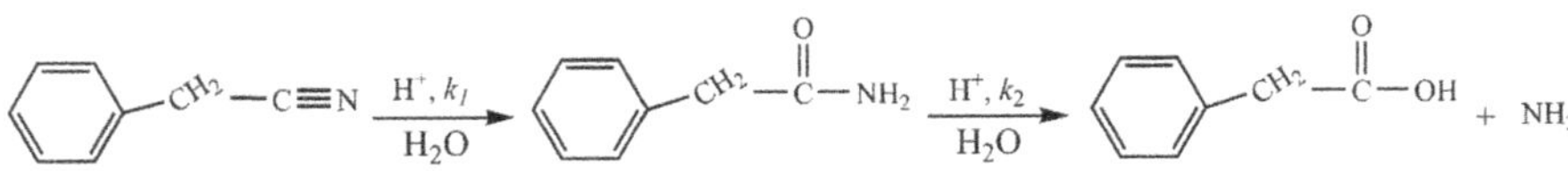

Scheme 1.6 The hydrothermal reaction of phenylacetonitrile

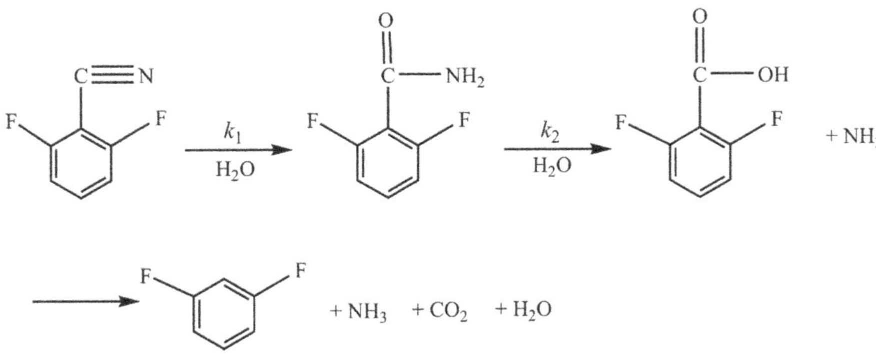

Scheme 1.7 The hydrothermal reaction of 2,6-difluorobenzonitrile

and Scheme 1.7 with the difference being that phenylacetic acid is the final product of phenylacetonitrile hydrolysis which is an autocatalytic reaction, while 2,6-difluorobenzoic acid is the intermediate product during DFBN hydrolysis and is unstable at high-temperatures. It is also found that the selectivity of phenylacetamide and phenylacetic acid in phenylacetonitrile hydrolysis system is close to 100%. Two electron-absorbing groups are added to the benzene ring of DFBN, which destabilizes the carboxyl group on the generated acid. Then, 2,6-difluorobenzoic continues to decarboxylate to produce m-difluorobenzene, thus leading to autocatalysis in later stages of the reaction and low yields of 2,6-difluorobenzoic. In addition, the introduction of two fluorine groups makes the hydrolysis reaction complicated [44].

1.4.2.4 Hydrolysis Reaction of Heterocyclic Nitriles in NCW

Pyridine is a compound in which a nitrogen atom replaces a carbon atom on benzene and has a planar structure. Pyridine-containing acrylonitrile compounds are similar to phenylnitrile in structure and are widely used in industrial production [40, 45]. The heterocyclic nitriles, 3-cyanopyridine and 4-cyanopyridine are typical compounds wherein both niacinamide (also referred to as nicotinamide) and niacin are the hydrolyzed products of 3-cyanopyridine and are widely used in the field of health care and medicine. It has been discovered that the hydrolysis of 3-cyanopyridine and 4-cyanopyridine are also typical chain reactions (Scheme 1.8 and Scheme 1.9) [46]. Although reaction temperature and time have little effect on the selectivity of hydrolyzed products of 3-cyanopyridine and 4-cyanopyridine, the

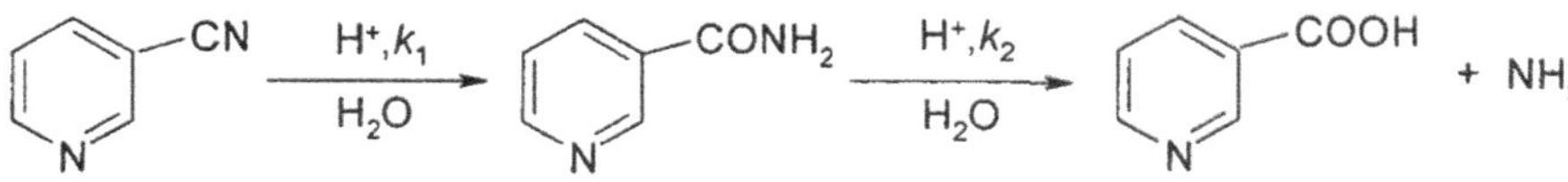

Scheme 1.8 The hydrothermal reaction of 3-cyanopyridine

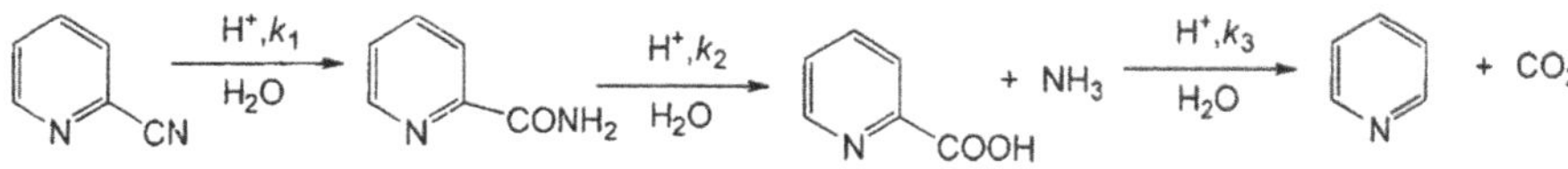

Scheme 1.9 The hydrothermal reaction of 4-cyanopyridine

yield of intermediate products during their hydrolysis is greatly affected by the conditions. The highest yield of nicotinamide gradually decreases as the reaction temperature increases. Therefore, desired products with high selectivity can be prepared by controlling the temperature of the hydrolysis reaction. For example, nicotinamide or nicotinic acid can be selectively prepared at relatively low or high reaction temperatures. Isoniacinamide also has a maximum yield value, but it is less affected by temperature and easily affected by reaction time. It is possible to selectively prepare isonicotinic acid and isonicotinamide in NCW with time as the main controlling factor.

1.4.2.5 Hydrolysis Reaction of Dicyan-Nitriles in NCW

The hydrolysis of dicyanyl compounds containing two cyano groups is more complicated than that of monocyanyl compounds due to multiple reaction pathways. Adiponitrile (ADN, $NC(CH_2)_4CN$) is an important member of dinitrile compounds and is mainly used to produce hexanediamine (the raw material of nylon 66). In addition, ADN is widely used in electronics, the light industry, and other organic synthesis fields. Hexanediamine obtained by catalytic hydrogenation of ADN not only is the main raw material of nylon −6,6 but can be used to produce nylon-610 resin, hexyl isocyanate and polyurethane, and plasticizers. ADN is a good solvent, for example, it can be used as a solvent for methyl acrylonitrile and methacrylate ternary copolymer and the mixture of ADN and tetrahydrofuran serve as a solvent for wet spinning and dry spinning of PVC fibers. ADN is widely used in the preparation of rocket fuels, as a solvent in electroplating industries, as washing additives, and in pesticides. ADN is used in analytical chemistry to determine the relative strength of organic acids and it serves as a medium for potentiometric titration.

Hydrolysis reactions of dicyanyl compounds such as adiponitrile are more complicated than that of monocyanyl groups. Duan et al. has studied a series of hydrolysis reactions of ADN and iminodiacetonitrile (IDAN) and their derivatives in the NCW system [45, 47, 48]. As shown in Scheme 1.10 and Scheme 1.11, many amides, carboxylic acids, nitrile, and other multi-group mixed products are produced

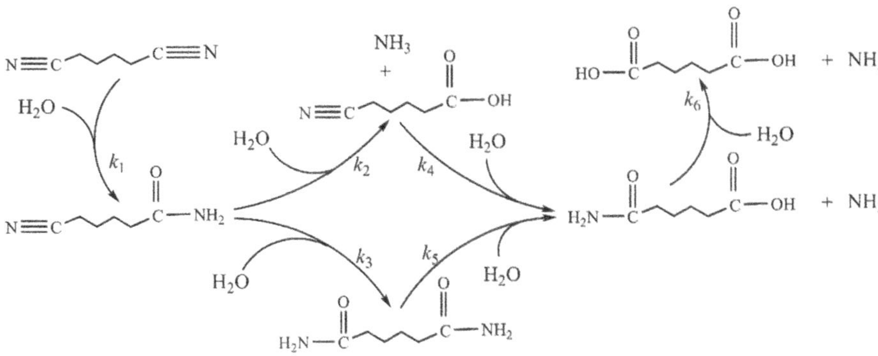

Scheme 1.10 Global reaction scheme for the hydrolysis of ADN in NCW

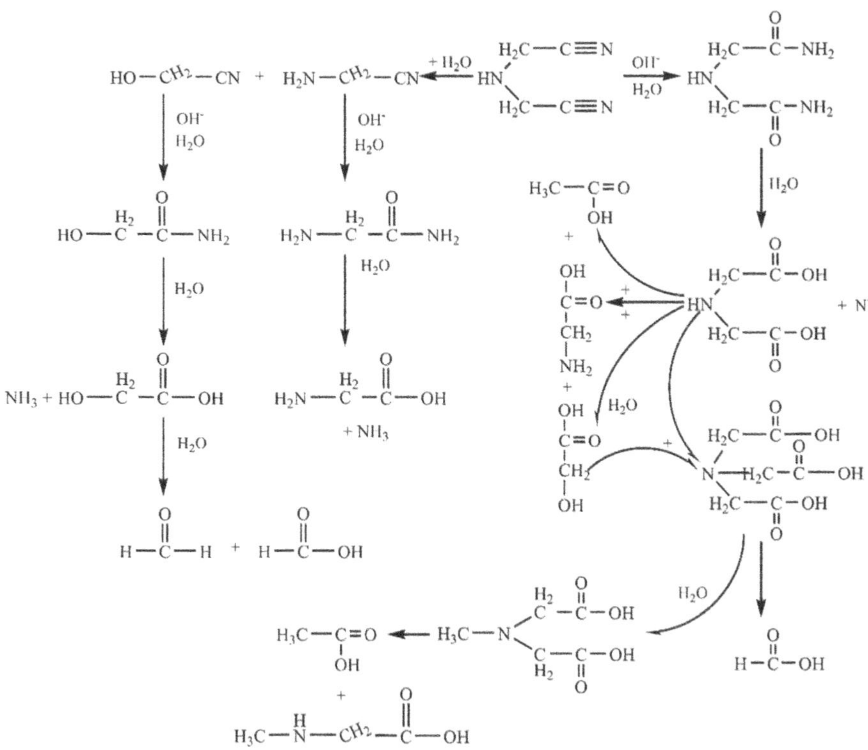

Scheme 1.11 General reaction pathway of IDAN in NCW

due to the complexity of dicyanogen hydrolysis. Taking the hydrolysis of adiponitrile in NCW as an example, the formation of 5-cyanopentamide by the action of a cyano group and water molecule occurs. Then, the other cyano group on 5-cyanopentamide is hydrolyzed to produce adipamide. The hydrolysis of adipamide also involves two of the same groups, and the hydrolysis of an amide group results in 5-cyanovalerate acid which continues to be hydrolyzed to form adipic acid. The concentration of ADN has a great effect on the yield of 5-cyanopentamide, with the lower the concentration of ADN being more favorable for the 5-cyanopentamide product. High concentrations of ADN and moderate temperature are conducive to the formation of adipamide. Both 5-cyanovaleric acid and adipamide can be hydrolyzed to form 5-amidovalerate acid. The formation rate of 5-cyanovaleric acid is less than its hydrolysis rate, resulting in a small amount of 5-cyanovaleric acid in the product.

1.4.3 Comparison of Hydrolysis Reactions of Different Nitrile Compounds in NCW

It is known that the structure of nitriles affects the reactivity and hydrolysis pathway in the NCW system. The primary reaction pathway for nitriles hydrolysis is the addition of water to the –CN groups. For the hydrolysis of saturated nitrile, 1,2 addition at the –CN carbon is the sole primary reaction to produce the corresponding amide and the hydrolysis rate of different saturated nitriles is related to their carbon chain length. For acrylonitrile, its unsaturated sites open additional hydrolysis pathways and provide a facile pathway for the formation of alcohols, amines, and ethers. This results in faster conversion of unsaturated acrylonitrile.

The hydrolysis of aliphatic and aromatic nitrile in NCW is different. In the case of acetonitrile (aliphatic nitrile) and benzonitrile (aromatic nitrile), although their dominant hydrolysis reaction is the addition of water to the electrophilic carbon to the same type of stable end products (corresponding carboxylic acid and ammonia) and there is autocatalytic kinetics for these reaction systems, acetonitrile has a higher reactivity. It may be because the steric effect of benzamide produced by the hydrolysis of benzonitrile may hinder its hydrolysis to benzoic acid, which results in lower acid (catalyst) concentrations and ultimately affects the autocatalytic effect.

In addition, the reactivity of phenyl nitrile with different substituents in near-critical water is different. For example, the hydrolysis rate of 2,6-difluorobenzonitrile is higher than that of cyanobenzene but lower than that of phenylacetonitrile, which is closely related to their molecular structure. Cyanobenzene is more stable than phenylacetonitrile due to the conjugation of the benzene ring, which results in high activation energy that lowers its hydrolysis rate. Two electron-absorbing groups introduced into the benzene ring of 2,6-difluoronitrile activates the cyanide group on the benzene ring and makes it easier for it to be hydrolyzed than cyanobenzene but also reduces the stability of the carboxyl groups on the resulting acid, making it easy to decarboxylate.

The main influencing factor of heterocyclic nitrile compounds is the position of the cyanide group. Considering the electron-drawing induction and electron-drawing conjugation effects of pyridine rings on cyanogen groups at different positions, the order of hydrolysis rate of heterocyclic nitrile compounds is ortho-position > meta-position > para-position.

For the hydrolysis of more complex nitrile with multiple cyanide groups, such as adiponitrile and iminodiacetonitrile, the reaction rate of dicyanyl compounds is significantly higher than that of mononitrile compounds due to the hydrolysis of both cyanide groups, but the distribution of its products is relatively more complex.

1.4.4 Factors Affecting Hydrolysis of Nitrile Compounds in NCW

1.4.4.1 Reaction Temperature and Time

The hydrolysis rate of nitrile compounds increases sharply in the region close to the critical point, indicating that temperature has a very strong effect on reactivity. Substrate molecules gain energy, and some molecules with lower energy are activated as the temperature rises, which speeds up the molecular motion rate and increases the number of effective collisions. Fig. 1.1 depicts the physical properties of NCW and SCW [49]. With an increase in temperature, the ionization constant of water increases sharply. Especially in the temperature range of 25–270 °C, the ionic product of water increases nearly 1000 times, which can provide abundant H^+ and OH^- and has acid/base catalytic functions. The hydrolysis of nitrile compounds is promoted by acid/base catalytic species, which greatly increases the hydrolysis rate. The yield of amide, which is an intermediate product of nitrile hydrolysis, increases first and then decreases with an increase in reaction time, and there is a maximum yield value. Therefore, controlling the reaction time has an important effect on the

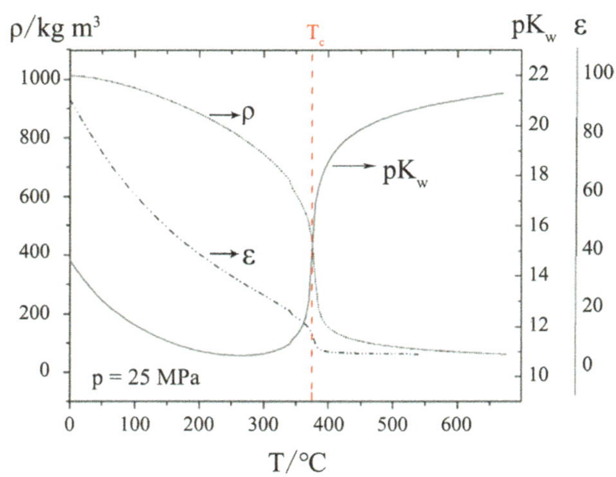

Fig. 1.1 Physical properties of near-critical water and supercritical water: dielectric constant (ε), ionization constant (pK_w), and density (ρ) as a function of temperature at a pressure of 25 MPa

selectivity of the product. The maximum yield values are closely related to reaction rates of the hydrolysis of nitrile to produce amide (k_1) and that of amide to produce carboxylate (k_2). When k_1 is much larger than k_2, amides can be accumulated rapidly in the system, and the yield increases. When k_1 is similar to or even less than k_2, little or no amide is accumulated in the system. Therefore, reaction temperature and time can be used as important control factors to adjust the selective production of amides and carboxylic acid compounds.

1.4.4.2 Reaction Pressure

The reaction pressure has an important effect on the physical properties of the near-critical water system. With an increase in pressure, the dielectric constant of the whole reaction system change, which has a certain effect on the hydrolysis reaction of nitrile compounds [50, 51]. Iyer et al. investigated the influence of a wide range of pressure systems on the hydrolysis kinetics of butanamide at 330 °C in the NCW system and found that the reaction rate increases significantly with an increase in pressure. The conversion rate of nitrile was 88% after 180 min under a pressure of 12.85 MPa and a water density of 641 kg/m^3, whereas a 99.5% conversion rate was achieved after just 30 min at a pressure of 253.9 MPa and a water density of 888 kg/m^3. In addition, the pressure affects the product yield. Both the time to reach the maximum yield of butanamide and butyric acid are shortened with increasing pressure, but the maximum yield of butanamide gradually decreases and that of butyric acid gradually increases. The maximum yield of butanamide at any pressure is generally at least four times lower than the corresponding maximum yield of butyric acid [52]. Changes in pressure are accompanied by concomitant changes in other solvent properties such as the dielectric constant and the ion product of NCW which affect the solvent polarity and rates of reaction.

1.4.4.3 Catalytic Additives

Nitrile compounds can be catalyzed by acids and bases at the same time to form amides and to be further hydrolyzed to produce carboxylic acids, but the reaction mechanisms are different. In NCW systems, the ionization constant of water as the reaction medium increases with temperature up to 250 °C, which provides abundant H^+ and OH^- and catalytic activity, so acid-base catalysis exists simultaneously. Although the concentration of H^+ and OH^- ionized is same in the system, the activation energy of alkali catalysis is lower than that of acid catalysis. Therefore, there is acid-base catalysis in the process of the hydrolysis of nitrile in the absence of additives in NCW, but the reaction mechanism of alkali catalysis is dominant.

When ammonia is added to the NCW hydrolysis system of phenylacetonitrile, ammonia is dissolved in NCW and ionized to produce OH^-, which is beneficial to alkali catalysis. The hydrolysis rate of phenylacetonitrile is significantly accelerated and the hydrolysis mechanism of the system is similar to Scheme 1.3, but the alkyl group is replaced by a benzene ring [53].

When CO_2 is added to the near-critical water system, the concentration of H^+ increases while OH^- decreases, and acid catalysis is significantly enhanced. However, the use of CO_2 is not enough to provide large amounts of H^+ to accelerate the hydrolysis rate of the whole reaction due to the limited acid strength of the system. On the other hand, when the Ea of acid catalysis is higher than that of base catalysis, the resulting hydrolysis rate of phenylacetonitrile becomes lower instead of higher [54]. Some acidic or basic inorganic salts or peroxides can also be added to the near-critical water system to catalyze the hydrolysis of nitriles [55, 56]. Similarly, these additives affect the hydrolysis rate by changing the H^+/OH^- concentration in the NCW system.

1.5 Conclusions and Future Outlook

The chapter gives a detailed outline of the hydrolysis of various nitrile compounds in near-critical water (NCW). Different types of nitrile compounds are discussed in the chapter which is useful for the production of important chemical products or intermediates. Individual water molecules can participate in reactions as a reactant or catalysts in NCW systems. Different nitrile compounds have different hydrolysis processes and products due to their different functional groups. Many nitrile compounds (such as acetonitrile, nitrile, and benzonitrile) can be completely hydrolyzed without additives to form the corresponding amide or carboxylic acid in NCW systems. Apart from the environmental pollution-related issues imposed by the conventional acid-base additives, the change of water polarity affected by the dielectric constant of water at different temperatures and pressures reduces mass transfer resistances of the reaction and also advantageously allows extraction and separation of the required products and lowers the amount of waste liquid. Moreover, the change of temperature and pressure in NCW systems affects the ionization constant of water, thus allowing control of the concentration of H^+ and OH^- in the system, as well as the apparent pH of the system, thus affecting hydrolysis rates and product yields. On the other hand, part of the N atom in nitrile compounds is converted to ammonia under high temperatures and high pressures. The gradually accumulated ammonia act as a base catalyst to promote the hydrolysis of the nitrile and finally transfer to the waste liquid. Hence, further research work on generated waste liquids in NCW systems is still required to achieve sustainable development of resources and green chemistry.

Acknowledgments We are grateful for the ffnancial support of the National Key Research and Development Project (2018YFC1902103) and the National Natural Science Foundation of China (21776063).

References

1. Poliakoff M, Fitzpatrick JM, Farren TR, Anastas PT. Green chemistry: science and politics of change. Science. 2002;297:807–10. https://doi.org/10.1126/science.297.5582.807.
2. Collins T. Toward sustainable chemistry. Science. 2001;291:48–9. https://doi.org/10.1126/science.291.5501.48.
3. Zhang Z, Xu J, Xie J, Zhu S, Li J, Ying G, Chen K. A low-energy and sustainable pulping technology for eucalyptus slabs using a deep eutectic solvent. Green Chem. 2023;25:3256–66. https://doi.org/10.1039/D3GC00589E.
4. Lubberink M, Finnigan W, Flitsch SL. Biocatalytic amide bond formation. Green Chem. 2023;25:2958–70. https://doi.org/10.1039/D3GC00456B.
5. Liu H, Chen BQ, Pan YJ, Fu CP, Kankala WSB, Chen AZ. Role of supercritical carbon dioxide ($scCO_2$) in fabrication of inorganic-based materials: a green and unique route. Sci Technol Adv Mater. 2021;22:695–717. https://doi.org/10.1080/14686996.2021.1955603.
6. Akiya N, Savage PE. Roles of water for chemical reactions in high-temperature water. Chem Rev. 2002;102:2725–50. https://doi.org/10.1021/cr000668w.
7. Lian S, Song C, Liu Q, Duan E, Ren H, Kitamura Y. Recent advances in ionic liquids-based hybrid processes for CO_2 capture and utilization. J Environ Sci (China). 2021;99:281–95. https://doi.org/10.1016/j.jes.2020.06.034.
8. Akizuki M, Fujii T, Hayashi R, Oshima Y. Effects of water on reactions for waste treatment, organic synthesis, and bio-refinery in sub- and supercritical water. J Biosci Bioeng. 2014;117:10–8. https://doi.org/10.1016/j.jbiosc.2013.06.011.
9. Eckert CA, Liotta CL, Bush D, Brown JS, Hallett JP. Sustainable reactions in tunable solvents. J Phys Chem B. 2004;108:18108–18. https://doi.org/10.1021/jp0487612.
10. Kruse A, Dinjus E. Hot compressed water as reaction medium and reactant. Properties and synthesis reactions. J Supercrit Fluids. 2007;39:362–80. https://doi.org/10.1016/j.supflu.2006.03.016.
11. Kim J, Kim HJ, Chang S. Synthesis of aromatic nitriles using nonmetallic cyano-group sources. Angew Chem Int Edit. 2012;51:11948–59. https://doi.org/10.1002/anie.201206168.
12. Rabinovitch BS, Winkler CA. The hydrolysis of acid amides in concentrated hydrochloric acid solutions. Can J Res. 1942;20b:10. https://doi.org/10.1139/cjr42b-032.
13. Katritzky AR, Nichols DA, Siskin M, Murugan R, Balasubramanian M. Reactions in high-temperature aqueous media. Chem Rev. 2001;101:837–92. https://doi.org/10.1021/cr960103t.
14. Ahlbom A, Maschietti M, Nielsen R, Lyckeskog H, Hasani M, Theliander H. Using isopropanol as a capping agent in the hydrothermal liquefaction of Kraft lignin in near-critical water. Energies. 2021;14(4):932. https://doi.org/10.3390/en14040932.
15. Lyu X, Xu M, Chen X, Xu L, Wang J, Deng S, Lu X. Beneficial effect of water on the catalytic conversion of sugars to methyl lactate in near-critical methanol solutions. Ind Eng Chem Res. 2019;58:12451–8. https://doi.org/10.1021/acs.iecr.9b02198.
16. Nekrasov EV, Tallon SJ, Vyssotski MV, Catchpole OJ. Extraction of lipids from New Zealand fern fronds using near-critical dimethyl ether and dimethyl ether–water–ethanol mixtures. J Supercrit Fluids. 2021;170:105137. https://doi.org/10.1016/j.supflu.2020.105137.
17. McConvey IF, Woods D, Lewis M, Gan Q, Nancarrow P. The importance of acetonitrile in the pharmaceutical industry and opportunities for its recovery from waste. Org Process Res Dev. 2012;16:612–24. https://doi.org/10.1021/op2003503.
18. Trangwachirachai K, Lin YC. Light hydrocarbon conversion to acrylonitrile and acetonitrile-a review. Dalton T. 2023;52:6211–25. https://doi.org/10.1039/d2dt03795e.
19. Yang Y, Li B, Zhang L. Design and synthesis of triphenylamine-malonitrile derivatives as solvatochromic fluorescent dyes. Sensors Actuators B Chem. 2013;183:46–51. https://doi.org/10.1016/j.snb.2013.03.108.
20. Hou Y, Dong X, Yang C, Hui Y, Xie Z. MCM-41 immobilized $H_3PW_{12}O_{40}$ catalyzed the addition-elimination reaction of imine with malonitrile in water. Chin J Org Chem. 2018;38:1749–54. https://doi.org/10.6023/cjoc201802006.

21. Dionisi EM, Binder JF, LaFortune JHW, Macdonald CLB. A remarkably stable acyclic phosphamethine cyanine dye. Dalton T. 2023;52:2448–54. https://doi.org/10.1039/d2dt04085a.
22. Suganya K, Maalmarugan J, Manikandan R, Nagaraj TS, Patel RP, Tamilarasi K, Vimalan M, Senthil Kannan K. Synthesis, studies of 2-benzyl-amino-4-p-tolyl-6,7-di-hydro 5H-cyclopenta–[b]pyridine-3 carbo-nitrile (BAPTDHCPCN) crystals for optical, photonic and mechano-electronic uses. J Mater Sci-Mater El. 2022;33:19320–30. https://doi.org/10.1007/s10854-022-08770-0.
23. Fang X, Yu P, Morandi B. Catalytic reversible alkene-nitrile interconversion through controllable transfer hydrocyanation. Science. 2016;351:832–6. https://doi.org/10.1126/science.aae042.
24. Huang CP, Chen CY. Toxicity of nitriles and their derivatives on raphidocelis subcapitata (Algae). Proceedings of the 9th international conference on environmental science and technology, vol A-Oral presentations, Pts A and B; 2005; Univ Aegean, Voulgaroktonou 30.
25. Xiao S, He Y, Xu G, Liu Q. Investigation on claisen rearrangement of allyl phenyl ethers in near-critical water. Res Chem Intermed. 2015;41:3299–305. https://doi.org/10.1007/s11164-013-1433-4.
26. Sinağ A, Gülbay S, Uskan B, Canel M. Biomass decomposition in near critical water. Energy Convers Manag. 2010;51:612–20. https://doi.org/10.1016/j.enconman.2009.11.009.
27. Lesutis HP, Gläser R, Liotta CL, Eckert CA. Acid/base-catalyzed ester hydrolysis in near-critical water. Chem Commun. 1999;20:2063–4. https://doi.org/10.1039/A906401J.
28. Azadi P, Otomo J, Hatano H, Oshima Y, Farnood R. Hydrogen production by catalytic near-critical water gasification and steam reforming of glucose. Int J Hydrog Energy. 2010;35:3406–14. https://doi.org/10.1016/j.ijhydene.2010.01.069.
29. Kang KY, Chun BS. Behavior of hydrothermal decomposition of silk fibroin to amino acids in near-critical water. Korean J Chem Eng. 2004;21:654–9. https://doi.org/10.1007/BF02705501.
30. Krieble VK, Noll CI. The hydrolysis of nitriles with acids. J Am Chem Soc. 1939;61:560–3. https://doi.org/10.1021/ja01872a005.
31. Bolton PD. Hydrolysis of amides II: substituent effects in dilute acid and alkali. Australian J Chem. 1966;19:1013–21. https://doi.org/10.1071/CH9661013.
32. Shi D, Zhu H, Han Y, Zhang Y, Zhao J. Hydrogenation of aliphatic nitriles to primary amines over a bimetallic catalyst Ni25.38Co18.21/MgO–0.75Al2O3 under atmospheric pressure. Catal Lett. 2021;151:2784–94. https://doi.org/10.1007/s10562-021-03532-9.
33. Rabinovitch BS, Winkler CA, Stewart ARP. The hydrolysis of propionitrile in concentrated hydrochloric acid solutions. Can J Res. 1942;20b(7):121–32. https://doi.org/10.1139/cjr42b-019.
34. Rabinovitch BS, Winkler CA. The hydrolysis of aliphatic nitrile in concentrated hydrochloric acid solutions. Can J Res. 1942;20b(10):221–30. https://doi.org/10.1139/cjr42b-032.
35. Rabinovitch BS, Winkler CA. Kinetics of the alkaline hydrolysis of propionitrile. Can J Res. 1942;20b(9):185–8. https://doi.org/10.1139/cjr42b-026.
36. Hinzmann A, Stricker M, Gröger H. Chemoenzymatic cascades toward aliphatic nitriles starting from biorenewable feedstocks. ACS Sustain Chem Eng. 2020;8:17088–96. https://doi.org/10.1021/acssuschemeng.0c04981.
37. Nguyen HVT, Bin Faheem AB, Kwak K, Lee KK. Propionitrile as a single organic solvent for high voltage electric double-layer capacitors. J Power Sources. 2020;463:228134. https://doi.org/10.1016/j.jpowsour.2020.228134.
38. Krikunova LI, Nikolayev AA, Porfiriev DP, Mebel AM. Reaction of propionitrile with methylidyne: a theoretical study. J Chin Chem Soc. 2022;70:439–50. https://doi.org/10.1002/jccs.202200461.
39. Izzo B, Klein MT, LaMarca C, Scrivner NC. Hydrothermal reaction of saturated and unsaturated nitriles: reactivity and reaction pathway analysis. Ind Eng Chem Res. 1999;38:1183–91. https://doi.org/10.1021/ie9803218.

40. Izzo B, Harrell CL, Klein MT. Nitrile reaction in high-temperature water: kinetics and mechanism. AICHE J. 1997;43:2048–58. https://doi.org/10.1002/aic.690430813.
41. Wei J, Shao W, Cao M, Ge J, Yang P, Chen L, Wang X, Kang L. Phenylacetonitrile in locusts facilitates an antipredator defense by acting as an olfactory aposematic signal and cyanide precursor. Sci Adv. 2019;5:eaav5495. https://doi.org/10.1126/sciadv.aav5495.
42. Perez VC, Zhao H, Lin M, Kim J. Occurrence, function, and biosynthesis of the natural auxin phenylacetic acid (PAA) in plants. Plan Theory. 2023;12:266. https://doi.org/10.3390/plants12020266.
43. Ivol F, Porcher M, Ghosh A, Jacquemin J, Ghamouss F. Phenylacetonitrile ($C_6H_5CH_2CN$) ionic liquid blends as alternative electrolytes for safe and high-performance supercapacitors. Molecules. 2020;25:2697. https://doi.org/10.3390/molecules25112697.
44. Ren HM, Lu XY. Kinetics of 2,6-difluorobenzonitrile hydrolysis in high temperature liquid water. J Chem Ind Eng (China). 2009;60:1435–41. https://doi.org/10.3969/j.issn.0438-1157.2011.07.016.
45. Duan PG, Wang X, Dai LY. Noncatalytic hydrolysis of iminodiacetonitrile in near-critical water-a green process for the manufacture of iminodiacetic acid. Chem Eng Technol. 2007;30:265–9. https://doi.org/10.1002/ceat.200600298.
46. Fu J, Ren HM, Zhu J, Lu XY. Hydrolysis kinetics of 2-pyridinecarboxamide, 3-pyridinecarboxamide and 4-pyridinecarboxamide in high-temperature water. Chin J Chem Eng. 2014;22:1005–8. https://doi.org/10.1016/j.cjche.2014.06.024.
47. Duan PG, Wang Y, Yang Y, Dai LY. Optimization of adiponitrile hydrolysis in subcritical water using an orthogonal array design. J Solut Chem. 2009;38:241–58. https://doi.org/10.1007/s10953-008-9362-3.
48. Duan PG, Niu YL, Wang YY, Dai LY. Hydrolysis of adiponitrile in near-critical water. Chin J Chem. 2008;26:1741–4. https://doi.org/10.1002/cjoc.200890315.
49. Kruse A, Vogel H. Heterogeneous catalysis in supercritical media: 2. Near-critical and supercritical water. Chem Eng Technol. 2008;31:1241–5. https://doi.org/10.1002/ceat.200800085.
50. Venardou E, Garcia-Verdugo E, Barlow SJ, Gorbaty YE, Poliakoff M. On-line monitoring of the hydrolysis of acetonitrile in near-critical water using Raman spectroscopy. Vib Spectrosc. 2004;35:103–9. https://doi.org/10.1016/j.vibspec.2003.12.003.
51. Chandler K, Deng F, Dillow AK, Liotta CL, Eckert CA. Alkylation reactions in near-rritical water in the absence of acid catalysts. Ind Eng Chem Res. 1997;36:5175–9. https://doi.org/10.1021/ie9702688.
52. Iyer SD, Klein MT. Effect of pressure on the rate of butyronitrile hydrolysis in high-temperature water. J Supercrit Fluids. 1997;10:191–200. https://doi.org/10.1016/S0896-8446(97)00004-1.
53. Ren HM, Lu XY. Reaction kinetics of phenylacetonitrile hydrolysis in NH_3-enriched high temperature liquid water. J Chem Ind Eng (China). 2011;62:1892–7. https://doi.org/10.3321/j.issn:0438-1157.2009.06.016.
54. Ren HM, Lu XY. Effect of carbon dioxide on phenylacetonitrile hydrolysis in high temperature liquid water. Fine Chemicals. 2012;29:86–90. doi: https://doi.org/CNKI:SUN:JXHG.0.2012-01-022.
55. Prihod'ko R, Sychev M, Kolomitsyn I, Stobbelaar PJ, EJM H, Santen RA. Layered double hydroxides as catalysts for aromatic nitrile hydrolysis. Microporous Mesoporous Mater. 2002;56:241–55. https://doi.org/10.1016/S1387-1811(02)00468-7.
56. Schmid TE, Herrera AG, Songis O, Sneddon D, Révolte A, Nahra F, Cazin CSJ. Selective NaOH-catalysed hydration of aromatic nitriles to amides. Cat Sci Technol. 2015;5:2865–8. https://doi.org/10.1039/c5cy00313j.

Chapter 2
Major Advances in Syntheses of Biomass Based Amines and Pyrrolidone Products by Reductive Amination Process of Major Bio-derived Platform Molecules

Tejas A. Gokhale and Bhalchandra M. Bhanage

Abstract This chapter focuses on advances in development of novel heterogeneous catalysts and their fundamentals of operation for reductive amination of these major platform molecules. Considerable efforts have been made in the last decade for conversion of these platform molecules to N-containing molecules. Mainly, transition metal catalysts have been explored for reductive amination of carbonyl compounds with ammonia or amine as nitrogen source and molecular hydrogen as the sole reductant, which have been proven to be the most effective and efficient. The catalytic systems developed in recent years for reductive amination of furfural (FUR), 5-hydroxymethylfurfural (HMF) will be discussed for syntheses of primary, secondary and tertiary amines. Catalytic systems used for reductive amination of levulinic acid (LA) will also be explored in detail. Primarily, the fundamentals of catalyst design in terms of respective catalysts being applied in mono-, bi- or tri-metallic setups, catalyst support and its synergistic effects, selectivity tuning based on the reactant concentrations and solvation effects have been discussed for these reactions. This chapter also discusses significant reports on reductive amination processes that aid in the understanding of amination reaction mechanisms.

Keywords Furfural · 5-hydroxymethyl furfural · Levulinic acid · Reductive amination · Nitrogen containing compounds · Catalysis · Platform chemicals · Value-added chemicals · Pyrrolidones · Amines

Abbreviations

0.75Co-phen@C-800-HCl	Cobalt-based single atom catalysts
Au/TiO_2-R	

T. A. Gokhale · B. M. Bhanage (✉)
Department of Chemistry, Institute of Chemical Technology, Mumbai, Maharashtra, India
e-mail: bm.bhanage@ictmumbai.edu.in

Z. Fang et al. (eds.), *Production of N-containing Chemicals and Materials from Biomass*, Biofuels and Biorefineries 12,
https://doi.org/10.1007/978-981-99-4580-1_2

	Single phase rutile titania supported gold
Au/ZrO_2-VS	Gold nanoparticles deposited on acid-tolerant ZrO_2
C-Au-Pd	Gold-Palladium alloy supported on carbon
CNF_x@Ni@CNTs	Porous-carbon-coated nickel catalysts supported on carbon nanotubes
Co@C-600-EtOH	Cobalt supported on graphitic carbon support
Co@NC-800	Cobalt supported on nitrogen doped carbon material pyrolyzed at 800 °C
CO-DRIFTS	CO adsorption Diffuse Reflectance Infrared Fourier Transform Spectroscopy
Co-terephthalic acid-piperazine@SiO_2-800	Cobalt nanoparticles embedded on MOF template of terephthalic acid-piperazine supported on silica
Co-Zr@Chitosan-20	Cobalt-Zirconium alloy supported on N-doped carbon-based material obtained from chitosan
Cu_{10}/AlB_3O	Boron Modified copper supported on alumina
$Cu_{15}Pr_3/Al_2O_3$	Praseodymium-copper alloy supported on alumina
$CuNiAlO_x$	Copper-Nickel alloy nanoparticles supported on alumina
DFT	Density Functional Theory
DOE	Department of Energy
DRIFTS-MS	Diffuse Reflectance Infrared Fourier Transform Spectroscopy- Mass Spectrometry
EXAFS	Extended X-ray absorption fine structure
FA	Furfural
Fcc	Face-centered cubic
GVL	Gamma valerolactone
HAADF-STEM	High-Angle Annular Dark-Field imaging-Scanning Transmission Electron Microscope
HMF	5-hydroxymethyl furfural
HZSM-5	H form of ZSM-5 zeolite
IEA	International Energy Agency

Ir/SiO_2-SO_3H	Iridium supported on sulfonic group modified silica surface
Ir-PVP	Iridium supported on polypyrrolidone
LA	Levulinic acid
MC/Ni	Nitrogen doped carbon support for loading of Ni nanoparticles
Ni/MMT	Nickel supported on Montmorillonite K10 clay
Ni@DS	Nickel supported on dendritic silica
NiMn(4: 1)/γ-Al_2O_3	Mn doped Nickel supported on gamma alumina
Pd/Al_2O_3	Palladium supported on alumina surface
Pd/ZrO_2	Zirconia-supported palladium nanoparticles
Pd	Palladium
Pt/P-TiO_2	Platinum supported on porous titania nanosheets
Pt/TiO_{2D}	Platinum supported on 2D titania nanorods
Pt/TiO_2-NT	Platinum supported on titania nanotubes
Pt	Platinum
Pt-MoO_X/TiO_2	Platinum embedded on molybdenum oxide supported on titania
Rh/TiO_2	Rhodium supported on titania
Rh	Rhodium
Ru/BNC	Ruthenium supported on boron and nitrogen doped carbon
Ru/BN-e	Layered boron nitride supported ruthenium catalyst
Ru/MMT	Ruthenium supported on Montmorillonite K10 clay
Ru/Nb_2O_5	Ruthenium supported on niobium pentoxide surface
Ru/SiO_2	Ruthenium supported on silica surface
Ru/TiO_2	Ruthenium supported on titania
Ru/TiP-100	Titanium phosphate support for Ru nanoparticles.
Ru	Ruthenium
Ru@GOIL	Ruthenium ion supported on ionic liquid immobilized into graphene oxide

Ru_1/NC-900-800NH_3	Ruthenium metal based single atom catalyst from ruthenium acetylacetonate and N/C precursors
Ru-HCP	Hexagonally packed ruthenium nanoparticles
Ru-PP/CNTs	Ruthenium based polymeric porphyrin-functionalized carbon nanotubes
Ru-PVP/HAP	Poly(N-vinyl-2-pyrrolidone)-capped ruthenium supported hydroxyapatite
THFA	Tetrahydrofurfurylamines
toe	Tonne of oil equivalent
UiO-67/PpPDA/Pd	Palladium supported on UiO-67 molecular organic framework coated with p-phenylenediamine polymer
XANES	X-ray Absorption Near Edge Structure
XPS	X-ray Photoelectron Spectroscopy

2.1 Introduction

Over the past few decades, the concerns regarding global energy demand have surged, with the aim of limiting the current dependence on non-renewable fossil fuel resources (i.e., coal, natural gas, crude petroleum) [1–3]. International Energy Agency (IEA) reported a 100% increase in global energy consumption close to the end of the current decade (358 Mtoe) as compared to the previous decade (188 Mtoe) [4]. As a result, this has led to overexploitation of non-renewable fossil resources, which had a great environmental and socio-economic impact affecting some developed and many developing nations, whilst affecting global energy supply. The scientific community has devoted considerable efforts to searching for alternative cost-effective and renewable technologies to combat the aforementioned issues. Presently, in general, biomass is the most effective replacement for these fossil fuel resources due to its abundance and renewable nature [5–7]. Specifically, lignocellulosic biomass holds promising avenues as a vital alternative feedstock source towards developing industry specific processes for biomass conversion and valorisation [8, 9]. The major components of lignocellulosic biomass namely cellulose, hemicellulose and lignin have been the major focus of the scientific community for developing novel and relevant processes to convert them to value-added chemicals [10–12].

Efficient extraction of different mono- or oligo-saccharides and further conversion to platform chemicals from lignocellulosic biomass warrants exploration of different treatments in the form of reaction media, solid catalysts, multiple phase reactions to get the most optimal results in terms of sustainability with expected highest yields. This research effectively comprises the other side of the coin to the carbon capture processes, which is certainly necessary for the global carbon cycle balance. This effectively creates a great need for efforts towards innovation for biomass extraction and valorisation processes. The different platform molecules extracted from these bio-sources are required to be studied in terms of raw materials for their conversion to chemicals of high commercial value. More than 30 of these platform chemicals namely levulinic acid (LA), furfural (FUR), 5-hydroxymethyl furfural (HMF), glycerol, sorbitol and many more have been listed in the US Department of Energy (DOE) report from the Office of Biomass Program for the last two decades having excellent industrial relevance for their conversion to value added chemicals and as a potential building block [13]. Extensive life cycle analysis has also shown these platform molecules to have encouraging industrial applicability, [14] however, extensive exploration of separate reaction pathways is currently required to unlock all the possibilities for these molecules.

FUR is considered to be one of the most important bio-based chemicals with current global market value at around US $850 million with a steady growth of its market expected for the next decade [15]. It is mainly manufactured with sugarcane bagasse and corncob as a raw material to extract xylose which later gets converted to FUR, involving mineral acids as a catalyst [16]. Countries such as China and South Africa being the majority contributors (~80%) to its production [17]. FUR and HMF having similar chemical structure do not hold much value as a fuel due to its low stability at ambient conditions, however, the commercial demand for FUR and its derivatives such as substituted tetrahydrofurans, lactones, furfuryl alcohol, cyclopentanone, furfurylamine and more has considerably grown over the years due to their diverse applications in polymer, paint, textile and pharmaceutical industries [18–20]. Similarly, HMF holds great potential in terms of tunability of the reaction pathways to convert to its different valuable derivatives being an excellent alternative for petroleum derived chemicals [13]. It almost has a symmetrical carbon skeleton which provides endless possibilities in terms of functionality tuning on both sides of the structure, as per the desired application. Its primary source of production involves acid catalyzed dehydration of simple sugars [21–24]. HMF conversion using simple oxidation and hydrogenation processes yields a variety of valuable chemicals such as 2,5-bishydroxymethyl furan, 2,5-bishydroxymethyl-tetrahydrofuran, 2,5-dimethyltetrahydrofuran, 2,5-Furandicarboxylic acid and 2,5-diformyl furan with each of them having numerous applications as industrial monomers, biofuel precursors and even industrial solvents [25–28].

LA is another promising and highly sought after bio-based platform feedstock with its excellent application possibilities as a building block and industrial raw material [14]. It has a wide range of reaction pathways which have been and currently being explored with different desired chemicals such as valerolactones, [29, 30] pentanediols, [31] maleic anhydride, [32] succinic acid, [33] *N*-substituted pyrrolidones [34, 35]. LA has been found to be versatile as its straight carbon chain provides many permutations of reactants that can be conceived through different

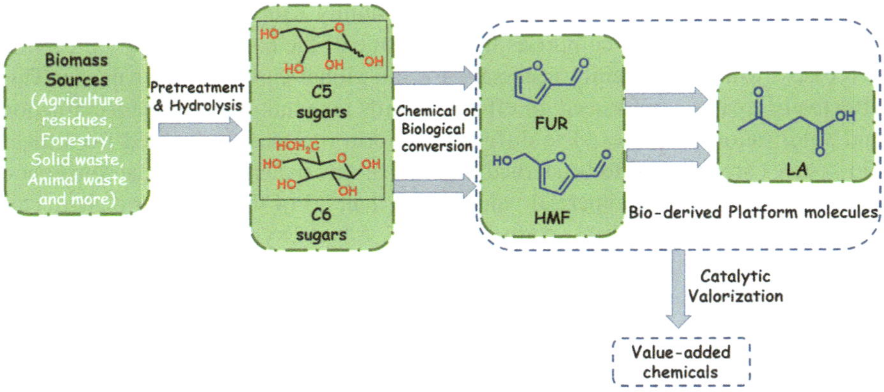

Fig. 2.1 Schematic representation of source and application of platform molecules

catalytic processes. LA and FUR share an interesting nexus as FUR can be easily converted to LA from selected catalytic processes [36, 37].

All three of these molecules have been extensively explored through experimental as well as predictive modelling and comprehensive life cycle analysis for anticipating their importance in the future, [5, 7, 14] shown in Fig. 2.1. One of the goals associated with these molecules is the development of processes for producing nitrogen containing biomass-based molecules. Nitrogen-containing compounds are an important class of chemicals which have been wide applications in pharmaceuticals, materials synthesis and even as high value chemicals [38–40].

Additionally, nitrogen containing compounds have wide applications in biochemical industry with focus on structures such as peptides, nucleic acids and amino acids [41, 42]. Around 75% of all pharmaceutical and agrochemical-based chemicals have nitrogen in the main structural skeleton [43, 44]. The global nitrogen containing chemicals market is around US $50 billion annually [45]. However, the main source of these chemicals is the use of fossil fuels as global feedstock for their production currently. These nitrogen containing compounds also are vital building blocks with applications in pharmaceuticals, food additives, household products and even agrochemicals [46–48]. Recently, the focus of the research community has shifted to major biomass derived platform molecules such as furfural (FA), [49] 5-hydroxymethylfurfural (HMF) [22] and levulinic acid (LA) [50] with the goal of insertion of nitrogen in its structure and their conversion to value-added chemicals such as *N*-substituted amines and pyrrolidones. These compounds have been shown to be synthesized by different processes such as *N*-alkylation, [51, 52] alcohol amination, [53] nitrile reduction, [54] hydroamination process [55] and the reductive amination process [56]. One of the processes that serve as sustainable alternative to traditional amination of fossil feedstocks is reductive amination of bio-based chemical feedstock. Reductive amination process holds several advantages for nitrogen containing compounds over traditional methods such as milder conditions, economical heterogeneous catalytic systems, and shorter reaction times. The use of molecular hydrogen as the sole reductant reaps more advantages such as higher atom economy, relatively lower waste generation, minimal reactor corrosion and maintenance. There are many reports involving reductive amination of bio-based platform chemicals using reductants other than hydrogen including formic acid, sodium borohydride, phenylsilane, ammonia-borane, dimethylamine borane. Reductive amination and the hydrogenation reaction are both competing processes, which can occur at the same time on the catalyst surface. The different parameters such as nitrogen source concentration, hydrogen concentration, temperature, time of the reaction and even catalyst surface properties play a pivotal role in steering the selectivity between these processes.

In this chapter, we will focus on the reductive amination of the main platform molecules namely furfural, 5-hydroxymethyl furfural and levulinic acid to bio-based amines and pyrrolidones. Different heterogeneous catalytic systems using different noble and non-noble metals for this process will be discussed with essential findings and fundamentals of operation of these catalysts. We will also discuss other significant reports of the reductive amination process using different catalytic systems in brief.

2.2 Reductive Amination of FUR and HMF

The reductive amination of furfural using ammonia or other amines has been quite the focus of the scientific community for the last decade as a great number of literature reports were published for this reaction. *N*-substituted or unsubstituted furfurylamine were the desired products with focus on maximizing of its yields and conversion with emphasis on the optimizing the catalyst parameters such as the type of metal, metallic nanoparticle size, metal support material and its surface properties. These reports would be discussed chronologically, which were published for conversion of FUR and HMF using the reductive amination process. As both these molecules share the same target functionality (aldehyde), common reports for both these molecules as well as any individual reports have been discussed separately.

All the reported heterogeneous catalysts for reductive amination process have been added to Tables 2.1 and 2.2 in chronological order. Initially, there were several reports for the reductive amination of specific aldehyde substrates such as benzaldehyde, [57, 58] butyraldehyde, [59] tropanone, [60] to their respective 1°, 2° and 3° amines, using noble metal catalyst and use of ammonium salts or amines as the nitrogen source. However, the first report for reductive amination of FUR was published by Kawanami and co-workers, [61] where they applied Ru, Rh, Pd, Pt based 13 heterogeneous catalysts available commercially for conversion of furfural to furfurylamine using liquor ammonia and molecular hydrogen. They were able to achieve an excellent selectivity of 92% for furfurylamine in just 2 h of reaction time at 80 °C using 5% wt. Rh/Al_2O_3 (Table 2.1 Entry 1). Additionally, 5% wt. Rh/Al_2O_3 gave a yield of 86.3% for 5-hydroxymethyl furfurylamine at identical conditions (Table 2.2 Entry 1). They also discussed the general mechanism for the process,

Table 2.1 Reductive amination of FUR using heterogeneous catalysts in chronological order

Entry No.	Catalyst	Product	Reactions conditions	Conv. (%)	Yield (%)	Ref.
1	Rh/Al_2O_3	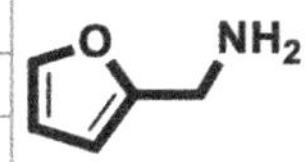	Cat. = 2 mg, sub = 200 mg, sub/NH_3 (mole ratio) = 0.03, T = 80 °C, t = 2 h, pH $_2$ = 20 bar	100	91.5	[61]
2	Pd/C			100	10.6	
3	Pd/Al_2O_3			100	18.4	
4	Pd/MCM			100	–	
5	Rh/C			100	27.3	
6	Rh/MCM			100	82.8	
7	Pt/C			100	26.9	
8	Pt/MCM			42.2	10.6	
9	Pt/Al_2O_3			55.6	12.4	
10	Ru/C			25.2	–	
11	Ru/Al_2O_3			33.5	–	
12	Ru/MCM			–	–	
13	Carbon-supported FeNi alloy	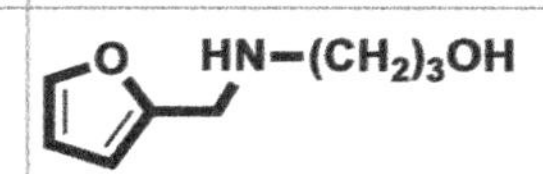	0.3 mL min^{-1}, WHSV = 24 h^{-1}, 10 bar H_2, 125 °C. aldehyde (50 mM), amine (50 mM) in ethanol	99	90	[62]
14	Carbon-supported FeNi alloy	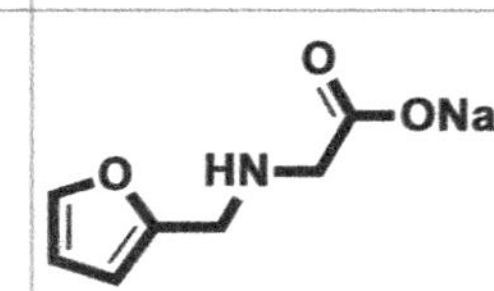	0.3 mL min^{-1}, WHSV = 24 h^{-1}, 10 bar H_2, 85 °C. aldehyde (50 mM), amino acid (50 mM) in ethanol	99	60	
15	Carbon-supported FeNi alloy	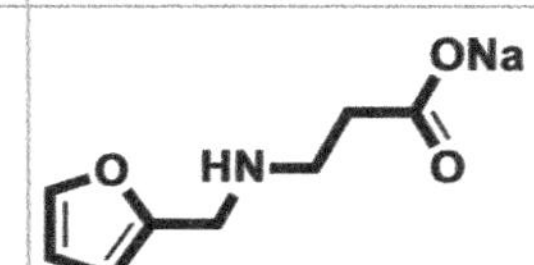	0.3 mL min^{-1}, WHSV = 24 h^{-1}, 10 bar H_2, 125 °C. aldehyde (50 mM), amino acid (50 mM) in ethanol	99	50	
16				90	10	

	Carbon-supported FeNi alloy	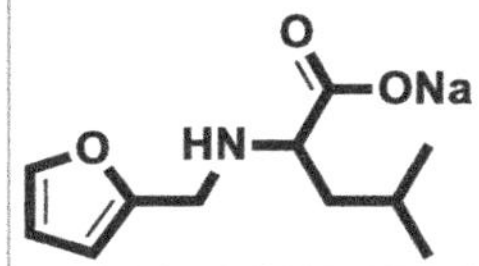	0.3 mL min^{-1}, WHSV = 24 h^{-1}, 10 bar H_2, 100 °C. aldehyde (50 mM), amino acid (50 mM) in ethanol			
17	5 wt% Ru-PVP/HAP	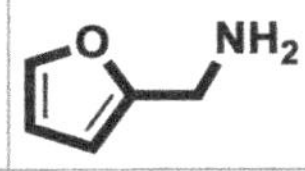	Catalyst (100 mg), furfural (1 mmol), 25% NH_3 aq. (3 mL), H_2 gas (4 bar), 373 K, 2 h, 500 rpm	100	60	[63]
18	Ru/ZrO_2	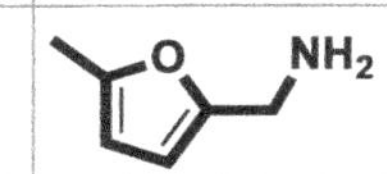	2 mmol sub., 2.0 mL 25 wt.% aq.NH_3, 0.1 g cat. (Ru loading = 5 wt.%), 85 °C, 12 h, 20 bar H_2	–	61	[64]
19	Ru/Nb_2O_5	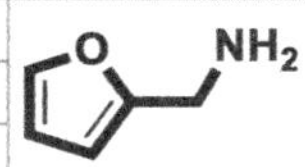	Catalyst (20 mg), 1a (0.5 mmol), MeOH (5 mL), NH_3 (1 bar), H_2 (40 bar), 90 °C, 4 h.	100	99	[65]
20	Ru/SiO_2			99	84	
21	Ru/TiO_2			99	72	
22	Ru/C			99	31	
23	Rh/Nb_2O_5			99	75	
24	Pd/Nb_2O_5			99	26	
25	Ru-NP	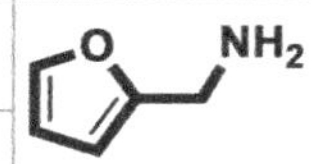	Catalyst (0.2 mg), sub (0.5 mmol), MeOH (5 mL), NH_3 (8 mmol), H_2 (20 bar), 90 °C, 2 h.	99	99	[66]
26	Ru-HCP		Catalyst (0.2 mg), sub (0.5 mmol), MeOH (5 mL), NH_3 (8 mmol), H_2 (20 bar), 90 °C, 5 h.	99	71	
27	Ru/Nb_2O_5		Catalyst (20 mg), sub (0.5 mmol), MeOH (5 mL), NH_3 (8 mmol), H_2 (20 bar), 90 °C, 4 h.	99	98	
28	Ru/SiO_2		Catalyst (20 mg), sub (0.5 mmol), MeOH (5 mL), NH_3 (8 mmol), H_2 (20 bar), 90 °C, 4 h.	99	91	
29	Ru/$Nb_2O_5 \cdot nH_2O$-300	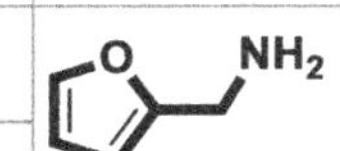	Catalyst (0.02 g), sub (0.5 mmol), NH_3 in MeOH (1.6 M, 5 mL), H_2 (40 bar), 70 °C, 4 h.	99	89	[67]
30	Ru/Nb_2O_5			99	76	
31	MC/Ni			100	88.4	[68]

(continued)

Table 2.1 (continued)

Entry No.	Catalyst	Product	Reactions conditions	Conv. (%)	Yield (%)	Ref.
		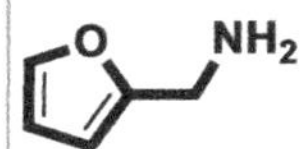	Sub (1 mmol), cata. (20 mg), MeOH (10 mL), $NH_3 \cdot H_2O$ (36 wt. %, 200 μL, 2.8 mmol), 80 °C, 12 h, 1 bar H_2			
32	Co@NC-800	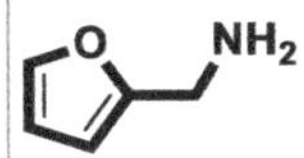	Sub (1 mmol), co@NC-800 (0.68 mol.% co), 130 °C, H_2 (10 bar), ethanol (8 mL), 26.5 wt% NH_3 aq. Soln. (2 mL), 12 h.	100	81.8	[69]
33	Ni_6AlOx	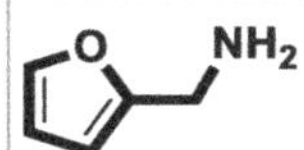	1 mmol substrate, 49: 1 NH_3/aldehyde, 50 mg of cat, and 3 mL of H_2O, 100 °C, 4 bar H_2, 5 h	99	90	[70]
34	Fe_3O_4@SiO_2-Ni	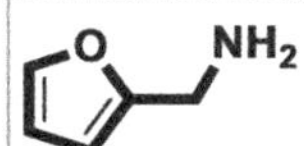	MW irradiation (3 h, 115 °C), 1 mmol sub., cat. (5 wt. % loading), H_2 (20 bar), 750 μL aq. NH_3 (25 wt %) and 75 μL of ethanol	100	100	[71]
35	1% Ru/Nb_2O_5-L	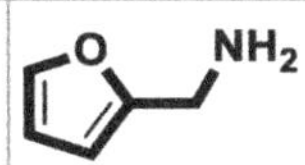	2 mmol sub., 5 mL NH_3-MeOH (2 M), 20 mg 1% Ru/Nb_2O_5-L, 90 ° C, 20 bar H_2, 8 h	99	60	[72]
36	Rh/C	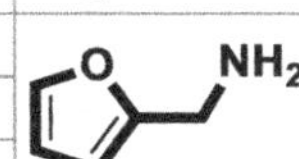	Sub. (0.5 g); *n*-dodecane, (0.1 g), CH_3OH, 20 mL, cat. (0.1 g), NH_3 1 bar, H_2 10 bar, 120 °C, 2 h	100	54	[73]
37	Raney Ni			100	65.1	
38	Raney co			100	98.9	
39	Raney co (0.05 g)			100	98.4	
40	Pd/C	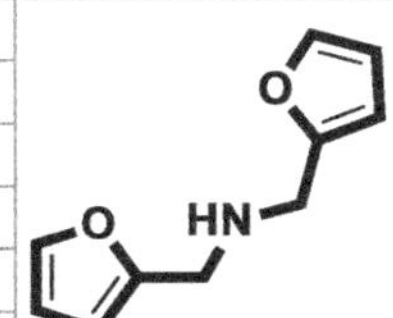	Schiff base 0.5 g, *n*-dodecane 0.1 g, CH_3OH, 20 mL, catalyst 0.1 g, 10 bar H_2, 120 °C, 2 h	100	99.1	[73]
41	Pt/C			100	99	
42	Ru/C			100	99	
43	Rh/C			100	99	
44	Raney Ni			100	99	
45	Raney co			94.4	78	

46	PdNPs	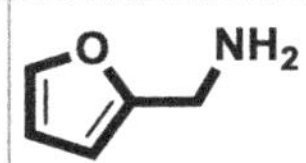	Sub. 0.05 mmol, 0.2 mmol, aq. NH_3, cat. 2.0 mg, 10 mL H_2O, H_2 balloon, 3 h, Rt., pH of reaction set at 2.0	–	97	[74]
47	Ru/HZSM-5(46)	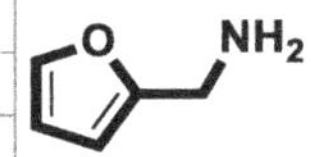	Sub. (2 mmol), cata. (0.15 g), 30 bar H_2, NH_3 (7 M in MeOH 6 mL), 100 °C, 15 min.	99	76	[75]
48	Pd/HZSM-5(46)			99	44	
49	Pt/HZSM-5(46)			99	47	
50	Ru/SiO_2	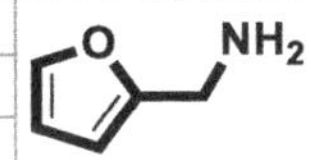	130 °C, 750 psi H_2, 10 h, (5:1 NH_3: Substrate), 0.1 M sub., 40:1 Dioxane:Water (40 mL), 15 mg cata., reduction conditions: 200 °C in 30 mL dioxane in batch reactor, 12 h, 750 psi H_2	100	72	[76]
51	Ni/SiO_2			98	51	
52	Co/SiO_2			100	70	
53	Pd/SiO_2	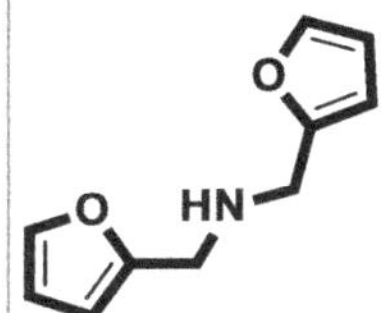		100	60	
54	Pd/Al_2O_3	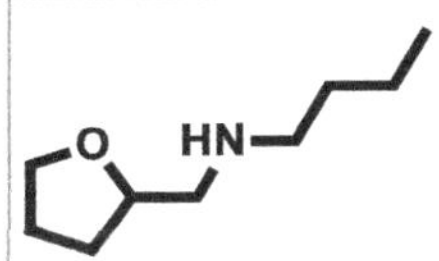	Sub. 100 mg, $BuNH_2$ 77 mg, EtOH 5 g, cat. (0.45 mol% to sub.), H_2 (10 bar), 80 °C, 90 min.	99	98	[77]
55	Ru/TiP-100	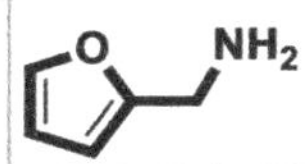	Sub. 1 mmol, methanol 2 mL, H_2 17 bar, NH_3 3 bar, cat. 0.2 mol.%, 30 °C, 24 h	99	91	[78]
56	Co_2P NRs	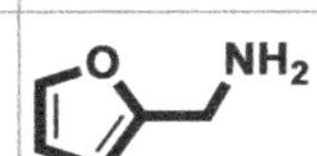	Cata. 4.0 mg, sub. 0.5 mmol, aq. NH_3 25% (3 mL), 5 bar H_2, 100 °C, 10 h.	93	90	[79]
57	10Ni/Al_2O_3	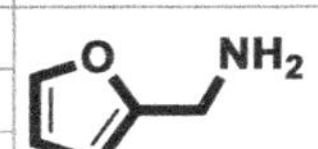	Sub. 2 mmol, cat. 0.1 g, 20 bar H_2, NH_3 (7 M in MeOH 6 mL), 100 °C, 2 h.	99	92	[80]
58	10Ni/ZrO_2			99	69	
59	10Ni/HZSM-5			99	48	

(continued)

Table 2.1 (continued)

Entry No.	Catalyst	Product	Reactions conditions	Conv. (%)	Yield (%)	Ref.
60	$10Ni/SiO_2$			99	54	
61	$10Ni/Nb_2O_5$			99	45	
62	$10Co/Al_2O_3$			99	69	
63	$5Pd/Al_2O_3$			99	46	
64	$5Ru/Al_2O_3$			99	86	
65	$5Ni/Al_2O_3$			99	82	
66	$20Ni/Al_2O_3$			99	89	
67	Ru_1/NC-900-800NH_3	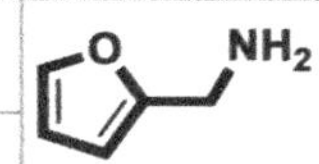	2 mmol sub., cata. Ru:Furfural =1:400, 3 g methanol, 5 bar NH_3, 20 bar H_2, 100 °C, 10 h	99	97	[81]
68	Ru/AC			99	82	
69	Ru/Nb_2O_5			99	91	
70	$Ru_1/NC/Nb_2O_5$		2 mmol sub., cata. Ru:Furfural =1:400, 3 g methanol, 5 bar NH_3, 20 bar H_2, 100 °C, 3 h	99	97	
71	2% Ru/MMT	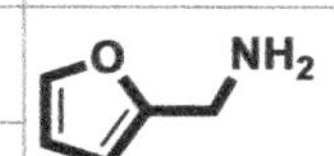	30 mg cat., 90 °C, 10 bar H_2 pressure, 0.5 mmol reactant, 4 mL 25% NH_3 aq. Soln	97	89	[82]
72	12.5% Ni/MMT		30 mg cat., 130 °C, 15 bar H_2 pressure, 0.5 mmol reactant, 4 mL 25% NH_3 aq. Soln.	96	84	
73	$BF_3{\cdot}Et_2O$		Sub. (1.0 mmol), aniline (1.1 mmol), cat. (0.5 mol%), FA (2.0 mmol), CH_3CN (2.0 mL), 85 °C, 1.5 h, air.	–	90	[83]
74		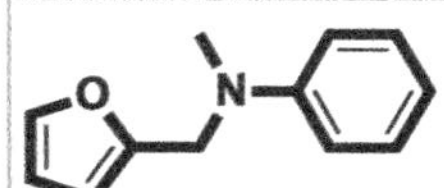	Same as above except time (5 h)	–	82	
75	Ru/BN-e			99	95	[84]

		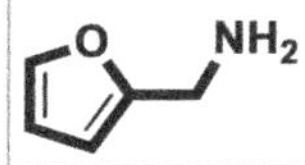	2.0 mmol sub., 2.2 mmol product, 2 eq. of 25% aq. NH_3, 30 mg cat. (Ru: 0.6 mol. %), 7.0 mL MeOH, 10 bar H_2, 90 °C.			
76	Co@C-600-EtOH	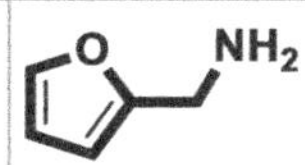	Sub. (0.05 mmol), 90 °C, 4 h, 10 mg cata., 7 M NH_3 solution, and 20 bar H_2	99	87	[85]
77	Ni@DS-7.5	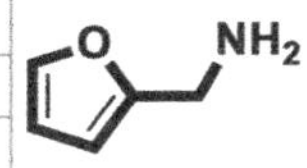	60 mg cat., 0.5 mmol sub., 50 °C, 3 bar NH_3, 9 bar H_2, 5 mL toluene, 6 h	–	88	[86]
78	Ni@DS-15			–	89	
79	Ni@DS-30			–	81	
80	1% Rh/TiO_2	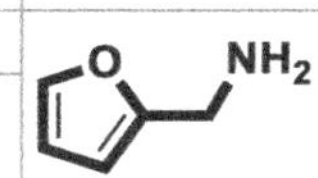	0.1 mL of furfural, 10 mL of NH_3 (33%), 20 bars of H_2, and 25 mg cat.	99	95–99	[87]
81	1% Ru/TiO_2			99	98	
82	Ru/BNC	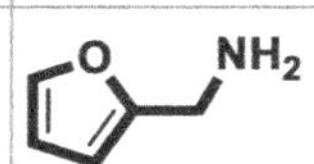	Sub. 0.12 mmol, cat. 10 mg, 1.8 wt% Ru, 1.1 mol% Ru of sub., $N_2H_4{\cdot}H_2O$ (0.24 M in 2.0 mL MeOH, 4 eq.), H_2 (20 bar), 80 °C, 16 h	99	99	[88]
83	NiAl-0-H_2	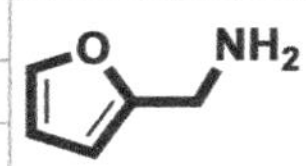	10 mg cat., 1 mmol sub., 10 mL methanol, 20 bar H_2, 5 bar NH_3, 80 ° C, 1 h.	–	99	[89]
84	NiAl-10			–	99	
85	Raney Ni			–	81	
86	Rh_2P/NC-1	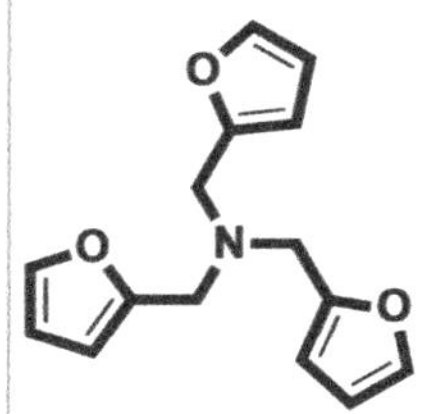	Cat. 20 mg, sub. 1.5 mmol, $HCOONH_4$ 0.3 mmol, EtOAc 3.0 mL, PH_2 30 bar, 60 °C, 24 h.	100	92	[90]

Table 2.2 Reductive amination of HMF using heterogeneous catalysts in chronological order

Entry No.	Catalyst	Product	Reaction condition	Conv. (%)	Yield (%)	Ref.
1	5% Rh/Al_2O_3	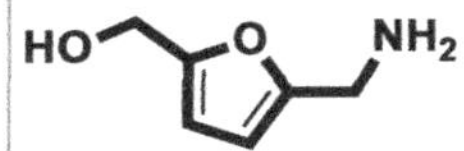	Cat. 2 mg, sub. 200 mg, sub./NH_3 (mole ratio) = 0.03, 80 °C, 4 h	100	86.3	[61]
2	Carbon supported FeNi alloy	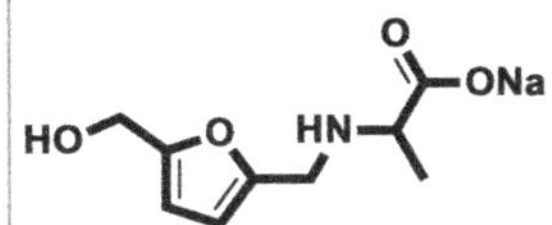	0.3 mL min^{-1}, WHSV = 24 h^{-1}, 10 bar H_2, 100 °C	99	78	[62]
3		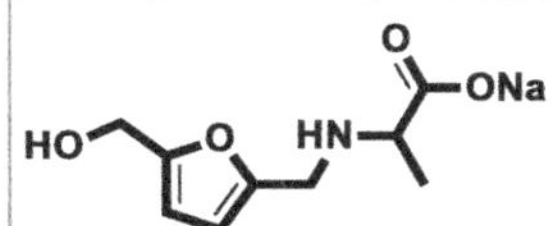	0.3 mL min^{-1}, WHSV = 24 h^{-1}, 10 bar H_2, 125 °C	99	77	
4	Au/TiO_2-R	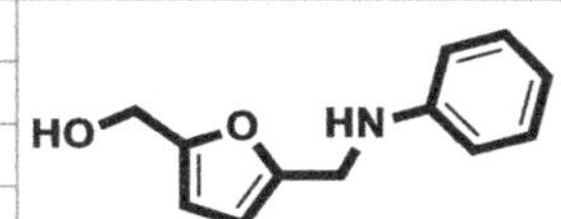	Sub. 0.5 mmol, amine 0.5 mmol, cat. (au 0.5 mol%), MeOH–H_2O 1:1, 3 mL, CO 20 bar, 60 °C, 2.5 h.	99	99	[91]
5	Au/TiO_2-P25			96	81	
6	Au/CeO_2			84	53	
7	Au/ZrO_2			79	25	
8	Au/TiO_2-R	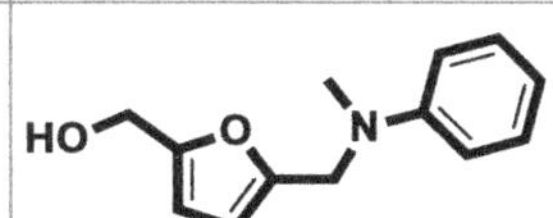	Sub. 0.5 mmol, amine 0.5 mmol, cat. (au 0.5 mol%), MeOH–H_2O 1:1, 3 mL, CO 20 bar, 60 °C, 6 h.	100	97	
9	Au/TiO_2-R	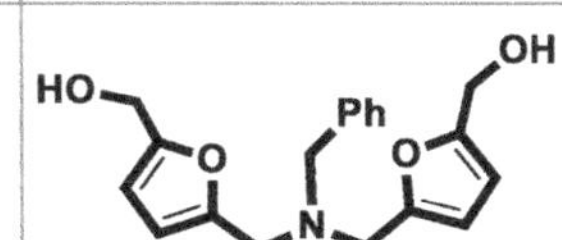	Sub. 0.5 mmol, amine 0.5 mmol, cat. (au 0.5 mol%), MeOH–H_2O 1:1, 3 mL, CO 40 bar, 90 °C, 10 h.	100	85	

10	Ru/Nb_2O_5	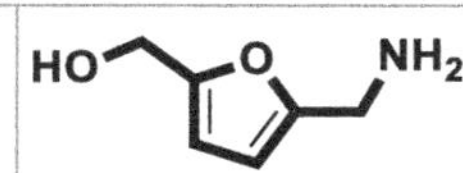	Cat. (0.02 g), sub. (0.5 mmol), MeOH (5 mL), NH_3 (1 bar), H_2 (40 bar), 90 °C, 4 h.	100	96	[65]
11	Ru-NPs	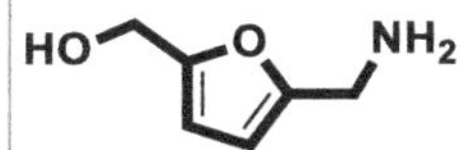	Catalyst (0.2 mg), sub (0.5 mmol), MeOH (5 mL), NH_3 (8 mmol), H_2 (20 bar), 90 °C, 2 h.	–	95	[66]
12	$Ru/Nb_2O_5 \cdot nH_2O$-300	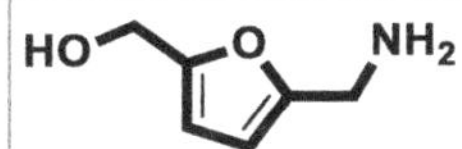	Cat. 20 mg, sub. 0.5 mmol, NH_3 in MeOH 1.6 M, 5 mL, H_2 30 bar 70 °C, 4 h.	99	80	[67]
13	Ni_1AlO_x	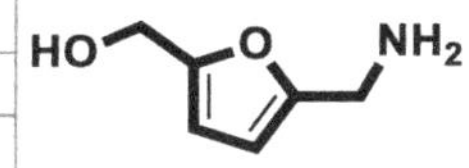	Sub. 1 mmol, 49:1 NH_3/HMF, 3 mL of H_2O, 50 mg of cat, 100 °C, 1 bar H_2, and 6 h.	65	32	[70]
14	Ni_2AlO_x			74	38	
15	Ni_4AlO_x			93	85	
16	Ni_6AlO_x			100	99	
17	Ni_8AlO_x			100	99	
18	$Ni_{10}AlO_x$			100	96	
19	NiO_x			91	33	
20	Raney-Ni			100	48	
21	Raney Ni	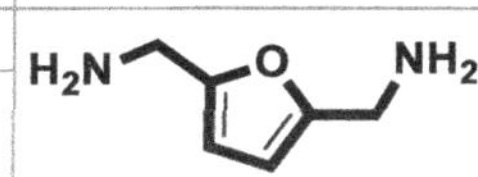	Sub. 0.5 g, THF 15 ml, cat. 0.25 g, NH_3 3.5 bar, H_2 10 bar, 160 °C, 1000 rpm, 12 h	100	60.7	[92]
22	Raney co			100	14.8	
23	Pd/C	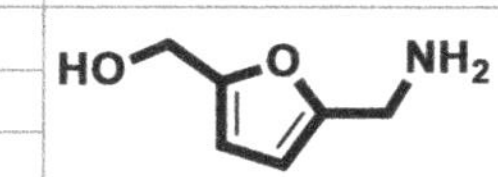		100	85.4	
24	Pt/C			100	85.1	
25	Ru/C			100	83.5	
26	UiO-67/Pd	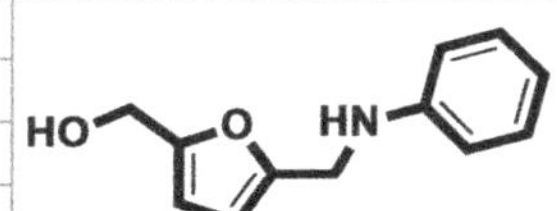	5 bar H_2, 50 °C, 10 mg cat. (0.4 wt% Pd), sub. 0.24 mmol, amine 0.24 mmol, 5 mL ethanol, 2 h.	100	93	[95]
27	UiO-67/PpPDA/Pd			100	94.5	
28	Fe-BTC/Pd			100	95	
29	Commercial Pd/C			100	82	

(continued)

Table 2.2 (continued)

Entry No.	Catalyst	Product	Reaction condition	Conv. (%)	Yield (%)	Ref.
30	$Cu_1Ni_4Al_4O_x$	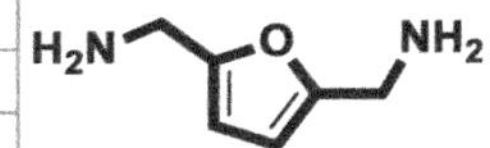	Sub. 79.4 mmol, NH_3 1352.9 mmol, cat. 2.0 g, Na_2CO_3 (8.6 mmol, 11 mol% to 5-HMF), H_2 (45 bar), dioxane 50 mL, 90 °C for 6 h then 210 °C for 18 h.	100	46.1	[96]
31	$Cu_1Ni_1Al_{1.6}O_x$			100	56.3	
32	$Cu_4Ni_1Al_4O_x$			100	75.2	
33	$Cu_6Ni_1Al_{5.6}O_x$			100	60.4	
34	$Cu_{19}Ni_1Al_{16}O_x$			100	36.6	
35	$Cu_1Al_1O_x$			100	36.9	
36	$Ni_1Al_4O_x$			100	45.5	
37	$Cu_4Ni_1Al_4O_x$		Sub. 79.4 mmol, NH_3 1352.9 mmol, cat. 2.0 g, Na_2CO_3 (8.6 mmol, 11 mol% to 5-HMF), H_2 (45 bar), dioxane 50 mL, 90 °C for 9 h then 210 °C for 18 h.	100	85.9	
38	$Cu_1Al_1O_x/Ni_1Al_4O_x$ (3.38: 1.51)		Sub. 79.4 mmol, NH_3 1352.9 mmol, cat. 2.0 g, Na_2CO_3 (8.6 mmol, 11 mol% to 5-HMF), H_2 (45 bar), dioxane 50 mL, 90 °C for 6 h then 210 °C for 18 h.	100	60.2	
39	Ru/TiP-100 (3.35 wt% Ru)	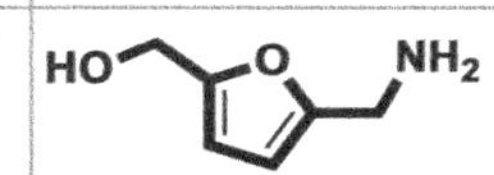	Sub. 1 mmol, methanol 2 mL, H_2 14 bar; NH_3, 6 bar, 0.2 mol % cat., 30 °C, 24 h	100	93	[78]
40	Co_2P NRs	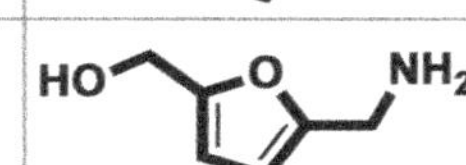	Cat. (4.0 mg), aldehyde (0.5 mmol), aq. NH_3 25% (3 mL), 5 bar H_2, 100 °C, 10 h.	95	87	[79]
41	Pt/C	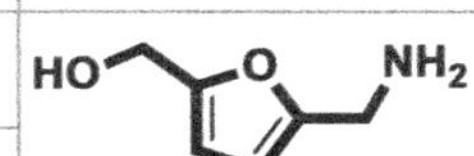	Sub. 0.5 g, THF 15 mL, cat. 0.25 g, 80 °C, H_2 10 bar, NH_3 3.5 bar, 2 h, stirring speed 1000 rpm	100	81.2	[93]
42	Raney Ni		Sub. 0.5 g, THF 15 mL, cat. 0.25 g, 120 °C, H_2 10 bar, NH_3 3.5 bar, 2 h, stirring speed 1000 rpm	100	80.7	
43	Raney co			100	99.5	

44	Raney Ni	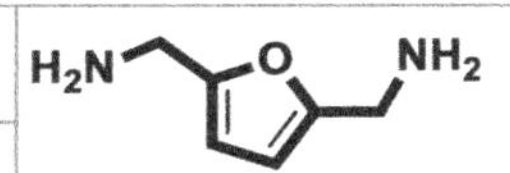	Sub. 0.5 g, THF 15 mL, cat. 0.25 g, 160 °C, H_2 10 bar, NH_3 2 bar, 12 h, stirring speed 1000 rpm	100	82.3	
45	Raney Ni		Sub. 0.5 g, THF 15 mL, cat. 0.25 g, reacting at 120 °C for the first 2 h and increasing temp. To 160 °C for the remaining 10 h (2 + 10 h), H_2 10 bar, NH_3 2 bar, stirring speed 1000 rpm	100	86.5	
46	10Ni/Al_2O_3	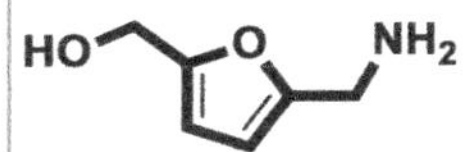	Sub. 0.2 g, cat., 0.1 g, 20 bar H_2, NH_3 (7 M in MeOH, 6 mL), 100 °C, and 30 min.	99	87	[80]
47	Ru_1/NC-900-800NH_3	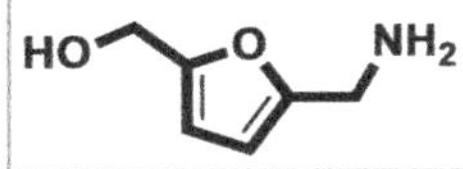	2 mmol sub., 22 mg cat., 0.25 mol% Ru (molar ratio of Ru: Substrate = 1:400), 3 g methanol, 5 bar NH_3, 20 bar H_2, 100 °C, 10 h, dodecane as an internal standard.	–	93	[81]
48		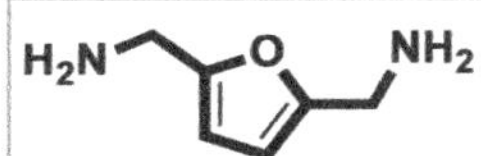		–	63	
49	2% Ru/MMT	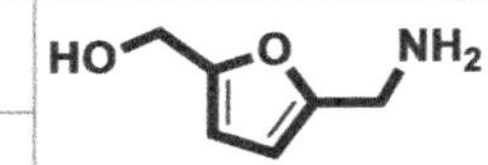	30 mg cat., 90 °C, 10 bar H_2, 0.5 mmol sub., 4 mL 25% NH_3 aq. Soln	98	86	[82]
50	12.5% Ru/MMT		30 mg cat., 130 °C, 15 bar H_2, 0.5 mmol sub., 4 mL 25% NH_3 aq. Soln.	94	79	
51	Ru/BN-e	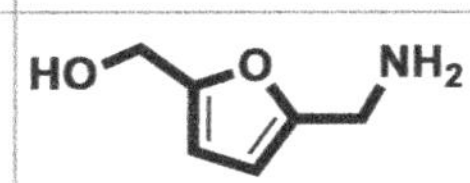	2.0 mmol sub., 2 equiv. of 25% aq. NH_3, 30 mg cat. (Ru 0.6 mol%), 7.0 mL MeOH, 1.0 MPa H_2, 90 °C, 3 h	–	94.5	[84]
52	Co-terephthalic acid-piperazine@SiO_2-800	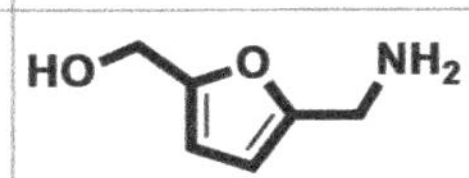	0.5 mmol sub., 5 bar NH_3, 25 mg cat. (5 mol% co), 10 bar H_2, 3 mL H_2O, 50 °C, 16 h.	99	94	[97]
53	Co-terephthalic acid-piperazine@SiO_2-800	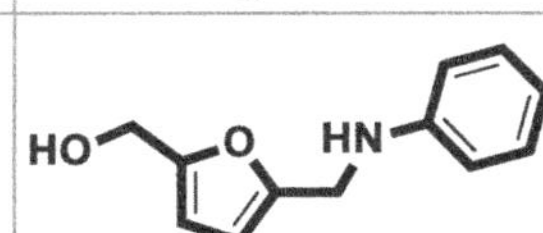	0.5 mmol sub., 0.5 mmol amine, 25 mg cat. 5 mol.% co, 20 bar H_2, 3 mL H_2O, 70 °C, 16 h.	99	95	

(continued)

Table 2.2 (continued)

Entry No.	Catalyst	Product	Reaction condition	Conv. (%)	Yield (%)	Ref.
54	Co@C-600-EtOH	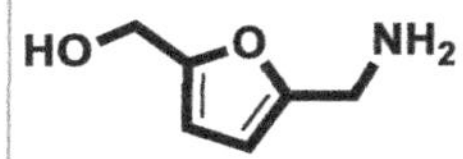	Sub. (0.05 mmol), 90 °C, 4 h, 10 mg cata., 7 M NH_3 solution, and 20 bar H_2	99	99	[85]
55	10Ni/γ-Al_2O_3	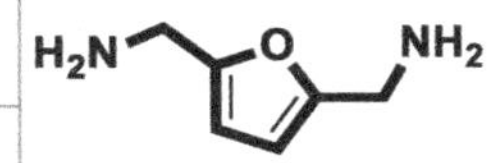	Sub. 0.5 g, THF 15 mL, NH_3 3.5 bar, H_2 10 bar, 160 °C, 1200 rpm, 36 h	–	86.3	[94]
56	10NiMn(4: 1)/γ-Al_2O_3		Sub. 0.5 g, THF 15 mL, NH_3 3.5 bar, H_2 10 bar, 160 °C, 1200 rpm, 24 h	–	82.1	
57	Ni@C/Al_2O_3-400	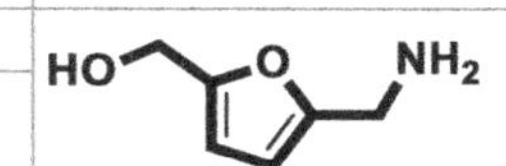	Sub. (0.5 mmol), cat. (100 mg), MeOH (5 mL), NH_3 (7 mmol), H_2 (20 bar), 30 °C, 16 h.	99	96	[98]
58	Ni@C/ZrO_2-400			99	60	
59	Ni@C/Al_2O_3-400	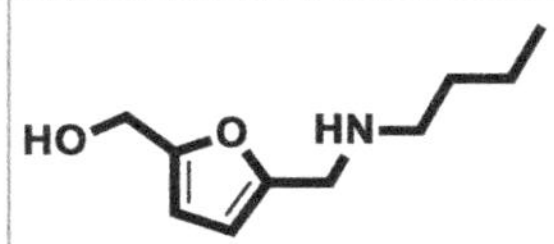	Cat. (100 mg), sub. (0.5 mmol), MeOH (5 mL), NH_3 (7 mmol), H_2 (20 bar), 30 °C, 16 h.	–	90	
60	Ru/BNC	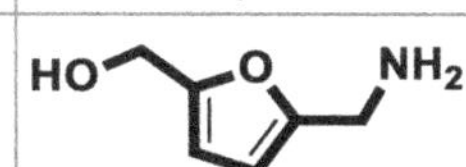	Sub. 0.12 mmol, cat. 10 mg, 1.8 wt% Ru, 1.1 mol% Ru of sub., $N_2H_4{\cdot}H_2O$ (0.24 M in 2.0 mL MeOH, 4 eq.), H_2 (20 bar), 80 °C, 16 h	99	94	[88]
61	NiAl-10	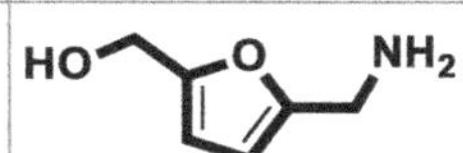	10 mg catalyst, 1 mmol substrate, 10 mL methanol, 1 h, 20 bar H_2, 5 bar NH_3, and 90 °C.	99	98	[89]

which involved formation of reversible imine intermediate from aldehyde and ammonia, which further underwent reduction to form furfurylamine. The catalysts which showed selectivity of FUR to furfurylamine were Pd/C, Pd/Al_2O_3, Rh/C, Rh/MCM, Rh/Al_2O_3, Pt/C, Pt/MCM and Pt/Al_2O_3 (Table 2.1 Entry 2–12). They further discussed that the reductive amination and hydrogenation reaction are the two main competing pathways at play on the catalytic active sites in this process. They showed it through extensive optimization of parameters such as H_2 pressure, NH_3 concentration and reaction time; and its necessity to achieve the desirable selectivity of the amine product. In the same year, Chieffi et al. [62] showed continuous reductive amination of bio-derived aldehydes over carbonised filter paper supported Fe-Ni alloy. The alloy nanoparticles have been shown to be enwrapped and stabilised by graphitic shells from the carbonised filter paper. This work represented the first report specifically targeted towards reductive amination of FUR and HMF using non-noble metal catalysts. They showed a wide variety of substrates shuffled with FUR and HMF with different amines, nitro compounds and even four different amino acids to produce the corresponding coupled secondary amine products in decent yields (Table 2.1 Entry 13–16 & Table 2.2 Entry 2–3).

Ebitani and group [63] reported a new catalyst namely Poly (N-vinyl-2-pyrrolidone)-capped ruthenium-supported hydroxyapatite (Ru-PVP/HAP) capable of exhibiting notable activity for the reductive amination of furfural. 5 wt.% Ru-PVP/HAP exhibited the highest yield of 60% for furfurylamine in liquor ammonia and hydrogen gas as the reductant (Table 2.1 Entry 17). They were able to confirm the Ru metal cluster as the catalytic active site in the Ru-supported HAP catalysts using material characterisation techniques such as Powder X-ray diffraction and TEM analysis. Ru particle size and product selectivity correlation was also established in this work.

Next, Zhang and group [64] reported an efficient catalyst system in the form of partly reduced Ru/ZrO_2 for reductive amination of bio-derived aldehydes and ketones using aq. ammonia and molecular hydrogen. Reductive amination of 5-methyl furfural using Ru/ZrO_2 catalyst gave a yield of 61% at optimized conditions (Table 2.1 Entry 18). They also showed a one pot approach to synthesis of ethanolamine from lignocellulosic biomass, demonstrating a yield of 10%. Moreover, through characterisation and experimental observations, a synergistic role of metallic Ru and RuO_2 species with being an active site for imine intermediate reduction and enhanced acidity respectively, necessary to accelerate the rate of the reaction.

The subsequent report by Hara and co-workers [65] further aided to clarify the role of Ru metal and support interactions needed for achieving higher efficiency for the reductive amination process. A highly selective synthesis of furfurylamine was reported over ruthenium nanoparticles supported on Nb_2O_5 as a reusable heterogeneous catalyst. The high selectivity of this process was attributed to the weak electron-donating nature of Ru nanoparticles on the Nb_2O_5 surface. They compared 14 different permutations of 8 support materials and 7 metals (5 noble and 2 non-noble metals) for this reaction, where 6 catalysts exhibited exceptional selectivity and % conversion for furfurylamine product (Table 2.1 Entry 19–24).

Ru/Nb_2O_5 also gave a substantial yield of 96% for conversion of HMF to 5-hydroxymethyl furfurylamine (Table 2.2 Entry 10). They also conducted free energy calculations for the reactants, products and intermediates involved in the reductive amination reaction through DFT study, which showed the primary amine (furfurylamine) product to be thermodynamically unfavorable over formation of secondary amine, which further highlights the catalytic ability of Ru/Nb_2O_5 catalyst. They further demonstrated the applicability of the Ru/Nb_2O_5 catalyst by employing it on 10 multi-functional diverse substrates.

Debraj et al. [66] further explored specific flat-shaped pristine fcc bare ruthenium nanoparticles with highly active {111} crystal planes exposed on the surface, as an efficient catalyst for having a large fraction of atomically active facets exposed on their flat surfaces have been developed that act as a highly selective and reusable heterogeneous catalyst. It produced a widely functionalized primary amine products with high reaction rates at mild reaction parameters. The phase selective synthesis of flat-shaped fcc Ru nanoparticles and its surface dwelling {111} facets achieved the much higher TOF compared to other Ru metal catalyst (Ru-HCP, Ru/Nb_2O_5, Ru/SiO_2) at 1850 h^{-1} for reductive amination of FUR to furfurylamine (Table 2.1 Entry 25–28). These templated Ru nanoparticles also exhibited excellent yield of 95% and TOF of 739 h^{-1} for reductive amination of HMF at identical conditions (Table 2.2 Entry 11).

Hara and co-workers [67] further contributed by reporting a novel catalyst namely $Ru/Nb_2O_5{\cdot}nH_2O$-300, comparing its catalytic performance for reductive amination of FUR with their previous catalyst, Ru/Nb_2O_5. It was observed that the nanocomposite catalyst reduced at 300 °C showed the highest catalytic performance among five other catalyst alternates. The superior catalytic activity was explained by high density acidic sites on support surface to assist in low temperature reductive amination of FUR. The $Ru/Nb_2O_5{\cdot}nH_2O$-300 catalyst reduced at 300 °C was able to achieve a furfurylamine yield of 89% while Ru/Nb_2O_5 achieved a yield of 76% at identical reaction conditions (Table 2.1 Entry 29 & 30). Similarly, HMF conversion to 5-hydroxymethyl furfurylamine with $Ru/Nb_2O_5{\cdot}nH_2O$-300 gave a yield of 80% (Table 2.2 Entry 12).

Zhang et al. [68] prepared a nitrogen doped carbon support for loading of Ni nanoparticles, which effectively catalyses reductive amination of carbonyl species to primary amines. The catalyst synthesis procedure involved use of organic monomers such as ethylene diamine, hexamethylenetetramine and 2,4-dihydroxybenzoic acid. It comprises three steps, which are hydrothermal polymerisation encapsulating Pluronic-123 support, surface ion exchange to Ni (II) ions and pyrolysis at 500 °C in inert atmosphere. MC/Ni catalyst achieved a yield of 88% at 80 °C at 1 bar hydrogen pressure (Table 2.1 Entry 31). In the immediate subsequent report, Ziliang et al. [69] reported applying Co@NC-800 (Cobalt supported on nitrogen doped carbon material pyrolyzed at 800 °C) catalyst for the reductive amination of furfural, exhibiting a decent yield of 81% (Table 2.1 Entry 32). The kinetic studies conducted showed formation of two pathway intermediates namely arylidene amine and imine, with both undergoing hydrogenation to form the primary amine product. The two following reports from Yuan et al. [70] and Manzoli et al. [71] showed Ni metallic nanoparticles on different catalytic supports namely Al_2O_3 and SiO_2 with magnetite respectively. Both the catalysts were able to produce substantial yields of 90% (Ni_6AlOx) with thermal approach and 100% ($Fe_3O_4@SiO_2$-Ni) with microwave irradiation approach for synthesis of furfurylamine (Table 2.1 Entry 33 and 34). Yuan et al. [70] tested 6 different variations of NiAlOx catalysts and Raney Ni for reductive amination of HMF producing a yield in the range of 30–90% (Table 2.2 Entry 13–20).

Ru/Nb_2O_5 catalyst was further studied by Guo et al., [72] where they made an interesting observation through extensive XPS and CO-DRIFTS analysis, that the catalytic performance for reductive amination reaction was not influenced by the electronic effect of the support material, rather the morphology of the support and dispersion of metallic nanoparticles played an important role in accelerating the rate of the reaction. They compared layered, hollow sphere and nanoflower surface morphologies of Nb_2O_5 for supporting Ru nanoparticles for conversion of FUR to furfurylamine. Although, support materials comprising of surface acid sites are definitely a basic necessity for reductive amination catalyst, metal-support electronic effect hardly played any role. Furthermore, Ru species was proven to be the activating species for carbonyl functionality through in situ DRIFTS-MS analysis. 1% Ru/Nb_2O_5-L exhibited a furfurylamine yield of 60% under optimized conditions (Table 2.1 Entry 35).

Xu and co-workers [73] tested Raney Co catalyst for reductive amination of furfural to synthesize primary and secondary amines. They also compared five other commercial catalysts with Raney Co such as Rh/C, Pd/C, Pt/C, Ru/C and Raney Ni. Raney cobalt achieved a furfurylamine yield of 98.4% with 100% conversion, while Rh/C and Raney Ni gave a yield of 54% and 65% respectively with 100% conversion (Table 2.1 Entry 36–39). They also tested secondary amine synthesis for this catalyst using furfural in absence of NH_3. All the catalysts except Raney Co gave quantitative yields close to 99%, with Raney Co getting a yield of 78% for secondary amine product (Table 2.1 Entry 40–45).

Wang and group [74] developed p-methylbenzylamine ligand stabilised Pd nanoparticles for reductive amination of carbonyl compounds at ambient temperature and hydrogen balloon pressure. These Pd nanoparticles achieved a yield of 97% for furfurylamine (Table 2.1 Entry 46) with good reusability up to 5 reaction runs. Subsequently, Dong et al. [75] incorporated HZSM-5 as catalytic support due to its high number of strong surface acid sites with different metallic nanoparticles of ruthenium, palladium and platinum. The highest yield for furfurylamine was achieved by Ru/HZSM-5(46) at 76% with 99% conversion of furfural (Table 2.1 Entry 47–49). They showed that the RuO_2 and metallic Ru in Ru/HZSM-5(46) catalyst showed synergistic effect in maximizing the yields for the reductive amination process.

In an attempt to improve the carbon balance in the reductive amination process, Gould et al. [76] applied different metals on silica support for reductive amination of furfural (Table 2.1 Entry 50–53). Ru/SiO_2 was among the best catalysts for reductive amination of furfural with furfurylamine yields reaching 72% with 77% carbon balance. The other catalysts involved Ni/SiO_2 and Co/SiO_2 exhibiting furfurylamine yields of 51% and 70% respectively with both having around 70% carbon balance. Pd/SiO_2 achieved decent yield of 60% for secondary amine product of furfural with 77% carbon balance.

For converting furfural into secondary and tertiary tetrahydrofurfurylamines; Pera-Titus, Vigier and co-workers [77] applied the commercial Pd/Al_2O_3 catalyst with THFA yields reaching 98% at 80 °C in just 90 min of reaction time (Table 2.1 Entry 54). They also showcased a wide scope library of secondary and tertiary tetrahydrofurfurylamines under mild conditions (25 °C/1 bar H_2) with excellent yields (>90%).

There were some attempts from the scientific community to apply catalysts which can reduce the temperature requirements of the reductive amination reaction to ambient conditions. Han and co-workers [78] employed naturally occurring phytic acid as a precursor for producing titanium phosphate support for Ru nanoparticles. Interestingly, the material obtained from this process had excellent catalytic performance for reductive amination of carbonyl compounds, where it was able to achieve almost quantitative furfurylamine yield of 91% using Ru/TiP-100 at 30 °C in 24 h of reaction time (Table 2.1 Entry 55). Ru/TiP-100 also gave a exceptional yield of 93% for 5-hydroxymethyl furfurylamine from HMF (Table 2.2 Entry 39). Ru (0), RuO_2 and TiP had a synergistic effect with balanced ratio of Ru/RuO_2 providing suitable electron density of metallic Ru and high surface acidity of TiP support.

As the earlier reports suggested that majority of the robust and widely reusable catalysts with excellent catalytic performance have mostly been noble metal catalysts. Non-noble metal catalysts lacked the stability essentially although the performance achieved by these catalysts is on par to the noble metal catalysts. To circumvent this limitation, Mitsudome and co-workers [79] synthesized single crystal Co_2P nanorods, which act as highly efficient catalyst for reductive amination of carbonyl compounds. They tested different nitrogen sources such as ammonia as gas and aqueous solution form and ammonium acetate with hydrogen pressure of just 1 bar. Co_2P nanorods exhibit metallic nature with Co-Co active sites, while being air stable and catalytically robust achieved through alloying with phosphorous in the structure. Co_2P nanorods gave a substantial yield of 90% and 87% for furfurylamine and 5-hydroxymethylfurfurylamine respectively (Table 2.1 Entry 56 & Table 2.2 Entry 39). On a similar note, Dong et al. [80] developed Ni nanoparticles supported on Ni nanoparticles supported on γ-Al_2O_3 as an active heterogeneous catalyst for synthesis of primary amines from the reductive amination process covering aromatic, branched, and purely aliphatic substrates with excellent yields under mild reaction conditions (100 °C and 20 bar H_2 pressure). In particular, the yield of primary amines for earth-abundant Ni catalysts was superior compared to noble metals catalysts (Ru, Rh, and Pd). Ni/Al_2O_3 catalyst achieved the highest furfurylamine yield of 92%, compared to nine other catalysts with different metals such as Co, Ru, Pd (Table 2.1 Entry 57–66). Moreover, $10Ni/Al_2O_3$ also achieved a yield of 87% for reductive amination of HMF to desired primary amine product (Table 2.2 Entry 46).

To further improve metal atom utilization efficiency for the reductive amination process, Qi et al. [81] reported a ruthenium metal based single atom catalyst synthesized through by pyrolysis of ruthenium acetylacetonate and N/C precursors at different elevated temperatures in nitrogen atmosphere followed by ammonia gas. The resultant Ru_1-N_3 structure exhibited great capability towards balancing the transamination and imine hydrogenation to yield primary amine. The single atom catalyst, Ru_1/NC-900-800NH_3 has been compared with three different variants of ruthenium catalysts including another single atom catalyst, Ru_1/NC/Nb_2O_5, which showed almost quantitative yields in the range of 80–97% (Table 2.1 Entry 67–70). This catalyst also showed encouraging yields for 5-hydroxymethyl furfurylamine and 2,5-bisaminomethyl furan from HMF and 2,5-diformyl furan producing 93% and 63% respectively (Table 2.2 Entry 47 and 48).

The previous reports over the years showed a keen focus two separate metallic nanoparticles as reassuring candidates for the reductive amination process, which were ruthenium and nickel metallic nanoparticles. So, our group [82] conducted a comparative assessment of these two metals for the reductive amination process using similar conditions. Montmorillonite clay K-10 was used as the support for both the metals to form nanocomposites, which were compared for the reductive amination of FUR and HMF. 2% Ru/MMT and 12.5% Ni/MMT produced yields of 89% and 84% for furfurylamine at separate optimized conditions (Table 2.1 Entry 71 and 72). Moreover, these catalysts achieved yields of 86% and 79% respectively for the reductive amination of HMF (Table 2.2 Entry 49 and 50). Nickel nanocomposites were able to achieve comparable yields to Ru-K10 nanocomposites, but they required slightly harsher reaction conditions.

The next report from Xu, Fan and group [83] covered a metal free catalyst namely, $BF_3 \cdot Et_2O$ for base-free direct reductive amination of aldehydes with amines using formic acid as a reductant. A wide scoped library of primary, secondary and tertiary amines was synthesized with decent functional group tolerance. FUR was used as a substrate to synthesize primary and tertiary amine using liq. Ammonia and N-methyl aniline as the nitrogen source, generating yields of 90 and 82% respectively (Table 2.1 Entry 73 and 74). In the same year, Gao et al. [84] showed a quantitative self-regulated control and usage of ammonia for the reductive amination process. The layered boron nitride supported ruthenium catalyst (Ru/BN-e) showed enhanced interfacial electronic effects of the metal with the support accelerating the hydrogenation step of the reductive amination process, which directly affected the use of ammonia in efficient capacity. This process was able to achieve yields of 95% and 94.5% for furfurylamine and 5-hydroxymethyl furfurylamine (Table 2.1 Entry 75 and Table 2.2 Entry 51).

Ma and co-workers [85] developed an eco-friendly, and highly effective procedure for reductive amination by applying a non-toxic Co based heterogeneous catalyst, whose catalytic performance was able to deliver substantial yields for furfural and widely functionalized aldehyde derivatives. The other advantages were that the Co@C-600-EtOH catalyst was magnetically separable with excellent reusability up to 8 consecutive runs. Through comprehensive characterisation of the catalyst, the authors concluded that the outer graphitic shell of the carbon support was activated by the electronic interaction, enabling the easy substitution of the –NH_2 moiety toward different substrates. Co@C-600-EtOH catalyst was able to produce a yield of 87% and 99% for reductive amination of FUR and HMF (Table 2.1 Entry 76 and Table 2.2 Entry 54).

After their two significant earlier reports, Hara and co-workers [86] developed a dendritic silica-based Ni catalyst for the reductive amination of carbonyl compounds at very mild conditions. It was inferred from detailed characterisation that the particle size of partially oxidized Ni nanoparticles at around 4.5 nm was responsible for the exceptional catalytic performance for reductive amination of carbonyl compounds. The three different variants developed namely Ni@DS-7.5, Ni@DS-15 and Ni@DS-30 were able to produce furfurylamine yields around ~85% (Table 2.1 Entry 77–79). They also applied this catalytic system on five different drug molecules and their precursors to showcase the practical applicability and the 'softness' of the process.

Rodríguez-Padron et al. [87] developed a highly active, selective and stable titania-based catalytic material from eco-friendly microwave-assisted approach. The catalytic materials namely 1% Rh/TiO_2 and 1% Ru/TiO_2 gave significant furfurylamine yields in the range of 95–99% for both the catalysts at optimized parameters (Table 2.1 Entry 80 and 81). Additionally, the materials showed remarkable recyclability, retaining high conversion and selectivity values after six reaction cycles. Subsequently, the majority reports focused on implementation of Ru and Ni metals with different catalytic supports as effective catalysts for the reductive amination process. Recently, Chen and group [88] developed a catalytic system comprised of Ru nanoparticles supported on boron/nitrogen doped carbon to be applied for the reductive amination of furfural using hydrazine as the nitrogen source. Mechanistic investigation revealed the explanation for the exceptional catalytic activity suggesting presence of Frustrated Lewis acid pairs (FLPs) enhancing the catalytic activity of the catalyst. This system was shown to be compatible with a variety of aldehydes with moderate to excellent yields. Ru/BNC catalyst was able to achieve yields of 99% and 94% for furfurylamine and 5-hydroxymethylfurfurylamine at the optimized parameters (Table 2.1 Entry 82 and Table 2.2 Entry 60).

The next report from Pan et al. [89] involved synthesis of two carbon-doped Ni catalysts distinguished over the Ni particle sizes, which aids steer the selectivity in the reductive amination of carbonyl compounds between primary amines and secondary imines. The smaller size of Ni provides higher selectivity for primary amines. The synthesis involves modification of confined pyrolysis of Ni-Al layered double hydroxides embedded on hollow polymer nanospheres. The two catalysts exhibit excellent stability up to 10 cycles. NiAl-10 catalyst was compared with Raney Ni, where it achieved a furfurylamine yield of 99% with the latter managing a yield of 81% (Table 2.1 Entry 83–85). This catalytic system was also applied to HMF, which afforded a yield of 98% (Table 2.2 Entry 61).

Most recently, Chen and co-workers [90] demonstrated a highly efficient strategy to various biomass-based tertiary amines by consecutive reductive amination of furfural with various amines in one-pot using rhodium phosphide (Rh_2P) catalyst. This acts as the first report for one pot synthesis of bio-derived tris(2-furanylmethyl) amine with yields reaching 92% at optimized conditions (Table 2.1 Entry 86).

Now, we will discuss the reports specifically focused towards the reductive amination process of HMF to 5-hydroxymethyl furfurylamine and 2,5-bisaminomethyl furan. There were multiple efforts in the last decade for improving the conversion and selectivity for reductive amination of HMF to different value added chemicals.

Firstly, Zhu et al. [91] reported a process involving reductive amination of HMF with CO gas and water as reductant. This was the first report with respect to molecular hydrogen or hydrogen surrogate free conversion of HMF to secondary and tertiary amines. Single phase rutile titania supported gold (Au/TiO_2-R) catalyst was shown to efficiently catalyze this molecular hydrogen free reductive amination process for varied library of aldehyde substrates. Au/TiO_2-R and 4 other variations of Au catalysts were compared for the reductive amination of HMF with aniline, N-methyl aniline, producing secondary and tertiary amines with yields in the range of 85–99% (Table 2.2 Entry 4–9).

Wei and group did significant amount of work on reductive amination of furfuryl alcohol, FUR and HMF with different metal catalysts. Raney Ni catalyst gave the highest furfurylamine yields from FUR at 100% at moderate conditions (Table 2.2 Entry 21–25, 93]. Raney Ni also produced a yield of 60.7% for 2,5-bisaminomethylfuran at 160 °C in 12 h of reaction time. DFT calculations were also carried out to understand each transition metal's catalytic potential for the reductive amination process. It was found that the adsorption energy difference for NH_3 and H_2 molecules was the lowest for Ni, which may aid efficient conversion of carbonyl compounds into desired amine products. Their group worked further on the same process, by applying the Raney Co catalyst [93]. Raney Co and Raney Ni achieved a yield of 99.5% and 80.7% for 5-hydroxymethyl furfurylamine respectively, while conversion of HMF to 2.5-bisaminomethyl furan yielded highest yield of 86.5% (Table 2.2 Entry 41–45). Following this work, the same group developed mono- and bi-metallic Ni-based catalysts with different permutations of 6 supports and 14 secondary metals for reductive amination of HMF to 2,5-bis(aminomethyl)-furan [94]. γ-Al_2O_3 and Mn as co-metal were the best candidates for enhancing the reactions yields. The catalysts namely 10Ni/γ-Al_2O_3 and 10NiMn(4: 1)/γ-Al_2O_3 produced 86.3% and 82.1% yields respectively for synthesis of 2,5-bisaminomethyl furan from HMF (Table 2.2 Entry 55 and 56).

Next, Karve et al. [95] developed a novel Pd metal supported on MOF/polymer composite for reductive amination of HMF to form secondary amines under mild conditions. These composite catalysts were able to exhibit exceptional catalytic activity with high TOF values of 302.4 h^{-1}, with great stability of the catalyst shown up to 15 reaction runs. UiO-67/PpPDA/Pd and Fe-BTC/Pd catalysts were able to produce yields of 94.5% and 95% respectively, compared with Pd/C (Table 2.2 Entry 26–29).

Shi and co-workers [96] developed a simple and highly efficient bifunctional $CuNiAlO_x$ catalyst for the one pot transformation of HMF into 2,5-bis (aminomethyl)furan. Different ratios of Cu, Ni, Al were used in stoichiometric amounts to synthesize nine different variations of $CuNiAlO_x$ catalyst. $Cu_4Ni_1Al_4O_x$

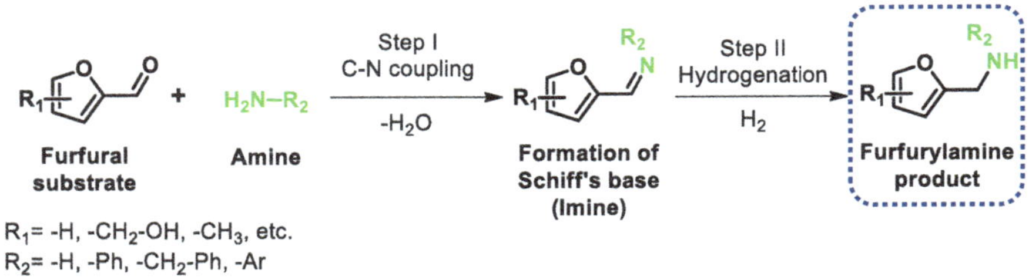

Fig. 2.2 Plausible mechanistic pathway for reductive amination of FUR and HMF based on literature reports

provided the best catalytic performance for synthesis of 2,5-bisaminomethyl furan product in 86% from HMF (Table 2.2 Entry 30–38).

Jagadeesh and group [97] synthesized cobalt nanoparticles embedded on MOF template of terephthalic acid-piperazine supported on silica, as a heterogeneous catalyst for reductive amination of HMF. The templated MOF aided to control the morphology and the particle size of the cobalt nanoparticles. The metal loading was achieved through immobilization of metal on the template and inert pyrolysis at different temperatures ranging from 400 °C to 1000 °C. Co-terephthalic acid-piperazine@SiO_2-800 achieved encouraging yields of 94% and 95% for primary and secondary amine (coupled with aniline) respectively (Table 2.2 Entry 52 and 53).

Finally, Hu et al. [98] reported an alumina supported carbon doped Ni catalyst developed by distinguishing pyrolysis reduction temperatures and absence of components in a controlled manner. Ni@C/Al_2O_3-400 was able to achieve a high yield of 96% for reductive amination of HMF to 5-hydroxymethyl furfurylamine using ammonia and molecular hydrogen at ambient temperatures (Table 2.2 Entry 58). Additionally, Ni@C/Al_2O_3-400 gave a decent yield of 90% for secondary amine from HMF and n-butylamine (Table 2.2 Entry 59).

Now, we discuss the mechanistic pathway for reductive amination of FUR and HMF as shown in Fig. 2.2. The first step involves the coupling of aldehyde functionality of the substrate and the respective amine used to form Schiff's base (imine). This step is catalyzed by material supports with acidic sites (Brønsted and Lewis), which is evident from the application of such materials as supports for all the reported catalysts. The second step however, requires both metal active sites as well as surface acidity (preferably Brønsted acidity) for completing the hydrogenation step. The reports for this reaction over the years have shown that multiple catalyst surface properties are collectively responsible for the desired selectivity tuning and achieving the desired product in substantial yields. Collectively, catalytic technologies for the reductive amination of FUR and HMF by heterogeneous catalysis pathway has seen significant development in the last decade. Different approaches to the choice of metal, metal oxidation state, choice of catalytic support, focused support material design and detailed parameter optimizations have been the focal point for improving the yields and conversion of FUR and HMF into corresponding amines. However, there have not been many noteworthy efforts towards upgrading this process to flow syntheses, which is a vital step towards commercialization of the reductive amination process for these bio-derived molecules. However, the reductive amination of LA has several reports for applying the reaction through flow synthesis pathway. We will explore this in the next section of this chapter.

2.3 Reductive Amination of LA

Pyrrolidones are an important class of compounds with multiple applications as organic solvents, ink applications, surfactants and even pharmaceutical drug design [99, 100]. The rational desired commercial synthesis pathway involves reductive amination of LA or its esters with corresponding alkyl or aryl amines [35]. In the last three decades, initial catalytic systems were patented for conversion of LA to respective pyrrolidone products [101, 102]. However, significant work in developing novel heterogeneous catalysts for this process was conducted in the last decade. The process development for the reductive amination of LA commenced much earlier as compared to the reductive amination of FUR and HMF. Now, we will discuss all the significant reports for the reductive amination of LA, in heterogeneous catalysis paradigm, in this section of the chapter.

Firstly, Du et al. [103] reported reductive amination of LA using gold nanoparticles deposited on acid-tolerant ZrO_2 (Au/ZrO_2-VS), where formic acid was used as the reductant. They covered a small but wide variety of aliphatic & aryl amines, anilines and even ammonia for demonstrating the catalytic activity of the catalyst for the reductive amination process. Au/ZrO_2-VS gave a great yield of 88% and 97% for aniline and benzylamine substrates respectively (Table 2.3 Entry 1 and 2). Next, the first catalyst free report came of Xiao and co-workers, [104] where formic acid and triethylamine were used as reagents for conversion of levulinic acid into *N*-substituted pyrrolidone products. The incorporation of triethylamine in the system was to balance the acidity of the system for enhancing product yields. This process was able to achieve a yield of 34% and 87% for aniline and benzylamine substrates respectively (Table 2.3 Entry 3 and 4), at much lower temperature of 100 °C in a catalyst free system.

The first report for reusable and widely applicable heterogeneous catalytic system for the reductive amination of LA, was from Touchy et al., [105] where Pt-MoO_X/TiO_2 catalyst was compared with 24 other catalysts in terms of its catalytic performance with a high TON value of 2150. It also achieved exceptional yields over a varied library of aliphatic and aryl amines at relatively mild conditions. Pt-MoO_X/TiO_2 exhibited a yield of 90% and 94% for aniline and benzylamine substrate respectively (Table 2.3 Entry 5 and 6).

Interestingly, another catalyst free process was reported by Andrioletti and group, [106] where the focus was on reducing the amount of waste generated, in terms of *E*-factor (amount of waste generated per kg of product). This process was compared with earlier reports, where this process only produced an *E*-factor of 0.2. The

Table 2.3 Reductive amination of LA using heterogeneous catalysts in chronological order

Entry No.	Catalyst	Product	Reaction conditions	Conv. (%)	Yield (%)	Ref.
1	Au/ZrO_2-VS	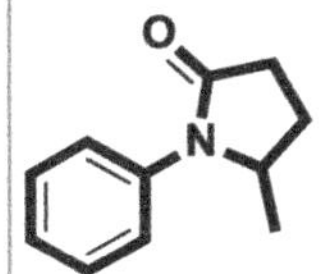	Sub. 8 mmol, amine 8 mmol, FA 8 mmol, au 0.05 mol.%, water 0.56 mL, 130 °C	91	88	[103]
2	Au/ZrO_2-VS	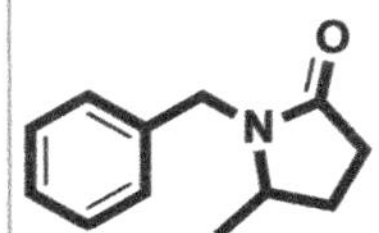		98	97	
3	Catalyst-free	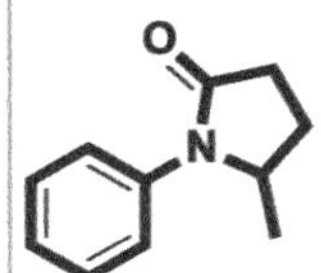	Sub. 1 mmol, amine 3 mmol, HCOOH 5 mmol, Et_3N 1 mmol, DMSO 3 mL, 100 °C, 12 h	–	34	[104]
4	Catalyst-free	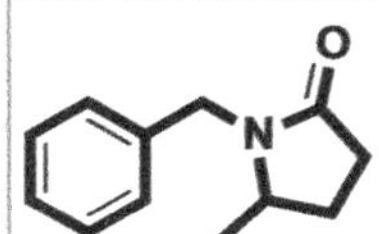	Sub. 1 mmol, amine 2 mmol, HCOOH 5 mmol, Et_3N 1 mmol, DMSO 3 mL, 100 °C, 12 h	–	87	
5	Pt-MoO_X/TiO_2	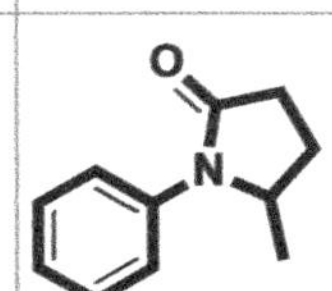	0.001 mmol cat., 1.0 mmol sub., 1.0 mmol amine, H_2 3 bar, 100 °C, 20 h, no solvent	–	90	[105]
6	Pt-MoO_X/TiO_2			–	94	

		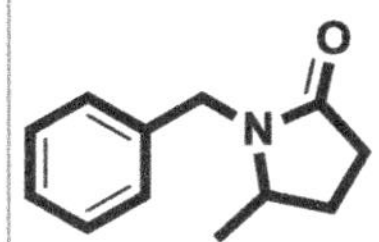				
7	Catalyst free	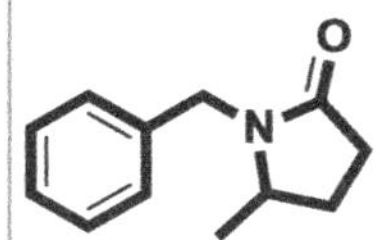	60–150 mmol scale, ratio LA/FA/sub.: 1: 1: 1. P = 25 to 35 bar, Tmax = 150–200 °C.	99	49	[106]
8	0.2% Pt/TiO_{2D}	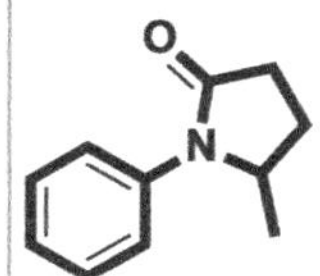	Sub. 2 mmol, amine 2 mmol, H_2 10 bar, 100 mg cat. (0.05 mol%), S/C = 2000, at 120 °C, neat	98	93	[107]
9	0.2% Pt/TiO_{2D}	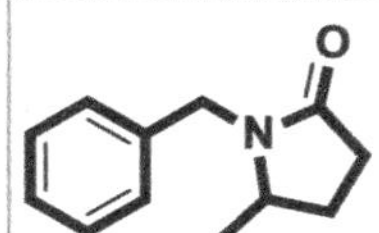		99	90	
10	Carbon-supported FeNi alloy	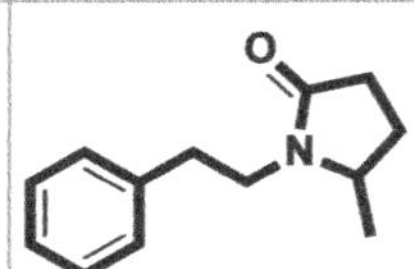	50 mM sub., 25 mM amine, 150 °C, H_2 85 bar, 0.3 mL min^{-1}, WHSV 25 h^{-1}	91	99	[62]
11	Pd/ZrO_2	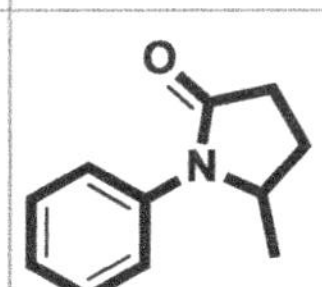	5 mmol sub., 5 mmol amine, S/C 1000, 12 h, 90 °C, H_2 5 bar.	94.4	82.7	[108]

(continued)

Table 2.3 (continued)

Entry No.	Catalyst	Product	Reaction conditions	Conv. (%)	Yield (%)	Ref.
12	0.2 wt.% Pt/TiO_2	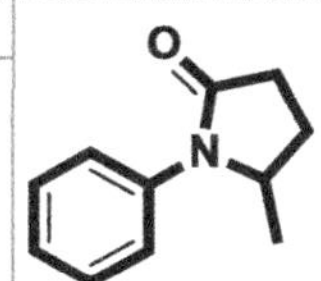	Sub. 2 mmol, amine 2 mmol, H_2 10 bar, cat. 0.05 mol. %, 120 °C, 6 h	20	10	[109]
13	0.2 wt.% Pt/TiO_2-NT		Sub. 2 mmol, amine 2 mmol, H_2 10 bar, cat. 0.05 mol. %, 120 °C, 48 h	94	92	
14	NHC-Ru solid 7	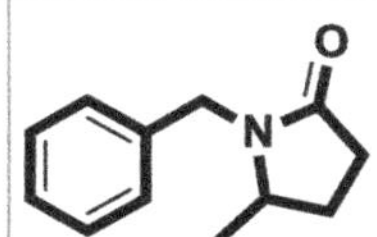	0.15 mol.% cat., amine precursor 2 mmol, ammonium formate 2.2 mmol & formic acid 4 mmol heated at 80 °C for 12 h, then sub. 2.1 mmol was added and stirred for 24 h	–	85	[110]
15	Ni@CNTs	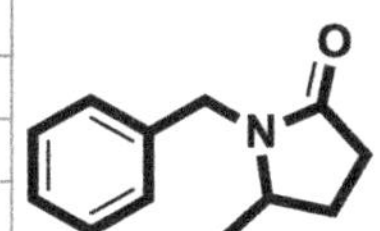	Sub. 10 mmol, amine 10 mmol, cat. 0.03 g, 10 wt.% Ni, γ-valerolactone 4 mL, 30 bar H_2, 130 °C 6 h.	–	99	[111]
16	CNF_{10}@Ni@CNTs			–	99	
17	CNF_{20}@Ni@CNTs			–	99	
18	CNF_{30}@Ni@CNTs			–	99	
19	CNF_{50}@Ni@CNTs			–	72	
20	CNF_{80}@Ni@CNTs			–	70	
21	Ni/CNTs			–	90	
22	Ni/C			–	99	
23	Ni/TiO_2			–	34	
24	Ni/$NbOPO_4$			–	42	
25	Ni/HZSM-5			–	21	
26	Ni/SAPO-34			–	32	
27	Ru-PP/CNT		Sub. 1.0 mmol, amine 1.2 mmol, cat. (50 mg), THF 5 mL, pH $_2$ 30 bar, 120 °C, 24 h	56.7	36.4	[108]

		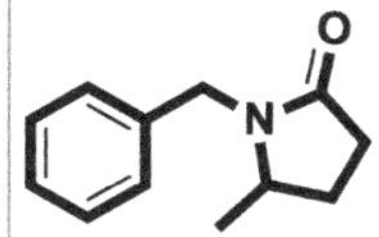				
28	Ir/ SiO_2-SO_3H	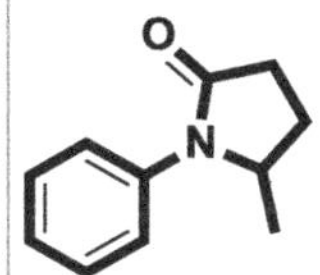	1000 rpm using 40 mL of 0.125 M aniline solution and 0.250 M levulinic acid solution, with ethyl Acetate, at a temperature of 100 °C, with H_2 34.4 bar, 0.1 g of cat.	–	40	[112]
29	Ir/ SiO_2-SO_3H	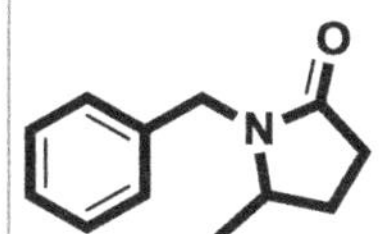		–	31	
30	Raney-Ni	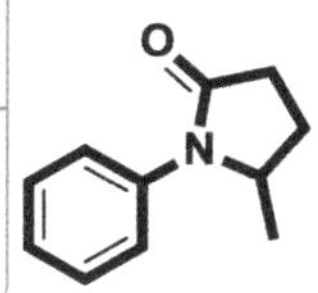	Sub. 1.0 mmol, amine 1.0 mmol, 1 mmol formic acid, 20 mg cat., 2 mL H_2O, 120 °C, 20 h.	62	55	[113]
31	Raney-Ni		Sub. 1.0 mmol, amine 1.0 mmol, 4 mmol formic acid, 20 mg cat., 2 mL H_2O, 180 °C, 6 h.	100	92	
32	$Cu_{15}Pr_3/Al_2O_3$	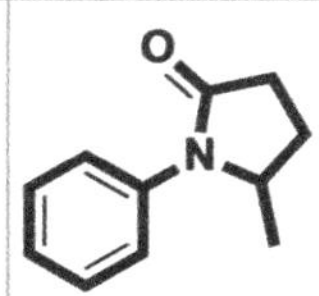	Sub./amine molar ratio = 1:1.5, cat. 5 mol%, 1, 4-dioxane, 175 °C, H_2 50 bar, 20 h.	–	66	[114]
33	$Cu_{15}Pr_3/Al_2O_3$	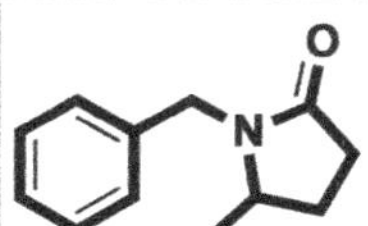		–	91	

(continued)

Table 2.3 (continued)

Entry No.	Catalyst	Product	Reaction conditions	Conv. (%)	Yield (%)	Ref.
34	Pt/P-TiO_2	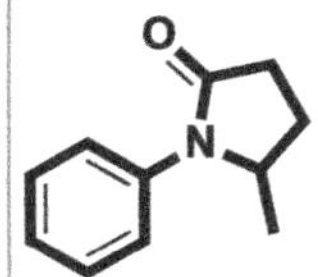	Sub. 1 mmol, amine 1 mmol, cat. 0.1 mol.% Pt, methanol 2 mL, 3 h, rt., H_2 balloon.	–	97	[115]
35	Pt/P-TiO_2	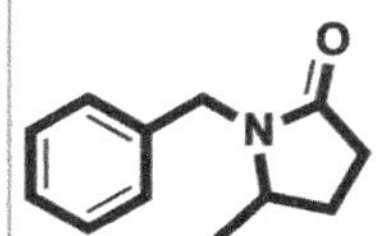		–	99	
36	C–$Au_{66}Pd_{34}$	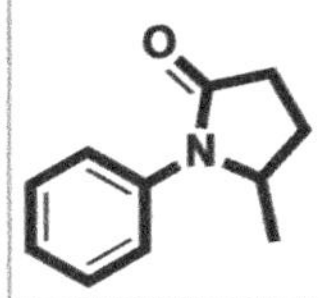	Amine 3 mmol, sub. 3 mmol, 0.3 mol.%, 85 °C, 12 h, 1 atm H_2.	13	10	[116]
37	C–$Au_{66}Pd_{34}$	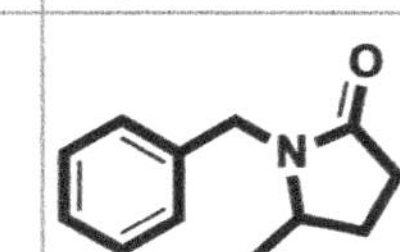		55	50	
38	5% Ru/C	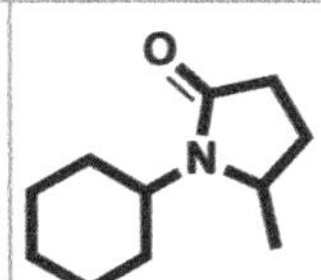	16 h, cat. 28 mg, i-octane 5 mL, 2 mmol sub., 1 mmol amine, 0.5 mL H_2O, 1 mmol FA, 150 °C, 35 bar H_2	96	88	[117]
39	Catalyst free			–	93	[118]

		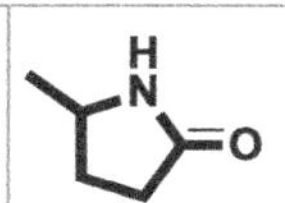	2 mmol sub., 6 mmol formic acid, 20 mmol NH_2CHO, 30 eq. H_2O, 160 °C, 1.5 h.			
40	Catalyst free	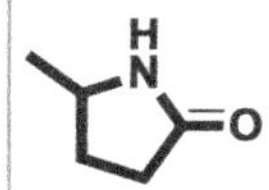	LA: Deionized water: $HCOONH_4$ (1:30:6), initial flow rate: 25 μL min^{-1}, 180 °C, adjusted flow rate: 1.5 μL min^{-1}, 1 h	100	90	[119]
41	Ru@GOIL	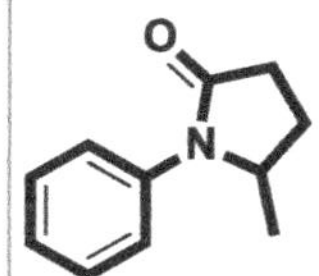	Sub. 1 mmol, amine 1 mmol, cat. 15 mg, 130 °C, 15 bar H_2, methanol 5 mL. 5 h	–	93	[120]
42	Ru@GOIL	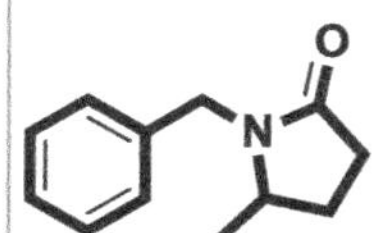		–	93	
43	0.95 wt% Pt@PW65S	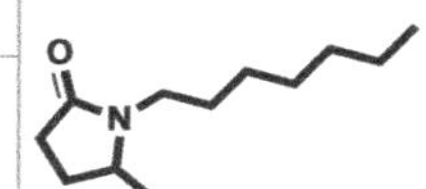	50 mg cat., 120 °C, H_2 4 bar, sub. 1 mmol in *p*-xylene 0.2 M, molar ratio (amine:Sub.) = 1.2, 4 h.	84.6	76.2	[121]
44	0.81 wt% Pt@PW79S-02%X-link			74.6	64.8	
45	Pt/c-C	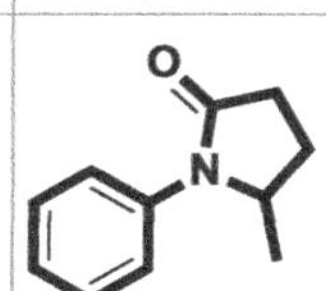	Amine 1 mmol, sub. 1 mmol, cat. 5 mg, methanol 1 mL, 30 °C, H_2 balloon, 3 h	–	96	[122]
46	Pt/c-C		Amine 1 mmol, sub. 1 mmol, cat. 5 mg, methanol 1 mL, 30 °C, H_2 balloon, 5 h	–	99	

(continued)

Table 2.3 (continued)

Entry No.	Catalyst	Product	Reaction conditions	Conv. (%)	Yield (%)	Ref.
		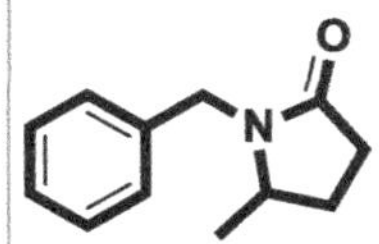				
47	5Pd/C(ox.) 250	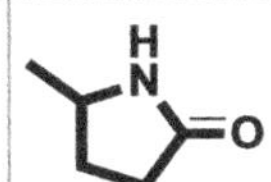	1.35 g sub., 1.5 g NH_4OH, 0.5 g H_2O, 50 bar H_2, 1 h, 200 °C	–	95.5	[123]
48	Ir-PVP	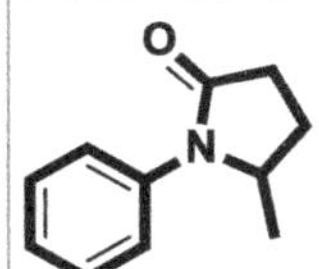	Amine 1 mmol, sub. 2 mmol, cat. (1.4 mol%), H_2 5 bar, neat, 30 °C, 24 h	99	95	[124]
49	Ir-PVP	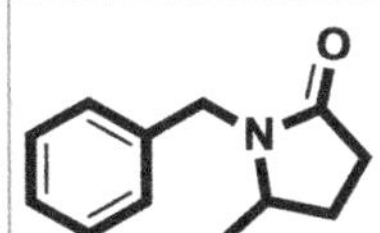		99	80	
50	Ni_2P/SiO_2	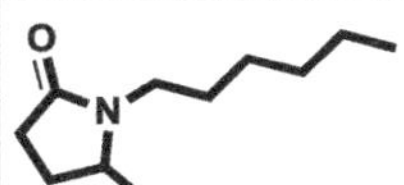	Sub. 0.04 M, amine 0.041 M in toluene, cat. 750 mg, 170 °C, 10 bar H_2, $V_{H2} = 30$ mL min^{-1}, V = 20 mL h^{-1}	98	94	[125]
51	Ni_2P/SiO_2			85	67	

		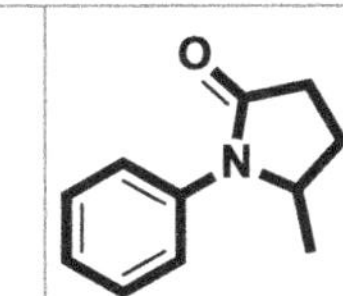				
52	Pt/C	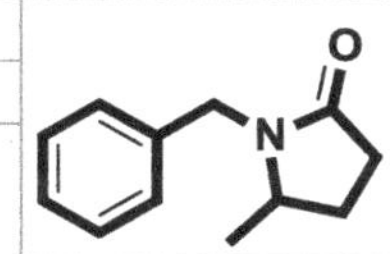	Sub. 4 mmol, nitrile 4.8 mmol, THF 5 mL, cat. 50 mg, H_2 20 bar, 80 °C, 4 h, 1000 rpm	76.8	54.7	[126]
53	Pd/C			100	78.8	
54	Raney Ni		Sub. 4 mmol, nitrile 4.8 mmol, THF 5 mL, cat. 50 mg, H_2 20 bar, 150 °C, 4 h, 1000 rpm	95.7	64.4	
55	Co@chitosan	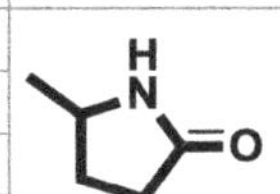	1 mL sub., 1 mL NH_4OH, 4.0 mL deionized water, 20 mol.% cat., 30 bar H_2, 130 °C, 24 h	97.2	74.1	[127]
56	Zr@chitosan			14.7	2.7	
57	Zr-co@chitosan			93.5	92.8	
58	Cu-co@chitosan			93.4	77.2	
59	Fe-co@chitosan			92.9	72.1	
60	Ni-co@chitosan			95.9	71.6	
61	Ce-co@chitosan			96.2	68.4	
62	La-co@chitosan			94.5	62.1	
63	In-co@chitosan			41.0	14.6	
64	12% Ni-MMT	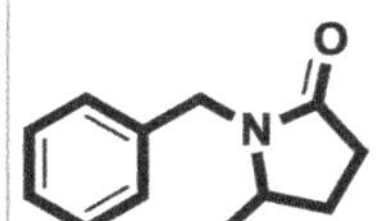	Aldehyde 0.5 mmol, 4 mL aq. NH_3, 50 mg Cat., 15 bar H_2, 2 mL MeOH, 7 h. then, sub. 0.7 mmol in 10 mL MeOH was added and H_2 pressure raised to 30 bar	100	86	[128]
65	12% Ni-MMT	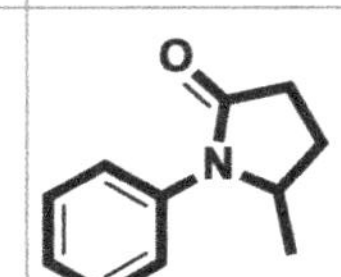	Amine 0.5 mmol, sub. 0.7 mmol, 750 mg Cat., 30 bar H_2, 2 mL MeOH, 9 h.	100	89	

(continued)

Table 2.3 (continued)

Entry No.	Catalyst	Product	Reaction conditions	Conv. (%)	Yield (%)	Ref.
66	0.75Co-phen@C-800-HCl	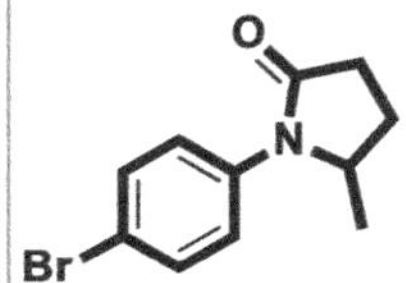	0.5 mmol sub., 0.5 mmol amine, 40 mg cat. (0.64 mmol% co), 20 bar H_2, 90 °C, 24 h, 2 mL, isopropanol.	–	91	[129]
67	0.75Co-phen@C-800-HCl	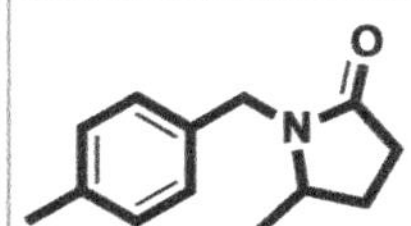		–	90	
68	0.75Co-phen@C-800-HCl	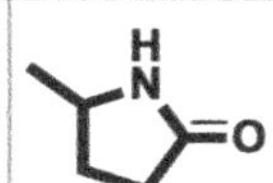	0.5 mmol sub., 0.5 mmol amine, 40 mg cat. (0.64 mmol% co), 5 bar NH_3, 20 bar H_2, 2 mL isopropanol, 90 °C, 24 h	–	96	
69	Cu_{10}/AlB_3O	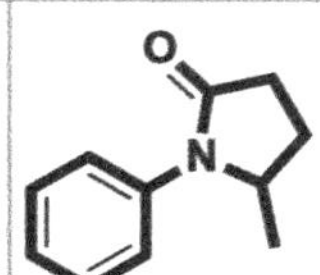	Sub.: Amine (1:1 molar ratio), H_2 30 bar, 1,4-dioxane, LHSV = 0.3 h^{-1}.	76	53	[130]
70	Cu_{10}/AlB_3O	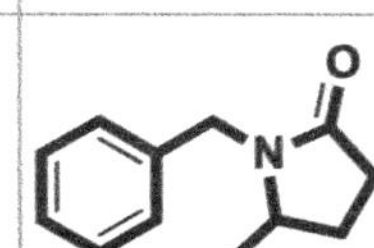		92	88	

reaction was scaled up to 60–150 mmol and tested at optimized conditions, which gave a moderate yield of 49% for benzylamine substrate with LA (Table 2.3 Entry 7).

Corma, Iborra and co-workers [107] attempted reductive amination of LA with multi-functional amines using 0.2% Pt/TiO_{2D} catalyst. They showed the 0.2% Pt/TiO_{2D} to be chemoselective by their functional group tolerance towards amine with hydrogenation susceptible groups such as cyano, carbonyl and vinyl groups. It was concluded from characterisation of the catalyst that the protonic acid sites on the support aided in hydrogen dissociation on the surface resulting in efficient hydrogenation second step for conversion of imine intermediate to the desired product. The optimized process with 0.2% Pt/TiO_{2D} catalyst was able to achieve a yield of 93% and 90% for aniline and benzylamine substrates respectively (Table 2.3 Entry 8 and 9).

Chieffi et al. [62] showcased the first report for the reductive amination of all the three bio-derived platform molecules using non-noble metal catalyst, Carbon-supported FeNi alloy. This work also featured a case study for commercial production of pyrrolidone with continuous flow experiments conducted for over 100 h. The selectivity of the pyrrolidone product kept increasing over time with absolutely no change to the catalyst's crystal structure even after 112 h of production. The highest yield achieved using 2-phenylethyl amine as the amine source was 99% with 91% conversion (Table 2.3 Entry 10).

Zheng, Xiao and co-workers [108] applied ZrO_2-supported Pd nanoparticles as a chemoselective, and robust catalyst for reductive amination of LA with molecular hydrogen at relatively mild conditions. Pd/ZrO_2 showed exceptional activity compared to conventional and commercial Pd catalysts. The enhanced catalytic activity was attributed to the plethora of Lewis acid sites available on the ZrO_2 support, which also aids reduce side products. Pd/ZrO_2 produced a decent yield of 82.7% for aniline substrate with LA (Table 2.3 Entry 11).

Another effort from Corma, Iborra and group [109] involved synthesis of *N*-substituted 5-methyl-2-pyrrolidones from ethyl levulinate and nitro compounds in the presence of Pt supported on TiO_2 nanotubes (Pt/TiO_2-NT). This one pot process comprised hydrogenation of nitro to amine moiety, formation of imine followed by reductive cyclization. 0.2 wt.% Pt/TiO_2-NT showed much better catalytic performance compared to Pt supported on TiO_2 due to poisoning of catalytic sites by nitro compounds in the first step, in turn reducing the rate of the reaction. 0.2 wt.% Pt/TiO_2-NT produced a yield of 92% of pyrrolidone with nitrobenzene (Table 2.3 Entry 12 and 13). Tu and co-workers [110] also came up with another sequential one pot process where they applied the NHC-Ru solid seven catalyst for a "dual reductive amination" process for conversion of carbonyl compounds (aldehydes and ketones) to primary amines, finally to pyrrolidone with LA. The solid catalyst was robust and reusability for 37 runs without obvious loss of activity. A high TON value of 6.7×10^4 was achieved in a molar-scale reaction with a catalyst loading at 0.001 mol%. An excellent yield of 85% was achieved with this process for benzaldehyde substrate with LA (Table 2.3 Entry 14).

The next non-noble metal catalyst advance for the reductive amination of LA was reported by Gao et al., [111] in which a facile synthesis of porous-carbon-coated Ni catalysts supported on carbon nanotubes (CNF_x@Ni@CNTs). Additionally, seven different variants of CNF_x@Ni@CNTs were synthesized and compared for reductive amination of LA with five other Ni composite materials. The best optimized CNF_{30}@Ni@CNTs catalyst gave 99% yield for pyrrolidone product while demonstrating excellent recyclability of up to 20 runs without any stability issues (Table 2.3 Entry 15–26). Based on experimental observations and density functional theory (DFT) calculations, it was inferred that Ni catalyzed reductive amination of LA followed an unconventional pathway via formation of amides followed by cyclization, dehydration, and subsequent hydrogenation, as opposed to noble metal imine intermediate mechanism.

Subsequently, Zhang et al. [108] developed a novel ruthenium based polymeric porphyrin-functionalized carbon nanotubes (Ru-PP/CNTs). Thin layer of Ru-PP/CNTs comprises cross-linked polymeric ruthenium porphyrin coating over the CNT surface. Ru-PP/CNTs exhibited excellent activity to promote reductive amination of levulinate esters for the synthesis of pyrrolidone derivatives, with yield of 36.4% for benzylamine substrate (Table 2.3 Entry 27).

Martinez, Rojas and co-workers [112] showed LA conversion to pyrrolidones using Ir/SiO_2-SO_3H catalyst in liquid phase. The modified acidity from the sulfonic groups on SiO_2 surface aided in the improving the product yield and suppressing side reactions within the process. Mechanistic investigations were also conducted through experimental and DFT calculations. The Ir/SiO_2-SO_3H catalytic system gave a yield of 40% and 31% for aniline and benzylamine substrate respectively at optimized conditions (Table 2.3 Entry 28 and 29). Amarasekara et al. [113] applied Raney Ni for reductive amination of LA using ammonium formate as a N and H source for synthesis of 5-methyl-2-pyrrolidone, showing decent reusability up to 4 cycles with only 10% loss in the yield. The process, however, did require an elevated temperature of 180 °C. Moreover, they also tested formic acid and library of other amines for synthesis of *N*-substituted pyrrolidones. Aniline as the amine source produced 55% and 92% using corresponding optimal parameters with ammonium formate and formic acid as H source (Table 2.3 Entry 30–31).

In majority, the emphasis for advancement in these catalytic systems has been on testing different transition metals and the effect of the support materials used, however, the reports utilizing lanthanide metals in these catalysts are scarce. One such work for reported by Zhao, Zhang and group [114] using praseodymium-copper alloy supported on alumina for the reductive amination of LA. Eleven different catalyst variants were developed using copper, alumina and lanthanide metals (La, Ce, Pr). A wide scope of amine substrates was explored for conversion to *N*-substituted pyrrolidones with moderate to substantial yields. $Cu_{15}Pr_3/Al_2O_3$ attained a yield of 66% and 91% for aniline and benzylamine respectively (Table 2.3 Entry 32–33).

The next two developments for this process were reported using noble metal catalysts. Firstly, Xie et al. [115] was able to report the reductive amination of LA with excellent yields at ambient conditions using Pt/P-TiO_2 catalyst (1.8% wt. Pt loading). P-TiO_2 denotes porous TiO_2 nanosheets which provides a high surface area with excellent metal dispersion capability in turn exponentially increasing the number of active sites on catalytic surface. The as-prepared Pt/P-TiO_2 showed impressive catalytic capability to produce *N*-substituted pyrrolidones with a large varied library of amines at ambient parameters. Pt/P-TiO_2 gave yields of 97% and 99% for aniline and benzylamine respectively at 1 bar hydrogen pressure and room temperature (Table 2.3 Entry 34–35). The second report was from Sun and co-workers, [116] which developed Au-Pd alloy nanoparticles for neat conversion of levulinate esters and amines to pyrrolidone products. C–$Au_{66}Pd_{34}$ catalyst showed decent catalytic performance with reusability up to 10 runs with no significant drop in yields. Primarily, this process exhibited exceptional performance for aliphatic amines (85–99%), however, moderate performance was observed for aryl amines (10–50%) (Table 2.3 Entry 36–37).

Selva and group [117] reported a multiphase solvent protocol for reductive amination of LA, using 5% Ru/C as a catalyst with formic acid as the hydrogen source. The process showed high flexibility for selectivity of products (GVL and pyrrolidone) based on relative volume of biphasic solvent, aqueous concentrations and formic acid concentration. The catalyst separation was also shown to be simplified, by isolation in the hydrocarbon organic phase. This system was able to attain a yield of 88% for cyclohexylamine substrate at optimal conditions (Table 2.3 Entry 38).

Earlier, the potential for the reductive amination of LA to be catalyst free was explored with formic acid and triethylamine system. The next report from Yang and co-workers [118] involved application of economical formamide and formic acid as the nitrogen and hydrogen sources respectively with an exceptionally short reaction time of 90 min at 160 °C. The role of formic acid as an acid and a H source through extensive labelling experiments and control experiments. Different keto acids including LA (Table 2.3 Entry 39) were tested using this process to get good to excellent yields. The second report was from the same group, [119] applying ammonium formate as the reagent and catalyst for conversion of LA and other keto acids to *N*-unprotected lactams. The driving force for this reaction was the use of ammonium formate and pressurized hot water as the solvent and promoter of hydrolysis of ammonium salt, with yields reaching up to 90% in just 60 min at 180 ° C (Table 2.3 Entry 40).

Raut et al. [120] was a contribution from our group, reporting a novel catalyst namely Ru@GOIL for synthesis of *N*-substituted pyrrolidones from LA and amines. Ru@GOIL denoted ruthenium ion supported on ionic liquid immobilized into graphene oxide, which was comprehensively characterised. This was the first example of a metal ion homogenized catalyst for the reductive amination process of LA with good recyclability up to 6 cycles with minor drop in product yield. A wide library of amines was explored for this process with yields ranging from 78% to 93% (Table 2.3 Entry 41–42) and with a great initial TOF of 187 h^{-1}. The next report also involved different bifunctional catalysts with immobilized metal nanoparticles and solid acid materials, from Barbaro et al. [121] for the reductive amination process. They tested different levulinate esters with primary amines with molecular hydrogen as a reductant. Pt loaded on solid acid Aquivion showed the best catalytic performance with excellent yield of 98.6% for *N*-heptyl-5-methyl-2-pyrrolidone (Table 2.3 Entry 43–44).

Liu and group [122] developed a Pt nanocatalyst with cellulose-derived carbon support, which was applied as a heterogeneous catalyst for ambient parameter synthesis of benzoxazoles from 2-nitroacylbenzenes and reductive amination of LA using primary amines with molecular hydrogen as the H source. It was shown that the carbon support had multiple functions such as surging the electron density of dispersed Pt nanoparticles and surface affinity for reactant molecules. Pt/c-C attained quantitative yields of 96% and 99% for aniline and benzylamine respectively, at optimized reaction conditions (Table 2.3 Entry 45–46).

Louven et al. [123] presented a two-step process, yielding *N*-vinyl-2-pyrrolidone monomer bio-derived keto acids namely itaconic and levulinic acid. A high TOF of 4000 $mol_{Pyr} \cdot mol_{Pd}$ $surface^{-1}$ h^{-1} was achieved for the reductive amination of LA by the designed palladium heterogeneous catalyst. Addtionally, vinylation of 2-pyrrolidone was carried out with decent yields up to 80%. 5Pd/C (ox.) 250 catalyst achieved a yield of 95.5% using NH_4OH as the N source with LA (Table 2.3 Entry 47). There were some polymeric metallic catalysts that were explored in the past reports with decent yields of pyrrolidones and showed exceptional robustness of these catalysts. One such work reported by Nagaoka, Chaudhari and co-workers [124] compared different noble metals (Ru, Pd, Pt, Ir, Rh) catalysts synthesized through stabilisation of metallic nanoparticles using polypyrrolidone. Ir proved to be the best performing metal from all the others with yields reaching up to 95% and 80% for aniline and benzylamine respectively to *N*-substituted pyrrolidones from LA (Table 2.3 Entry 48 and 49). Furthermore, separate functionalities namely amines, nitro and nitrile compound substrate were used to showcase the exceptional catalytic activity of Ir-PVP catalyst for their conversion to corresponding pyrrolidone products.

The subsequent report from Wang et al. [125] got attention for being one of first report of flow synthesis of pyrrolidones from using a heterogeneous catalyst (Ni_2P/SiO_2). Different parameters such as solvation effect, temperature, pressure, hydrogen flow rates were precisely optimized to achieve significant yields for alkyl pyrrolidones. Ni_2P/SiO_2 catalyst was able to provide an excellent yield of 94% for hexylamine and a moderate yield of 67% for aniline substrate at optimized parameters (Table 2.3 Entry 50–51).

Commercial catalysts hold great value in industry as their availability, ease of procuring and their versatility is widely accepted. It is assumed that tuning any process for using commercial catalysts to achieve good yields can be highly sought out endeavour. Interestingly, Wei and associates [126] reported a one-pot synthesis of *N*-substituted-5-methyl-2-pyrrolidones for reductive amination of LA with nitriles as the N source. Pd/C proved to be the best choice of catalyst with yields of pyrrolidones ranging from 50% to 79% at relatively mild conditions than earlier reports (Table 2.3 Entry 52–54).

Another example of metal-support interaction and highly tunable carbon-based support material applied to different transition and lanthanide metals was reported by

Wu et al., [127] where cobalt mixed with other transition metal (i.e., Zr, Cu, Fe, Ni, Ce, La) coupled with N-doped carbon-based material obtained from chitosan. Each distinct interactions of components such as formation of metal alloys, metal-carbon bonds, metal-nitrogen bonds and metal oxides formed on the surface were shown through characterisation techniques. The addition of second metal showed a significant enhancement in the yield due to surge in number of active sites and the deviations in the Fermi energy level by presence of the other metal atoms. Co-Zr@Chitosan-20 showed superior catalytic performance for reductive amination of levulinic acid to pyrrolidone product with 92.8% yield (Table 2.3 Entry 55–63).

The next report was published from our group by Gokhale et al., [128] which comprised exploration of cascade and sequential one pot syntheses pathways for production of *N*-substituted pyrrolidones from aldehydes and LA. These processes were optimized on nickel nanoparticles embedded on montmorillonite K10 clay (12% Ni-MMT). The parameter optimization yielded interesting results, with sequential one pot pathway being the better of the two due to higher atom economy, lesser amount of side products. Some mechanistic studies were also conducted by the help of control experiments that the reaction mechanism in case of 12% Ni-MMT catalyst proceeds through formation of imine intermediate between the amine and ketonic functionality of LA, which showed another avenue of observations than Gao et al. [111]. This can be explained by the stronger metal-support interaction of K10 clay and Ni nanoparticles, which were shown to be capable of activating the ketonic group of LA for formation of imine intermediate. 12% Ni-MMT showed impressive yields of pyrrolidone products with yields ranging between 70% and 90% using the sequential one pot pathway (Table 2.3 Entry 64–65).

Recently, Beller, Jagadeesh and group [129] reported reusable cobalt-based single atom catalysts (SAC) for reductive amination of LA and keto acids. Co-SACs were found to be extremely active catalyst for the reductive amination process using amines, nitriles, nitro moieties. These catalysts were also extensively characterized, with techniques such as HAADF-STEM, XANES, EXAFS, XPS. 0.75Co-phen@C-800-HCl was found to be the most optimal version of the catalyst with yields of pyrrolidone products reaching 80–98% for most substrates (Table 2.3 Entry 66–68). The latest report from *Zeng* et al., [130] involved synthesis of a class of boron doped Cu/Al_2O_3 catalysts developed for the continuous reductive amination of LA. These catalysts showed highly enhanced activity compared to the Cu/Al_2O_3, which may have been due to better dispersion of Cu with enhanced surface acidity. Cu_{10}/AlB_3O exhibited great resilience for over 200 h of continuous flow synthesis of pyrrolidone products. Cu_{10}/AlB_3O gave an impressive yield of 88% and moderate yield of 53% in continuous flow synthesis using benzylamine and aniline substrates respectively (Table 2.3 Entry 69 and 70).

Unlike reductive amination of FUR and HMF, there have been many studies proving the mechanistic pathways for the reductive amination of LA in the past literature. There are two mechanisms shown in Fig. 2.3, which have extensive experimental data proving their validity. There are three steps involved in the reductive amination of LA, Step I include formation of Schiff's base (imine) [a] at the ketonic site or formation of amide species [b] at the carboxylic site of LA. It was

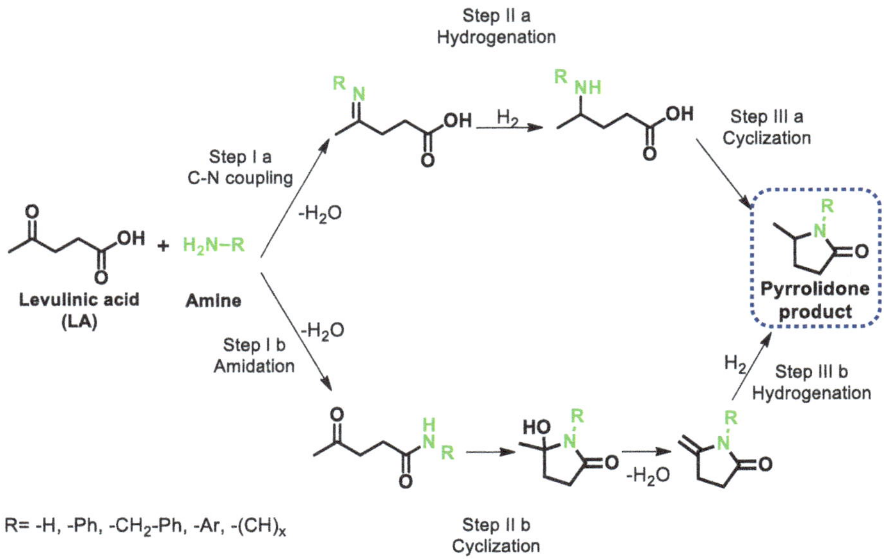

Fig. 2.3 Accepted mechanistic pathways for reductive amination of LA based on literature reports

found by Gao et al. [111] that the non-noble metals showed the amidation pathway for reductive amination of LA and noble metal catalysts exhibited the imine pathway. However, based on the recent work in our group by Gokhale et al. [128] that multiple catalytic factors may contribute to the mechanistic pathway for the reaction as our group showed Ni catalyst favoring the imine pathway, which would be due to multiple factors such as surface acidity of catalyst supports and even the strength of metal-support interaction. The next steps in both the reaction pathways are similar with different sequence. It generally involves hydrogenation followed by cyclization to form the desired pyrrolidone product.

2.4 Conclusions and Future Outlook

Overall, it can be seen that the focus of the scientific community can be expressed based on the number of significant reports that were published in the last decade with a special focus on valorisation of these three bio-derived platform molecules. Reductive amination of LA accounted for some of major breakthroughs in terms of catalyst design, flow synthesis, cascade reactions and continuous syntheses pathways. Reductive amination of FUR and HMF also observed many significant reports in terms of catalyst design, however, lacks exploration of other aspects of process development and mechanistic investigations. This topic is still attractive as much more investigation is required for these processes to scaled up to commercial phase and it will be interesting to see the interesting works that will be carried out in the coming future.

As for the future prospects related to upgradation of the reductive amination of FUR, HMF and LA to the commercial phase, there are many vital points, which need to studied and confirmed before the next step. Firstly, for report FUR and HMF, many catalytic supports were reported to provide excellent performance such as N-doped carbon, Al_2O_3 and Nb_2O_5 based on unique surface properties. However, commercial availability and cost can be the restricting roadblocks. These factors also affect the metal, which will be used for synthesis of these catalysts. Several non-noble metals such as Co, Ni, Cu have been shown to provide similar level of catalytic activity to noble metals. However, the long-term stability studies of these non-noble metal catalysts are absolutely necessary. Additionally, appropriate choice of solvents for this reaction is another factor, which needs to be considered. Based on the literature, polar protic solvents provide the best results for these reactions. However, more rigorous exploration of their solvation effects, hydrogen gas solubility of these solvents is the need of the hour. Another segment which needs to be explored for the reductive amination of FUR and HMF, is its potential of being applied to flow syntheses paradigm. Many novel reactions, which require heterogeneous catalysts for operation are being explored for flow syntheses, which creates a window of opportunity for this reaction to be studied under similar conditions.

On the other hand, reductive amination of LA has been explored in all batch, continuous and flow synthesis scenarios with encouraging results. However, it will still require detailed level of optimization and process parameter studies to be scaled up to the commercial phase.

Acknowledgements Tejas Gokhale appreciates funding provided by IMRCD, Department of Science & Technology (DST), Govt. of India, under the DST-BRICS-ICP-BIO research project sanctioned for DST/IMRCD/BRICS/PilotCall2/ICP-BIO/2018 (G) project call.

References

1. Sudarsanam P, Zhong R, Van den Bosch S, Coman SM, Parvulescu VI, Sels BF. Functionalised heterogeneous catalysts for sustainable biomass valorisation. Chem Soc Rev. 2018;47(22):8349–402. https://doi.org/10.1039/c8cs00410b.
2. Gallezot P. Conversion of biomass to selected chemical products. Chem Soc Rev. 2012;41: 1538–58. https://doi.org/10.1039/C1CS15147A.
3. Saxena RC, Adhikari DK, Goyal HB. Biomass-based energy fuel through biochemical routes: a review. Renew Sust Energ Rev. 2009;13:167–78. https://doi.org/10.1016/j.rser.2007.07.011.
4. Sajid M, Farooq U, Bary G, Azim MM, Zhao X. Sustainable production of levulinic acid and its derivatives for fuel additives and chemicals: progress, challenges, and prospects. Green Chem. 2021;23(23):9198–238. https://doi.org/10.1039/d1gc02919c.
5. Nong D, Escobar N, Britz W, Börner J. Long-term impacts of bio-based innovation in the chemical sector: a dynamic global perspective. J Clean Prod. 2020;272:122738. https://doi.org/10.1016/j.jclepro.2020.122738.

6. Troiano D, Orsat V, Dumont M-J. Status of biocatalysis in the production of 2, 5-Furandicarboxylic acid. ACS Catal. 2020;10:9145–69. https://doi.org/10.1021/acscatal.0c02378.
7. Dheskali E, Koutinas AA, Kookos IK. Risk assessment modeling of bio-based chemicals economics based on Monte-Carlo simulations. Chem Eng Res Des. 2020;163:273–80. https://doi.org/10.1016/j.cherd.2020.09.011.
8. Li H, Yang S, Saravanamurugan S, Riisager A. Glucose isomerization by enzymes and chemo-catalysts: status and current advances. ACS Catal. 2017;7:3010–29. https://doi.org/10.1021/acscatal.6b03625.
9. Zhang N, Zou Y, Tao L, Chen W, Zhou L, Liu Z, Zhou B, Huang G, Lin H, Wang S. Electrochemical oxidation of 5-hydroxymethylfurfural on nickel nitride/carbon nanosheets: reaction pathway determined by in situ sum frequency generation vibrational spectroscopy. Angew Chem. 2019;131(44):16042–50. https://doi.org/10.1002/ange.201908722.
10. Desir P, Saha B, Vlachos DG. Ultrafast flow chemistry for the acid-catalyzed conversion of fructose. Energy Environ Sci. 2019;12:2463–75. https://doi.org/10.1039/C9EE01189G.
11. Lee SY, Kim HU, Chae TU, Cho JS, Kim JW, Shin JH, Kim DI, Ko Y-S, Jang WD, Jang Y-S. A comprehensive metabolic map for production of bio-based chemicals. Nat Catal. 2019;2(1):18–33. https://doi.org/10.1038/s41929-018-0212-4.
12. Sudarsanam P, Peeters E, Makshina EV, Parvulescu VI, Sels BF. Advances in porous and nanoscale catalysts for viable biomass conversion. Chem Soc Rev. 2019;48(8):2366–421. https://doi.org/10.1039/c8cs00452h.
13. Werpy T, Petersen G. Top value added chemicals from biomass: volume I -- results of screening for potential candidates from sugars and synthesis gas. Golden: US Department of Energy; 2004. https://doi.org/10.2172/15008859.
14. Dunn JB, Adom F, Sather N, Han J, Snyder S, He C, Gong J, Yue D, You F. Life-cycle analysis of bioproducts and their conventional counterparts in GREET. Office of Scientific and Technical Information (OSTI); 2015. https://doi.org/10.2172/1250468.
15. Grand View Research (2022) Furfural market size, share & trends analysis report by process (quaker batch process, chinese batch process), by raw material (corncob, sugarcane bagasse), by application, by end use, and segment forecasts, 2022–2030, 978-1-68038-301-0.
16. Dutta S, De S, Saha B, Alam MI. Advances in conversion of hemicellulosic biomass to furfural and upgrading to biofuels. Cat Sci Technol. 2012;2:2025. https://doi.org/10.1039/c2cy20235b.
17. Schöppe H, Kleine-Möllhoff P, Epple R. Energy and material flows and carbon footprint assessment concerning the production of HMF and furfural from a cellulosic biomass. PRO. 2020;8:119. https://doi.org/10.3390/pr8010119.
18. Kumar S, Devi MM, Kansal SK, Saravanamurugan S. Untangling the active sites in the exposed crystal facet of zirconium oxide for selective hydrogenation of bioaldehydes. Cat Sci Technol. 2020;10:7016–26. https://doi.org/10.1039/D0CY01259A.
19. Li X, Jia P, Wang T. Furfural: a promising platform compound for sustainable production of C_4 and C_5 chemicals. ACS Catal. 2016;6:7621–40. https://doi.org/10.1021/acscatal.6b01838.
20. Streiff S, Jérôme F. Hydroamination of non-activated alkenes with ammonia: a holy grail in catalysis. Chem Soc Rev. 2021;50:1512–21. https://doi.org/10.1039/C9CS00873J.
21. Huber GW, Iborra S, Corma A. Synthesis of transportation fuels from biomass: chemistry, catalysts, and engineering. Chem Rev. 2006;106:4044–98. https://doi.org/10.1021/cr068360d.
22. van Putten R-J, van der Waal JC, de Jong E, Rasrendra CB, Heeres HJ, de Vries JG. Hydroxymethylfurfural, a versatile platform chemical made from renewable resources. Chem Rev. 2013;113(3):1499–597. https://doi.org/10.1021/cr300182k.
23. Rosatella AA, Simeonov SP, Frade RFM, Afonso CAM. 5-Hydroxymethylfurfural (HMF) as a building block platform: biological properties, synthesis and synthetic applications. Green Chem. 2011;13:754. https://doi.org/10.1039/c0gc00401d.

24. Alam MI, De S, Singh B, Saha B, Abu-Omar MM. Titanium hydrogenphosphate: an efficient dual acidic catalyst for 5-hydroxymethylfurfural (HMF) production. Appl Catal A Gen. 2014;486:42–8. https://doi.org/10.1016/j.apcata.2014.08.019.
25. Feriani A, Gaviraghi G, Toson G, Mor M, Barbieri A, Grana E, Boselli C, Guarneri M, Simoni D, Manfredini S. Cholinergic agents structurally related to Furtrethonium. 2. synthesis and antimuscarinic activity of a series of N-[5-[(1'-substituted-acetoxy)methyl]-2-furfuryl] dialkylamines. J Med Chem. 1994;37(25):4278–87. https://doi.org/10.1021/jm00051a004.
26. Plitta B, Adamska E, Giel-Pietraszuk M, Fedoruk-Wyszomirska A, Naskręt-Barciszewska M, Markiewicz WT, Barciszewski J. New cytosine derivatives as inhibitors of DNA methylation. Eur J Med Chem. 2012;55:243–54. https://doi.org/10.1016/j.ejmech.2012.07.024.
27. Xu Z, Yan P, Xu W, Jia S, Xia Z, Chung B, Zhang ZC. Direct reductive amination of 5-hydroxymethylfurfural with primary/secondary amines via Ru-complex catalyzed hydrogenation. RSC Adv. 2014;4(103):59083–7. https://doi.org/10.1039/c4ra10349a.
28. Cukalovic A, Stevens C, v. Production of biobased HMF derivatives by reductive amination. Green Chem. 2010;12:1201. https://doi.org/10.1039/c002340j.
29. Omoruyi U, Page S, Hallett J, Miller PW. Homogeneous catalyzed reactions of levulinic acid: to γ-valerolactone and beyond. ChemSusChem. 2016;9:2037–47. https://doi.org/10.1002/cssc.201600517.
30. Dutta S, Yu IKM, Tsang DCW, Ng YH, Ok YS, Sherwood J, Clark JH. Green synthesis of gamma-valerolactone (GVL) through hydrogenation of biomass-derived levulinic acid using non-noble metal catalysts: a critical review. Chem Eng J. 2019;372:992–1006. https://doi.org/10.1016/j.cej.2019.04.199.
31. Mizugaki T, Nagatsu Y, Togo K, Maeno Z, Mitsudome T, Jitsukawa K, Kaneda K. Selective hydrogenation of levulinic acid to 1,4-pentanediol in water using a hydroxyapatite-supported Pt–Mo bimetallic catalyst. Green Chem. 2015;17(12):5136–9. https://doi.org/10.1039/c5gc01878a.
32. Chatzidimitriou A, Bond JQ. Oxidation of levulinic acid for the production of maleic anhydride: breathing new life into biochemicals. Green Chem. 2015;17:4367–76. https://doi.org/10.1039/C5GC01000D.
33. Kawasumi R, Narita S, Miyamoto K, Tominaga K-I, Takita R, Uchiyama M. One-step conversion of levulinic acid to succinic acid using I(2)/t-BuOK system: the Iodoform reaction revisited. Sci Rep. 2017;7(1):17967. https://doi.org/10.1038/s41598-017-17116-4.
34. Xue Z, Yu D, Zhao X, Mu T. Upgrading of levulinic acid into diverse N-containing functional chemicals. Green Chem. 2019;21:5449–68. https://doi.org/10.1039/C9GC02415H.
35. Bukhtiyarova MV, Bukhtiyarova GA. Reductive amination of levulinic acid or its derivatives to pyrrolidones over heterogeneous catalysts in the batch and continuous flow reactors: a review. Renew Sust Energ Rev. 2021;143:110876. https://doi.org/10.1016/j.rser.2021.110876.
36. Wang S, Zhao Y, Lin H, Chen J, Zhu L, Luo Z. Conversion of C5 carbohydrates into furfural catalyzed by a Lewis acidic ionic liquid in renewable γ-valerolactone. Green Chem. 2017;19(16):3869–79. https://doi.org/10.1039/c7gc01298e.
37. Dixit RJ, Bhattacharyya K, Ramani VK, Basu S. Electrocatalytic hydrogenation of furfural using non-noble-metal electrocatalysts in alkaline medium. Green Chem. 2021;23:4201–12. https://doi.org/10.1039/D1GC00579K.
38. Signor A, Guerrato A, Signor G. Process for the preparation of furosemide. WIPO; 1998. WO/1996/012714
39. Roose P, Eller K, Henkes E, Rossbacher R, Höke H. Amines, aliphatic. In: Ley C, Elvers B, editors. Ullmann's Encyclopedia of industrial chemistry. Weinheim: Wiley-VCH Verlag Gmb H & Co. KGaA; 2015. p. 1–55.
40. Afanasyev OI, Kuchuk E, Usanov DL, Chusov D. Reductive amination in the synthesis of pharmaceuticals. Chem Rev. 2019;119:11857–911. https://doi.org/10.1021/acs.chemrev.9b00383.

41. Deng W, Wang Y, Zhang S, Gupta KM, Hülsey MJ, Asakura H, Liu L, Han Y, Karp EM, Beckham GT, Dyson PJ, Jiang J, Tanaka T, Wang Y, Yan N. Catalytic amino acid production from biomass-derived intermediates. Proc Natl Acad Sci USA. 2018;115(20):5093–8. https://doi.org/10.1073/pnas.1800272115.
42. Qi Y, Wang J, Kou Y, Pang H, Zhang S, Li N, Liu C, Weng Z, Jian X. Synthesis of an aromatic N-heterocycle derived from biomass and its use as a polymer feedstock. Nat Commun. 2019;10(1):2107. https://doi.org/10.1038/s41467-019-10178-0.
43. Hülsey MJ, Yang H, Yan N. Sustainable routes for the synthesis of renewable heteroatom-containing chemicals. ACS Sustain Chem Eng. 2018;6:5694–707. https://doi.org/10.1021/acssuschemeng.8b00612.
44. Li H, Guo H, Su Y, Hiraga Y, Fang Z, Hensen EJM, Watanabe M, Smith RL Jr. N-formyl-stabilizing quasi-catalytic species afford rapid and selective solvent-free amination of biomass-derived feedstocks. Nat Commun. 2019;10(1):699. https://doi.org/10.1038/s41467-019-08577-4.
45. Xu L, Shi C, He Z, Zhang H, Chen M, Fang Z, Zhang Y. Recent advances of producing biobased N-containing compounds via thermo-chemical conversion with ammonia process. Energy Fuels. 2020;34(9):10441–58. https://doi.org/10.1021/acs.energyfuels.0c01993.
46. Yan N, Wang Y. Catalyst: is the amino acid a new frontier for biorefineries? Chem. 2019;5:739–41. https://doi.org/10.1016/j.chempr.2019.03.016.
47. Maschmeyer T, Luque R, Selva M. Upgrading of marine (fish and crustaceans) biowaste for high added-value molecules and bio (nano)-materials. Chem Soc Rev. 2020;49:4527–63. https://doi.org/10.1039/C9CS00653B.
48. Li H, Bunrit A, Li N, Wang F. Heteroatom-participated lignin cleavage to functionalized aromatics. Chem Soc Rev. 2020;49:3748–63. https://doi.org/10.1039/D0CS00078G.
49. Mariscal R, Maireles-Torres P, Ojeda M, Sádaba I, López Granados M. Furfural: a renewable and versatile platform molecule for the synthesis of chemicals and fuels. Energy Environ Science. 2016;9(4):1144–89. https://doi.org/10.1039/c5ee02666k.
50. Yan K, Jarvis C, Gu J, Yan Y. Production and catalytic transformation of levulinic acid: a platform for speciality chemicals and fuels. Renew Sust Energ Rev. 2015;51:986–97. https://doi.org/10.1016/j.rser.2015.07.021.
51. Jung HS, Yun T, Cho Y, Jeon HB. Simple and convenient copper-catalyzed amination of aryl halides to primary arylamines using NH4OH. Tetrahedron. 2016;72:5988–93. https://doi.org/10.1016/j.tet.2016.07.021.
52. Fu M-C, Shang R, Cheng W-M, Fu Y. Boron-catalyzed N-alkylation of amines using carboxylic acids. Angew Chem Int Ed. 2015;54:9042–6. https://doi.org/10.1002/anie.201503879.
53. Huy PH, Motsch S, Kappler SM. Formamides as Lewis base catalysts in S_N reactions-efficient transformation of alcohols into chlorides, amines, and ethers. Angew Chem Int Ed. 2016;55:10145–9. https://doi.org/10.1002/anie.201604921.
54. Elangovan S, Topf C, Fischer S, Jiao H, Spannenberg A, Baumann W, Ludwig R, Junge K, Beller M. Selective catalytic hydrogenations of nitriles, ketones, and aldehydes by well-defined manganese pincer complexes. J Am Chem Soc. 2016;138(28):8809–14. https://doi.org/10.1021/jacs.6b03709.
55. Davies J, Svejstrup TD, Fernandez Reina D, Sheikh NS, Leonori D. Visible-light-mediated synthesis of Amidyl radicals: transition-metal-free Hydroamination and <i>N</i>-Arylation reactions. J Am Chem Soc. 2016;138(26):8092–5. https://doi.org/10.1021/jacs.6b04920.
56. Nayal OS, Bhatt V, Sharma S, Kumar N. Chemoselective reductive amination of carbonyl compounds for the synthesis of tertiary amines using $SnCl_2 \cdot 2H_2O$/PMHS/MeOH. J Org Chem. 2015;80:5912–8. https://doi.org/10.1021/acs.joc.5b00156.
57. Heinen AW, Peters JA, van Bekkum H. The reductive amination of benzaldehyde over Pd/C catalysts: mechanism and effect of carbon modifications on the selectivity. Eur J Org Chem. 2000;13:2501–6. https://doi.org/10.1002/1099-0690(200007)2000:13<2501::AID-EJOC2501>3.0.CO;2-S.

58. Gomez S, Peters JA, Van der Waal C, Maschmeyer T. High-throughput experimentation as a tool in catalyst design for the reductive amination of benzaldehyde. Appl Catal A Gen. 2003;254(1):77–84. https://doi.org/10.1016/s0926-860x(03)00278-3.
59. Bódis J, Lefferts L, Müller TE, Pestman R, Lercher JA. Activity and selectivity control in reductive amination of butyraldehyde over noble metal catalysts. Catal Lett. 2005;104(1–2): 23–8. https://doi.org/10.1007/s10562-005-7431-4.
60. Allegretti M, Berdini V, Cesta MC, Curti R, Nicolini L, Topai A. One-pot, new stereoselective synthesis of endo-tropanamine. Tetrahedron Lett. 2001;42(25):4257–9. https://doi.org/10.1016/s0040-4039(01)00663-3.
61. Chatterjee M, Ishizaka T, Kawanami H. Reductive amination of furfural to furfurylamine using aqueous ammonia solution and molecular hydrogen: an environmentally friendly approach. Green Chem. 2016;18:487–96. https://doi.org/10.1039/C5GC01352F.
62. Chieffi G, Braun M, Esposito D. Continuous reductive amination of biomass-derived molecules over carbonized filter paper-supported FeNi alloy. ChemSusChem. 2015;8:3590–4. https://doi.org/10.1002/cssc.201500804.
63. Nishimura S, Mizuhori K, Ebitani K. Reductive amination of furfural toward furfurylamine with aqueous ammonia under hydrogen over Ru-supported catalyst. Res Chem Intermed. 2016;42:19–30. https://doi.org/10.1007/s11164-015-2334-5.
64. Liang G, Wang A, Li L, Xu G, Yan N, Zhang T. Production of primary amines by reductive amination of biomass-derived aldehydes/ketones. Angew Chem Int Ed. 2017;56(11):3050–4. https://doi.org/10.1002/anie.201610964.
65. Komanoya T, Kinemura T, Kita Y, Kamata K, Hara M. Electronic effect of ruthenium nanoparticles on efficient reductive amination of carbonyl compounds. J Am Chem Soc. 2017;139(33):11493–9. https://doi.org/10.1021/jacs.7b04481.
66. Chandra D, Inoue Y, Sasase M, Kitano M, Bhaumik A, Kamata K, Hosono H, Hara M. A high performance catalyst of shape-specific ruthenium nanoparticles for production of primary amines by reductive amination of carbonyl compounds. Chem Sci. 2018;9(27):5949–56. https://doi.org/10.1039/c8sc01197d.
67. Deng D, Kita Y, Kamata K, Hara M. Low-temperature reductive amination of carbonyl compounds over Ru deposited on $Nb_2\ O_5\ n\ H_2O$. ACS Sustain Chem Eng. 2019;7:4692–8. https://doi.org/10.1021/acssuschemeng.8b04324.
68. Zhang Y, Yang H, Chi Q, Zhang Z. Nitrogen-doped carbon-supported nickel nanoparticles: a robust catalyst to bridge the hydrogenation of nitriles and the reductive amination of carbonyl compounds for the synthesis of primary amines. ChemSusChem. 2019;12:1246–55. https://doi.org/10.1002/cssc.201802459.
69. Yuan Z, Liu B, Zhou P, Zhang Z, Chi Q. Preparation of nitrogen-doped carbon supported cobalt catalysts and its application in the reductive amination. J Catal. 2019;370:347–56. https://doi.org/10.1016/j.jcat.2019.01.004.
70. Yuan H, Li J-P, Su F, Yan Z, Kusema BT, Streiff S, Huang Y, Pera-Titus M, Shi F. Reductive amination of Furanic aldehydes in aqueous solution over versatile Ni(y) AlO(x) catalysts. ACS Omega. 2019;4(2):2510–6. https://doi.org/10.1021/acsomega.8b03516.
71. Manzoli M, Gaudino EC, Cravotto G, Tabasso S, Baig RBN, Colacino E, Varma RS. Microwave-assisted reductive amination with aqueous ammonia: sustainable pathway using recyclable magnetic nickel-based Nanocatalyst. ACS sustainable chemistry & Engineering. 2019;7(6):5963–74. https://doi.org/10.1021/acssuschemeng.8b06054.
72. Guo W, Tong T, Liu X, Guo Y, Wang Y. Morphology-tuned activity of Ru/Nb ₂ O ₅ catalysts for ketone reductive amination. ChemCatChem. 2019;11(16):4130–8. https://doi.org/10.1002/cctc.201900335.
73. Zhou K, Chen B, Zhou X, Kang S, Xu Y, Wei J. Selective synthesis of furfurylamine by reductive amination of furfural over Raney cobalt. ChemCatChem. 2019;11(22):5562–9. https://doi.org/10.1002/cctc.201901269.
74. Jv X, Sun S, Zhang Q, Du M, Wang L, Wang B. Efficient and mild reductive amination of carbonyl compounds catalyzed by dual-function palladium nanoparticles. ACS sustainable

chemistry & amp. Engineering. 2019;8(3):1618–26. https://doi.org/10.1021/acssuschemeng.9b06464.
75. Dong C, Wang H, Du H, Peng J, Cai Y, Guo S, Zhang J, Samart C, Ding M. Ru/HZSM-5 as an efficient and recyclable catalyst for reductive amination of furfural to furfurylamine. Mol Catal. 2020;482:110755. https://doi.org/10.1016/j.mcat.2019.110755.
76. Gould NS, Landfield H, Dinkelacker B, Brady C, Yang X, Xu B. Selectivity control in catalytic reductive amination of furfural to furfurylamine on supported catalysts. ChemCatChem. 2020;12(7):2106–15. https://doi.org/10.1002/cctc.201901662.
77. Jiang S, Muller E, Jerôme F, Pera-Titus M, De Oliveira VK. Conversion of furfural to tetrahydrofuran-derived secondary amines under mild conditions. Green Chem. 2020;22(6):1832–6. https://doi.org/10.1039/d0gc00119h.
78. Xie C, Song J, Hua M, Hu Y, Huang X, Wu H, Yang G, Han B. Ambient-temperature synthesis of primary amines via reductive amination of carbonyl compounds. ACS Catal. 2020;10(14):7763–72. https://doi.org/10.1021/acscatal.0c01872.
79. Sheng M, Fujita S, Yamaguchi S, Yamasaki J, Nakajima K, Yamazoe S, Mizugaki T, Mitsudome T. Single-crystal cobalt phosphide Nanorods as a high-performance catalyst for reductive amination of carbonyl compounds. JACS Au. 2021;1(4):501–7. https://doi.org/10.1021/jacsau.1c00125.
80. Dong C, Wu Y, Wang H, Peng J, Li Y, Samart C, Ding M. Facile and Efficient Synthesis of primary amines via reductive amination over a Ni/Al₂O₃ catalyst. ACS sustain Chem Engin. 2021;9(21):7318–27. https://doi.org/10.1021/acssuschemeng.1c01456.
81. Qi H, Yang J, Liu F, Zhang L, Yang J, Liu X, Li L, Su Y, Liu Y, Hao R, Wang A, Zhang T. Highly selective and robust single-atom catalyst Ru(1)/NC for reductive amination of aldehydes/ketones. Nat Commun. 2021;12(1):3295. https://doi.org/10.1038/s41467-021-23429-w.
82. Gokhale TA, Raut AB, Bhanage BM. Comparative account of catalytic activity of Ru- and Ni-based nanocomposites towards reductive amination of biomass derived molecules. Mol Catal. 2021;510:111667. https://doi.org/10.1016/j.mcat.2021.111667.
83. Luo Z, Pan Y, Yao Z, Yang J, Zhang X, Liu X, Xu L, Fan Q-H. BF₃·et₂O as a metal-free catalyst for direct reductive amination of aldehydes with amines using formic acid as a reductant. Green Chem. 2021;23(14):5205–11. https://doi.org/10.1039/d1gc01468d.
84. Gao M, Jia X, Ma J, Fan X, Gao J, Xu J. Self-regulated catalysis for the selective synthesis of primary amines from carbonyl compounds. Green Chem. 2021;23(18):7115–21. https://doi.org/10.1039/d1gc01275d.
85. Zhuang X, Liu J, Zhong S, Ma L. Selective catalysis for the reductive amination of furfural toward furfurylamine by graphene-co-shelled cobalt nanoparticles. Green Chem. 2022;24:271–84. https://doi.org/10.1039/D1GC03578A.
86. Bhunia MK, Chandra D, Abe H, Niwa Y, Hara M. Synergistic effects of earth-abundant metal–metal oxide enable reductive amination of carbonyls at 50 °C. ACS Appl Mater Interfaces. 2022;14(3):4144–54. https://doi.org/10.1021/acsami.1c21157.
87. Ronda-Leal M, Espro C, Lazaro N, Selva M, Perosa A, Osman SM, Pineda A, Luque R, Rodríguez-Padrón D. Efficient and stable titania-based nanocatalytic materials for the reductive amination of furfural. Mater Today Chem. 2022;24:100873. https://doi.org/10.1016/j.mtchem.2022.100873.
88. Zou H, Chen J. Efficient and selective approach to biomass-based amine by reductive amination of furfural using Ru catalyst. Appl Catal B. 2022;309:121262. https://doi.org/10.1016/j.apcatb.2022.121262.
89. Pan Z, Zhang Q, Wang W, Wang L, Wang G-H. Size-Tunable carbon-doped Ni nanoparticles for switchable reductive amination of biomass-derived carbonyl compounds to primary amines and secondary imines. ACS Sustain Chem Engin. 2022;10(11):3777–86. https://doi.org/10.1021/acssuschemeng.2c00800.

90. Lin C, Zhou J, Zheng Z, Chen J. An efficient approach to biomass-based tertiary amines by direct and consecutive reductive amination of furfural. J Catal. 2022;410:164–79. https://doi.org/10.1016/j.jcat.2022.04.016.
91. Zhu M-M, Tao L, Zhang Q, Dong J, Liu Y-M, He H-Y, Cao Y. Versatile CO-assisted direct reductive amination of 5-hydroxymethylfurfural catalyzed by a supported gold catalyst. Green Chem. 2017;19(16):3880–7. https://doi.org/10.1039/c7gc01579h.
92. Zhou K, Liu H, Shu H, Xiao S, Guo D, Liu Y, Wei Z, Li X. A comprehensive study on the reductive amination of 5-hydroxymethylfurfural into 2,5-bisaminomethylfuran over Raney Ni through DFT calculations. ChemCatChem. 2019;11(11):2649–56. https://doi.org/10.1002/cctc.201900304.
93. Wei Z, Cheng Y, Zhou K, Zeng Y, Yao E, Li Q, Liu Y, Sun Y. One-step reductive amination of 5-hydroxymethylfurfural into 2,5-Bis(aminomethyl)furan over Raney Ni. ChemSusChem. 2021;14(11):2308–12. https://doi.org/10.1002/cssc.202100564.
94. Wei Z, Cheng Y, Huang H, Ma Z, Zhou K, Liu Y. Reductive amination of 5-hydroxymethylfurfural to 2,5-Bis(aminomethyl)furan over alumina-supported Ni-based catalytic systems. ChemSusChem. 2022;15(13):e202200233. https://doi.org/10.1002/cssc.202200233.
95. Karve VV, Sun DT, Trukhina O, Yang S, Oveisi E, Luterbacher J, Queen WL. Efficient reductive amination of HMF with well dispersed Pd nanoparticles immobilized in a porous MOF/polymer composite. Green Chem. 2020;22(2):368–78. https://doi.org/10.1039/c9gc03140e.
96. Yuan H, Kusema BT, Yan Z, Streiff S, Shi F. Highly selective synthesis of 2,5-bis(aminomethyl)furan via catalytic amination of 5-(hydroxymethyl) furfural with NH(3) over a bifunctional catalyst. RSC Adv. 2019;9(66):38877–81. https://doi.org/10.1039/c9ra08560b.
97. Chandrashekhar VG, Natte K, Alenad AM, Alshammari AS, Kreyenschulte C, Jagadeesh RV. Reductive amination, hydrogenation and hydrodeoxygenation of 5-hydroxymethylfurfural using silica-supported cobalt- nanoparticles. ChemCatChem. 2021;14(1):e202101234. https://doi.org/10.1002/cctc.202101234.
98. Hu Q, Jiang S, Wu Y, Xu H, Li G, Zhou Y, Wang J. Ambient-temperature reductive amination of 5-hydroxymethylfurfural Over Al ₂ O ₃ -supported carbon-doped nickel catalyst. ChemSusChem. 2022;15(13):e202200192. https://doi.org/10.1002/cssc.202200192.
99. Roy S, Reiser O. A catalytic multicomponent approach for the stereoselective synthesis of *cis* −4,5-Disubstituted Pyrrolidinones and Tetrahydro-3 *H* -pyrrolo[3,2- *c*]quinolines. Angew Chem Int Ed. 2012;51:4722–5. https://doi.org/10.1002/anie.201107831.
100. Sarkar R, Mukhopadhyay C. Admicellar catalysis in multicomponent synthesis of polysubstituted pyrrolidinones. Tetrahedron Lett. 2013;54:3706–11. https://doi.org/10.1016/j.tetlet.2013.05.017.
101. Todd W, Frye John G Jr, Wang Y, Zacher AH. Methods of making pyrrolidones. Battelle Memorial Institute; 2003. p. 6670483.
102. Manzer Leo Ernest (2005) Production of 5-methyl-N-aryl-2-pyrrolidone and 5-methyl-N-alkyl-2-pyrrolidone by reductive amination of levulinic acid esters with nitro compounds., US6818593B2.
103. Du X-L, He L, Zhao S, Liu Y-M, Cao Y, He H-Y, Fan K-N. Hydrogen-independent reductive transformation of carbohydrate biomass into γ-valerolactone and pyrrolidone derivatives with supported gold catalysts. Angew Chem Int Ed. 2011;50(34):7815–9. https://doi.org/10.1002/anie.201100102.
104. Wei Y, Wang C, Jiang X, Xue D, Liu Z-T, Xiao J. Catalyst-free transformation of levulinic acid into pyrrolidinones with formic acid. Green Chem. 2014;16(3):1093–6. https://doi.org/10.1039/c3gc42125b.
105. Touchy AS, Hakim Siddiki SMA, Kon K, Shimizu K. Heterogeneous Pt catalysts for reductive amination of levulinic acid to pyrrolidones. ACS Catal. 2014;4:3045–50. https://doi.org/10.1021/cs500757k.

106. Ledoux A, Sandjong Kuigwa L, Framery E, Andrioletti B. A highly sustainable route to pyrrolidone derivatives – direct access to biosourced solvents. Green Chem. 2015;17:3251–4. https://doi.org/10.1039/C5GC00417A.
107. Vidal JD, Climent MJ, Concepcion P, Corma A, Iborra S, Sabater MJ. Chemicals from biomass: chemoselective reductive amination of ethyl levulinate with amines. ACS Catal. 2015;5(10):5812–21. https://doi.org/10.1021/acscatal.5b01113.
108. Zhang T, Ge Y, Wang X, Chen J, Huang X, Liao Y. Polymeric ruthenium Porphyrin-functionalized carbon nanotubes and graphene for levulinic ester transformations into γ-valerolactone and pyrrolidone derivatives. ACS Omega. 2017;2(7):3228–40. https://doi.org/10.1021/acsomega.7b00427.
109. Vidal JD, Climent MJ, Corma A, Concepcion DP, Iborra S. One-pot selective catalytic synthesis of pyrrolidone derivatives from ethyl levulinate and nitro compounds. ChemSusChem. 2016;10(1):119–28. https://doi.org/10.1002/cssc.201601333.
110. Sun Z, Chen J, Tu T. NHC-based coordination polymers as solid molecular catalysts for reductive amination of biomass levulinic acid. Green Chem. 2017;19:789–94. https://doi.org/10.1039/C6GC02591A.
111. Gao G, Sun P, Li Y, Wang F, Zhao Z, Qin Y, Li F. Highly stable porous-carbon-coated Ni catalysts for the reductive amination of levulinic acid via an unconventional pathway. ACS Catal. 2017;7(8):4927–35. https://doi.org/10.1021/acscatal.7b01786.
112. Martínez JJ, Silva L, Rojas HA, Romanelli GP, Santos LA, Ramalho TC, Brijaldo MH, Passos FB. Reductive amination of levulinic acid to different pyrrolidones on Ir/SiO 2 -SO 3 H: elucidation of reaction mechanism. Catal Today. 2017;296:118–26. https://doi.org/10.1016/j.cattod.2017.08.038.
113. Amarasekara AS, Lawrence YM. Raney-Ni catalyzed conversion of levulinic acid to 5-methyl-2-pyrrolidone using ammonium formate as the H and N source. Tetrahedron Lett. 2018;59:1832–5. https://doi.org/10.1016/j.tetlet.2018.03.087.
114. Cao P, Ma T, Zhang H-Y, Yin G, Zhao J, Zhang Y. Conversion of levulinic acid to N-substituted pyrrolidinones over a nonnoble bimetallic catalyst Cu15Pr3/Al2O3. Catal Commun. 2018;116:85–90. https://doi.org/10.1016/j.catcom.2018.07.016.
115. Xie C, Song J, Wu H, Hu Y, Liu H, Zhang Z, Zhang P, Chen B, Han B. Ambient reductive amination of levulinic acid to pyrrolidones over Pt nanocatalysts on porous TiO₂ nanosheets. J Am Chem Soc. 2019;141(9):4002–9. https://doi.org/10.1021/jacs.8b13024.
116. Muzzio M, Yu C, Lin H, Yom T, Boga DA, Xi Z, Li N, Yin Z, Li J, Dunn JA, Sun S. Reductive amination of ethyl levulinate to pyrrolidones over AuPd nanoparticles at ambient hydrogen pressure. Green Chem. 2019;21(8):1895–9. https://doi.org/10.1039/c9gc00396g.
117. Bellè A, Tabanelli T, Fiorani G, Perosa A, Cavani F, Selva M. A multiphase protocol for selective hydrogenation and reductive amination of levulinic acid with integrated catalyst recovery. ChemSusChem. 2019;12(14):3343–54. https://doi.org/10.1002/cssc.201900925.
118. Wu H, Dai W, Saravanamurugan S, Li H, Yang S. Quasi-catalytic approach to N-unprotected lactams via transfer hydro-amination/cyclization of biobased Keto acids. ACS Sustain Chem Engin. 2019;7(12):10207–13. https://doi.org/10.1021/acssuschemeng.9b00412.
119. Wu H, Yu Z, Li Y, Xu Y, Li H, Yang S. Hot water-promoted catalyst-free reductive cycloamination of (bio-)keto acids with HCOONH4 toward cyclic amides. J Supercrit Fluids. 2020;157:104698. https://doi.org/10.1016/j.supflu.2019.104698.
120. Raut AB, Shende VS, Sasaki T, Bhanage BM. Reductive amination of levulinic acid to N-substituted pyrrolidones over RuCl3 metal ion anchored in ionic liquid immobilized on graphene oxide. J Catal. 2020;383:206–14. https://doi.org/10.1016/j.jcat.2020.01.020.
121. Barbaro P, Liguori F, Oldani C, Moreno-Marrodán C. Sustainable catalytic synthesis for a bio-based alternative to the reach-restricted *N* -methyl-2-pyrrolidone. Adv Sustain Syst. 2020;4:1900117. https://doi.org/10.1002/adsu.201900117.
122. Wu Y, Zhao Y, Wang H, Zhang F, Li R, Xiang J, Wang Z, Han B, Liu Z. Ambient reductive synthesis of N-heterocyclic compounds over cellulose-derived carbon supported Pt

nanocatalyst under H₂atmosphere. Green Chem. 2020;22(12):3820–6. https://doi.org/10.1039/d0gc01177k.

123. Louven Y, Haus MO, Konrad M, Hofmann JP, Palkovits R. Efficient palladium catalysis for the upgrading of itaconic and levulinic acid to 2-pyrrolidones followed by their vinylation into value-added monomers. Green Chem. 2020;22(14):4532–40. https://doi.org/10.1039/d0gc01043j.
124. Chaudhari C, Shiraishi M, Nishida Y, Sato K, Nagaoka K. One-pot synthesis of pyrrolidones from levulinic acid and amines/nitroarenes/nitriles over the Ir-PVP catalyst. Green Chem. 2020;22(22):7760–4. https://doi.org/10.1039/d0gc01725f.
125. Wang Y, Nuzhdin AL, Shamanaev IV, Bukhtiyarova GA. Flow synthesis of N-alkyl-5-methyl-2-pyrrolidones over Ni2P/SiO2 catalyst. Mol Catal. 2021;515:111884. https://doi.org/10.1016/j.mcat.2021.111884.
126. Liu Y, Zhang K, Zhang L, Wang Y, Wei Z. One-pot synthesis of pyrrolidone derivatives via reductive amination of levulinic acid/ester with nitriles over Pd/C catalyst. React Kinet Mech Catal. 2021;134(2):777–92. https://doi.org/10.1007/s11144-021-02073-x.
127. Wu P, Li H, Fang Z. Synergistic catalysis of co-Zr/CN $_x$ bimetallic nanoparticles enables reductive amination of biobased levulinic acid. Adv Sustain Syst. 2022;6:2100321. https://doi.org/10.1002/adsu.202100321.
128. Gokhale TA, Raut AB, Chawla SK, Bhanage BM. Insights into cascade and sequential one-pot pathways for reductive amination of aldehydes paired with bio-derived levulinic acid to *N* -substituted pyrrolidones using molecular hydrogen. React Chem Eng. 2022;7:1005–13. https://doi.org/10.1039/D1RE00384D.
129. Gao J, Feng L, Ma R, Su B-J, Alenad AM, Liu Y, Beller M, Jagadeesh RV. Cobalt single-atom catalysts for domino reductive amination and amidation of levulinic acid and related molecules to N-heterocycles. Chem Catal. 2022;2(1):178–94. https://doi.org/10.1016/j.checat.2021.12.009.
130. Zeng Y, Wang B, Yan F, Xu W, Bai G, Li Y, Yan X, Chen L. Boron modified cu/al ₂ o ₃ catalysts for the selective reductive amination of levulinic acid to n-substituted pyrrolidinones. ChemCatChem. 2022;14(13):e202200311. https://doi.org/10.1002/cctc.202200311.

Chapter 3
Producing N-Heterocyclic Compounds from Lignocellulosic Biomass Feedstocks

Ananda S. Amarasekara

Abstract Major components of lignocellulosic biomass are cellulose, hemicellulose, lignin and these bio-macromolecules are composed of C, H and O atoms. In the production of N-heterocyclic compounds from biomass, N atoms are incorporated into intermediate molecules after depolymerization of bio-macromolecules. Depolymerization of lignocellulosic biomass results in common intermediates such as monosaccharides, furfural, 5-hydroxymethylfurfural, levulinic acid, succinic anhydride and vanillin used in producing N-heterocyclic compounds which are discussed in the introduction of this chapter. Nitrogen atoms are generally introduced into intermediate molecules via ammonia, amines or amine derivatives. The synthesis of N-heterocyclic compounds from lignocellulosic biomass feedstocks are arranged according to increasing ring size and number of nitrogen atoms in heterocyclic system. Synthesis of five-membered N-heterocyclic compounds: pyrrolidines, pyrrolidin-2-ol, pyrrolidones, bicyclic and fused pyrrolidones, pyrroles, pyrazoles, imidazoles and tetrazoles are discussed. Synthesis of six-membered N-heterocyclic compounds: pyridinium salts: pyridazin-3(2H)-one, pyridazines and pyrazines are also discussed. Synthesis of lesser known compounds: quinolines, benzodiazepinones, pyrido[2,3-d]pyrimidines, 1,2,4-triazines, quinoxalines and pyrazolo[3,4-*b*]quinoxalines from lignocellulosic biomass feedstocks are reviewed. The potential of selected reactions for large scale processes is presented. Furthermore, catalyst development needs for transformation of polysaccharides to N-heterocyclic compounds with polyhydric side chains or glycosidic groups are discussed throughout the text, highlighting their potential in drug discovery.

Keywords Lignocellulosic biomass · Monosaccharides · Levulinic acid · Pyrrolidines · Pyrrolidones · Pyrroles · Pyridazines · Pyrazines · Quinolines

---

A. S. Amarasekara (✉)
Department of Chemistry, Prairie View A&M University, Prairie View, TX, USA
e-mail: asamarasekara@pvamu.edu

Z. Fang et al. (eds.), *Production of N-containing Chemicals and Materials from Biomass*, Biofuels and Biorefineries 12,
https://doi.org/10.1007/978-981-99-4580-1_3

3.1 Introduction

With the declining petroleum resources and climate change concerns researchers have focused their attention to the development of technologies for producing chemical feedstocks and synthetic polymer precursors form renewable carbon resources. Lignocellulosic biomass is in the forefront of this challenge as the most abundant organic substance on earth for processing into renewable replacements for petroleum products. The lignocellulosic biomass in plant materials mainly composed of cellulose, hemicellulose, and lignin, and the typical percentage composition by dry weight are 35–50% cellulose, 20–35% hemicellulose, and 5–30% lignin [1]. Thus, cellulose is the most abundant biopolymer; however, the main carbohydrate polymers cellulose and hemicelluloses are tightly bound to the lignin, by hydrogen and covalent bonds making them difficult to access. Furthermore, cellulose is an unbranched polymer of glucose, linked together through β-1,4 links of the glucose monomer unit [2, 3]. The complex structure of cellulose is composed of stiff polymeric molecular chains with close packing via numerous strong, inter and intramolecular hydrogen bonding, making it extremely difficult for solvent molecules to penetrate the structure. The second most abundant polysaccharide in lignocellulosic biomass is hemicellulose, which is a branched polymer composed of diverse sugars. In general, hemicellulose consists of the five-carbon sugar xylose as the major component and another five-carbon sugar arabinose as well as six-carbon sugars, glucose, mannose and galactose that are present in a lesser extent. As a result of this molecular architecture, it is hard to dissolve cellulose and hemicellulose in water and in most common organic solvents. This property causes difficulties in improving the processability, fusibility, functionality and hydrolysis of polysaccharides. Considerable efforts are still being devoted to improve the solubility and processability of cellulose [4, 5]. However, recent development in use of enzyme systems, acid and ionic liquid catalysts have paved the way for more efficient depolymerizations of cellulose and hemicellulose to its monomers C6 and C5 sugars [6, 7]. Further processing of these sugars can lead to a series of intermediate biomass derived feedstocks such as monohydric alcohols, methanol, ethanol as well as polyhydric alcohols; carbonyl compounds, including furfural, 5-hydroxymethylfurfural and levulinic acid. Then depolymerization of the lignin fraction can produce polyhydric alcohols as well as phenolic compounds such as vanillin. A representative sample of lignocellulosic biomass derived feedstocks with the potential to be converted to N-heterocyclic compounds is shown in Fig. 3.1.

In this complex biomass refinery process, depolymerization of the major polysaccharide cellulose to give glucose is the entry point to majority of renewable feedstocks. Additionally, in this important downstream process, acid catalyzed dehydration of glucose with the loss of three molecules of water yields 5-hydroxymethylfurfural. In further reactions of 5-hydroxymethylfurfural involving a rehydration can lead to levulinic acid or 4-oxopentanoic acid as shown in Fig. 3.2 [8–10]. This C5 keto-acid is an important biomass derived feedstock as well as renewable carbon based intermediate in new generation feedstock compound for

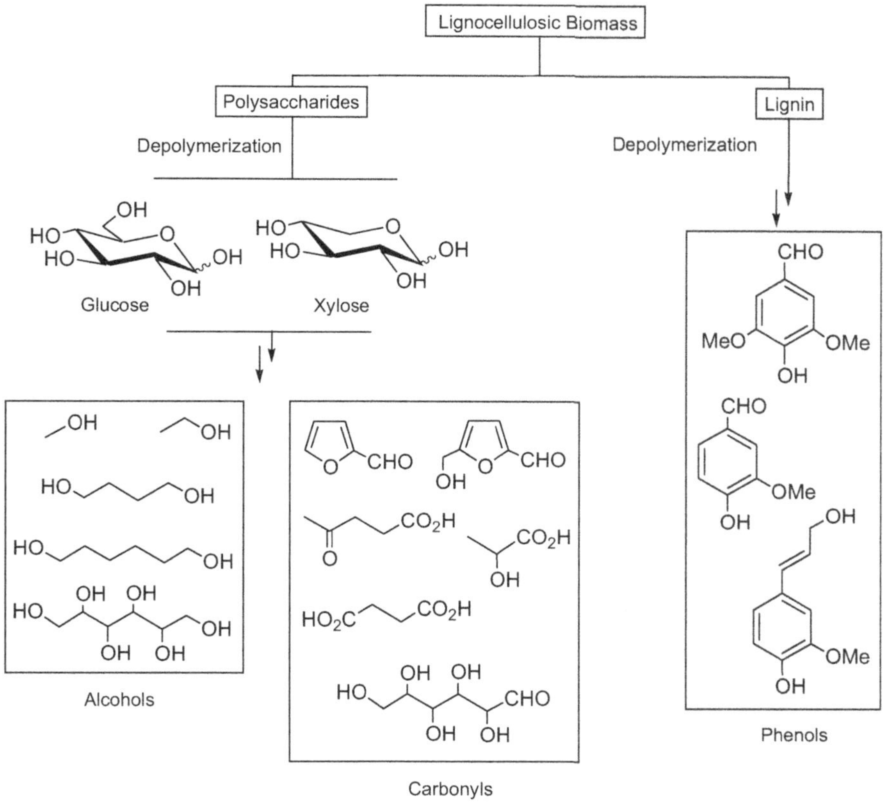

Fig. 3.1 Lignocellulosic biomass components and intermediates for producing N-heterocyclic compounds

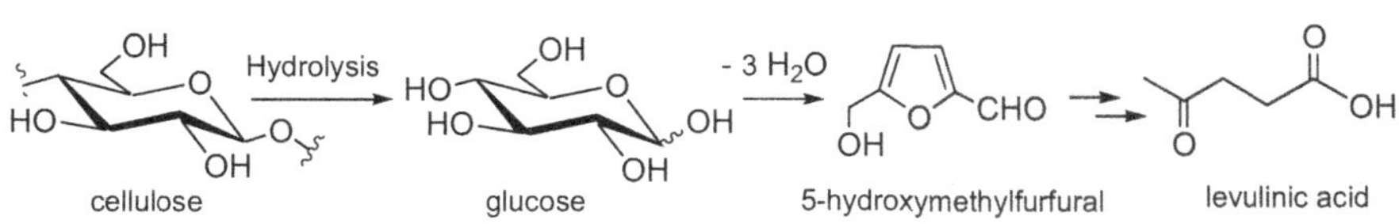

Fig. 3.2 Levulinic acid production from cellulose

polymers and many other applications such as pharmaceuticals and biofuels [3, 9, 11]. In addition, levulinic acid was listed as one of the top 12 most promising value added feedstock chemicals from biomass by the Biomass Program of the US Department of Energy in 2004 [12] and also continues to rank highly in more recent reviews of major biorefinery target products [13]. Even though the current world production of levulinic acid is around 2600 tons/year, it is predicted to increase in the next 2–5 years, mainly due to the introduction of the dilute acid catalyzed rapid hydrolysis technique known as Biofine process. In this patented reactor configuration refined biomass is converted into four products: levulinic acid, formic acid, furfural, and biochar, where levulinic acid can be separated from the other products. As a result of the success of the novel Biofine process levulinic acid production cost is expected to decrease as low as US $0.22 per kg making this keto acid an attractive intermediate for many biobased chemicals [14].

The bifunctional levulinic acid can be used in the synthesis of a number of useful chemicals such as γ-valerolactone, 2-methyl tetrahydrofuran, acrylic acid, 1,4-pentanediol, β-acetylacrylic acid, α-angelica lactone, δ-amino levulinic acid, etc. [14–21]. Similar to these oxygenated derivatives, introduction of a nitrogen function to this C5 keto-acid has been explored for the production of 5-methyl-2-pyrrolidone and its related compounds and these developments are presented in this chapter.

The lignin fraction of biomass can also be depolymerized and processed to produce feedstocks for the synthesis of N-heterocyclic compounds. Lignin is a three-dimensional amorphous macromolecule composed of phenyl propane units that arise from the copolymerization of three primary precursors: coniferyl alcohol, sinapyl alcohol and *p*-coumaryl alcohol. The lignin content of biomass varies greatly on the plant species and many other factors such as age and part of the plant. Lignin from softwoods is principally based on structural units derived from the coniferyl alcohol (guaiacyl units), which are the precursors for vanillin yield [22, 23]. On the other hand, hardwood lignins present structures with much broader chemical compositions with various guaiacyl, syringil ratios [24, 25]. Some of the prominent monomers derived from depolymerized lignin are phenol, methoxyphenols, 4-ethylphenol, catechol, pyrocatechol, cresol, resorcinol, eugenol, syringol, coniferol, guaiacol, propenylguaiacol, 4-propylguaiacol, 4-methylguaiacol and vanillin [26, 27].

Lignin derived from wide variety of plants can be depolymerized under oxidative conditions to obtain phenolic compounds. The processing of lignin to produce value added aromatic compounds is a well established field and is discussed in a number of recent reviews [22, 28–31], Among many aromatic value added product possible through depolymerization of lignin, the production of vanillin is probably the most widely studied process [32–34]. The use of vanillin as a building block for the chemical industry and polymer industry is a current high priority interest and this stable phenolic aldehyde can undergo a wide variety of reactions at its aldehyde and phenol functions as well as in the reactive polarized aromatic ring [35–37]. Currently vanillin is the only molecular phenolic compound manufactured on an industrial scale from biomass, and many research groups have explored the potential of vanillin as a sustainable polymer feedstock [38–42]. Poly-functionalized vanillin can be an excellent renewable feedstock for the synthesis of N-heterocyclic compounds as well. However, in comparison to the widely studied polymer feedstock application, lignin aromatics are not a well explored avenue for the synthesis of N-heterocyclic compounds and N-containing polymers. Nevertheless, Gendron has recently discussed some of the applications of vanillin as a building block for N-heterocyclic compounds in a short review titled "Vanillin: A Promising Biosourced Building Block for the Preparation of Various Heterocycles" [43].

The production of N-heterocyclic compounds from biomass feedstocks is partially reviewed in a few recent reviews [44–46]. Literature on upgrading of levulinic acid into N-containing functional chemicals is discussed in a 2019 review by Xue and co-workers [45]. The widely studied levulinic acid derived five-membered N-heterocyclics are well reviewed in this publication [45]. In another review, Zhang et al. discussed the applications of levulinic acid for in drug synthesis highlighting the potential of biomass derived feedstocks in pharmaceutical industry [46]. In a more recent 2020 review Li et al. discussed the cycloamination strategies for renewable N-heterocycles, where a more critical approach was taken in comparison of synthesis from renewable feedstock and classical strategies for preparation of these heterocyclic compounds [47]. However, these journal articles are more focused on a selected group of starting compounds, ring systems or a specific application like drug development [48, 49]. This book chapter reviews recent developments in the synthesis of common N-heterocyclic ring systems starting from lignocellulosic biomass-based feedstocks.

3.2 Five-Membered N-Heterocyclic Compounds

3.2.1 Pyrrolidines

Pyrrolidines are five-membered saturated heterocyclic compounds with one nitrogen atom. This common heterocyclic moiety can be found in amino acid proline, a number of natural products such as nicotine and pharmaceuticals [50, 51]. Industrially pyrrolidine is prepared by NiO/Al_2O_3 catalyzed cyclo-condensation of 1,4-butanediol and ammonia under high pressure and temperature [52]. Interestingly, 1,4-butanediol can be obtained from renewable sources and the same technology can be used in synthesis of bio-based pyrrolidine. Carbohydrate derived C5 levulinic acid is the feedstock chemical of choice for many five-membered N-heterocyclic compounds. In this case, condensation of levulinic acid with amines under strong reductive conditions is known to give renewable carbon based 2-methyl-pyrrolidines. $InI_3/PhSiH_3$ is such a strong reducing agent and Ogiwara et al. have shown that various pyrrolidines could be synthesized from levulinic acid and amines in toluene. at 120 °C [53]. In this process, pyrrolidines were produced through the formation of pyrrolidones and the subsequent over-reduction of the carbonyl groups as shown in mechanism in Fig. 3.3. Furthermore, Ogiwara et al. explained the reason to form pyrrolidines rather than pyrrolidones as the higher acidity of InI_3 than that of $In(OAc)_3$, which was used to selectively synthesize pyrrolidones [53].

A few other strongly reducing catalyst systems such as $RuCl_3/PhSiH_3$ may also used for the reaction. Additional examples for the synthesis of N-aryl/alkyl-2-methyl-pyrrolidines from condensation of levulinic acid with amines under strong reductive conditions are shown in Table 3.1.

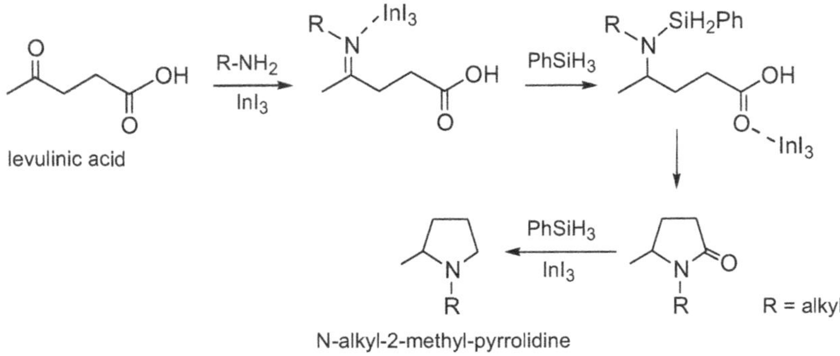

Fig. 3.3 Possible reaction pathway for the formation of N-alkyl-2-methyl-pyrrolidines by reductive amination of levulinic acid using $PhSiH_3/InI_3$ system

Table 3.1 Preparation of pyrrolidines by condensation of levulinic acid with amines under strong reductive conditions

Reaction conditions	Product (yield %)	Reference
Ar-NH_2, $RuCl_3$. $3H_2O$ (1 mol%), $PhSiH_3$, 45 °C	N-alkyl-2-methyl-pyrrolidine (85–96%)	[54]
Aryl/alkyl-NH_2, Cs_2CO_3 (5 mol%), $PhSiH_3$, 25 °C	N-alkyl/aryl-2-methyl-pyrrolidine (81–93%)	[55]
Ar-NH_2, [$Fe(CO)_4$(IMes)] (B, IMes = 1,3-bis(2,4,6-trimethylphenyl) imidazol-2-ylidene), $PhSiH_3$, 100 °C	N-aryl-2-methyl-pyrrolidine (80–99%)	[56]

3.2.2 *1-Ethyl-2-(Ethylideneamino)-5-Methylpyrrolidin-2-Ol*

In one unusual nitrogen source example, Rodríguez-Padrón and co-workers reported the use acetonitrile as the N-source where acetonitrile is reduced in situ to a mixture of ethyl amine and an imine under strongly reducing reaction conditions. In this instance, N-Ethyl, 5-methyl-2-pyrrolidone was formed as the primary product and further reaction of this with imine resulted 1-ethyl-2-(ethylideneamino)-5-methylpyrrolidin-2-ol. Ru deposited on TiO_2 was used as the catalyst system, where under the optimized conditions 79% conversion and 85% selectivity to 1-ethyl-2-(ethylideneamino)-5-methylpyrrolidin-2-ol was achieved in experiments using 3 wt% Ru, homogeneously deposited on the titania surface as the catalyst. Furthermore, they have proposed a reasonable reaction mechanism for the formation of this functionalized 5-methylpyrrolidin-2-ol product in good yields as shown in Fig. 3.4 [57].

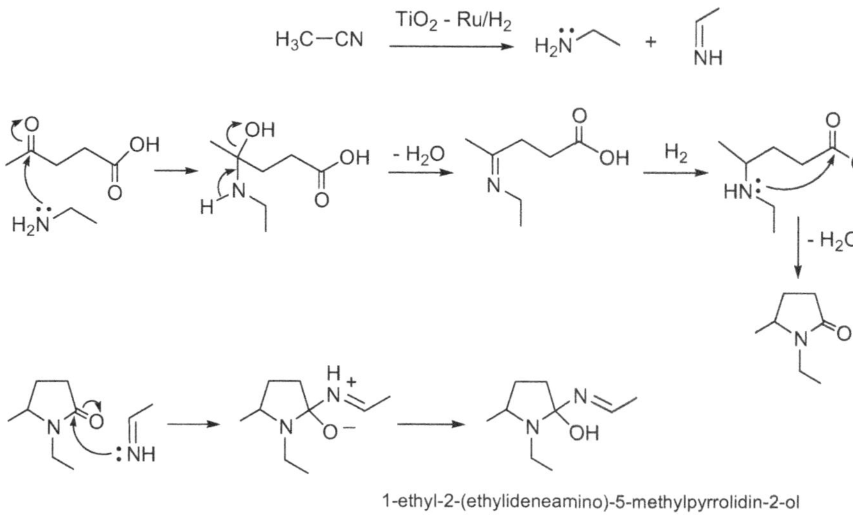

Fig. 3.4 Proposed mechanism for the formation of 1-ethyl-2-(ethylideneamino)-5-methylpyrrolidin-2-ol in the reaction of acetonitrile with levulinic acid under hydrogen using Ru/TiO_2 catalyst [57]. Reprinted with permission from [57]. Copyright 2018, American Chemical Society

3.2.3 Pyrrolidones

Reductive amination of levulinic acid to pyrrolidones is one of the most widely studied syntheses of N-heterocyclic compounds from lignocellulosic biomass feedstocks. A large portion of these studies have been focused on the noble metal catalyzed process for the conversion of levulinic acid to give 5-methyl-2-pyrrolidone using a mixture of hydrogen and ammonia gases. In the analogous process using primary alkyl amines gives N-alkyl-5-methyl-2-pyrrolidones. These N-alkylated pyrrolidones have a wide diversity of applications, such as the use as an alternative to common solvent N-methyl-2-pyrrolidone, surfactants and important intermediates in the synthesis of agricultural bioactive compounds and pharmaceuticals [58, 59]. In addition, N-methyl-2-pyrrolidone is used as an important constituent in many cleaning agents, refrigerants, air conditioning lubricants, inks and in aerosol formulations [60]. The nitrogen function introduction is known via a wide range of N-sources such as ammonia, ammonium hydroxide, ammonium salts, and primary amines. The condensation—cyclization process to make 2-pyrrolidones from levulinic acid is normally carried out under reduction conditions in the presence of a catalyst. A selected list of representative synthesis of 5-methyl-2-pyrrolidone and derivatives from levulinic acid and its derivatives are shown in Table 3.2.

Most techniques use excess of ammonia and hydrogen under high pressure, for instance, the patented method for producing N-methyl-2-pyrrolidone from levulinic acid using H_2-NH_3 mixture and noble metal catalysts have several drawbacks such as the requirement of a high excess of H_2 gas, expensive noble metals, hazardous

Table 3.2 Synthesis of 5-methyl-2-pyrrolidone (5MP) and derivatives from levulinic acid (LA) and its derivatives

Starting material	Reaction conditions	Product (yield %)	Reference
Levulinic acid	H_2, NH_4OH, Pd, water; 2 h, 50 bar, 200 °C	5-methyl-2-pyrrolidone (96%)	[61]
Levulinic acid	$NH_3{\cdot}BH_3$, methanol; 120 °C	5-methyl-2-pyrrolidone (95%)	[62]
Levulinic acid	H_2, NH_3, Pt, methanol; 72 h, 14 atm, 25 °C	5-methyl-2-pyrrolidone (89%)	[63]
Levulinic acid	H_2, NH_3, Ru, methanol; 4 h, 3 MPa, 70 °C	5-methyl-2-pyrrolidone (90%)	[64]
Levulinic acid	HCO_2NH_4, water; 3 h, 160 °C	5-methyl-2-pyrrolidone (97%)	[65]
Levulinic acid	NH_4OH, ZrO_2, Chloroauric acid, water; rt.; 6 h, pH 9.0, rt	5-methyl-2-pyrrolidone (85%)	[66]
Levulinic acid	HCO_2H, NH_3, Bis(dichloro(η-*p*-cymene)ruthenium), tri-tert-butylphosphonium tetrafluoroborate, water; 12 h, 80 °C	5-methyl-2-pyrrolidone (70%)	[67]
^{n}-Butyl-levulinate	HCO_2H, NH_3, water; 4 h, 180 °C	5-methyl-2-pyrrolidone (93%)	[68]
Ethyl-levulinate	NH_4HCO_3, H_2, Ru (polyporphyrin complexes), 1,4- Benzenedicarboxaldehyde, polymer with 1H-pyrrole, methanol, rt.; 24 h, 120 °C	5-methyl-2-pyrrolidone (96%)	[69]
Ethyl-levulinate	$HCONH_2$, HCO_2H; 300 min, 160 °C	5-methyl-2-pyrrolidone (93%)	[70]
Ethyl-levulinate	Benzonitrile, H_2, Pd, THF, Dodecane; 4 h, 20 bar, 80 °C	N-benzyl, 5-methyl-2-pyrrolidone (70%)	[71]
Levulinic acid	N-Methylformamide, Water, 4 h, 160 °C	N-methyl, 5-methyl-2-pyrrolidone (98%)	[65]
Levulinic acid	H_2, Ni_3N, acetonitrile, 60 min, 50 bar, 100 °C	N-ethyl, 5-methyl-2-pyrrolidone (68%)	[72]
Levulinic acid	Benzyl amine, H_2, Cobalt catalyst, Toluene, 24 h, 20 bar, 70 °C; 70 °C → rt	N-benzyl, 5-methyl-2-pyrrolidone (68%)	[73]
Levulinic acid	Benzaldehyde, H_2, NH_3, Ni, methanol-water, Rt. → 140 °C; 7 h, 15 bar	N-benzyl, 5-methyl-2-pyrrolidone (88%)	[74]
Levulinic acid	R-NH_2, R-CN, R-NO_2, H_2, 0.75Co-phen@C-800, 90–120 °C, *i*-PrOH	N-alkyl, 5-methyl-2-pyrrolidone (89%)	[75]

organic solvents, and limited selectivity for the formation of the expected product when aryl amines are used in place of ammonia [58, 59]. The most widely used noble metal catalysts in this technique are Pt/C, Pd, Ru [76], Ru/C [59], Ir [77], and [Ru

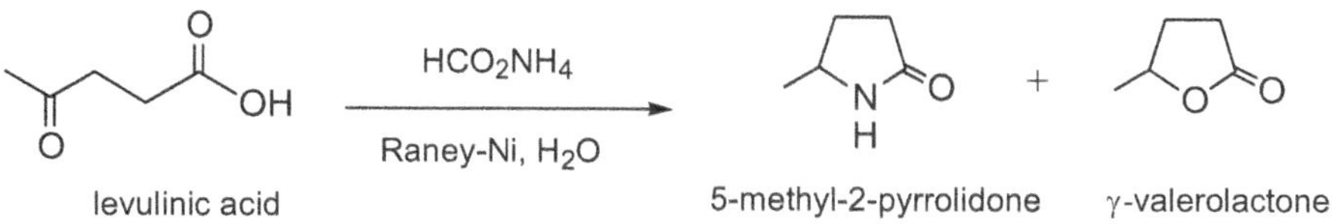

Fig. 3.5 Raney-Ni catalyzed conversion of levulinic acid to 5-methyl-2-pyrrolidone and γ-valerolactone using ammonium formate as the hydrogen and nitrogen source. Reprinted with permission from [80]. Copyright 2018, Elsevier

(*p*-cymene)$Cl_2]_2$ [67, 78]. The polymeric ruthenium porphyrin-functionalized carbon nanotubes (Ru-PP/CNTs) prepared by the metalation of polymeric porphyrin-functionalized carbon nanotubes with $Ru_3(CO)_{12}$, has also been used in the synthesis of pyrrolidone derivatives from ethyl levulinate, primary amines and hydrogen gas [69]. The use of inexpensive non-noble metal catalysts and improvements in reusability are potential future research directions for this process.

A few research groups have studied the possibility of reductive amination of levulinic acid using transfer hydrogenation, without using hydrogen gas as the H atom source, especially using Ru and Au based catalysts [66, 67]. Du and co-workers developed the use of gold deposited on ZrO_2 as the catalyst, where they reported a one pot process requiring only formic acid, ammonia or primary amines, and the Au catalyst to convert levulinic acid directly into the corresponding pyrrolidones [66]. In another example of transfer hydrogenation, Sun et al. used Ru based catalyst fabricated via self-assembly from a *p*-phenylene-bridged *bis*-benzimidazolium salt with metal precursors, where ammonium formate was used to prepare 5-methyl-2-pyrrolidone at a 0.15 mol% catalyst loading [79].

In another example, Amarasekara and Lawrence used ammonium formate in transfer hydrogenation, which is both H and N source where Raney-Ni was the catalyst. This reaction gives both 5-methyl-2-pyrrolidone and γ-valerolactone, as shown in Fig. 3.5. However, levulinic acid: HCO_2NH_4 1: 4 ratio and under the reaction conditions of oven heating at 180 °C, 3 h, produced only 5-methyl-2-pyrrolidone in excellent yield as shown in Table 3.3. Microwave heating also gave similar results with somewhat lower yield. Furthermore, Amarasekara and Lawrence explained the formation of 5-methyl-2-pyrrolidone and γ-valerolactone products in a reaction scheme as shown in Fig. 3.6 [80]. Ammonium formate is expected to decompose generating hydrogen and ammonia and ammonium levulinate formed can dehydrate to the amide. Cyclization of the amide followed by H_2/Raney-Ni reduction may lead to 5-methyl-2-pyrrolidone as shown in Fig. 3.6.

During this work Amarasekara and Lawrence extended the Raney-Ni catalyzed process to synthesize N-alkyl/aryl-5-methyl-2-pyrrolidones as well as shown in Fig. 3.7. In these experiments alkyl/aryl amines were used together with formic acid, which decomposed to hydrogen and carbondioxide during the reaction [80].

Table 3.3 The levulinic acid (LA): HCO_2NH_4 ratio used, catalyst loading, reaction condition, LA conversion (%), product yield based on LA reacted (%) and carbon balance (%) in Raney-Ni catalyzed conversion of LA to 5-methyl-2-pyrrolidone (MPD) and γ-valerolactone (GVL). 1.0 mmol levulinic acid and 2.00 mL of water were used in all experiments [80]

Entry	LA: HCO_2NH_4	Raney-Ni catalyst (mg/mmol of LA)	Reaction condition	LA conversion[a] (%)	Product yield based on LA reacted (%)[b]		Carbon balance[c] (%)
					MPD	GVL	
1	1: 1	20	Oven, 120 °C, 20 h	40	48	36	84
2	1: 2	20	Oven, 120 °C, 20 h	63	52	30	82
3	1: 4	20	Oven, 160 °C, 3 h	100	60	28	88
4	1: 4	20	Oven, 180 °C, 3 h	100	94	–	94
5	1: 4	–	Oven, 180 °C, 3 h	15	68	23	91
6	1: 4	20	MW (920 W), 5 × 10 s	11	80	–	80

[a]Levulinic acid conversion % = (levulinic acid used—levulinic acid unreacted)/levulinic acid used × 100
[b]Yield was calculated using: Yield % = (products detected/levulinic acid reacted) × 100
[c]Carbon balance determined on the basis of 5-methyl-2-pyrrolidone (MPD) and γ-valerolactone (GVL) products. Reprinted with permission from [80]. Copyright 2018, Elsevier

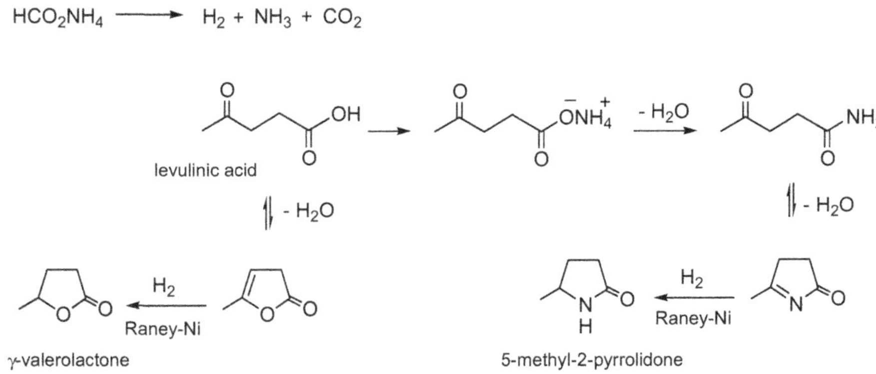

Fig. 3.6 Proposed reaction pathways for the Raney-Ni catalyzed conversion of levulinic acid to 5-methyl-2-pyrrolidone and γ-valerolactone using ammonium formate [80]. Reprinted with permission from [80]. Copyright 2018, Elsevier

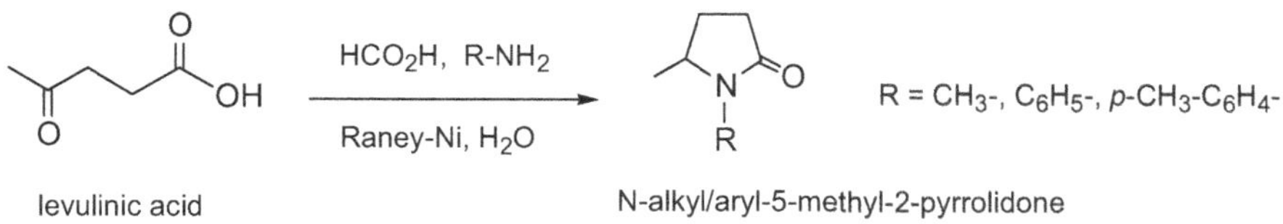

Fig. 3.7 Raney-Ni catalyzed conversion of levulinic acid to N-alkyl/aryl-5-methyl-2-pyrrolidone using ammonium formate [80]. Reprinted with permission from [80]. Copyright 2018, Elsevier

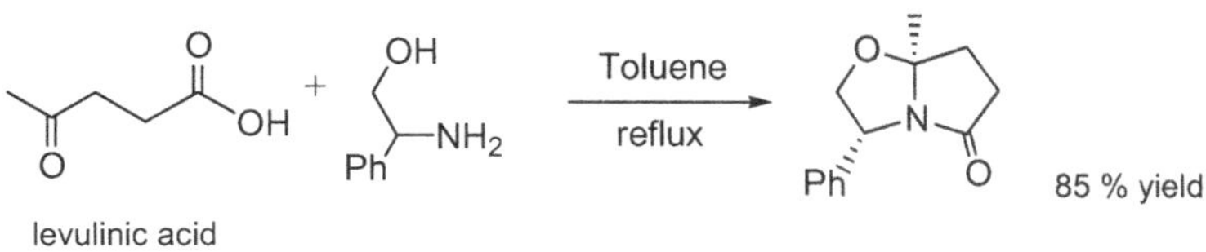

Fig. 3.8 Condensation of 2-amino, 2-phenyl ethanol with levulinic acid in refluxing toluene to give bicyclic fused pyrrolidone [84]. Reprinted with permission from [84]. Copyright 1991, American Chemical Society

3.2.4 *Bicyclic and Fused Pyrrolidones*

Synthesis of a number of bicyclic and fused lactams are possible with reactions of levulinic acid with a wide range amino alcohols, amino acids and diamines. The use of chiral amines has been used to the advantage of synthesizing chiral products and many of these bicyclic and fused lactams are known to have wide applications in therapeutics for different diseases. Zhang et al. have recently reviewed the use of levulinic acid derived compounds for application in drug synthesis and most of these active compounds are N-heterocyclics [46]. A wide range of medicinal applications are known in these compounds, including the use of 4-aminoquinoline c-and d-lactams as anti-parasitic drugs [81]. Furthermore, Bouchmaa and co-workers have synthesized a series of new pyridazine-3(2H)-one derivatives as breast cancer drugs, and reported their activity against human triple-negative breast cancer (MDA-MB-468) [82, 83].

The condensation of levulinic acid with 1,2-amino alcohols are known to give bicyclic lactams. Burgess and Meyers have shown that 2-amino, 2-phenyl ethanol with levulinic acid in refluxing toluene gives the lactam in 85% yield as shown in Fig. 3.8 [84]. Later the same group extended this reaction for the preparation of a series of chiral bicyclic lactams [85, 86].

In continuation of the synthesis of bio-active N-heterocyclic compounds from biomass derived compounds, Sharma and co workers used levulinic acid based chiral bicyclic lactams as central scaffolds [87]. These compounds were synthesized from (R)-phenylalaninol, levulinic acid and 3-(2-nitrophenyl) levulinic acid [87]. The diversity-oriented synthesis library was generated by either direct substitution with o-nitrobenzylbromide at the carbon α- to the amide functionality or by conversion to fused pyrroloquinolines. Upon screening this diverse library for their

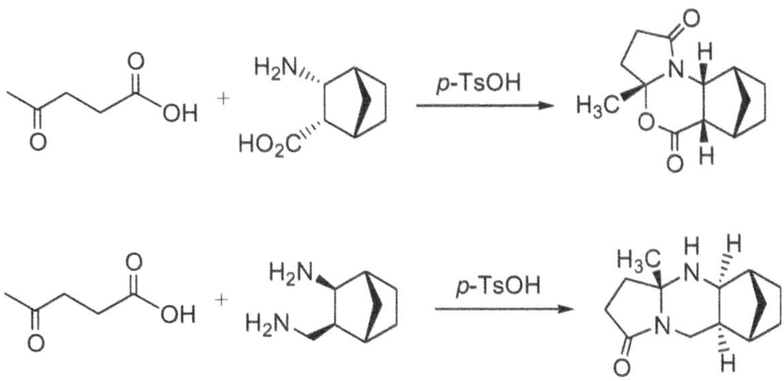

Fig. 3.9 Condensation of levulinic acid with norbornane amino acids and diamines to produce fused pyrrolidones

anti-malarial activity, a dinitro/diamine substituted bicyclic lactam was found to demonstrate exceptional activity of >85% inhibition at 50 μM concentration across different strains of P. falciparum with no toxicity against mammalian cells. In addition, loss of mitochondrial membrane potential, mitochondrial functionality and apoptosis was observed in parasite treated with diamine-substituted bicyclic lactams [87].

Stájer and co-workers have studied the condensation reactions of levulinic acid with norbornane/ene amino acids and diamines and shown that norbornane/ene fused bicyclic lactams can be obtained in moderate yields in reactions carried out under *p*-toluene sulfonic acid catalysis [88]. During this work they have found that cyclizations of di-*endo*-3-aminobicyclo[2.2.1]heptane-2-carboxylic acid and di-*endo*-3-aminobicyclo[2.2.1]hept-5-ene-2-carboxylic acid with levulinic acid yield methanodioxopyrrolo[1,2-*a*][1,3] benzoxazines [88]. Furthermore, levulinic acid reacts with di-*exo*-3-aminobicyclo[2.2.1]heptane-2-methanamine and di-*exo*-3-aminobicyclo[2.2.1]hept-5-ene-2-methanamine to give the methylene-bridged pyrrole[2,1-*b*]quinazolinones. A representative example is shown in Fig. 3.9 [88].

In another synthesis of bicyclic lactams Verardo et al. have reported that α-amino acid phenylhydrazides readily react with levulinic acid to produce the imidazolidin-4-ones [89]. In further reaction, these intermediates undergo a second ring closure to afford the 3-substituted dihydro-1-phenylamino-1*H*-pyrrolo[1,2-*a*]imidazole-2,5 (3*H*,6*H*)diones [89] as shown in Fig. 3.10. Furthermore, it has been established that the solvent polarity has a great influence on the rate of the second condensation reaction, but not on the first. In addition, Verardo et al. proposed a mechanism, supported by experimental evidence to explain how the intermediates give a single isomer for the bicyclic dihydro-1*H*-pyrrolo[1,2-*a*]imidazole-2,5-dione derivatives and the absolute stereochemistry of these compounds were also determined by X-ray crystallographic analysis [89]. This type of imidazolidin-4-one derivatives are known to have interesting biological activities such β-secretase (BACE-1) inhibition activity, and has been identified as a promising drug target for disease-modifying

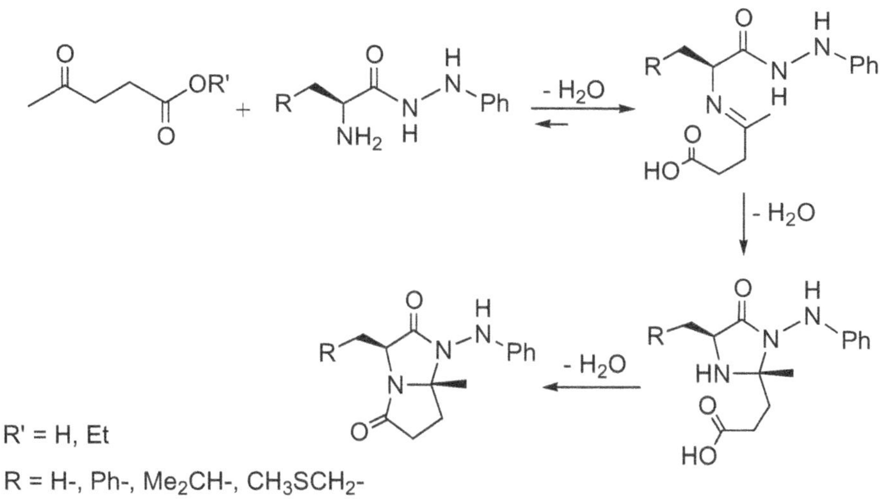

Fig. 3.10 Reaction of α-amino acid phenylhydrazides with levulinic acid to produce fused pyrrolidones

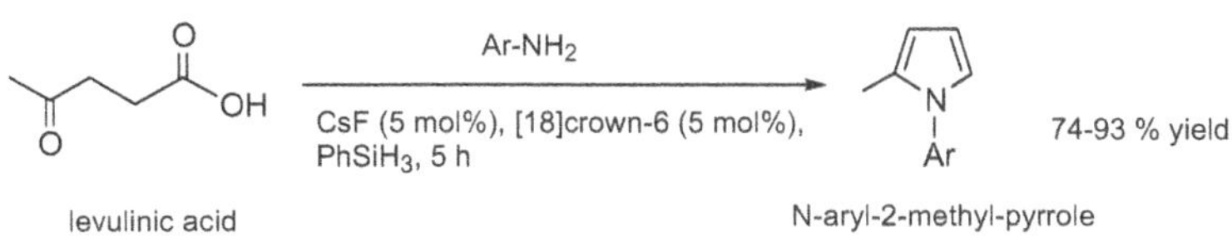

Fig. 3.11 Synthesis of N-substituted aryl pyrroles via reductive amination/cyclization of levulinic acid with primary aromatic amines using phenylsilane in the presence of CsF catalyst

therapy and has attracted significant attention from the medicinal chemistry community [90, 91].

3.2.5 Pyrroles

Recently Wu and co-workers reported the synthesis of N-substituted aryl pyrroles via reductive amination/cyclization of levulinic acid with primary aromatic amines and hydrosilanes such as polymethylhydrosiloxane over CsF catalyst [92]. This transformation is known only with a limited class of catalysts and hydrogen donors, however, during this work a series of N-substituted aryl pyrroles could be obtained in good to excellent yields at 120 °C as shown in Fig. 3.11. Furthermore, Wu et al. proposed a mechanism involving two steps [92]; in which, the cyclization between amine and levulinic acid was proposed for the first step to form intermediate N-alkyl-5-methyl-1,3-dihydro-2H-pyrrolones and their isomers, and then the chemo and

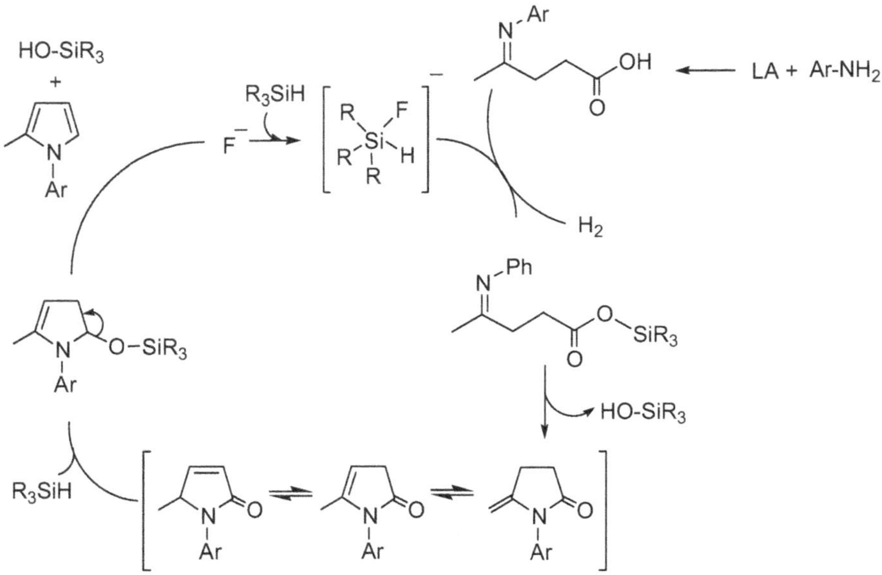

Fig. 3.12 A possible reaction pathway for the reductive amination—cyclization of levulinic acid to N-aryl pyrroles

regio-selective reduction of intermediates take place to produce the final product as shown in Fig. 3.12.

Another approach to pyrrole is via biomass derived furans. 5-Hydroxymethylfurfural derived from lignocellulosic biomass can be converted to the symmetrical 2,5-bis-hydroxymethylfurn by reduction of the aldehyde group [93–95]. The furan ring opening of –OH group protected 2,5-bis-hydroxymethylfurn is known with a range of reagents including singlet oxygen, bromine in water–methanol [96] or aqueous acetone [97], and 3-chloroperbenzoic acid [98]. However, 3-chloroperbenzoic acid or singlet oxygen is preferred for oxidative ring opening in furans, as the slightly acidic conditions allowed the isolation of the *cis*-hexenediones in high yields (>90%). The enediones were then reduced using $TiCl_3$ or zinc/acetic acid producing the desired hexane-2,5-diones, which were then condensed with various amines to produce pyrroles in good yields as shown in Fig. 3.13 [98].

3.2.6 *Pyrazoles*

Pyrazoles are five-membered aromatic compounds with two adjacent nitrogen atoms. Pyrazole derivatives have a long history of application in pharmaceutical industry and as agrochemicals. The pyrazole containing pharmacophore are reported to have a number of biological activities, including: anti-inflammatory [99, 100],

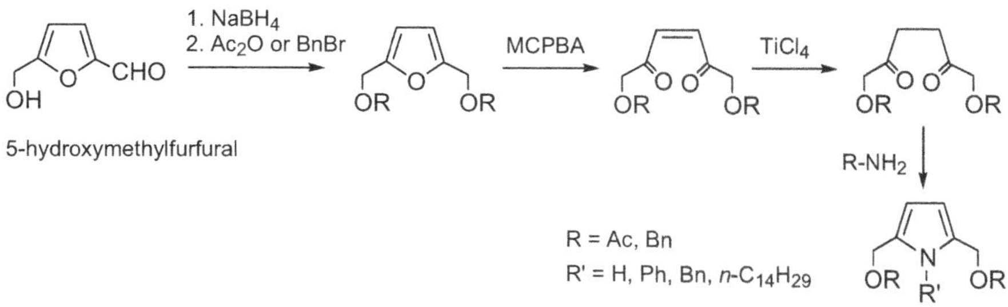

Fig. 3.13 Conversions of 5-hydroxymethylfurfural into 2,5-disubstituted pyrroles [98]. Reprinted with permission from [98]. Copyright 1999, Royal Society of Chemistry

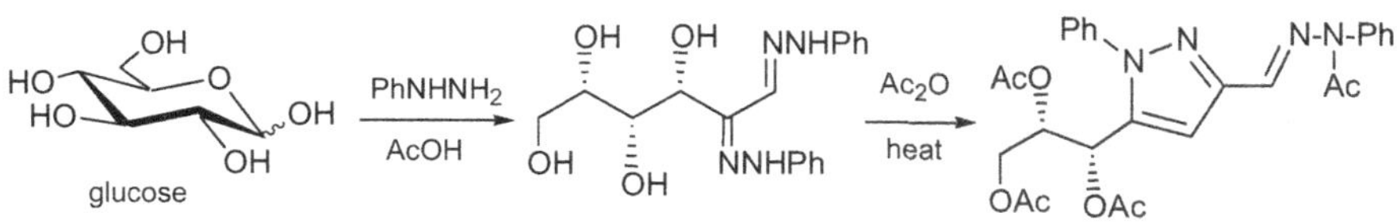

Fig. 3.14 Synthesis of 3,5-disubstituted pyrazoles from D-glucose [107]. Reprinted with permission from [107]. Copyright 1964, American Chemical Society

anti-bacterial [101], and anticonvulsant [102] activities making them an important ring system in medicinal chemistry. Furthermore, pyrazole ring system is common in widely used medicine, such as fezolamine, celecoxib and similar COX-2 inhibitors zaleplon and betazole [103]. In addition the pyrazole ring is found within a variety of agrochemicals: insecticides, pesticides, fungicides and herbicides [104].

One of the earliest synthesis of a pyrazole derivative from a biomass derived feedstock was reported by El Khadem et al., where glucose was reacted with phenyl hydrazine in the presence of acetic acid to give a phenylosazone as shown in Fig. 3.14 [105]. Then reacting with acetic anhydride produced the pyrazole derivative [106, 107]. Furthermore, microwave-assisted reaction is also known to produce respective osazones derivatives and pyrazoles from galactose, arabinose, and xylose in good yields [108, 109].

Levulinic acid with two active functional groups is a popular intermediate for many N-heterocyclic compounds from lignocellulosic biomass feedstocks. Flores et al. developed a Synthesis for bis-heterocycles of 1-[(5-Hydroxy-5-trifluoromethyl-3-substituted-4,5-dihydro-1*H*-pyrazol-1-yl)-3-(5-trifluoromethyl-1*H*-pyrazol-3-yl)propan-1-ones starting from levulinic acid [110]. This method involved cyclocondensation of 3-(5-trifluoromethyl-1*H*-pyrazol-3-yl)propanoyl hydrazide obtained from levulinic acid, with 1,1,1-trifluoro-4-methoxy-3-alken-2-ones proceeding regiospecifically to 1-[(5-trifluoromethyl-5-hydroxy-3-substituted-4,5-dihydro-1*H*-pyrazol-1-yl)-3-(5-trifluoro-methyl-1*H*-pyrazol-3-yl)propan-1-one derivatives as shown in Fig. 3.15 [111].

The same group have systematically investigated the use of methyl 7,7,7-trifluoro-4-methoxy-6-oxo-4-heptenoate methyl ester for the synthesis of a number

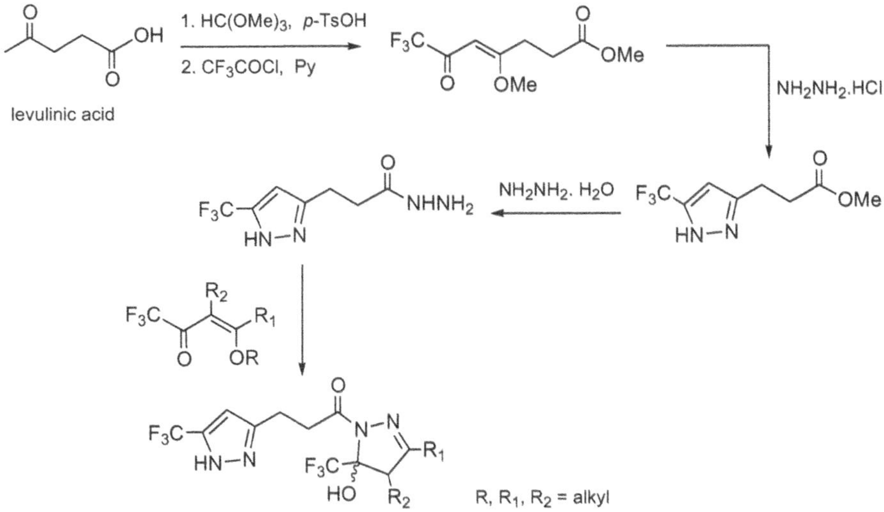

Fig. 3.15 Synthesis of bis-heterocyclic systems with dihydro-pyrazole and pyrazole moieties starting from levulinic acid

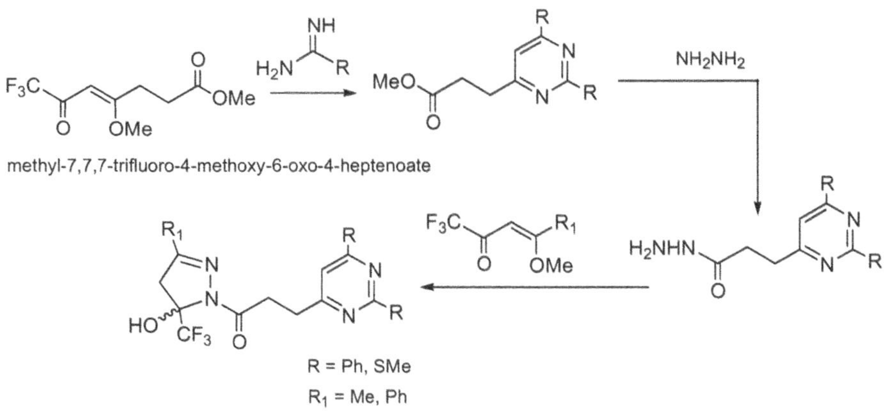

Fig. 3.16 Synthesis of bis-heterocyclic systems with dihydro pyrazole and pyridazine moieties from methyl-7,7,7-trifluoro-4-methoxy-6-oxo-4-heptenoate derived from levulinic acid

of bis-heterocyclic systems [112, 113]. This approach uses the versatility of the trifluoromethyl-substituted 1,3-dielectrophilic moiety for heterocycle diversification in the 3-position of the propanoic chain and the ability to transform the methyl ester into a hydrazide. In an extension of this strategy, the use of 2-substituted formamidine resulted the six-membered pyridazine as the first heterocycle moiety. Later the ester group was converted to hydrazide and to build the dihydro pyrazole as the second heterocycle moiety as shown in Fig. 3.16 [113].

3.2.7 Imidazoles

Imidazoles are five-membered aromatic compounds with two nitrogen atoms in 1,3-arrangement. Like the isomeric pyrazole moiety, imidazoles are also well known for their wide spectrum of biological activities including: anti-tuberculosis [114], antifungal and antimycobacterial activities [115], antimicrobial, antiviral [116], and antileishmanial [117] activities. The condensation of 1,2-dicarbonyls with ammonia and aldehydes via the Radziszewski reaction is currently used common industrial route to simple imidazoles [118]. One of the earliest examples of preparation of a N heterocyclic compound from a biomass derived feedstock is Darby and co-workers 1942 synthesis of 4(5)-hydroxymethylimidazole from D-fructose $CuCO_3$ or $Cu(OH)_2$ catalyzed reaction as shown in Fig. 3.17 [119]. In this experiment a mixture of formaldehyde, concentrated ammonia and fructose or glucose was heated in the presence of Cu(II) catalyst to produce 4(5)-hydroxymethylimidazole, which was isolated in the form of its picrate. During this process, the C6 sugar most likely underwent a retro-aldol fission in the presence of the basic $CuCO_3$ or $Cu(OH)_2$ catalyst, producing dihydroxyacetone and glyceraldehyde, which then reacted with ammonia to yield the imidazole [119].

Later on, following the 4(5)-hydroxymethylimidazole synthesis example, Brust and Cuny showed that when glucose is heated with formamidinium acetate and liquid ammonia, the retro-aldol fission is remarkably inhibited and instead affords 4-tetrahydroxy-butyl imidazole. Furthermore, they have extended this reaction to a series of mono and disaccharides, showing that various carbohydrates can be condensed with amidines in a ammonium carbonate melt, yielding tetrahydroxybutyl substituted imidazoles with a variable α/β-D-Glc/Gal-glycosylation depending on used starting carbohydrate [120] as shown in Fig. 3.18. The series of reducing mono and disaccharides used for producing tetrahydroxybutyl substituted imidazoles with amidines in molten ammonium carbonate and product yields are shown in Table 3.4.

The proposed mechanism involves an imine formation with the carbonyl moiety of the reducing sugar, followed by Amadori rearrangement [121, 122] under formation of an adjacent carbonyl moiety as shown in Fig. 3.19. This intermediate then undergoes a cyclization and tautomeric aromatization to yield the imidazole product [120].

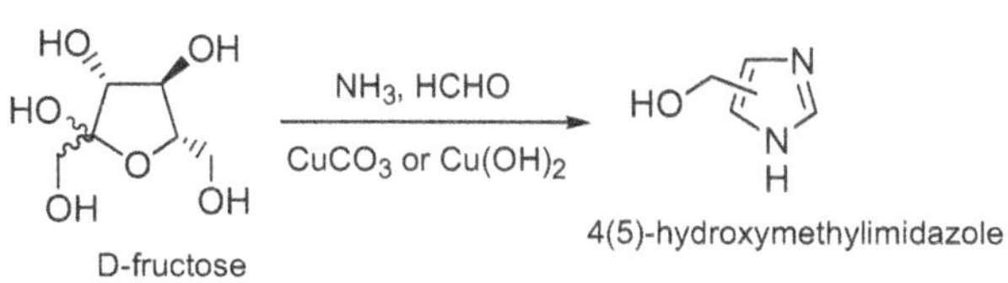

Fig. 3.17 Synthesis of 4(5)-hydroxymethylimidazole from D-fructose, ammonia and formaldehyde using copper carbonate or copper hydroxide as the catalyst [119]. Reprinted with permission from [119]. Copyright 1942, American Chemical Society

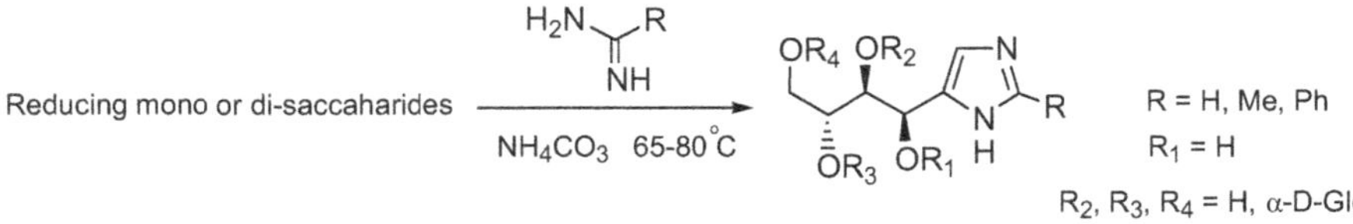

Fig. 3.18 One pot condensation of reducing mono and disaccharides with amidines in molten ammonium carbonate, yielding tetrahydroxybutyl substituted imidazoles with a variable α/β-D-Glc/Gal-glycosylation [120]

Table 3.4 Product yields obtained from one pot transformation of reducing mono and disaccharides into tetrahydroxybutyl substituted imidazoles using amidines in molten ammonium carbonate [120]. Reprinted with permission from [120]. Copyright 1999, Royal Society of Chemistry

Reducing mono/disaccharide	R	R_4	R_3	R_2	R_1	Yield [%]
Fructose	H	H	H	H	H	50
Glucose	H	H	H	H	H	47
Isomaltulose	H	α-D-Glc	H	H	H	49
Isomaltulose	Me	α-D-Glc	H	H	H	30
Isomaltulose	Ph	α-D-Glc	H	H	H	5
Melibiose	H	α-D-Glc	H	H	H	38
Melibiose	Me	α-D-Glc	H	H	H	27
Leucrose	H	H	α-D-Glc	H	H	38
Leucrose	Me	H	α-D-Glc	H	H	26
Maltose	H	H	H	α-D-Glc	H	28
Cellobiose	H	H	H	α-D-Glc	H	25
Lactose	H	H	H	α-D-Glc	H	40

3.2.8 Tetrazoles

Vanillin has attracted the attention in building heterocyclic moieties at the aldehyde function through hydrazone and Schiff base formation reaction. In this example, vanillin was first reacted with succinic anhydride to obtain an ester [123]. This di-ester was reacted at aldehyde groups with 2,4-Di-nitro-phenyl hydrazine or 4-chloroaniline to produce a new hydrazone or a Schiff base. Next the hydrazone or Schiff base was reacted with sodium azide to produce bis-tetrazole as shown in Fig. 3.20 [123].

3.3 Six-Membered N-Heterocyclic Compounds

3.3.1 Pyridinium Salts

Pyridinium salts are a widely studied class of compounds with a range of applications as ionic liquids, anti-microbial, anti-cancer, anti-malarial drugs and

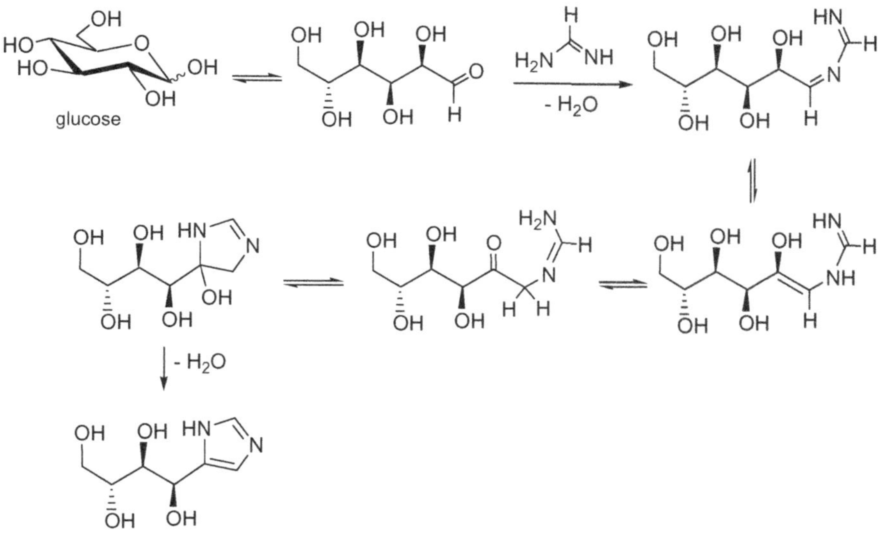

Fig. 3.19 Proposed mechanism for 4-tetrahydroxy-butyl imidazole formation from condensation of D-glucose and formamidine [120]. Reprinted with permission from [120]. Copyright 1999, Royal Society of Chemistry

anti-cholinesterase inhibitors, as well as surfactants. The synthesis of 6-(hydroxymethyl) pyridinium salts derivatives from D-fructose was first reported for a reaction between 5-hydroxymethylfurfural and L-alanine. In this early example, Villard and co-workers first converted D-fructose to 5-hydroxymethylfurfural by an acid catalyzed dehydration reaction. Which was then used in the synthesis of a novel taste enhancer N-(1-carboxyethyl)-6-(hydroxymethyl)pyridinium-3-ol, called alapyridaine, as a racemic mixture and as pure (+)-(S) and (−)-(R) enantiomers, as shown in Fig. 3.21 [124]. Alternatively, reductive amination of 5-(hydroxymethyl)-2-furaldehyde with Raney-Ni/hydrogen and L- or D-alanine followed by mild oxidation also could be used to prepare the same pyridinium salt [124]. This pyridinium salt, N-(1-carboxyethyl)-6-(hydroxymethyl)pyridinium-3-ol is known to form while heating sugar—L-alanine mixtures; as well as found in beef bouillon and has been shown to exhibit general taste-enhancing activities.

Later Kirchhecker et al. further developed this reaction to produce a series of pyridinium ionic liquids via the hydrothermal decarboxylation of pyridinium zwitterions derived from furfural and amino acids [125]. The functionality of the resulting ionic liquid could be tuned by choice of different amino acids as well as different natural carboxylic acids as counterions. In addition, a representative member of this new class of ionic liquids have been successfully used for the synthesis of ionogels and as a solvent for the Heck coupling reactions [125].

The N-alkylpyridinium salt synthesis by condensation of amino acids and 5-hydroxymethylfurfural has been further extended to aliphatic amines as well.

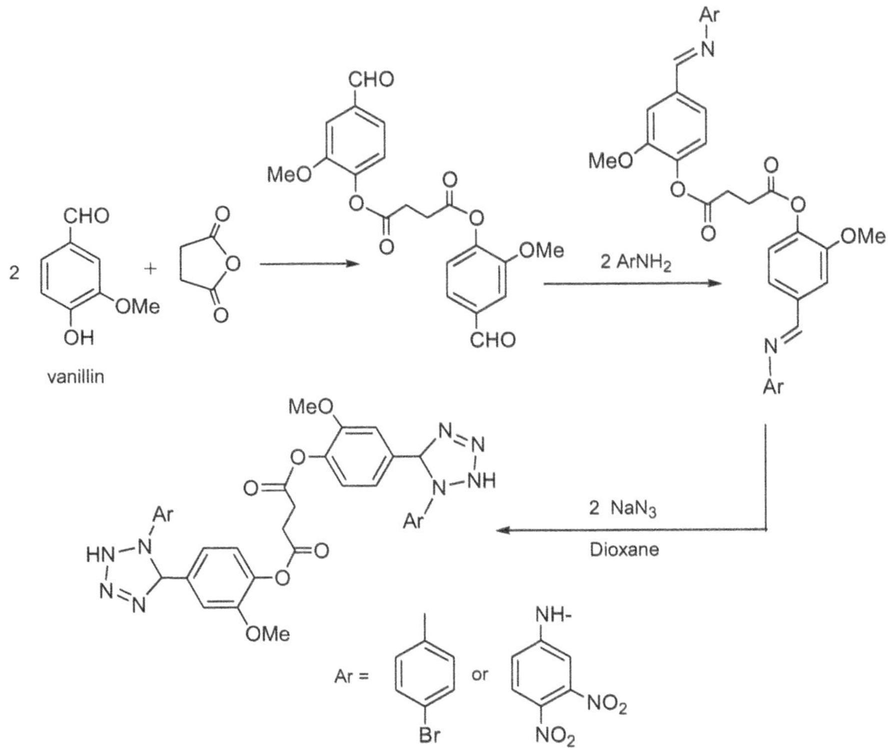

Fig. 3.20 Synthesis of bis-tetrazole heterocycles from vanillin

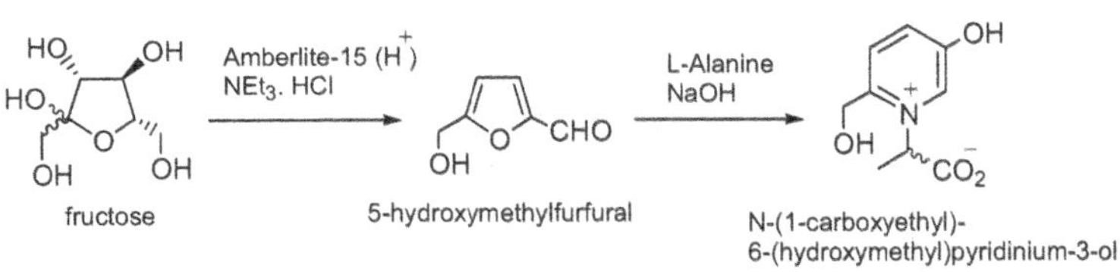

Fig. 3.21 Synthesis of N-(1-carboxyethyl)-6-(hydroxymethyl) pyridinium-3-ol from D-fructose via 5-hydroxymethylfurfural

Where, Sowmiah and co-workers developed a general method for the synthesis of N-alkylpyridinium salts from biomass derived 5-hydroxymethylfurfural and alkyl amines using organo catalysts as shown in Fig. 3.22. During this work Sowmiah et al. optimized the condensation reaction; identified formic acid as a suitable catalyst and aqueous ethanol medium as a convenient solvent for the reaction [126]. Further this protocol was extended to various diamines providing the exclusive formation of mono-N-alkylpyridinium salts. In addition, the mechanism for the formation of pyridinium salts was studied by DFT methods and using $H_2{}^{18}O$ isotope

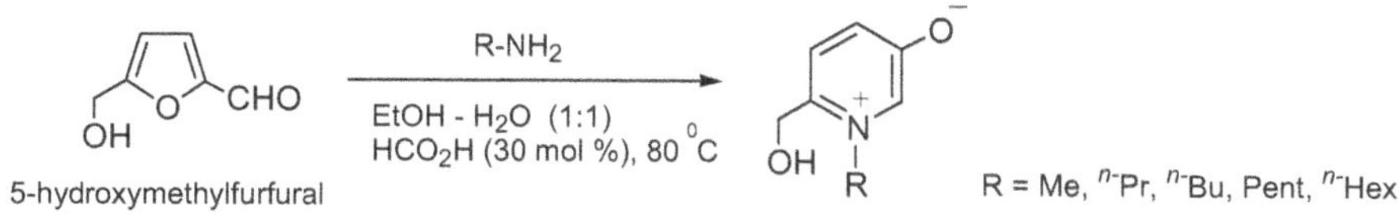

Fig. 3.22 Synthesis of N-alkyl pyridinium salts from 5-hydroxymethylfurfural

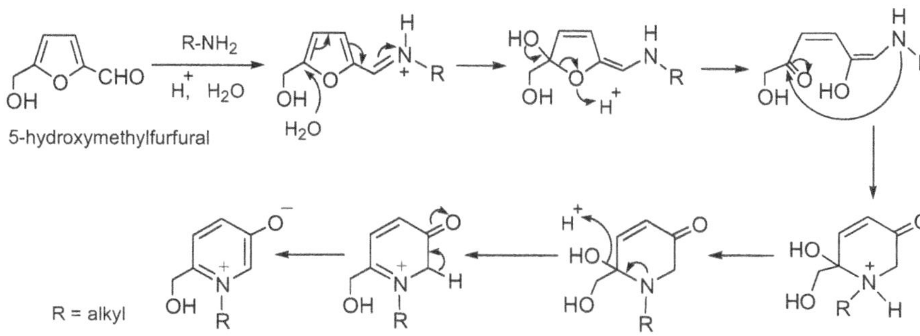

Fig. 3.23 Proposed mechanism for the reaction of producing N-alkylpyridinium salts from 5-hydroxymethylfurfural and alkyl amines in the presence of formic acid as catalyst [126]. Reprinted with permission from [126]. Copyright 2015, American Chemical Society

labeled experiments [126]. The mechanism proposed by Sowmiah and co-workers is shown in Fig. 3.23.

3.3.2 *Pyridazin-3(2H)-One*

The synthesis of pyridazine and its derivatives from biomass derived feedstock is rare. However, synthesis of fully bio-based aromatic N-heterocycle 6-(4-hydroxy-3-methoxyphenyl)pyridazin-3(2H)-one was reported as a part of a broader project on renewable feedstock-based polymers. In this project, lignin derived guaiacol and succinic anhydride were used as starting compounds shown in Fig. 3.24 [127]. In the first step guaiacol was coupled with succinic anhydride through Friedel–Crafts acylation to yield an intermediate keto-acid 3-(4-hydroxy-3-methoxybenzoyl) propanoic acid, next this intermediate was reacted with hydrazine to yield N-heterocyclic compound 6-(4-hydroxy-3-methoxyphenyl)-4,5-dihydro-2H-pyridazin-3-one. Which was later dehydrogenated to obtain the polymer precursor 6- (4-hydroxy-3-methoxyphenyl) pyridazin-3(2H)-one [127].

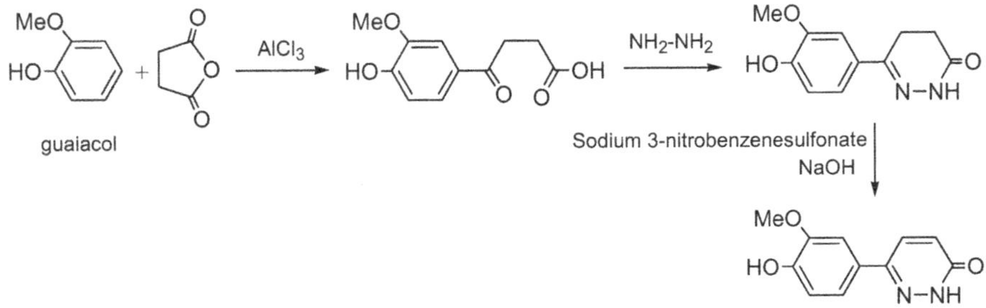

6-(4-hydroxy-3-methoxyphenyl)-4,5-dihydro-2H-pyridazin-3-one

Fig. 3.24 Synthesis of pyridazin-3(2H)-one from guaiacol and succinic anhydride

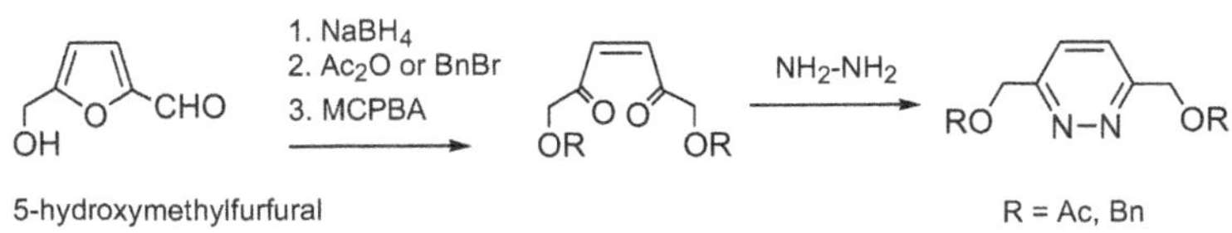

Fig. 3.25 Conversions of 5-hydroxymethylfurfural into 3,6-disubstituted pyridazines [98]. Reprinted with permission from [98]. Copyright 1999, Royal Society of Chemistry

3.3.3 Pyridazines

Pyridazines are six-membered aromatic compounds with two adjacent nitrogen atoms. Substituted and fused pyridazines are well known for their antimicrobial as well as a number of other biological activities [128, 129]. In classical synthesis they are prepared by cyclocondensation of 1,4-dicrbonyls with hydrazine followed by aromatization [130]. Synthesis of pyridazines from biomass feedstocks is not common; Lichtenthaler and co-workers have reported a simple synthesis of these compounds following the classical synthesis as shown in Fig. 3.25. In this route 5-hydroxymethyl is first converted to an enedione, then condensation with hydrazine gave the 3,6-disustituted pyridazines in good yields [98].

3.3.4 Pyrazines

Pyrazines are six-membered aromatic compounds with two nitrogen atoms in 1,4-positions in the ring. This group of aromatics is well known for a wide range of biological activities and is extensively applied in commercial medicines and in the polymer and coatings industries. The classical routes to pyrazines are intermolecular dehydrogenative coupling of ethylenediamine with 1,2-propanediol and dehydrogenative self-coupling of 2-amino alcohols [131, 132]. A few methods for producing pyrazines from biomass derived carbon compounds are known, and

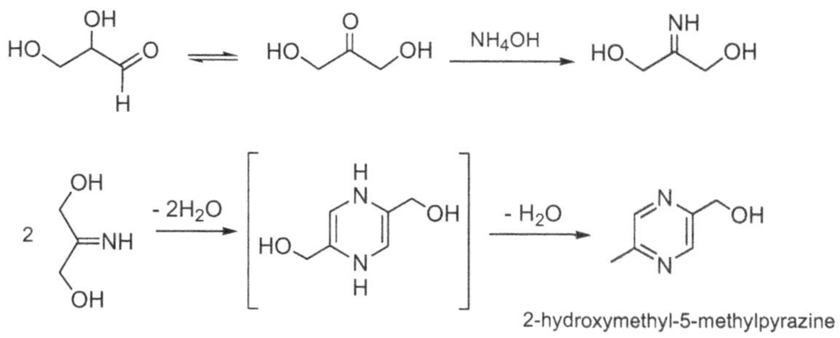

Fig. 3.26 Possible pathways for the formation of 2-hydroxymethyl-5-methylpyrazine from 1,3-dihydroxyacetone and diammonium phosphate under hydrothermal conditions

among them the synthesis of 2-hydroxymethyl-5-methylpyrazine from biomass derived 1,3-dihydroxyacetone and di-ammonium phosphate via a one-pot reaction is one of the most efficient and green approaches [133]. As reported by Song et al. the product yield as high as 72% could be achieved under optimized conditions of pH = 8.0–9.1 at 90 °C for 1 h by using a dioxane and water mixture as the solvent [133]. In addition, a possible reaction mechanism was also proposed where 2-imino-1,3-propanediol was identified to be a key reaction intermediate; which might be formed by ketimine condensation of 1,3-dihydroxyacetone with NH_3 generated in situ from diammonium phosphate, followed by cyclization and dehydration to yield the pyrazine product as shown in Fig. 3.26 [133].

In another interesting pyrazine synthesis Chen et al. reported that glucose can be converted to 2-methyl pyrazine in the presence of ammonium metatungstate as catalyst in 25% yield in aqueous ammonia at 180 °C [134]. Furthermore, they have proposed a reaction pathway involving a tandem fragmentation and cyclization for this single pot process as shown in Fig. 3.27. In the proposed mechanism, β-D-glucopyranosylamine was identified as an important intermediate and was formed as a result of condensation between glucose and NH_3. In addition, they have noted that the reaction is facilitated by tungsten clusters such as $[HW_2O_7]^-$ and $[W_4O_{13}]^{2-}$ in the catalyst; furthermore, the role of different tungsten oxide clusters would be an interesting topic for further explorations. A series of monosaccharides including fructose, xylose, glucosamine as well as common disaccharides and cellulose were tested; producing 2-methyl pyrazine in 7.2–23.3% moderate yields [134]. However, considering the simplicity of operation, sustainability and the cost of starting compounds, this is a fascinating process and further research into tandem fragmentation and cyclization methodologies to improve the yields would be highly desirable.

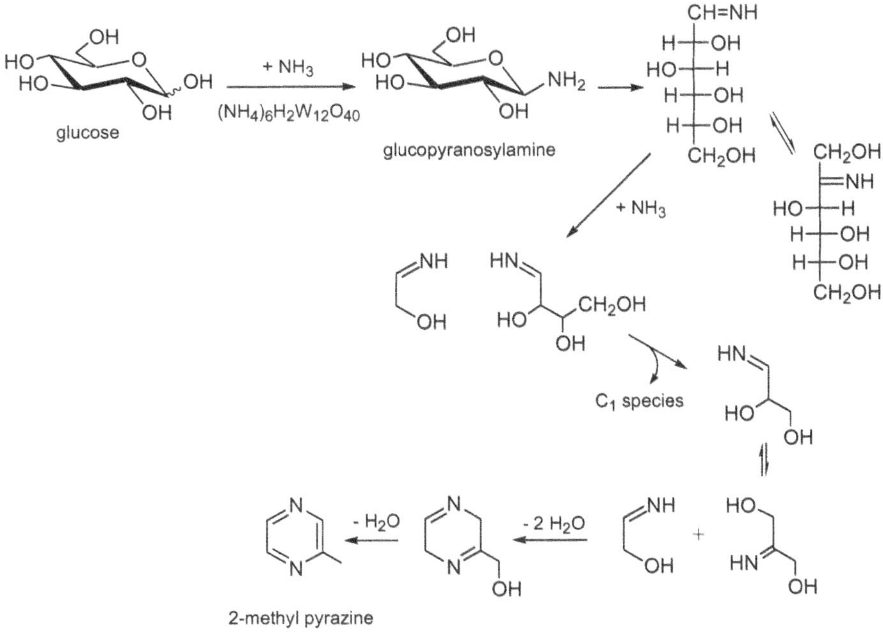

Fig. 3.27 Proposed pathway for the conversion of D-glucose to 2-methyl pyrazine in the presence of ammonium metatungstate catalyst in 25% aqueous ammonia at 180 °C [134]. Reprinted with permission from [134]. Copyright 2017, American Chemical Society

3.4 Quinolines

Quinolines are an important class of aromatic N-heterocyclic compounds with many important biological activities [135, 136]. While studying the reductive amination of levulinic acid with a variety of aromatic amines to produce pyrrolidones Ortiz-Cervantes and co-workers discovered an interesting route to produce quinolones from levulinic acid and aromatic amines [110]. Through this work they have found the formation of 2-(2,4-dimethyl-quinolin-3-yl) acetic acid as a condensation product of levulinic acid and 2-ethynylaniline in 85% yield, after 3 h at 80 °C [110]. Additionally, as an extension of this method, they reported the synthesis of a series of substituted quinolines through condensation reactions between levulinic acid and different 2-alkynylanilines promoted by *p*-toluenesulfonic acid. Proposed reaction pathway for the formation of substituted quinolines through a condensation reaction between levulinic acid and 2-alkynylanilines is shown in Fig. 3.28 [110].

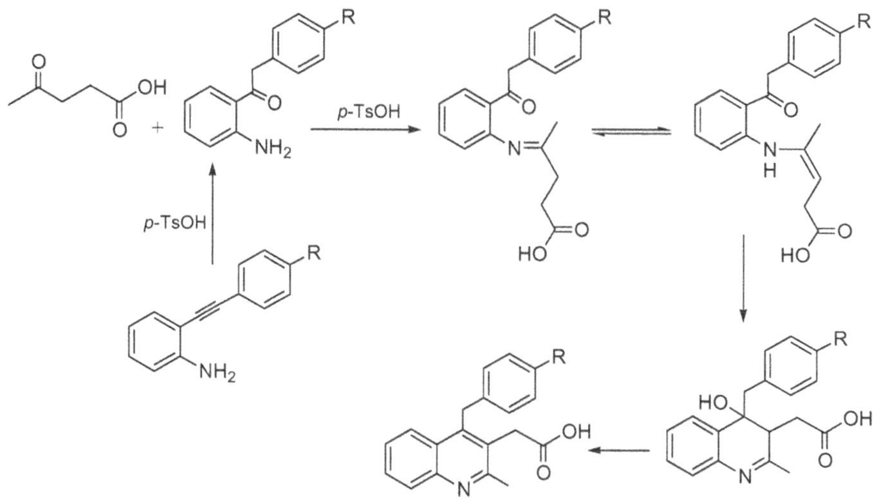

Fig. 3.28 Proposed reaction pathway for the formation of substituted quinolines through a condensation reaction between levulinic acid 2-alkynylanilines [110]. Reprinted with permission from [110]. Copyright 2018, Elsevier

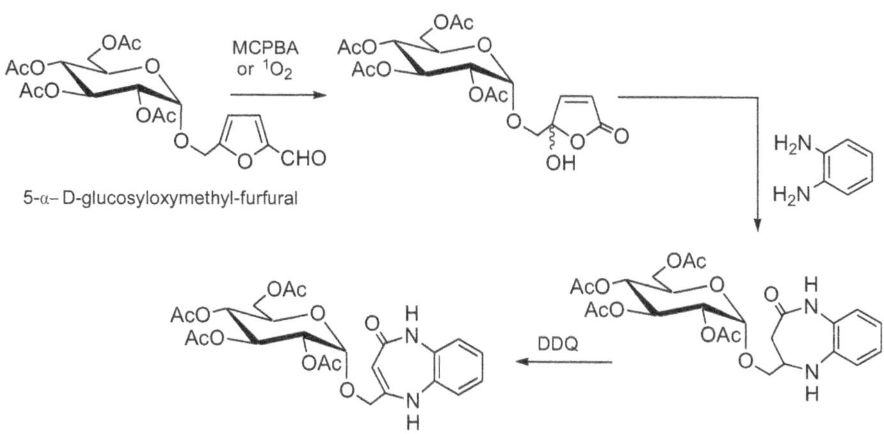

Fig. 3.29 Synthesis of a benzodiazepinone derivative from 5-α-D-glucosyloxymethyl-furfural [98]. Reprinted with permission from [98]. Copyright 1999, Royal Society of Chemistry

3.5 Benzodiazepinones

Benzodiazepinones are another interesting class of N-heterocycles that can be prepared from biomass derived compounds. In this instance, 5-α-D-glucosyloxymethylfurfural was used as the starting compound as shown in Fig. 3.29 [98].

In the first step *m*-chloroperbenzoic acid (MCPBA) oxidation of furan aldehyde moiety results an unsaturated lactone. Next, condensation of this intermediate with 1,2-diaminobenzene resulted the dihydro-benzodiazepinone, which was oxidized with 2,3-dichloro-5,6-dicyano-1,4-benzoquinone (DDQ) to produce the final benzodiazepinone as shown in Fig. 3.29.

3.6 Pyrido[2,3-d]Pyrimidines

Multi-component condensation of amines and carbonyl compounds is a convenient technique to build complex molecules and this method is particularly attractive in the synthesis of compound libraries for testing biological activities. For example, a large collection of pyrimidine-fused heterocycles has been prepared as illustrated in Fig. 3.30 for evaluation of their biological activities. In these experiments a series of monosaccharides: glucose, galactose, arabinose, xylose, maltose, mannose and lactose were combined separately with an amine and barbituric acid or thio-barbituric acid to form pyrimidine-fused heterocycles as shown in Fig. 3.30 [137, 138]. Excellent yields have been reported in these one-pot multi-component condensation reactions in the presence of an acid catalyst such as, *p*-toluenesulfonic acid or nanocrystalline cellulose sulfuric acid under mild conditions of 50 °C [137–139]. Furthermore, an addition of another component, malononitrile together with glucosamine, aldehyde, and barbituric acid, allowed a four-component condensation reaction as well, producing polyhydroxy-substituted pyrido[2,3-d]pyrimidines in 89–94% good yields, catalyzed by *p*-toluenesulfonic acid in ethanol at 50 °C [140].

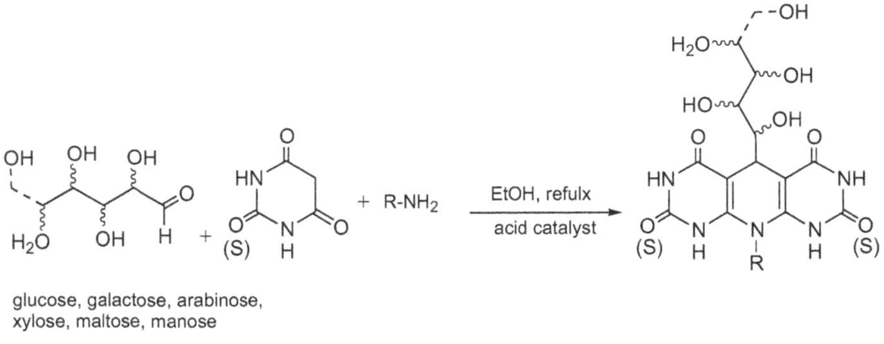

Fig. 3.30 Synthesis of pyrimidine-fused heterocycles via multi-component condensation of monosaccharide, amine, and barbituric acid or thio-barbituric acid [137]. Reprinted with permission from [137]. Copyright 2017, Elsevier

3.7 1,2,4-Triazine, Quinoxaline and Pyrazolo[3,4-*b*] Quinoxaline

In continuation of conversion of carbohydrates to nitrogen heterocycles, Burst and Cuny reported the conversion of reducing disaccharides into industrially relevant 1,2,4-triazine-, quinoxaline-, and pyrazolo[3,4-*b*]quinoxaline- type heterocycles as shown in Fig. 3.31 [120, 141]. In these reactions, heterocycle formation was facilitated by chemical conversion of reducing sugars into 1,2-dicarbonyl intermediates and their subsequent cyclization with nitrogen bis-nucleophiles. Throughout this work a range of disaccharides were converted into quinoxalines carrying a diverse glycosylation patterns on the poly-hydroxy alkyl side chain. Interestingly, all transformations were performed without the need of protecting group chemistry [141].

3.8 Conclusion and Future Outlook

Environmental concerns and decline in fossil carbon resources has guided the current interest in renewable resource-based chemical feedstocks and fuels. Production of N-heterocyclic compounds from new renewable resources-based platform chemicals for the replacement of petroleum-based compounds has taken off during

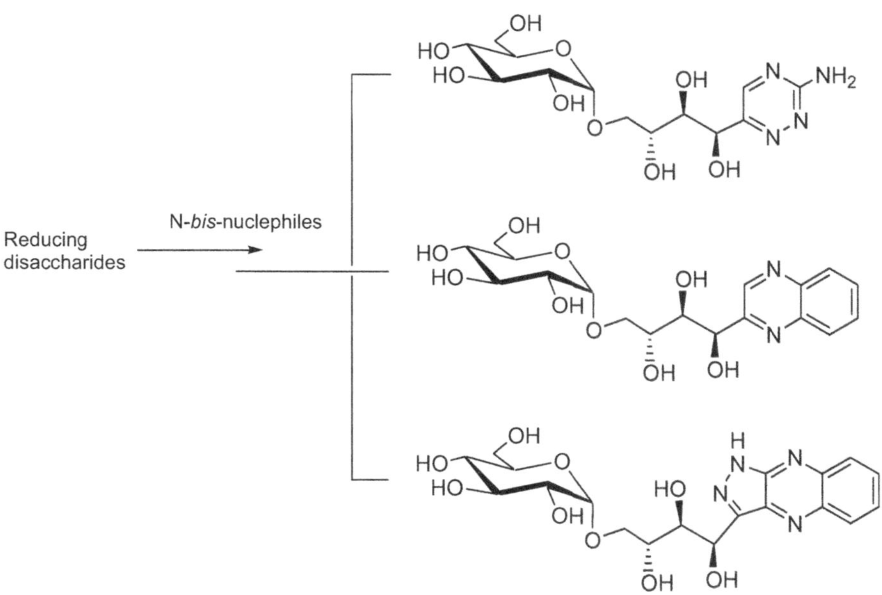

Fig. 3.31 Synthesis of 1,2,4-triazine, quinoxaline and pyrazolo[3,4-*b*]quinoxaline compounds, starting from reducing disaccharides

the last two decades and some of these methods are now in the verge of becoming industrial manufacturing processes. However, only a handful of lignocellulosic biomass-derived compounds such as glucose, 5-hydroxymethylfurfural, levulinic acid and vanillin have been widely studied and used in preparation of N-heterocyclic compounds. Levulinic acid has received most of the attention in this area as this elegant five-carbon keto acid is well set for the preparation of five and six membered heterocyclic compounds with one or two nitrogen atoms. Even though levulinic acid can be derived from the major fraction of the renewable carbon on earth, there are major challenges associated with large scale industrial production of levulinic acid, like the choice of carbohydrate source and issues related to dissolution and efficient depolymerization of cellulose, especially with direct use of lignocellulosic biomass as the raw material. Nevertheless, with the new developments like bio-fine process and reusable ionic liquid catalysts, the cost of levulinic acid is expected to come down rapidly in coming years, paving the way to industrial production of some of the levulinic acid-based N-heterocyclics and other feedstock chemicals.

One-pot transformation of monosaccharides such as glucose or fructose to five- or six- membered N-heterocyclic compounds is a class of reactions with great potential. These reactions require retro-aldol type fragmentations of the carbohydrate chain to C2–C4 range aldehyde-alcohol intermediates and re-assembling into heterocyclic rings with ammonia, amines or other nitrogen sources. Even though the current methods for these reactions gives poor to moderate yields, there are opportunities to improve the yields with the development of new catalytic systems or methods for these one-pot multi-step transformations. In many examples like the use of 1,4-dicarbonyls derived from biofurans, bio-based succinic acid or succinic anhydride the nitrogen insertion is possible through classical reactions to give simple N-heterocyclics. However, there are no parallels to the direct use of disaccharides and complex carbohydrates in synthesis of N-heterocyclic compounds. This type of multi-step transformations allows the preparation of highly decorated N-heterocyclic ring systems with side chains containing chiral hydroxyl groups or glycosides through few simple reactions. The synthesis of these complex and chiral group functionalized heterocyclic systems in a few operations is a very attractive proposition for the pharmaceutical industry as many drugs have similar motifs. Therefore, direct synthesis of complex N-heterocyclic systems from mono and disaccharides is a practical technique for generating compound libraries required for drug development. Some of the examples of levulinic acid-based N-heterocyclic synthesis and imidazole synthesis from fructose discussed are known since 1940s and the renaissance in the interest of producing N-heterocyclic compounds from lignocellulosic biomass feedstocks is reflected in the rapidly increasing number of publications in this field in last few years. Therefore, as highlighted in this review, large scale production of N-heterocyclic chemicals and feedstocks from renewable biomass is very likely be an economically viable industry in the near future.

References

1. Lynd LR, Van Zyl WH, McBride JE, Laser M. Consolidated bioprocessing of cellulosic biomass: an update. Curr Opin Biotechnol. 2005;16(5):577–83. https://doi.org/10.1016/j.copbio.2005.08.009.
2. Petrus L, Noordermeer MA. Biomass to biofuels, a chemical perspective. Green Chem. 2006;8 (10):861–7. https://doi.org/10.1039/b605036k.
3. Amarasekara AS, editor. Handbook of cellulosic ethanol. All Rights Reserved Wiley; 2013. vol Copyright (C) 2019 American Chemical Society (ACS)
4. Zoghlami A, Paës G. Lignocellulosic biomass: understanding recalcitrance and predicting hydrolysis. Front Chem. 2019;7:874. https://doi.org/10.3389/fchem.2019.00874.
5. Yang B, Dai Z, Ding S-Y, Wyman CE. Enzymatic hydrolysis of cellulosic biomass. Biofuels. 2011;2(4):421–49. https://doi.org/10.4155/bfs.11.116.
6. Amarasekara AS, Owereh OS. Hydrolysis and decomposition of cellulose in bronsted acidic ionic liquids under mild conditions. Ind Eng Chem Res. 2009;48. (Copyright (C) 2019 American Chemical Society (ACS). All Rights Reserved.):10152–5. https://doi.org/10.1021/ie901047u.
7. Lynd LR, Beckham GT, Guss AM, Jayakody LN, Karp EM, Maranas C, McCormick RL, Amador-Noguez D, Bomble YJ, Davison BH. Toward low-cost biological and hybrid biological/catalytic conversion of cellulosic biomass to fuels. Energy Environ Sci. 2022;15(3): 938–90. https://doi.org/10.1039/d1ee02540f.
8. Mukherjee A, Dumont MJ, Raghavan V. Review: sustainable production of hydroxymethylfurfural and levulinic acid: challenges and opportunities. Biomass Bioenergy. 2015;72:143–83. https://doi.org/10.1016/j.biombioe.2014.11.007.
9. Pileidis FD, Titirici MM. Levulinic acid biorefineries: new challenges for efficient utilization of biomass. ChemSusChem. 2016;9(6):562–82. https://doi.org/10.1002/cssc.201501405.
10. Liu C, Lu X, Yu Z, Xiong J, Bai H, Zhang R. Production of levulinic acid from cellulose and cellulosic biomass in different catalytic systems. Catalysts. 2020;10(9):1006. https://doi.org/10.3390/catal10091006.
11. Kang S, Fu J, Zhang G. From lignocellulosic biomass to levulinic acid: a review on acid-catalyzed hydrolysis. Renew Sustain Energy Rev. 2018;94:340–62. https://doi.org/10.1016/j.rser.2018.06.016.
12. Werpy T, Petersen G, Aden A, Bozell J, Holladay J, White J, Manheim A, Eliot D, Lasure L, Jones S. Top value added chemicals from biomass. Volume 1-results of screening for potential candidates from sugars and synthesis gas. DTIC document. 2004. doi:https://doi.org/10.2172/15008859.
13. Bozell JJ, Petersen GR. Technology development for the production of biobased products from biorefinery carbohydrates-the US department of energy's "Top 10" revisited. Green Chem. 2010;12(4):539–54. https://doi.org/10.1039/b922014c.
14. Bozell JJ, Moens L, Elliott DC, Wang Y, Neuenscwander GG, Fitzpatrick SW, Bilski RJ, Jarnefeld JL. Production of levulinic acid and use as a platform chemical for derived products. Resour Conserv Recycl. 2000;28(3–4):227–39. https://doi.org/10.1016/s0921-3449(99)00047-6.
15. Rackemann DW, Doherty WO. The conversion of lignocellulosics to levulinic acid. Biofuels Bioprod Biorefin. 2011;5(2):198–214. https://doi.org/10.1002/bbb.267.
16. Amarasekara AS, Hawkins SA. Synthesis of levulinic acid–glycerol ketal–ester oligomers and structural characterization using NMR spectroscopy. Eur Polym J. 2011;47(12):2451–7. https://doi.org/10.1016/j.eurpolymj.2011.09.007.
17. Zhang Y, Wu L, Li F, Li B. Lactonization and condensation polymerization of 5-hydroxylevulinic acid. Huagong Xuebao/J Chem Ind Eng. 2006;57(4):992–6. https://doi.org/10.1007/s10924-008-0082-y.

18. Zhang Y, Guo Z, Cheng J, Fang Z. Synthesis and characterization of biodegradable poly (5-hydroxylevulinic ACID-co-L-lactic acid). Acta Polym Sin. 2009;2:180–6. https://doi.org/10.3724/sp.j.1105.2009.00180.
19. Amarasekara AS, Singh TB, Larkin E, Hasan MA, Fan H-J. NaOH catalyzed condensation reactions between levulinic acid and biomass derived furan-aldehydes in water. Ind Crop Prod. 2015;65:546–9. https://doi.org/10.1016/j.indcrop.2014.10.005.
20. Liang G, Wang A, Zhao X, Lei N, Zhang T. Selective aldol condensation of biomass-derived levulinic acid and furfural in aqueous-phase over MgO and ZnO. Green Chem. 2016;18(11):3430–8. https://doi.org/10.1039/c6gc00118a.
21. Faba L, Díaz E, Ordóñez S. Base-catalyzed condensation of levulinic acid: a new biorefinery upgrading approach. ChemCatChem. 2016;8(8):1490–4. https://doi.org/10.1002/cctc.201600064.
22. Liu X, Bouxin FP, Fan J, Budarin VL, Hu C, Clark JH. Recent advances in the catalytic depolymerization of lignin towards phenolic chemicals: a review. ChemSusChem. 2020;13(17):4296–317. https://doi.org/10.1002/cssc.202001213.
23. Wendisch VF, Kim Y, Lee J-H. Chemicals from lignin: recent depolymerization techniques and upgrading extended pathways. Curr Opin Green Sustain Chem. 2018;14:33–9. https://doi.org/10.1016/j.cogsc.2018.05.006.
24. Chio C, Sain M, Qin W. Lignin utilization: a review of lignin depolymerization from various aspects. Renew Sustain Energy Rev. 2019;107:232–49. https://doi.org/10.1016/j.rser.2019.03.008.
25. Patil V, Adhikari S, Cross P, Jahromi H. Progress in the solvent depolymerization of lignin. Renew Sustain Energy Rev. 2020;133:110359. https://doi.org/10.1016/j.rser.2020.110359.
26. Deepa AK, Dhepe PL. Lignin depolymerization into aromatic monomers over solid acid catalysts. ACS Catal. 2015;5(1):365–79. https://doi.org/10.1021/cs501371q.
27. Wang H, Tucker M, Ji Y. Recent development in chemical depolymerization of lignin: a review. J Appl Chem. 2013;2013:9. https://doi.org/10.1155/2013/838645.
28. Zhou N, Thilakarathna W, He Q, Rupasinghe H. A review: depolymerization of lignin to generate high-value bio-products: opportunities, challenges, and prospects. Front Energy Res. 2022;9:758744. https://doi.org/10.3389/fenrg.2021.758744.
29. Raikwar D, Majumdar S, Shee D. Effects of solvents in the depolymerization of lignin into value-added products: a review. Biomass Convers Biorefin. 2021:1–34. https://doi.org/10.1007/s13399-021-02030-7.
30. Nguyen LT, Phan D-P, Sarwar A, Tran MH, Lee OK, Lee EY. Valorization of industrial lignin to value-added chemicals by chemical depolymerization and biological conversion. Ind Crop Prod. 2021;161:113219. https://doi.org/10.1016/j.indcrop.2020.113219.
31. Ye K, Liu Y, Wu S, Zhuang J. A review for lignin valorization: challenges and perspectives in catalytic hydrogenolysis. Ind Crop Prod. 2021;172:114008. https://doi.org/10.1016/j.indcrop.2021.114008.
32. Ciriminna R, Fidalgo A, Meneguzzo F, Parrino F, Ilharco LM, Pagliaro M. Vanillin: the case for greener production driven by sustainability megatrend. ChemistryOpen. 2019;8(6):660–7. https://doi.org/10.1002/open.201900083.
33. Khwanjaisakun N, Amornraksa S, Simasatitkul L, Charoensuppanimit P, Assabumrungrat S. Techno-economic analysis of vanillin production from kraft lignin: feasibility study of lignin valorization. Bioresour Technol. 2020;299:122559. https://doi.org/10.1016/j.biortech.2019.122559.
34. Wongtanyawat N, Lusanandana P, Khwanjaisakun N, Kongpanna P, Phromprasit J, Simasatitkul L, Amornraksa S, Assabumrungrat S. Comparison of different kraft lignin-based vanillin production processes. Comput Chem Eng. 2018;117:159–70. https://doi.org/10.1016/j.compchemeng.2018.05.020.
35. Fache M, Boutevin B, Caillol S. Vanillin, a key-intermediate of biobased polymers. Eur Polym J. 2015;68:488–502. https://doi.org/10.1016/j.eurpolymj.2015.03.050.

36. Fache M, Darroman E, Besse V, Auvergne R, Caillol S, Boutevin B. Vanillin, a promising biobased building-block for monomer synthesis. Green Chem. 2014;16(4):1987–98. https://doi.org/10.1039/c3gc42613k.
37. Fache M, Boutevin B, Caillol S. Vanillin production from lignin and its use as a renewable chemical. ACS Sustain Chem Eng. 2016;4(1):35–46. https://doi.org/10.1021/acssuschemeng.5b01344.
38. Savonnet E, Grau E, Grelier S, Defoort B, Cramail H. Divanillin-based epoxy precursors as DGEBA substitutes for biobased epoxy thermosets. ACS Sustain Chem Eng. 2018;6(8):11008–17. https://doi.org/10.1021/acssuschemeng.8b02419.
39. Amarasekara AS, Hasan MA. Vanillin based polymers: III. electrochemical dimerization of vanillin revisited and synthesis of hydrovanilloin-formaldehyde polymer. Pol Sci Ser B. 2016;58. (Copyright (C) 2018 American Chemical Society (ACS). All Rights Reserved.):307–12. https://doi.org/10.1134/s1560090416030015.
40. Amarasekara AS, Wiredu B, Razzaq A. Vanillin based polymers: I. an electrochemical route to polyvanillin. Green Chem. 2012;14. (Copyright (C) 2018 American Chemical Society (ACS). All Rights Reserved.):2395–7. https://doi.org/10.1039/c2gc35645g.
41. Amarasekara AS, Garcia-Obergon R, Thompson AK. Vanillin-based polymers: IV. hydrovanilloin epoxy resins. J Appl Polym Sci. 2019;136(4):47000. https://doi.org/10.1002/app.47000.
42. Shibata M, Ohkita T. Fully biobased epoxy resin systems composed of a vanillin-derived epoxy resin and renewable phenolic hardeners. Eur Polym J. 2017;92:165–73. https://doi.org/10.1016/j.eurpolymj.2017.05.007.
43. Gendron D. Vanillin: a promising biosourced building block for the preparation of various heterocycles. Front Chem. 2022;10:949355. https://doi.org/10.3389/fchem.2022.949355.
44. Lichtenthaler FW. Unsaturated O-and N-heterocycles from carbohydrate feedstocks. Acc Chem Res. 2002;35(9):728–37. https://doi.org/10.1021/ar010071i.
45. Xue Z, Yu D, Zhao X, Mu T. Upgrading of levulinic acid into diverse N-containing functional chemicals. Green Chem. 2019;21(20):5449–68. https://doi.org/10.1039/c9gc02415h.
46. Zhang M, Wang N, Liu J, Wang C, Xu Y, Ma L. A review on biomass-derived levulinic acid for application in drug synthesis. Crit Rev Biotechnol. 2022;42(2):220–53. https://doi.org/10.1080/07388551.2021.1939261.
47. Li H, Guo H, Fang Z, Aida TM, Smith RL. Cycloamination strategies for renewable N-heterocycles. Green Chem. 2020;22(3):582–611. https://doi.org/10.1039/c9gc03655e.
48. Nasrollahzadeh M, Ghasemzadeh M, Gharoubi H, Nezafat Z. Progresses in polysaccharide and lignin-based ionic liquids: catalytic applications and environmental remediation. J Mol Liq. 2021;342:117559. https://doi.org/10.1016/j.molliq.2021.117559.
49. Corma A, Iborra S, Velty A. Chemical routes for the transformation of biomass into chemicals. Chem Rev. 2007;107(6):2411–502. https://doi.org/10.1021/cr050989d.
50. Olivier WJ, Smith JA, Bissember AC. Synthesis of pyrrolidine-and γ-lactam-containing natural products and related compounds from pyrrole scaffolds. Chem Rec. 2022;22(3):e202100277. https://doi.org/10.1002/tcr.202100277.
51. Li Petri G, Raimondi MV, Spanò V, Holl R, Barraja P, Montalbano A. Pyrrolidine in drug discovery: a versatile scaffold for novel biologically active compounds. Top Curr Chem. 2021;379(5):1–46. https://doi.org/10.1007/s41061-021-00347-5.
52. Chedid RB, Melder J-P, Dostalek R, Pastre J, Tan AM.Process for the preparation of pyrrolidine, EP2872494B1. 2013. https://patents.google.com/patent/EP2872494B1/en.
53. Ogiwara Y, Uchiyama T, Sakai N. Reductive amination/cyclization of keto acids using a hydrosilane for selective production of lactams versus cyclic amines by switching of the indium catalyst. Angew Chem. 2016;128(5):1896–9. https://doi.org/10.1002/ange.201509465.
54. Wu C, Luo X, Zhang H, Liu X, Ji G, Liu Z, Liu Z. Reductive amination/cyclization of levulinic acid to pyrrolidones versus pyrrolidines by switching the catalyst from AlCl3 to RuCl3 under mild conditions. Green Chem. 2017;19(15):3525–9. https://doi.org/10.1039/c7gc00999b.

55. Li H, Zhao W, Dai W, Long J, Watanabe M, Meier S, Saravanamurugan S, Yang S, Riisager A. Noble metal-free upgrading of multi-unsaturated biomass derivatives at room temperature: silyl species enable reactivity. Green Chem. 2018;20(23):5327–35. https://doi.org/10.1039/c8gc02934b.
56. Wei D, Netkaew C, Darcel C. Iron-catalysed switchable synthesis of pyrrolidines vs pyrrolidinones by reductive amination of levulinic acid derivatives via hydrosilylation. Adv Synth Catal. 2019;361(8):1781–6. https://doi.org/10.1002/adsc.201801656.
57. Rodríguez-Padrón D, Puente-Santiago AR, Balu AM, Romero AA, Muñoz-Batista MJ, Luque R. Benign-by-design orange peel-templated nanocatalysts for continuous flow conversion of levulinic acid to N-heterocycles. ACS Sustain Chem Eng. 2018;6(12):16637–44. https://doi.org/10.1021/acssuschemeng.8b03896.
58. Werpy T, Frye JG, Wang Y, Zacher AH. Methods of making pyrrolidones, US20030120087A1. Google patents. 2003. https://patents.google.com/patent/US20030120087A1/en.
59. Manzer LE. Production of 5-methyl-N-aryl-2-pyrrolidone and 5-methyl-N-alkyl-2-pyrrolidone by reductive amination of levulinic acid with nitro compounds. Google patents. 2004. https://patents.google.com/patent/US6818593B2/en.
60. Du XL, He L, Zhao S, Liu YM, Cao Y, He HY, Fan KN. Hydrogen-independent reductive transformation of carbohydrate biomass into γ-valerolactone and pyrrolidone derivatives with supported gold catalysts. Angew Chem. 2011;123(34):7961–5. https://doi.org/10.1002/ange.201100102.
61. Louven Y, Haus MO, Konrad M, Hofmann JP, Palkovits R. Efficient palladium catalysis for the upgrading of itaconic and levulinic acid to 2-pyrrolidones followed by their vinylation into value-added monomers. Green Chem. 2020;22(14):4532–40. https://doi.org/10.1039/d0gc01043j.
62. Zhao W, Meier S, Yang S, Riisager A. Ammonia borane enabled upgrading of biomass derivatives at room temperature. Green Chem. 2020;22(18):5972–7. https://doi.org/10.1039/d0gc02372h.
63. Xie C, Song J, Wu H, Hu Y, Liu H, Zhang Z, Zhang P, Chen B, Han B. Ambient reductive amination of levulinic acid to pyrrolidones over Pt nanocatalysts on porous TiO2 nanosheets. J Am Chem Soc. 2019;141(9):4002–9. https://doi.org/10.1021/jacs.8b13024.
64. Deng D, Kita Y, Kamata K, Hara M. Low-temperature reductive amination of carbonyl compounds over Ru deposited on Nb2O5·nH2O. ACS Sustain Chem Eng. 2018;7(5):4692–8. https://doi.org/10.1021/acssuschemeng.8b04324.
65. Li H, Wu H, Zhang H, Su Y, Yang S, Hensen EJ. A facile direct route to N-(un) substituted lactams by cycloamination of oxocarboxylic acids without external hydrogen. ChemSusChem. 2019;12(16):3778–84. https://doi.org/10.1002/cssc.201901780.
66. Du XL, He L, Zhao S, Liu YM, Cao Y, He HY, Fan KN. Hydrogen-independent reductive transformation of carbohydrate biomass into γ-valerolactone and pyrrolidone derivatives with supported gold catalysts. Angew Chem Int Ed. 2011;50(34):7815–9.
67. Huang YB, Dai JJ, Deng XJ, Qu YC, Guo QX, Fu Y. Ruthenium-catalyzed conversion of levulinic acid to pyrrolidines by reductive amination. ChemSusChem. 2011;4(11):1578–81. https://doi.org/10.1002/cssc.201100344.
68. Liu Y, Wang Y, Cheng Y, Wei Z. Sustainable efficient synthesis of pyrrolidones from levulinic acid over Pd/C catalyst. ChemistrySelect. 2022;7(26):e202201191. https://doi.org/10.1002/slct.202201191.
69. Zhang T, Ge Y, Wang X, Chen J, Huang X, Liao Y. Polymeric ruthenium porphyrin-functionalized carbon nanotubes and graphene for levulinic ester transformations into γ-valerolactone and pyrrolidone derivatives. ACS Omega. 2017;2(7):3228–40. https://doi.org/10.1021/acsomega.7b00427.
70. Wu H, Dai W, Saravanamurugan S, Li H, Yang S. Quasi-catalytic approach to N-unprotected lactams via transfer hydro-amination/cyclization of biobased keto acids. ACS Sustain Chem Eng. 2019;7(12):10207–13. https://doi.org/10.1021/acssuschemeng.9b00412.

71. Liu Y, Zhang K, Zhang L, Wang Y, Wei Z. One-pot synthesis of pyrrolidone derivatives via reductive amination of levulinic acid/ester with nitriles over Pd/C catalyst. React Kinet Mech Catal. 2021;134(2):777–92. https://doi.org/10.1007/s11144-021-02073-x.
72. Defilippi C, Rodríguez-Padrón D, Luque R, Giordano C. Simplifying levulinic acid conversion towards a sustainable biomass valorisation. Green Chem. 2020;22(9):2929–34. https://doi.org/10.1039/c9gc04306c.
73. Pan Y, Luo Z, Yang J, Han J, Yang J, Yao Z, Xu L, Wang P, Shi Q. Cobalt-catalyzed selective transformation of Levulinic Acid and amines into pyrrolidines and pyrrolidinones using hydrogen. Adv Synth Catal. 2022;364(16):2830–6. https://doi.org/10.1002/adsc.202200578.
74. Gokhale TA, Raut AB, Chawla SK, Bhanage BM. Insights into cascade and sequential one-pot pathways for reductive amination of aldehydes paired with bio-derived levulinic acid to N-substituted pyrrolidones using molecular hydrogen. React Chem Eng. 2022;7(4):1005–13. https://doi.org/10.1039/d1re00384d.
75. Song J. Co single-atom catalysts enable efficient reductive amination of biomass-derived levulinic acid. Chem Catal. 2022;2(1):16–8. https://doi.org/10.1016/j.checat.2021.12.022.
76. Touchy AS, Hakim Siddiki SMA, Kon K, Shimizu K-i. Heterogeneous Pt catalysts for reductive amination of levulinic acid to pyrrolidones. ACS Catal. 2014;4(9):3045–50. https://doi.org/10.1021/cs500757k.
77. Wang S, Huang H, Bruneau C, Fischmeister C. Selective and efficient iridium catalyst for the reductive amination of levulinic acid into pyrrolidones. ChemSusChem. 2017;10(21):4150–4. https://doi.org/10.1002/cssc.201701299.
78. Ledoux A, Sandjong Kuigwa L, Framery E, Andrioletti B. A highly sustainable route to pyrrolidone derivatives - direct access to biosourced solvents. Green Chem. 2015;17(6):3251–4. https://doi.org/10.1039/c5gc00417a.
79. Sun Z, Chen J, Tu T. NHC-based coordination polymers as solid molecular catalysts for reductive amination of biomass levulinic acid. Green Chem. 2017;19(3):789–94. https://doi.org/10.1039/c6gc02591a.
80. Amarasekara AS, Lawrence YM. Raney-Ni catalyzed conversion of levulinic acid to 5-methyl-2-pyrrolidone using ammonium formate as the H and N source. Tetrahedron Lett. 2018;59. (Copyright (C) 2019 American Chemical Society (ACS). All Rights Reserved.):1832–5. https://doi.org/10.1016/j.tetlet.2018.03.087.
81. Musonda CC, Gut J, Rosenthal PJ, Yardley V, Carvalho de Souza RC, Chibale K. Application of multicomponent reactions to antimalarial drug discovery. Part 2: new antiplasmodial and antitrypanosomal 4-aminoquinoline γ- and δ-lactams via a 'catch and release' protocol. Bioorg Med Chem. 2006;14(16):5605–15. https://doi.org/10.1016/j.bmc.2006.04.035.
82. Bouchmaa N, Tilaoui M, Boukharsa Y, Jaâfari A, Mouse HA, Oukerrou A, Taoufik J, Zyad A. In vitro antitumor activity of newly synthesized pyridazin-3 (2H)-one derivatives via apoptosis induction. Pharm Chem J. 2018;51(10):893–901. https://doi.org/10.1007/s11094-018-1712-x.
83. Bouchmaa N, Ben Mrid R, Boukharsa Y, Nhiri M, Ait Mouse H, Taoufik J, Ansar M, Zyad A. Cytotoxicity of new pyridazin-3 (2H)-one derivatives orchestrating oxidative stress in human triple-negative breast cancer (MDA-MB-468). Arch Pharm. 2018;351(12):1800128. https://doi.org/10.1002/ardp.201800128.
84. Burgess LE, Meyers A. Asymmetric synthesis of. gamma.,. gamma.-dialkyl-. gamma.-aminobutyric acid analogs and 2, 2-disubstituted pyrrolidines. Control of stereochemistry in aminal ring opening by varying the extent of allylic 1, 3 strain. J Am Chem Soc. 1991;113(26):9858–9. https://doi.org/10.1021/ja00026a026.
85. Meyers AI, Seefeld MA, Lefker BA, Blake JF. Origin of stereochemistry in simple pyrrolidinone enolate alkylations. J Am Chem Soc. 1997;119(19):4565–6. https://doi.org/10.1021/ja970173x.
86. Meyers AI, Seefeld MA, Lefker BA, Blake JF, Williard PG. Stereoselective alkylations in rigid systems. Effect of remote substituents on π-facial additions to lactam enolates.

Stereoelectronic and steric effects. J Am Chem Soc. 1998;120(30):7429–38. https://doi.org/10.1021/ja980614s.
87. Sharma V, Agarwal S, Madurkar SM, Datta G, Dangi P, Dandugudumula R, Sen S, Singh S. Diversity-oriented synthesis and activity evaluation of substituted bicyclic lactams as antimalarial against plasmodium falciparum. Malar J. 2014;13(1):467. https://doi.org/10.1186/1475-2875-13-467.
88. Stájer G, Szabó Angela E, Csámpai A, Sohár P. Reactions of levulinic acid with norbornane/ene amino acids and diamines. Eur J Org Chem. 2004;2004(6):1318–22. https://doi.org/10.1002/ejoc.200300604.
89. Verardo G, Geatti P, Merli M, Castellarin Elena E. Synthesis of 3-substituted dihydro-1-phenylamino-1H-pyrrolo[1,2-a]imidazole-2,5(3H,6H)-diones from α-amino acid phenylhydrazides and levulinic acid. Eur J Org Chem. 2004;2004(13):2833–9. https://doi.org/10.1002/ejoc.200400112.
90. Barrow JC, Rittle KE, Ngo PL, Selnick HG, Graham SL, Pitzenberger SM, McGaughey GB, Colussi D, Lai MT, Huang Q. Design and synthesis of 2, 3, 5-substituted imidazolidin-4-one inhibitors of BACE-1. ChemMedChem. 2007;2(7):995–9. https://doi.org/10.1002/cmdc.200700038.
91. Akdag A, McKee ML, Worley SD. Mechanism of formation of biocidal imidazolidin-4-one derivatives: an Ab initio density-functional theory study. Chem Eur J. 2006;110(24):7621–7. https://doi.org/10.1021/jp060879q.
92. Wu C, Lou M, Sun M, Wang H, Li Z, Qiu J, Wang J, Liu Z. Selectively reductive amination of levulinic acid with aryl amines to N-substituted aryl pyrroles. Green Energy Environ. 2021;8(2):438–43. https://doi.org/10.1016/j.gee.2021.04.010.
93. Amarasekara AS. 5-hydroxymethylfurfural based polymers. Scrivener Publishing; 2012. p. 381–428. https://doi.org/10.1002/9781118217689.ch9.
94. Xu C, Paone E, Rodríguez-Padrón D, Luque R, Mauriello F. Recent catalytic routes for the preparation and the upgrading of biomass derived furfural and 5-hydroxymethylfurfural. Chem Soc Rev. 2020;49(13):4273–306. https://doi.org/10.1039/d0cs00041h.
95. Kong X, Zhu Y, Fang Z, Kozinski JA, Butler IS, Xu L, Song H, Wei X. Catalytic conversion of 5-hydroxymethylfurfural to some value-added derivatives. Green Chem. 2018;20(16):3657–82. https://doi.org/10.1039/c8gc00234g.
96. Williams PD, LeGoff E. Enedione-functionalized macrocycles via oxidative ring opening of furans. J Org Chem. 1981;46(21):4143–7. https://doi.org/10.1021/jo00334a007.
97. Jurczak J, Pikul S. Novel oxidative opening of 2,5-disubstituted furan ring. Tetrahedron Lett. 1985;26(25):3039–40. https://doi.org/10.1016/S0040-4039(00)98612-X.
98. Lichtenthaler FW, Brust A, Cuny E. Sugar-derived building blocks. Part 26. Part 25. See ref. 1. Hydrophilic pyrroles, pyridazines and diazepinones from d-fructose and isomaltulose. Green Chem. 2001;3(5):201–9. https://doi.org/10.1039/b105630c.
99. Yewale SB, Ganorkar SB, Baheti KG, Shelke RU. Novel 3-substituted-1-aryl-5-phenyl-6-anilinopyrazolo[3,4-d]pyrimidin-4-ones: docking, synthesis and pharmacological evaluation as a potential anti-inflammatory agents. Bioorg Med Chem Lett. 2012;22(21):6616–20. https://doi.org/10.1016/j.bmcl.2012.08.119.
100. Russo F, Guccione S, Romeo G, Monsu'Scolaro L, Pucci S, Caruso A, Cutuli V, Roxas MA. Synthesis and pharmacological properties of pyrazolotriazolopyrimidine derivatives. Eur J Med Chem. 1992;27(1):73–80. https://doi.org/10.1016/0223-5234(92)90064-8.
101. Ragavan RV, Vijayakumar V, Kumari NS. Synthesis and antimicrobial activities of novel 1, 5-diaryl pyrazoles. Eur J Med Chem. 2010;45(3):1173–80. https://doi.org/10.1016/j.ejmech.2009.12.042.
102. Abdel-Aziz M, Abuo-Rahma GE-DA, Hassan AA. Synthesis of novel pyrazole derivatives and evaluation of their antidepressant and anticonvulsant activities. Eur J Med Chem. 2009;44(9):3480–7. https://doi.org/10.5958/0974-360x.2019.00325.1.

103. Karrouchi K, Radi S, Ramli Y, Taoufik J, Mabkhot YN, Al-Aizari FA, Ansar M. Synthesis and pharmacological activities of pyrazole derivatives: a review. Molecules. 2018;23(1):134. https://doi.org/10.3390/molecules23010134.
104. Giornal F, Pazenok S, Rodefeld L, Lui N, Vors J-P, Leroux FR. Synthesis of diversely fluorinated pyrazoles as novel active agrochemical ingredients. J Fluor Chem. 2013;152:2–11. https://doi.org/10.1016/j.jfluchem.2012.11.008.
105. El Khadem H, Mohammed-Aly M. 943. New anhydrophenylosazones. J Chem Soc (Resumed). 1963:4929–32. https://doi.org/10.1039/jr9630004929.
106. Khadem HE. Dianhydrophenylosazone acetates from isomaltose and gentiobiose. J Org Chem. 1964;29(10):3072–4. https://doi.org/10.1021/jo01033a507.
107. Khadem HE, El-Shafei Z, Mohammed-Ali M. The action of acetic anhydride on osazones and hydrazones. J Org Chem. 1964;29(6):1565–7. https://doi.org/10.1021/jo01029a070.
108. Diehl V, Cuny E, Lichtenthaler FW. Preparative scale conversion of D-xylose into hydrophilically functionalized pyrazoles. Heterocycles. 1998;6(48):1193–201. https://doi.org/10.3987/com-98-8139.
109. Oikawa N, Müller C, Kunz M, Lichtenthaler FW. Hydrophilically functionalized pyrazoles from sugars. Carbohydr Res. 1998;309(3):269–79. https://doi.org/10.1016/s0008-6215(98)00137-2.
110. Ortiz-Cervantes C, Flores-Alamo M, García JJ. Synthesis of pyrrolidones and quinolines from the known biomass feedstock levulinic acid and amines. Tetrahedron Lett. 2016;57(7):766–71. https://doi.org/10.1016/j.tetlet.2016.01.018.
111. Flores AF, Malavolta JL, Frigo LM, Doneda M, Flores DC. From levulinic acid to biheterocycles: synthesis of 1-[(5-Hydroxy-5-trifluoromethyl-3-substituted-4, 5-dihydro-1 H-pyrazol-1-yl)-3-(5-trifluoromethyl-1 H-pyrazol-3-yl) propan-1-ones]. Synth Commun. 2015;45(10):1198–204. https://doi.org/10.1080/00397911.2014.1003352.
112. Malavolta JL. Síntese de pirimidinas derivadas de 7, 7, 7-trialo-4-metoxi-6-oxo-4-heptenoatos de metila e bi-heterociclos trifluormetilados. Santa Maria: Universidade Federal de Santa Maria; 2012.
113. Flores AF, Goularte RB, Malavolta JL. Synthesis of bis-heterocycles from levulinic acid; Sintese de bis-heterociclos a partir do acido levulinico. 2011. doi:https://doi.org/10.47749/t/unicamp.2003.291265.
114. Gupta P, Hameed S, Jain R. Ring-substituted imidazoles as a new class of anti-tuberculosis agents. Eur J Med Chem. 2004;39(9):805–14. https://doi.org/10.1016/j.ejmech.2004.05.005.
115. Zampieri D, Mamolo MG, Vio L, Banfi E, Scialino G, Fermeglia M, Ferrone M, Pricl S. Synthesis, antifungal and antimycobacterial activities of new bis-imidazole derivatives, and prediction of their binding to P45014DM by molecular docking and MM/PBSA method. Bioorg Med Chem. 2007;15(23):7444–58. https://doi.org/10.1016/j.bmc.2007.07.023.
116. Sharma D, Narasimhan B, Kumar P, Judge V, Narang R, De Clercq E, Balzarini J. Synthesis, antimicrobial and antiviral evaluation of substituted imidazole derivatives. Eur J Med Chem. 2009;44(6):2347–53. https://doi.org/10.1016/j.ejmech.2008.08.010.
117. Bhandari K, Srinivas N, Marrapu VK, Verma A, Srivastava S, Gupta S. Synthesis of substituted aryloxy alkyl and aryloxy aryl alkyl imidazoles as antileishmanial agents. Bioorg Med Chem Lett. 2010;20(1):291–3. https://doi.org/10.1016/j.bmcl.2009.10.117.
118. Munoz J, Junior J, Silva F. Radziszewski reaction: an elegant, easy, simple and efficient method to synthesise imidazoles. Curr Org Synth. 2014;11(6):824–34. https://doi.org/10.2174/1570179411666140623223611.
119. Darby WJ, Lewis HB, Totter JR. The preparation of 4 (5)-hydroxymethyl-imidazole. J Am Chem Soc. 1942;64(2):463–4. https://doi.org/10.1021/ja01254a502.
120. Brust A, Cuny E. Conversion of reducing carbohydrates into hydrophilic substituted imidazoles. Green Chem. 2013;15(10):2993–8. https://doi.org/10.1039/c3gc41203b.
121. Hodge JE. The amadori rearrangement. In: Advances in carbohydrate chemistry, vol. 10. Elsevier; 1955. p. 169–205. https://doi.org/10.1016/s0096-5332(08)60392-6.

122. Cui H, Yu J, Zhai Y, Feng L, Chen P, Hayat K, Xu Y, Zhang X, Ho C-T. Formation and fate of amadori rearrangement products in Maillard reaction. Trends Food Sci Technol. 2021;115: 391–408. https://doi.org/10.1016/j.tifs.2021.06.055.
123. Abd Al Hussain HAA, Abbas SK. The synthesis of some new heterocyclic compounds from vanillin derivatives. Res J Pharm Technol. 2019;12(4):1943–6. https://doi.org/10.5958/0974-360x.2019.00325.1.
124. Villard R, Robert F, Blank I, Bernardinelli G, Soldo T, Hofmann T. Racemic and enantiopure synthesis and physicochemical characterization of the novel taste enhancer N-(1-carboxyethyl)-6-(hydroxymethyl) pyridinium-3-ol inner salt. J Agric Food Chem. 2003;51(14):4040–5. https://doi.org/10.1021/jf034246+.
125. Kirchhecker S, Tröger-Müller S, Bake S, Antonietti M, Taubert A, Esposito D. Renewable pyridinium ionic liquids from the continuous hydrothermal decarboxylation of furfural-amino acid derived pyridinium zwitterions. Green Chem. 2015;17(8):4151–6. https://doi.org/10.1039/c5gc00913h.
126. Sowmiah S, Veiros LF, Esperanca JM, Rebelo LP, Afonso CA. Organocatalyzed one-step synthesis of functionalized n-alkyl-pyridinium salts from biomass derived 5-hydroxymethylfurfural. Org Lett. 2015;17(21):5244–7. https://doi.org/10.1021/acs.orglett.5b02573.
127. Qi Y, Wang J, Kou Y, Pang H, Zhang S, Li N, Liu C, Weng Z, Jian X. Synthesis of an aromatic N-heterocycle derived from biomass and its use as a polymer feedstock. Nat Commun. 2019;10(1):1–9. https://doi.org/10.1038/s41467-019-10178-0.
128. Kandile NG, Mohamed MI, Zaky H, Mohamed HM. Novel pyridazine derivatives: synthesis and antimicrobial activity evaluation. Eur J Med Chem. 2009;44(5):1989–96. https://doi.org/10.1016/j.ejmech.2008.09.047.
129. Tucaliuc R-A, Cotea VV, Niculaua M, Tuchilus C, Mantu D, Mangalagiu II. New pyridazine–fluorine derivatives: synthesis, chemistry and biological activity. Part II. Eur J Med Chem. 2013;67:367–72. https://doi.org/10.1016/j.ejmech.2013.04.069.
130. Hilmy Elnagdi M, Al-Awadi NA, Abdelshafy Abdelhamid I. Chapter 1 Recent developments in pyridazine and condensed pyridazine synthesis. In: Katritzky AR, editor. Advances in heterocyclic chemistry, vol. 97. Academic; 2009. p. 1–43. https://doi.org/10.1016/S0065-2725(08)00201-8.
131. Mortzfeld FB, Hashem C, Vranková K, Winkler M, Rudroff F. Pyrazines: synthesis and industrial application of these valuable flavor and fragrance compounds. Biotechnol J. 2020;15(11):2000064. https://doi.org/10.1002/biot.202000064.
132. Ong KT, Liu Z-Q, Tay MG. Review on the synthesis of pyrazine and its derivatives. Borneo J Resour Sci Technol. 2017;7(2):60–75. https://doi.org/10.33736/bjrst.591.2017.
133. Song L, Zheng M, Pang J, Sebastian J, Wang W, Qu M, Zhao J, Wang X, Zhang T. One-pot synthesis of 2-hydroxymethyl-5-methylpyrazine from renewable 1, 3-dihydroxyacetone. Green Chem. 2017;19(15):3515–9. https://doi.org/10.1039/c7gc00578d.
134. Chen X, Yang H, Hülsey MJ, Yan N. One-step synthesis of N-heterocyclic compounds from carbohydrates over tungsten-based catalysts. ACS Sustain Chem Eng. 2017;5(11):11096–104. https://doi.org/10.1021/acssuschemeng.7b03048.
135. Weyesa A, Mulugeta E. Recent advances in the synthesis of biologically and pharmaceutically active quinoline and its analogues: a review. RSC Adv. 2020;10(35):20784–93. https://doi.org/10.1039/d0ra03763j.
136. Chung P-Y, Bian Z-X, Pun H-Y, Chan D, Chan AS-C, Chui C-H, Tang JC-O, Lam K-H. Recent advances in research of natural and synthetic bioactive quinolines. Future Med Chem. 2015;7(7):947–67. https://doi.org/10.4155/fmc.15.34.
137. Gupta S, Khare NK. Multi-component synthesis and computational studies of three novel thiobarbituric acid carbohydrate derivatives. J Mol Struct. 2017;1127:309–13. https://doi.org/10.1016/j.molstruc.2016.07.063.

138. Nourisefat M, Panahi F, Khalafi-Nezhad A. Carbohydrates as a reagent in multicomponent reactions: one-pot access to a new library of hydrophilic substituted pyrimidine-fused heterocycles. Org Biomol Chem. 2014;12(46):9419–26. https://doi.org/10.1039/c4ob01791a.
139. Nikoofar K, Heidari H, Shahedi Y. Nano crystalline cellulose sulfuric acid (s-NCC): a novel green nanocatalyst for the synthesis of polyhydroxy pyrimidine-fused heterocyclic compounds (PPFHs). Cellulose. 2018;25(10):5697–709. https://doi.org/10.1007/s10570-018-1942-9.
140. Khajeh Dangolani S, Panahi F, Tavaf Z, Nourisefat M, Yousefi R, Khalafi-Nezhad A. Synthesis and antioxidant activity evaluation of some novel aminocarbonitrile derivatives incorporating carbohydrate moieties. ACS Omega. 2018;3(8):10341–50. https://doi.org/10.1021/acsomega.8b01124.
141. Brust A, Cuny E. Reducing disaccharides and their 1, 2-dicarbonyl intermediates as building blocks for nitrogen heterocycles. RSC Adv. 2014;4(11):5759–67. https://doi.org/10.1039/c3ra47349j.

Chapter 4
Waste Shell Biorefinery: Sustainable Production of Organonitrogen Chemicals

Yue Zheng, Yudi Wang, and Xi Chen

Abstract This chapter focuses on sustainable production of organonitrogen chemicals from the concept of a waste shell biorefinery. Chitin waste is naturally nitrogenous and can be converted into small molecule, high value-added organic compounds. In aqueous media or organic solvents/co-solvents, chitin can be depolymerized/deacetylated to acetylglucosamine or glucosamine. In addition, chitin can also be converted to important platform chemicals such as amino acids and furans by oxidation and dehydration. Finally, this chapter introduces conversion strategies for chitin, such as hydrothermal methods, hydrogenation reactions and condensation reactions and demonstrates the types and routes of conversion of chitin to organic nitrogenous chemicals.

Keywords Chitin · Hydrolysis · Amino acids · Dehydration · Shell biorefinery

4.1 Waste Shell Biorefinery

4.1.1 Global Shell Waste Generation

With over 100 million annual global fishery production (178 million tons in 2021, expected to expand to 204 million metric tons in 2030), it is estimated that more than 8 million of crustacean shell waste (e.g. shrimp shells, lobster shells and crab shells, etc.) are generated each year [1]. Crustacean shells milling valorization has low value, for example the dried shrimp shell powders sell for US $100–120 per ton. The low monetary value leads to direct disposal or landfilling of these shell waste, which brings about environmental issues and an increase in carbon emissions. Yet shell waste is undervalued, since it is composed of valuable chemicals such as proteins and chitin (a polymer similar to cellulose), showing the potential to be raw materials for processing useful chemicals. Thanks to the concept of "Shell Biorefinery"

Y. Zheng · Y. Wang · X. Chen (✉)
China-UK Low-Carbon College, Shanghai Jiao Tong University, Shanghai, China
e-mail: chenxi-lcc@sjtu.edu.cn

Z. Fang et al. (eds.), *Production of N-containing Chemicals and Materials from Biomass*, Biofuels and Biorefineries 12,
https://doi.org/10.1007/978-981-99-4580-1_4

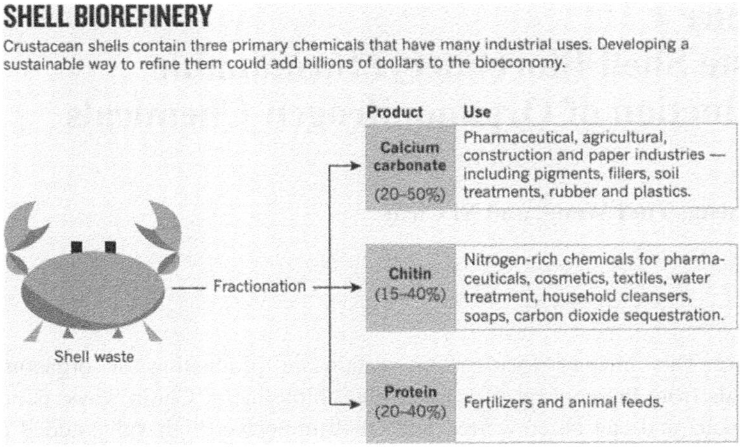

Fig. 4.1 Shell biorefinery model (Adapted with permission from [2], Copyright © 2015, Elsevier)

(Fig. 4.1) proposed in 2015, shell waste can be expected to be efficiently converted into value added materials in a green way. Utilization of waste shells is favorable from both environmental and economic points and there is enormous scope to employ green chemistry methods in sustainable upgrading of waste shells [3].

4.1.2 The Ocean-Based Chitin Biomass

Chitin is an earth abundant natural polymer next to cellulose with bio-fixed nitrogen that is found in fungi, organisms, plankton and the exoskeletons of insects and crustaceans. Chitin was systematically isolated and named by a French naturalist, Odier [4]. The most common way to extract chitin is by chemical processing of ocean-based crustaceans, such as crabs, shrimps and lobsters. The weight content of chitin in crustaceans varied from 15% to 40%, depending on their species [5]. A large amount of shell waste of crustaceans from fishing industries provides a reliable resource for chitin production. The structure of chitin is similar to that of cellulose, consisting of 2-amino-2-deoxy-D-glucose (GlcN) and 2-acetylamino-2-deoxy-D-glucose (GlcNAc; sometimes also called as "NAG") units linked by β-(1–4) bonds (usually higher proportion of GlcNAc than GlcN). When the content of GlcN falls to below 50% degree of deacetylation in chitin, the compound is referred to as chitosan. Chitin can be converted into chitosan through deacetylation treatments with their properties (e.g., solubility) varying greatly based on the degree of deacetylation. Chitosan is soluble in most dilute acids, while chitin is unable to be dissolved by dilute acids [6]. Chitin and chitosan have wide applications in life and industry, for instance, they can be used as drug carriers and tissue scaffolds in biomedical fields, food additives in food processing, absorbents in water treatment, supports for catalysts, materials for textiles and especially as raw materials for production of N-containing chemicals [7–9]. Conversion of chitin biomass into value-added N-containing chemicals poses great commercial potential and is necessary for sustainable development.

4.1.3 Chitin Extraction from Shell Waste

Chitin is commercially processed from shell waste by separating and removing the other components of shells, including the 20–50% carbonates and the 20–40% proteins [10]. After pretreatment into small dried pigments, chitin is obtained from waste shells in three basic steps (Fig. 4.2), namely demineralization, deproteination and decolouration [1].

Currently there are different methods available for chitin extraction and the relevant merits and drawbacks are listed in Table 4.1. The investigation of chitin extraction and separation dates back to the last century [12]. Traditional methods

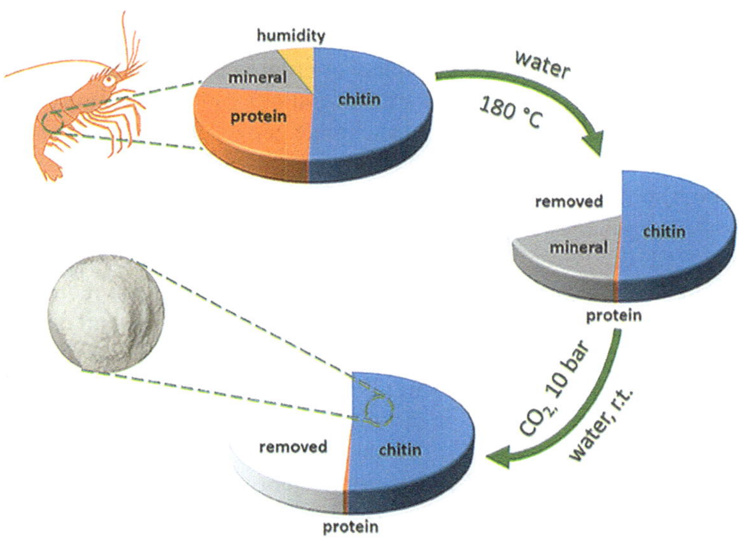

Fig. 4.2 Chitin fractionation through hot water-carbonic acid (HOW-CA) process from raw shrimp shells. (Adapted with permission from [11] Copyright © 2019, Elsevier)

Table 4.1 Comparison of chitin fractionation methods

Method	Advantages	Disadvantages
Traditional	Mass processing Short reaction time	Environmentally hazardous Requirements for erosion-resistant equipment
Biological	Mild reaction conditions	Long cultivation time
Solvent extraction	Environmentally friendly High efficiency	Lab-scale Costly

include chemical treatment and bioprocessing. In general, chemical methods outperform bioprocessing due to short reaction periods and wide reaction conditions, though they still need to be improved from the perspective of environmental friendliness [2]. For chemical methods, pretreated shells are firstly demineralized using strong acids or organic acids, such as hydrochloric acid (HCl), sulfuric acid (H_2SO_4), nitric acid (HNO_3) or acetic acid (HOAc) and formic acid (FA), to remove carbonate-based minerals [13–15]. In the second step, demineralized shells are soaked in basic solutions upon heating to remove proteins. In addition to typical sodium hydroxide (NaOH), sodium sulfide or phosphate (Na_2S or Na_3PO_4), potassium hydroxide or carbonate (KOH or K_2CO_3) and calcium hydrogen sulfite (Ca (HSO_3)$_2$) or sodium phosphate are also employed as alkaline reagents [16]. Demineralization and deproteination leads to raw chitin product, which is yellowish because the presence of astaxanthin, a type of carotenoid pigment that remains bonded with chitin after alkaline treatments. Normally astaxanthin is combined with proteins to give macromolecular complexes called crustacyanins which make the heat-treated shells orange-red. To obtain white/colorless chitin, the raw chitin needs to decolorized employing bleaching agents such as hydrogen peroxide (H_2O_2) and potassium permanganate ($KMnO_4$). Astaxanthin can also be extracted as a high-value byproduct in the very beginning of the process. In biological methods, chitin is separated from waste shells by a combination of fermentation and enzymatic treatment [15]. The enzymatic treatment for demineralized shells was reported as early as 1968, when Broussignac et al. found that chitin could be deproteinized by proteases such as trypsin and pepsin after acid demineralization [17]. The library of chitin depolymerization enzymes was expanded under exploration. Papain, tuna proteinase and a bacterial proteinase have been shown to be effective, with ca. 5% protein remaining [18]. Gagne and Simpson used the proteolytic enzymes to improve the deproteinization efficiency [19]. The remaining protein content was reduced to 1.3% and 2.8% over chymotrypsin and papain respectively under optimum conditions. A lactic acid (LAc) fermentation method was developed to avoid acid demineralization which is hazardous to the environment [20]. Demineralization and deproteinization were simultaneously completed in a bioreactor by a culture of the lactic acid bacterium over 5 days in which 61.0% calcium and 77.5% protein were dissolved to give the facile separation of chitin, with liquor as a by-product showing the potential for animal feed supplements. The types of microorganism greatly determined the deproteinization and demineralization efficiencies. Teng et al. removed 72% minerals and 84% proteins using *Bacillus subtilis* (a type of roteolytic *Aspergillus niger*) [21], while Sedaghat et al. removed 79% minerals and 75% proteins using *Pseudomonas aeruginosa* (a type of proteases-producing bacteria) [22]. Duan et al. isolated a fungus, namely epiphytic *Lactobacillus acidophilus* SW01, from shrimp waste and applied it for the fermentation of waste shells [23]. The deproteinization and demineralization efficiencies were significantly enhanced, while the fermentation period was remarkably shortened, removing more than 99% minerals and ~92% proteins within 48 h. Bioprocessing of shell waste was conducted under mild condition, but had long reaction period, with enzymatic treatment lasting more than 10 h and the fermentation step lasting several days.

Generally, chemical methods outperform bioprocessing in terms of reaction time and potential for mass production [2]. But there are a lot of contaminants generated along with strong acid and base treatments during the chemical extraction process and the requirements for costly erosion-resistant equipment results in high price of chitin and thus limits its promotion and application.

Novel methods such as solvent extraction have been proposed that improve environmental friendliness and economic feasibility. Ionic liquids (ILs) are popular solvents for biomass separation and purification processes due to their unique physical and chemical properties [24, 25]. Inspired by the cellulose dissolution in ILs, Xie et al. demonstrated chitin dissolution in 1-butyl-3-methyl-imidazolium chloride ([Bmim]Cl) IL [26]. Semi-clear and viscous solutions were obtained after 5 h at 110 °C. Then, chitin was separated and regenerated by adding water or methanol into the mixture.

Several types of ILs including 1-allyl-3-methylimidazolium bromide ([Amim]Br) [27] and 1-butyl-3-methylimidazolium acetate ([Bmim]Oac) [28] have been reported that have the ability to dissolve chitin upon heating (over 100 °C). Qin et al. applied ILs to directly dissolve waste shells for the first time [29]. Chitin could conveniently be extracted or made into fibers from the dissolution of crustacean shells in 1-ethyl-3-methyl-imidazolium acetate ([Emim]Oac). Yet the mineral components of shells (mainly $CaCO_3$) were also dissolved and remained in ILs during the process. Setoguchi et al. proposed adding a demineralization step with citric acid following chitin extraction by [Amim]Br, which acted as a chelating agent so that residual $CaCO_3$ was removed [30]. Deep eutectic solvents, such as choline chloride– thiourea, chloro-choline chloride–urea, and betaine hydrochloride–urea also achieved chitin dissolution at elevated temperatures [31]. Deep eutectic solvents have been extensively employed to extract biomass due to their low toxicity, low cost, easy degradability, and other excellent characteristics [32]. Energy intensive measure including microwave and ultrasonication can effectively reduce dissolution temperatures and shorten reaction times. Cheap and green solvents are highly desirable to improve industrial practicality of solvent extraction methods. It was reported the pretreatments of waste shells in hot water or glycerol gave facile and fast separation of chitin [33]. Yan's group established a hot water-carbonic acid (HOW-CA) process for shell waste fractionation [11]. Following the deproteinization in hot water (180–220 °C) and demineralization by pressured CO_2 in room-temperature aqueous solution, highly pure (>90%) chitin with less deacetylation was produced from shrimp shells. The system has advantages of simple equipment, low cost and environmental friendliness, showing great potential for scale-up shell fractionation. A natural deep eutectic solvent system for preparing chitin form insects was reported by Zhou et al. [34]. Sun et al. designed a natural deep eutectic solvents system consisting of LAc and choline chloride to extract highly pure chitin from shell waste [35]. Chitin with a purity of up to 99.3% was successfully prepared under optimum conditions. Such new and green solvent systems are expected to further enhance efficiency and reduce costs of chitin fractionation.

4.2 Chitin Hydrolysis into the Amino- or Amide-Sugar Products

4.2.1 Water Solvent Systems

The low solubility of chitin/chitosan limits direct application as biomass in the downstream upgrading of organonitrogen chemicals. Hydrolysis of chitin/chitosan into soluble oligosaccharides and monomers is the primary requirement for realizing chitin biorefineries [36–39]. Through the hydrolysis step (see Fig. 4.3), amino- or amide-sugars (GlcN or GlcNAc) which are chito-monomers can be generated [41]. These products have great potential in biomedicine, food, agriculture, cosmetics and chemical syntheses [42–45]. Taking GlcN as an example, it is the besta nutrients for joint disease and its market size is estimated to be valued at US $763.7 million in 2020 [46].

Conventionally, GlcN and GlcNAc are obtained by chemical hydrolysis and/or deacetylation of chitin with concentrated acids (e.g., HCl, H_2SO_4, H_3PO_4) upon heating [47–53]. Usually, reactions last for several hours. For instance, the chitin degradation reaction to GlcN using concentrated HCl requires 4 h as conducted by Mojarrad and co-workers [54]. Innovative assisted heating technologies such as microwave irradiation, ultrasonication and subcritical water have been introduced to shorten reaction times and increase selectivities [55–59], yet these methods are

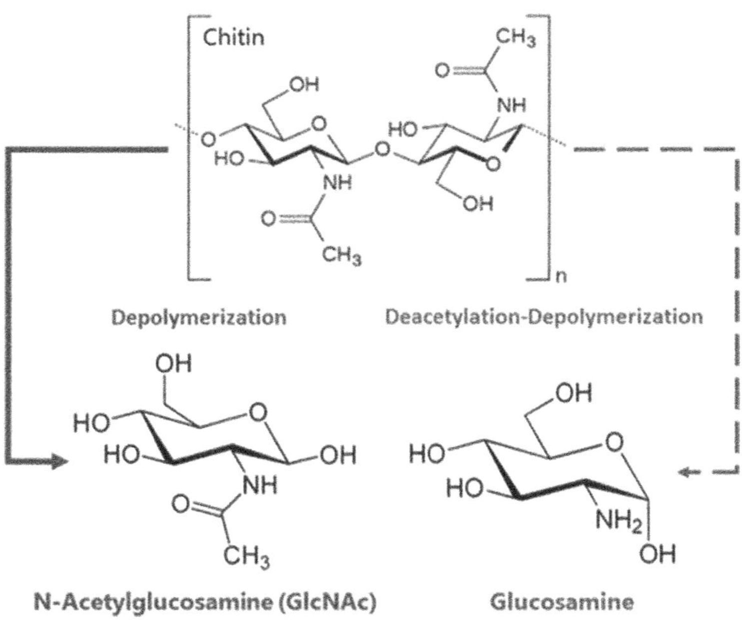

Fig. 4.3 Chitin hydrolysis reaction formula: chitin hydrolysis to N-acetylglucosamine and glucosamine. (Adapted with permission from [40], Copyright © 2021, Elsevier)

energy intensive and lack suitability for industrial production. Enzymatic degradation of chitin has also been explored [60–76], but has weaker ability for chitin depolymerization compared with chemical means.

Mechanochemical methods have been proposed to enhance degradation efficiency of chitin, posing the advantages of low catalyst usage, synergistic effects and solvent-free [77–81]. Combined physical and chemical forces lead to the decomposition of the hydrogen bonding network and cleavage of β-(1–4) bonds, resulting in appreciable yields of water-soluble products [82]. Fukuoka's group reported on the mechanochemical depolymerization of chitin with almost complete retention of acetamido groups (CH_3CONH-) [83]. In that research, chitin powder was impregnated in ether with a catalytic amount of H_2SO_4, followed by evaporation of solvent. H_2SO_4 was evenly attached on the polymer chain of chitin after acid-impregnation. Then acid-chitin was ball-milled at 500 rpm for 6 h at solvent-free conditions, leading to 100% soluble compounds. Control experiments showed that either acid-impregnation or ball-milling alone could not give completely water-soluble products, indicating a synergistic effect between mechanical and chemical forces. A yield of 53% GlcNAc was obtained by further hydrolyzing ball-milled mixtures in water at 170 °C for 1 h. The large retention degree of acetamido groups was due to tensile stress loading on main chains of chitin polymers rather than polymer side chains. Margoutidis and co-workers used kaolinite, a natural clay, as catalyst for mechanochemical degradation of chitin [84]. Considerable amounts of water-soluble compounds (75.8%) were obtained after 6 h of ball milling. The mixture was mainly composed of chito-oligomers with a degree of polymerization (DP) ranging from one to five. Solid acid catalysts instead of mineral acids were adopted in the method not only bringing the convenience of catalyst-product separation, but also posing the potential of catalyst recycling. Solid base catalysts have also been applied in the mechanochemical method [85]. Depolymerization and deacetylation occurred simultaneously employing solid base catalysts, leading to fully water-soluble low molecular weight chitosan (LMWC) products, and whose molecular weights are around 1.2 kDa to 13 kDa. Narrower product distribution was achieved while much less amount of base was consumed in the solid base catalyzed mechanochemical process compared with hydrolysis without ball-milling. This method could be directly applied to raw shrimp shells, recovering 20.9 wt% LMWC product with a degree of deacetylation of 80.5% and after ball milling at 700 rpm for 3 h.

4.2.2 *Organic or Co-solvent Systems*

The insolubility of chitin in water inhibits its hydrolysis [86]. Organic solvent systems were proposed by Yan's group to enhance the hydrolysis accessibility of chitin [87]. Over 75% of chitin was liquefied in ethylene glycol (EG) at 165 °C within 90 min using 8 wt% H_2SO_4 as the catalyst with major products being identified as hydroxyethyl-2-amino-2-deoxyhexopyranoside (HADP) and

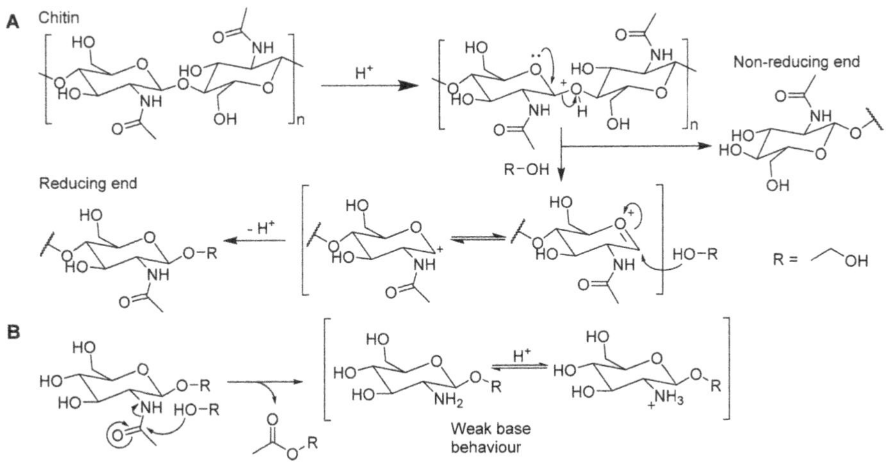

Fig. 4.4 Liquefaction mechanism of chitin in EG solvent. (Adapted with permission from [87], Copyright © 2014, Elsevier)

hydroxyethyl-2-acetamido-2-deoxyhexopyranoside (HAADP) by GC-MS and confirmed by NMR. HAADP was initially formed during the reaction and deacetylated into HADP in subsequence, implying that hydrolysis of the acetyl amide group occurred after depolymerization of chitin backbones. The characterization results indicated that EG helped to destroy the crystalline structure of chitin due to the interactions between hydroxyl group (−OH) and the polymer chain. EG also played a role in stabilizing the EG-derived chitin monomers. The alcoholysis reaction opened up a new route for chitin conversion and upgrading (see Fig. 4.4). However, this acid-catalyzed system deactivated after 90 min caused by accumulation of basic amine group ($-NH_2$). To address demerits in the EG -H_2SO_4 system, a novel system employing formic acid (FA) as the both the solvent and catalyst was proposed [88]. Ball-milled chitin and raw shrimp shells were completely degraded at 100 °C in the presence of FA, obtaining soluble formylated chitin oligomers and various monomers (recovery >60%). They eventually turned into a single compound, 5-(formyloxymethyl)furfural (FMF), in a yield of 35% after sufficient reaction time. Through product evolution monitored by electrospray ionization mass spectrometry (ESI-MS), it was revealed that the reaction could be divided into three stages: (1) partial formylation of the −OH in chitin side chains generated soluble polymeric derivatives, (2) FA catalyzed the polymer chain breakage to form monomers and oligomers, (3) water consumed in the hydrolysis and rehydration step was replenished via chitin formylation to make a self-sustained reaction cycle. Thus, chitin was transformed into monomers and eventually FMF as the reaction time was prolonged. The FA system had the advantages of having few side reactions because of weaker acidity, direct use of raw shrimp shells as the starting material and feasibility of catalyst separation and reutilization.

Solvent optimization has a significant effect on the efficiency and selectivity of chitin hydrolysis [89, 90]. Ionic liquids (ILs) have gained increasing attention as the solvents for catalytic systems. Zhang et al. devised a [Bmim]Cl IL solvent system for chitosan hydrolysis, recovering 60% amino sugars using HCl as catalyst at 100 °C for 7 h reaction time [89]. Later, microwave irradiation technology was applied to a sulfonic acid-functionalized IL system for improving reaction efficiency that increased total yield of reducing sugars to 90% [91]. Yan's group put forward an organic-water cosolvent system to obtain GlcN with high purity from chitin [92]. Ten cosolvents were screened and the aprotic diethylene glycol diethyl ether (DGDE) cosolvent showed the best performance. An 80% yield of GlcN was obtained after heating the system at 175 °C for 1 h, which used ball-milled chitin as reactant and 0.1 M H_2SO_4 as catalyst. The presence of aprotic solvent promoted hydrolysis and especially the deacetylation of chitin, thus notably improved GlcN selectivity. It was pointed that the basicity and solvating ability of cosolvent were dominant factors for the improved catalytic performance. An acidified lithium halide molten salt hydrate (AMSH) system was developed by the same group, encouraged by the enhanced ability of LiBr AMSH for cellulose hydrolysis [93]. Nine halide salts (i.e. LiBr, LiCl, LiI, $ZnCl_2$, KCl, KBr, $MgCl_2$, NaCl and NaI) combined with diluted acids (HCl or H_2SO_4) were screened and their influence on cleavage of the β-1,4-glycosidic bond were investigated. The AMSH system consisting of 60 wt% LiBr and 40 mM HCl was identified as the most effective system for chitin hydrolysis, offering 71.5% GlcNAc yield under optimum conditions (120 °C, 30 min). Kinetic studies indicated that the Li-containing salts substantially altered the hydrogen bonding network of chitin, thus achieving a remarkable enhancement in chitin dissolution. The high H^+ acidity resulted from the deshielding effect by the concentrated salt medium efficiently contributed to the improvement of depolymerization performance. At the same time, the deacetylation reaction was inhibited due to the presence of Li^+, leading to the high selectivity to GlcNAc. More novel hydrolysis systems are expected to be constructed and developed for downstream chitin upgrading.

4.3 Chitin Oxidation into Amino Acids

4.3.1 Amino Acids

Amino acids are a class of highly-valued organonitrogen chemicals with wide applications, many of which are produced by reduction of fossil-based chemicals using ammonia [94–98]. Production of amino acids from renewable resources with green methods is highly desirable. With the proposal of the "Shell biorefinery" concept, transforming chitin with its bio-fixed nitrogen into amino acids has drawn increasing attention [99, 100]. By oxidizing chitin monomers with a C6 backbone, many kinds of amino acids can be obtained. For example, in the oxidation of GlcN, the deacetylated monomer of chitin, leads to formation of glucosaminic acid

(GlcNA), which has potential economic value [101]. As a sugar acid, GlcNA is regarded as a potent sweeter and condiment in the food industry [102] and as an effective additive in cosmetics for preventing skin aging [103]. As a chiral molecule, GlcNA is a precursor for chemical preparation [104–106] and a starting material for pharmaceutical syntheses [107, 108]. As an important cationic ligand chelator with many metals [109], GlcNA can be employed as an adsorbent for heavy metals in water treatment [110]. Since it is biocompatible and nontoxic, GlcNA has potential to be used as a carrier in drug delivery systems and as a support for tissue engineering in clinical medicine [111–114]. Chen's group reported the air oxidation of chitin-derived GlcN to GlcNA at 35 °C using Au nanoparticles prepared by deposition-precipitation method as a carrier. The Au/ZnO catalyst prepared by the deposition-precipitation method reduces the E_a of the reaction, more than the Au/ZnO catalyst prepared by the deposition-reduction method. Comparative experiments and structural characterization results show that the small diameter of Au nanoparticles and the abundance of surface oxygen vacancy sites are the key factors responsible for their outstanding catalytic oxidation ability. This study provides a new idea for the conversion of GlcN to GlcNA [115].

4.3.2 *Oxidation of Chitin Monomers to Produce Amino Acid Sugars*

The simplest way to obtain amino acids from chitin biomass is via direct oxidization of the monomers of chitin. When the aldehyde group (–CHO) is oxidized into a carboxyl group (–COOH) on amino/amido sugars, corresponding amino acids can be obtained. Normally the oxidation is favored by the addition of a basic compounds, while the reported studies on chitin biomass oxidation into amino acid sugars could be achieved without base additives.

Studies on the oxidation of GlcN into GlcNA can be traced back to the last century. A 55% yield of GlcNA was recovered in hydrogen sulfide (H_2S) atmosphere using yellow mercuric oxide as oxidizing agent [116]. Optimization of this method led to the reduction of mercuric oxide dosage while increasing the recovery of GlcNA up to 62% [117]. Toxic agents in the classical oxidation method for producing GlcNA limited its application in medicines and foods. Electrochemical methods have been studied to selectively oxidize GlcN into GlcNA using gold-modified carbon felt electrodes and gold-plated electrode [118]. Recent studies have used O_2 gas, in the presence of heterogeneous metal catalysts. Gu and Xia recovered 70% GlcNA from GlcN using active charcoal-supported Pd-Bi catalyst (Pd-Bi/AC) with continuous flush of oxygen [119]. Ohmi et al. demonstrated that supported nano gold catalysts could effectively and selectively oxidize chitin-derived amino sugars into corresponding α-amino acids [120]. GlcN with yields of 89% and 93% was obtained over hydrotalcite (HT) and magnesium oxide supported gold catalysts (Au/HT and Au/MgO) respectively by purging O_2 at 40 °C for 3 h. Using MgO instead of

HT as the basic support not only enhanced activity but also significantly improved the reusability of catalysts, which possibly benefitted from strong interactions between Au nanoparticles and basic sites on MgO. Yan's group employed the Au/MgO catalyst with 0.1% Au loading to oxidize GlcN into GlcNA with 5 bar O_2 at 40 °C, achieving a yield of 88% within 3 h [121]. Heterogeneous catalytic systems in aqueous solutions employing O_2 as oxidant have the advantage of being green, environmentally-friendly and allowing facile separation of products.

GlcNAc was also the potential starting material for the synthesis of valuable amino acids. The oxidation of GlcNAc led to a new amino acid, *N*-acetyl-glucosaminic acid (GlcNAcA) [120]. As presented in Ohmi's work, near quantitative conversion of GlcNAc to GlcNAcA was achieved by using Au/HT catalyst at 40 °C for 5 h in the presence of O_2 flush.

4.3.2.1 Oxidative Cleavage of Chitin Monomer to Produce Short-chain Amino Acids

The C–C bond tends to break under harsh conditions, which leads to the formation of short-chain small molecules. Fukuoka's group reported acetylglycine (AcGly) production from GlcNAc using carbon supported ruthenium catalyst (Ru/C) via two steps (Fig. 4.5) [122]. In the first step, GlcNAc undergoes retro-aldol and hydrogenation reactions to generate *N*-acetylmonoethanolamine (AMEA) over Ru/C under H_2 pressure. In the second step, AcGly is generated by the oxidation of AMEA by combining Ru/C and $NaHCO_3$ catalysts in the presence of pressurized O_2. Introduction of H_2 pressure inhibited hydrogenation of GlcNAc to ADS and resonance stabilization of the acetamido group of GlcNAc avoided its isomerization, which was advantageous to the further improvement of AcGly yield. Mechanistic investigation revealed that Ru species was present on carbon as hydrous RuO_2 with hydroxo, aqua, μ-hydroxo, and μ-oxo ligands in the presence of O_2. The small hydrous RuO_2 clusters were inferred to be the active species for AMEA oxidation to produce AcGly.

4.3.2.2 Oxidation of Chitin/Chitosan into Amino Acids

Directly obtaining amino acids from low-cost chitin or chitosan is highly desired. Yan's group constructed a "one pot-two step" reaction system for the direct

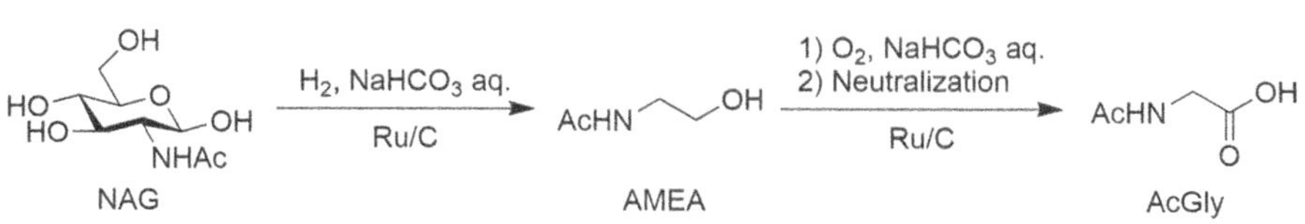

Fig. 4.5 Conversion of NAG to AcGly over Ru/C. (Adapted with permission from [122], Copyright © 2018, Elsevier)

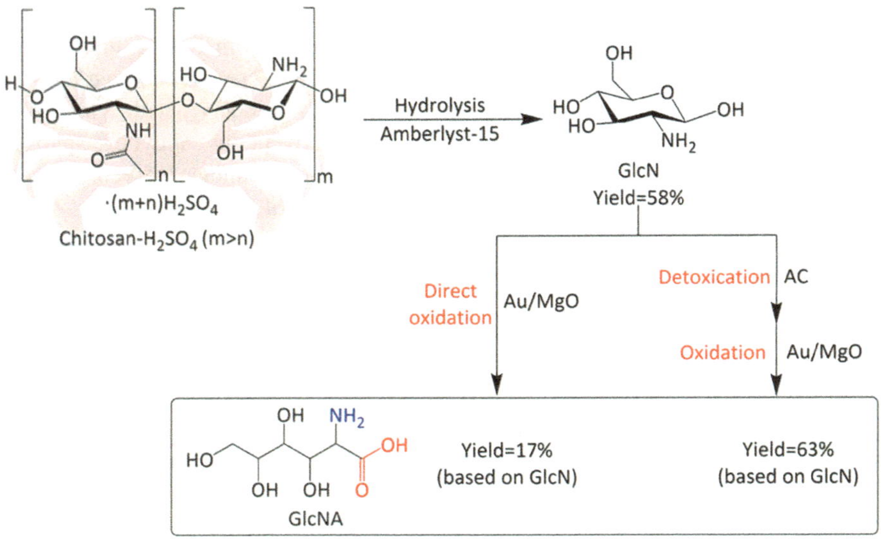

Fig. 4.6 Catalytic conversion of chitosan to GlcNA by a two-step chemical process (Adapted with permission from [121], Copyright © 2019, Elsevier)

synthesis of GlcNA from cheaper chitosan, following the tandem hydrolysis of chitosan and the oxidation of GlcN [121]. The chitosan was firstly hydrolyzed into GlcN over Amberlyst-15 catalyst at 160 °C. Then the sugar monomer was oxidized into GlcNA over Au/MgO catalyst at 40 °C with 5 bar pressurized O_2. By using activated carbon as the adsorbent to remove by-products such as humins which probably poisoned the Au/MgO catalyst, the yield of GlcNA was significant enhanced from 17% to 63%. The overall yield of GlcNA from chitosan attained 36% under optimum conditions (see Fig. 4.6). This study paved the way for direct utilization and industrial processing of chitin biomass. By referring to the cases of "one-step" conversion of cellulose, bifunctional catalyst especially solid acid supported metal catalyst can be developed to realize tandem hydrolysis and oxidation, with solid acid and metals acting as the active centers for hydrolyzing and oxidizing substrates, respectively.

Chitin has potential as starting material for the synthesis of amino acids. Yan's group developed an integrated biorefinery process to upgrade shrimp shell waste (SSW) derived-chitin into tyrosine and levodopa (L-DOPA, generally used as a drug in the treatment of Parkinson's disease) [123]. Two aromatic N-containing chemicals from chitin could be produced through a single chemical process step. At the beginning of the integrated process, the dried SSW was treated with concentrated HCl at room temperature for demineralization. Then the HCl-treated SSW was ball-milled with H_2SO_4 for dissolution and partial depolymerization, with nearly 100% chitin hydrolysates recovered. The unpurified chitin hydrolysates were directly put into a one-pot fermentation system, with the residual proteins being the nutrient source for microorganisms. 0.91 g/L tyrosine or 0.41 g/L L-DOPA was successfully produced from 22.5 g/L unpurified SSW-derived chitin hydrolysates using the engineered *E. coli* strains. This integrated biorefinery process demonstrated the feasibility of upgrading renewable chitin-containing waste into value-added amino acids.

4.4 Chitin Dehydration into Furanic Amide (3A5AF)

4.4.1 Potential of 3A5AF as a Building Block Chemical

The compound, 3-acetamido-5-acetylfuran (3A5AF) is an important dehydration product of GlcNAc and is obtained by removal of three molecules of water. Analogous to 5-hydroxymethylfurfural (5-HMF) which is the dehydration product of cellulose [124], 3A5AF is a versatile platform chemical affording a number of valuable derivatives (see Fig. 4.7). Modification of acetamido group or furan ring on 3A5AF gives useful N-containing heterocyclic compounds and is worth exploring. Liu demonstrated that 3A5AF hydrolysis over NaOH generated amino-substituted furan, 2-acetyl-4-aminofuran, affording a 57% yield at 80 °C for 1 h [125]. Changing the catalyst from NaOH to Ir based complex led to the reduction of the acetyl group, forming of 3-acetamido-5-(1-hydroxylethyl)furan, which was apt to dehydrated to the corresponding alkene in the presence of tert-butoxide. Pham et al. obtained a novel dihydrodifuropyridine scaffold from chitin-derived 3A5AF and ketones over

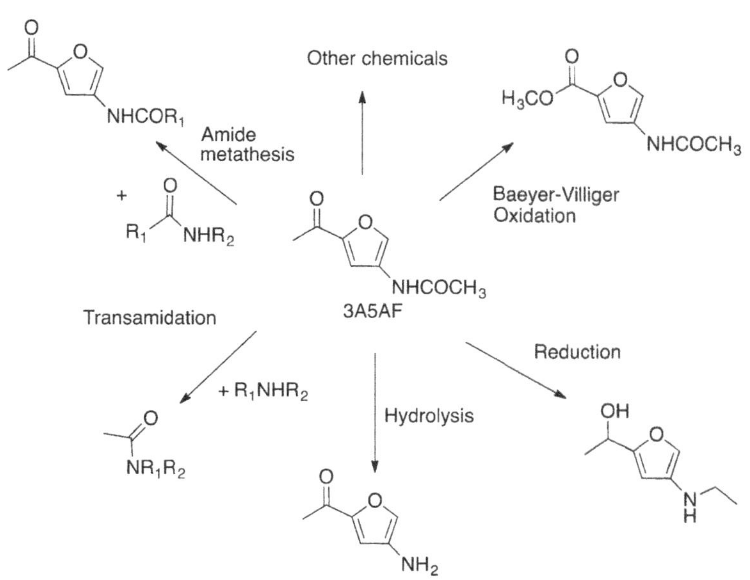

Fig. 4.7 Useful derivatives from 3A5AF (Adapted with permission from [45], Copyright © 2018, Elsevier)

10% HCl [126]. Conversion of 3A5AF to generate 3-aminofuran/3-amidofuran chemicals was studied by Pham and co-workers [127, 128]. Incorporation of non-Haber-Bosch nitrogen into value-added N-containing compounds has great potential. Apart from furan derivatives, 3A5AF can also be converted into heterocyclic compounds that do not contain a furan ring. Pham subsequently reported *N*-acetyl-L-rednose (Red-NAc) production from 3A5AF in considerable yield via Noyori reduction-Achmatowicz rearrangement sequence, which is a rare and valuable chiral amino sugar found in natural anthracycline and angucycline antibiotics [128]. Other work showed that Red-NAc was a versatile platform chemical for obtaining stereochemically pure 2-amino sugars. Up to nine kinds of fine N-containing compounds were obtained by Pham and co-workers [129], demonstrating the high reactivity and transformation possibility of 3A5AF.

In pharmaceutical syntheses, 3A5AF can act as a starting material. Proximicin A is a useful agent for cancer treatment, which can be synthesized from 3A5AF via multiple steps [130, 131], but the reaction involves many toxic agents such as organic lithium, azide compounds and methyl chloroformate. Sadiq et al. improved the procedure by using greener solvents (i.e., ILs, methanol, dimethyl carbonate) and making full utilization of nitrogen [132]. The potential value of 3A5AF is still under exploration through establishing and developing more novel conversion routes and systems. Efficient and affordable production of 3A5AF from chitin is one of the prerequisites for downstream upgrading.

4.4.2 Chitin Monomer Dehydration to 3A5AF

Dehydration of chitin monomer generates N-containing heterocyclic compounds, including 2-acetamido-2,3-dideoxy-D-erythro-hex-2-enofuranose (Chromogen I), 3-acetamido-5-(1',2'-dihydroxyethyl) furan (Chromogen III), 3-acetamido-5-acetylfuran (3A5AF) and their derivatives. These chemicals have potential applications in food and medicine fields [133, 134], especially 3A5AF as a building block chemical able to synthesize useful materials as mentioned above (Fig. 4.8).

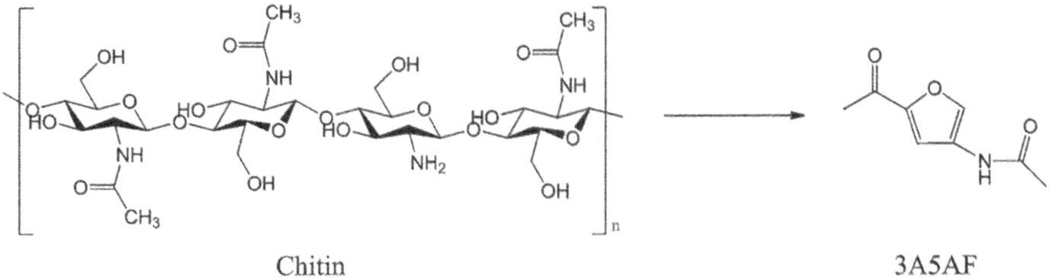

Fig. 4.8 Reaction of chitin conversion into 3A5AF product. (Adapted with permission from [135], Copyright © 2015, Elsevier)

Chromogen I, Chromogen III and 3A5AF are the dehydration products of GlcNAc by removing one, two and three molecules of water, respectively.

Dehydration of GlcNAc was firstly reported by Ogata et al. in 2010, using borate buffer (0.4 mol/L, pH = 7.0) as both the solvent and catalyst [136]. Chromogen I and its isomerization products were generated in a total yield of ~80% after reaction for 2 h at 100 °C. The non-catalytic degradation of GlcNAc in high-temperature water was studied by Osada and co-workers [137, 138]. A 23.0% yield of Chromogen I and 23.1% of Chromogen III were recovered from GlcNAc in water at (120–220) °C and 25 MPa within 40 s [137]. The plausible reaction routes were afterwards investigated and reported, giving of basis for following research work on hydrothermal treatment of chitin [138]. It was revealed that the dehydration occurred between–H on C2 and –OH on C3 and followed by the ring closure between C1 and C4, which was different from glucose dehydration to 5-HMF due to the presence of acetylamino group on GlcNAc [139]. Without introducing catalysts or additives, H^+ and OH^- in hot water were deemed as the active species that can efficiently promote the dehydration process.

Further dehydration to 3A5AF (removing three H_2O molecules in total) was completed back in the last century by non-catalytic pyrolysis at high temperatures (400 °C), but giving 2% yield at most [140, 141]. Kerton's group constructed an efficient catalytic system in organic solvents to convert GlcNAc to 3A5AF [142]. Dipolar aprotic solvents (i.e., *N,N*-dimethylformamide (DMF) and *N,N*-dimethylacetamide (DMA)) instead of esters like tert-butyl acetate were shown to be efficient for GlcNAc dehydration. With 1 equivalent $B(OH)_3$ and 2 equivalent NaCl as the additives, a maximum yield of 58% was reached after 15 min with microwave heating at 220 °C in DMA. Increasing the amount of NaCl to 4 equivalents gave a maximum yield of 62%. By conducting control experiments and ICP-MS analysis, it was found that the Cl and B elements played important roles in the formation of 3A5AF. Subsequently, the researchers combined Cl-containing ILs as solvent and $B(OH)_3$ as additive for effective dehydration of GlcNAc [142]. Optimum conditions were 180 °C oil-bath heating for 60 min and 180 °C microwave heating for 3 min in 1-butyl-3-methylimidazolium chloride ([Bmim]Cl) which gave a 60% yield of 3A5AF. The employment of microwave energy remarkably shortened the reaction time. Though $B(OH)_3$ displayed positive effects for GlcNA dehydration, it is not a green additive. In 2015, Yan's group reported on the treatment of chitin by ball-milling, steam explosion, alkaline treatment, phosphoric acid and ionic liquid dissolution/reprecipitation. The structure of chitin was intensively studied through a series of analytical techniques, and the reactivity after each treatment was evaluated for dehydration and liquefaction reactions. After parallel studies, it was concluded that dry ball-milling was the most efficient method, and that crystal size and hydrogen bond networks were two key factors to improve reactivity. After being pulverized by a ball mill, the yield of 3A5AF obtained by dehydration of chitin reached 28.5%, while the yield of untreated chitin was 7.5% at most [143]. An organic solvent system not involving $B(OH)_3$ was developed by Wang and co-workers [144]. The highest yield of 3A5AF reached 53% over glycine chloride ([Gly]Cl) IL catalyst in DMA solvent at 200 °C for 10 min. Adding $CaCl_2$

further increased the yield to 53%. The stable [Gly]Cl catalyst gave facile separation and reliable reuse for production of 3A5AF, with only ~6% decrease of yield after eight runs. Fukuoka's group conducted GlcNAc dehydration in DMF solvent using $AlCl_3 \cdot 6H_2O$ as catalyst, which is cheap, green and easy to separate [145]. This simple and facile reaction system afforded a 30% yield of 3A5AF as the major product at 120 °C for 30 min. This method was applicable over a wide range of temperatures and substrate concentrations, exhibiting the potential for industrial scale-up. The non-catalytic conversion of GlcNAc to 3A5AF in pyridinium-based IL solvent was put forward by Zang et al. [146]. Without introducing any additives, the yield of 3A5AF reached 37.5% in 1-carboxymethyl pyridinium chloride ([CMPy]Cl) at 180 °C for 60 min. Adding B_2O_3 and $CaCl_2$ increased the optimum yield to 67.4% in 20 min, highlighting the promoting roles of B and Cl elements on the generation of 3A5AF. Biobased deep eutectic solvent (DES) system was developed by Wu et al. to transform GlcNAc and chitin into 3A5AF [147]. Screening test results indicated that the DES composed of choline chloride/citric acid (CCCA) had the best performance for GlcNAc dehydration. A maximum yield of 47.1% was obtained in DMA solvent at 220 °C with $CaCl_2 \cdot 2H_2O$ as an additive. Recycling tests showed CCCA was easy to separate and reuse, with acceptable losses of activity after five runs.

Chemo-enzymatic methods have been applied to convert chitin into 3A5AF. Zhang's group reported on efficient production of 3A5AF by a chemo-enzymatic method using chitin as raw material. First, chitin was converted to GlcNAc by chitin hydrolase using an integrated affinity adsorption-enzymatic reaction method. Then, 3A5AF was prepared by using the obtained GlcNAc as the reactant and ammonium thiocyanate (NH_4SCN) as the screened catalyst. At 180 °C, in DMA solvent, the ratio of GlcNAc, NH_4SCN and $CaCl_2 \cdot 2H_2O$ was 2:1:2, a molar yield of 56.7% 3A5AF was obtained after 10 min. The NH_4SCN applied in this method has good recyclability for 5 cycles [148].

4.4.3 Chitin Polymer Dehydration to 3A5AF

Direct conversion of chitin/chitosan to 3A5AF is attractive but challenging. In 2014, Yan's group demonstrated direct synthesis of 3A5AF from chitin [149]. Six organic solvents, different types of additives including metal chlorides, and acids were screened. Under optimum conditions including *N*-methyl-2-pyrrolidone (NMP) as the solvent, boric acid and alkaline chlorides as additives, and reaction temperature at 220 °C, the yield of 3A5AF nearly reached 7.5% in 2 h with ca. 50% GlcNAc converted. Levoglucosenone, 4-(acetylamino)-1,3-benzenediol, HOAc and chitin-humins were obtained as byproducts (see Fig. 4.9). Consistent with previous research, the combination of $B(OH)_3$ and NaCl significantly promoted formation of 3A5AF. The boron–GlcN complex, which was the active species, was observed by NMR spectroscopy. Solvent optimization was also conducted [135] and it was shown that the IL solvent ([Bmim]Cl) accelerated the reaction rate and allowed

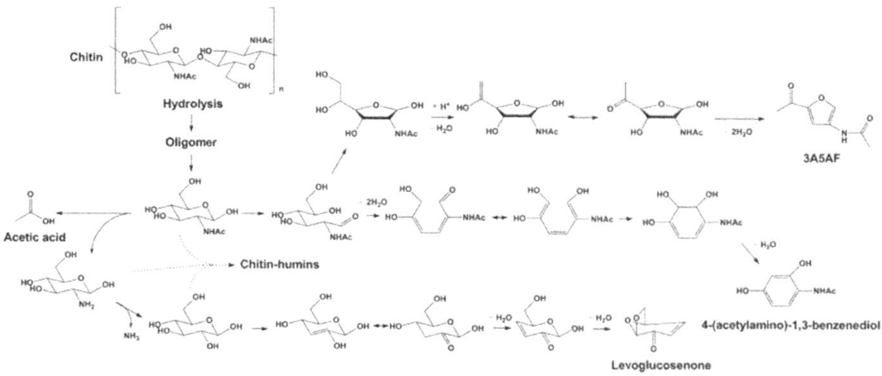

Fig. 4.9 Chitin dehydration reaction pathway: Chitin to 3A5AF, HOAc, levoglucosenone, 4-(acetylamino)-1,3-benzenediol. (Adapted with permission from [149], Copyright © 2014, the Royal Society of Chemistry)

lower reaction temperatures (from 215 °C to 180 °C) compared with organic solvents. Chitosan hydrolysis in the presence of Lewis acids with $SnCl_4{\cdot}5H_2O$ led to the production of levulinic acid and 5-HMF [150], whereas 23.9 wt% levulinic acid was recovered from 100 mg chitosan using 0.24 mmol $SnCl_4{\cdot}5H_2O$ in 4 mL water, assisted with microwave irradiation at 200 °C for 30 min reaction time.

Chitin offers an alternative source for the synthesis of 3A5AF and gives comparable yields [151]. Kinetic studies indicate that depolymerization of chitin polymer is the rate-determining step owing to its high crystallinity. To further enhance the efficiency of simultaneous depolymerization and dehydration, pretreatments including ball-mill grinding, steam explosion, alkaline treatment, phosphoric acid treatment, and IL dissolution/reprecipitation are necessary [135]. Ball-mill grinding exhibits the best performance among these techniques. A notably improved 29% yield of 3A5AF was achieved within 1 h, using ball-milled chitin as the reactant in [Bmim]Cl with $B(OH)_3$ and HCl as additives at 180 °C. Mechanistic study indicated that the mechanical force could effectively reduce the crystallite size and destroy the strong hydrogen bond network of chitin polymer, thus improving the reactivity of chitin. Solvent optimization is regarded as the key strategy to realize the selective and highly efficient synthesis of 3A5AF. Inspired by the satisfactory results of hydrothermal treatment of GlcNAc, the hot water system was employed to directly convert chitin powder [152]. Higher temperatures (290–390 °C) were required to enhance the reaction accessibility of chitin. By employing a semi-batch reactor to control the reaction temperature and time, a 90% yield of water-soluble products was attained in 3 h in which Chromogen I was the major product while most others were not identified. The maximum yield was limited at 2.6% under optimum conditions. The EDS system was also applied to the direct production of 3A5AF from chitin [147]. Under optimum conditions at 220 °C in *N,N*-dimethylacetamide, the recovery of 3A5AF attained 5 mol%. Because no satisfactory recovery was gained from direct hydration from chitin polymer, it was a prior task to establish an effective catalytic system for chitin depolymerization. Zhang's group reported on the preparation of ethanolamine ionic liquid with inexpensive raw materials to catalyze the conversion of GlcNAc to 3A5AF. Using triethanolamine hydrochloride ([TEA]Cl) ionic liquid as catalyst, under the optimal reaction conditions of 170 °C and 20 min, the yield of 3A5AF can reach 62%, and the ionic liquid has good reusability and can be recycled. It still has good catalytic activity after five times use [153].

4.5 Other Transformation Strategies

4.5.1 Hydrothermal Methods

Hydrothermal methods are highly favored for use of water as a green solvent and low-cost features. Without introducing any catalyst or additive, the products can be easily separated and recovered. In addition to 3A5AF [137, 138, 152], some other N-heterocyclic compounds can be obtained by hydrothermal reaction of chitin. Direct hydrothermal treatment at 300 °C in NaOH aqueous solution was applied to degradation of chitin. This system gave HOAc as the main product while generating a small amount of N-heterocyclic compounds as well, including pyrrole, pyrazine, and pyridine derivatives [154]. Substituting the reactant from chitin to shrimp shell increased the yield of HOAc from 38.1% to 47.9% at optimum conditions, including 1.25 equivalent (mole of copper oxide (CuO) catalyst to the mole of carbon in chitin in the substrate) ball-milled CuO as catalyst and 5 bar O_2. It was speculated that some carbon in the protein may be advantageous to the formation of HOAc. However, the highest yield of pyrrole was limited to 12% even when external nitrogen sources like ammonia were introduced. But this result was exciting compared to the 3% recovery from naphtha through an industrial chemical route. Valuable findings that such pyrrole formation is favored at high temperatures could provide a perspective for increasing the yield of pyrrole in follow-up studies. The hydrothermal carbonization of chitin was investigated by Liang et al. [155]. The combination of mechanical activation and $FeCl_3$ addition was adopted to pretreat chitin, and significantly enhanced the hydrothermal carbonization efficiency. Qi et al. demonstrated that hydrothermal conversion of GlcNAc and ball-milled chitin into valuable organic acids over vanadium pentoxide (V_2O_5) catalyst in the presence of pressured O_2 [156]. Multiple organic acids including formic acid (FA), glycolic acid, LAc and HOAc were generated from chitin in base-free water at 200 °C after 2–3 h. It can be suggested that the high concentration of H^+ and OH^- available from water at high temperatures (ion product) promoted the decomposition of GlcNAc.

4.5.2 Hydrogenation/Hydrogenolysis Reactions

Similar to polyol production from glucose or cellulose, polyol amines/amides can be obtained through hydrogenation/hydrogenolysis of chitin biomass. Pressure hydrogen, water solvent and noble metal catalysts such as Pt/C, Pd/C, Ru/C, and Rh/C are usually employed in these type of reactions [157]. Among the noble metal catalysts, Ru based catalysts performed the best for higher product yields and carbon balances. The quantitative conversion of GlcN can be obtained over Ru/C at 80 °C, with 2-acetoamido-2-deoxy-D-sorbitol (ADS) as the dominant product. Elevating the reaction temperature leads to the formation of smaller C2 to C4 amines/amides though hydrogenolysis. The chitin-derived ADS can be further dehydrated into prospective precursors for organonitrogen chemicals, namely 2-acetamide-2-deoxyisosorbide (ADI)(see Fig. 4.10) [158–160].

As for using chitin/chitosan as the reactant, higher temperatures are required to destroy the crystalline structure and enhance the reactivity of polymer. Around 8% yield of C2 products as well as 10% gaseous products are obtained at 260 °C. A "one-pot, two-step" hydrolytic hydrogenation of chitin polymer was proposed by Fukuoka's group [161]. First, acid treatment of ball-milled chitin gave water-soluble GlcN and oligomer in a total yield of ~83% at 175 °C for 12 min. Then $NaHCO_3$ was added to change solution pH from 2 to 3 ~ 4, providing favorable condition for the following hydrogenation reaction. Fifty two percent yield of ADS was finally reached at 120 °C with 40 bar H_2 over Ru/TiO_2 catalyst. Theoretical investigation indicated that pH adjustment effectively inhibits side reactions thus leads to the high selectivity of ADS.

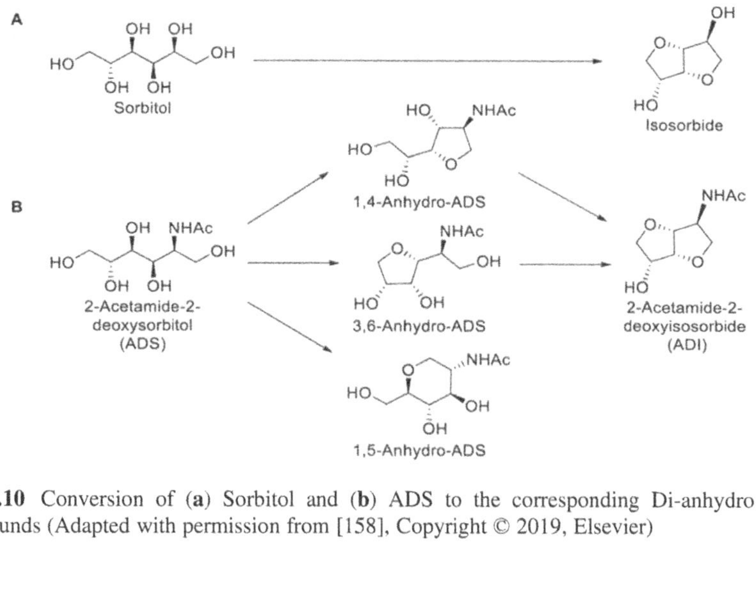

Fig. 4.10 Conversion of (**a**) Sorbitol and (**b**) ADS to the corresponding Di-anhydro cyclic compounds (Adapted with permission from [158], Copyright © 2019, Elsevier)

4.5.3 Condensation Reactions

Condensation of chitin monomers can produce valuable heterocyclic compounds. The chitin dimer, *N,N′*-diacetylchitobiose $(GlcNAc)_2$, tends to become hydrated into 4-*O*-β-2-acetamido-2-deoxy-D-glucopyranosyl 2-acetamido-2,3-dideoxydidehydro-glucopyranose (GND) in hot water [162]. A maximum yield of 24.7% was achieved at (120–220) °C and 25 MPa with a reaction time of (8–39) s. Limited conversions of chitin disaccharide were probably because the glycosidic bonds were hard to break at H^+ insufficient conditions, without any catalysts or additives.

Dehydration of two molecules of GlcN over acid or base catalysts can obtain potential pharmaceutical agent, deoxyfructosazine (DOF) and fructosazine (FZ). Fujii et al. reported on the epimerization and self-condensation of GlcN to FZ with a 24.8% yield at 70 °C in methanol in the presence of metallic sodium [163]. Recent studies focus on the mild conversion of GlcN to DOF using boron-based catalysts such as phenylboronic acid and boric acid [164]. The DOF with a yield of 58% has been identified to be the sole product after the reaction of 3 h at room temperature. NMR analysis reveals the promotion effect of B-based catalyst: the B atom in the catalyst interacted with the with the three groups of the GlcN in an open chain (the two –OH groups on the C3 and C4 positions and the $–NH_2$ group on the C2 position) to form a B-GlcN intermediate, which transforms into a pyrazine ring, and subsequently dehydrates and isomerizes, and ultimately affords DOF. The waste acid and waste alkali generated during the reaction poses a risk for the environment. A novel green system using [Bmim]OH IL as both solvent and catalyst was proposed by Jia et al. [165]. The effect of reaction parameters including substrate concentration, reaction temperature, reaction time, additives and co-solvents on the yields of products were studied. Under optimum conditions (120 °C, 180 min), a 35% yield of DOF was obtained in [Bmim]OH solvent. Adding NaOH or KOH had negligible effects on improving the recovery of DOF, while introducing the cosolvent of dimethyl sulfoxide (DMSO) remarkably enhances the reaction activity, achieving a maximum DOF yield of 49% within 3 h. A plausible reaction mechanism was also investigated and proposed. Substituting [Bmim]OH into the basic 1-ethyl-3-methylimidazolium acetate ([Emim]OAc) efficiently converts the GlcN into DOF and FZ simultaneously [166]. A mechanistic investigation by NMR indicated that [Emim]OAc with strong hydrogen bonding basicity could coordinate with the hydroxyl and amino groups of GlcN, thus activating the substrate and catalyzing the product formation at the same time. Hydrogen bonding as an activation force contributes to selectivity and rate enhancement of the condensation reaction, however the total yields of DOF and FZ are still unsatisfactory, limited to 15%. Introducing $B(OH)_3$ as the additive significantly promoted the yield of DOF to 40.2% [167]. It was revealed by the combining the mechanistic investigation with the ESI-MS analyses that a mixed 1:1 boron complex was the major active species. DFT calculations indicated that the coordination effects of B atoms make the ring-opening and subsequent dehydration reaction more favorable from both thermodynamic and kinetic aspects. When adding a trace amount of hydrogen peroxide

(H_2O_2) as oxidant, the proceeding direction changed and the major product became FZ in a 25.3% yield.

4.6 Concluding Remarks and Future Outlook

Chitin biomass is an available and abundant alternative resource for synthesis of organonitrogen chemicals. Hydrolysis (and deacetylation) of chitin and chitosan leads to platform chitin monomers, GlcN and GlcNAc. A series of organonitrogen chemicals can be obtained from GlcN/GlcNAc via multiple reaction pathways, including oxidation to amino acids, dehydration into furanic amides, hydrogenation/hydrogenolysis to polyol amines/amide and condensation reaction to N-heterocyclic compounds. Satisfactory yields can be obtained for chitin monomer conversion, demonstrating the implement ability of "Shell Biorefinery". But the cost of chitin monomer upgrading is costly at present due to lack of chemistry for accessing GlcN/GlcNAc from chitin biomass. Directly using chitin polymer which avoids hydrolysis and separation steps to prepare valuable N-containing chemicals would be preferred. The strong crystallinity and extensive hydrogen bond network impedes reaction of chitin polymer. Pretreatment methods such as grinding, solvent optimization, innovative assisted technologies such as microwave heating and ultrasonic irradiation, and the mechanochemical method have been shown to be effective for enhancing the reactivity of chitin. With research efforts during the past few decades, it has been demonstrated that value-added chemicals can be obtained from either chitin or chitosan.

Direct conversion pathways that avoid the separation of intermediates are preferred for biorefineries, despite recovery aspects and costs being challenging. Establishing effective and efficient reaction systems are the very first step. To achieve this, the following avenues should be given attention. Novel solvent systems are required to effectively dissolve and activate the chitin polymer. Though organic solvents and ILs have good performance, their toxicity and high cost have to be considered. Some novel means such as microwave heating or ultrasonication can significantly shorten the reaction, but these methods tend to be energy-intensive. Mechanochemical methods are worth to be addressed as they are cost-effective and applicable. Heterogeneous catalytic systems are favored in biomass conversion owing to the advantages of facile separation and reuse of catalyst. It is urgently required to develop green and stable catalysts to enhance reaction efficiency. Catalyst design along with reaction systems can realize efficient chemical production from chitin biomass in a mild and environmentally-friendly route. Given the structural similarity between chitin and cellulose, some strategies for cellulose refinery can be used for chitin, along with the consideration of differences brought by the acetyl/acetylamino group on chitin/chitosan. The present price of pure chitin still remains high such that further efforts should be devoted into the process improvement and innovation of chitin extraction from raw shells for cost reduction. Based on the present achievements, the aim of "Waste Shell Biorefinery" will come into reality with more studies being conducted and findings reported. Organonitrogen chemicals are expected to be produced from chitin biomass with the incorporation of non-Haber-Bosch nitrogen.

References

1. Kerton FM, Liu Y, Omari KW, Hawboldt K. Green chemistry and the ocean-based biorefinery. Green Chem. 2013;15(4):860–71. https://doi.org/10.1039/c3gc36994c.
2. Beaney P, Lizardi-Mendoza J, Healy M. Comparison of chitins produced by chemical and bioprocessing methods. J Chem Technol Biotechnol. 2005;80(2):145–50. https://doi.org/10.1002/jctb.1164.
3. Chen X, Song S, Li H, Gözaydın G, Yan N. Expanding the boundary of biorefinery: organonitrogen chemicals from biomass. Acc Chem Res. 2021;54(7):1711–22. https://doi.org/10.1021/acs.accounts.0c00842.
4. Stephen AM, Phillips GO, Williams PA. Food polysaccharides and their applications. CRC/Taylor & Francis; 2006.: 733 p. https://doi.org/10.1201/9781420015164.
5. Bastiaens L, Soetemans L, D'Hondt E, Elst K. Sources of chitin and chitosan and their isolation. 2020:1–34. https://doi.org/10.1002/9781119450467.ch1.
6. Sogias IA, Khutoryanskiy VV, Williams AC. Exploring the factors affecting the solubility of chitosan in water. Macromol Chem Phys. 2010;211(4):426–33. https://doi.org/10.1002/macp.200900385.
7. Kumar M. A review of chitin and chitosan applications. React Funct Polym. 2000;46(1):1–27. https://doi.org/10.1016/S1381-5148(00)00038-9.
8. Liu L, Zhu L, Zhang S, Ma Y, Wang L, Wang H, Niu X. Preparation and properties of chitosan-based bacteriostatic agents and their application in strawberry bacteriostatic preservation. J Food Sci. 2021;86(10):4611–27. https://doi.org/10.1111/1750-3841.15912.
9. Aranaz I, Acosta N, Civera C, Elorza B, Mingo J, Castro C, Gandía MDlL, Caballero AH. Cosmetics and cosmeceutical applications of chitin, chitosan and their derivatives. Polymers. 2018;10(2):213. https://doi.org/10.3390/polym10020213.
10. Yan N, Chen X. Sustainability: don't waste seafood waste. Nature. 2015;524(7564):155–7. https://doi.org/10.1038/524155a.
11. Yang H, Gözaydın G, Nasaruddin RR, Har JRG, Chen X, Wang X, Yan N. Toward the shell biorefinery: processing crustacean shell waste using hot water and carbonic acid. ACS Sustain Chem Eng. 2019;7(5):5532–42. https://doi.org/10.1021/acssuschemeng.8b06853.
12. Shahidi F, Synowiecki J. Isolation and characterization of nutrients and value-added products from snow crab (Chionoecetes opilio) and shrimp (Pandalus borealis) processing discards. J Agric Food Chem. 2002;39(8):1527–32. https://doi.org/10.1021/jf00008a032.
13. Cheung PCK, Leung AYH, Ang PO. Comparison of supercritical carbon dioxide and Soxhlet extraction of lipids from a brown seaweed, sargassum hemiphyllum (Turn.) C Ag. J Agric Food Chem. 1998;46(10):4228–32. https://doi.org/10.1021/jf980346h.
14. Chevolot L, Foucault A, Chaubet F, Kervarec N, Sinquin C, Fisher A-M, Boisson-Vidal C. Further data on the structure of brown seaweed fucans: relationships with anticoagulant activity. Carbohydr Res. 1999;319(1–4):154–65. https://doi.org/10.1016/s0008-6215(99)00127-5.
15. Naylor RL, Hardy RW, Bureau DP, Chiu A, Elliott M, Farrell AP, Forster I, Gatlin DM, Goldburg RJ, Hua K, Nichols PD. Feeding aquaculture in an era of finite resources. Proc Natl Acad Sci U S A. 2009;106(36):15103–10. https://doi.org/10.1073/pnas.0905235106.
16. Charoenvuttitham P, Shi J, Mittal GS. Chitin extraction from black tiger shrimp (Penaeus monodon) waste using organic acids. Sep Sci Technol. 2006;41:1135–53. https://doi.org/10.1080/01496390600633725.

17. Broussignac P. Un haut polymère naturel peu connu dans l'industrie, Le chitosane. Chim Ind Genie Chim. 1968;99:1241–7.
18. Lezica RP, Quesada-AlluÉ L. In: Chitin PMD, editor. Carbohydrates. Academic; 1990. p. 443–81. https://doi.org/10.1016/b978-0-12-461012-5.50019-7.
19. Gagné N, Simpson BK. Use of proteolytic enzymes to facilitate the recovery of chitin from shrimp wastes. Food Biotechnol. 1993;7(3):253–63. https://doi.org/10.1080/08905439309549861.
20. Zakaria Z, Hall GM, Shama G. Lactic acid fermentation of scampi waste in a rotating horizontal bioreactor for chitin recovery. Process Biochem. 1998;33(1):1–6. https://doi.org/10.1016/s0032-9592(97)00069-1.
21. Teng WL, Khor E, Tan TK, Lim LY, Tan SC. Concurrent production of chitin from shrimp shells and fungi. Carbohydr Res. 2001;332(3):305–16. https://doi.org/10.1016/s0008-6215(01)00084-2.
22. Sedaghat F, Yousefzadi M, Toiserkani H, Najafipour S. Chitin from penaeus merguiensis via microbial fermentation processing and antioxidant activity. Int J Biol Macromol. 2016;82:279–83. https://doi.org/10.1016/j.ijbiomac.2015.10.070.
23. Duan S, Li L, Zhuang Z, Wu W, Hong S, Zhou J. Improved production of chitin from shrimp waste by fermentation with epiphytic lactic acid bacteria. Carbohydr Polym. 2012;89(4):1283–8. https://doi.org/10.1016/j.carbpol.2012.04.051.
24. Swatloski RP, Spear SK, Holbrey JD, Rogers RD. Dissolution of cellulose [correction of cellose] with ionic liquids. J Am Chem Soc. 2002;124(18):4974–5. https://doi.org/10.1021/ja025790m.
25. Siankevich S, Fei Z, Yan N, Dyson PJ. Application of ionic liquids in the downstream processing of lignocellulosic biomass. Chimia (Aarau). 2015;69(10):592–6. https://doi.org/10.2533/chimia.2015.592.
26. Xie, H., S. Zhang, and S. Li Chitin and chitosan dissolved in ionic liquids as reversible sorbents of CO2. Green Chem, 2006, 8(7): 630–633.doi:https://doi.org/10.1039/b517297g.
27. Yamazaki S, Takegawa A, Kaneko Y, Kadokawa J-i, Yamagata M, Ishikawa M. An acidic cellulose–chitin hybrid gel as novel electrolyte for an electric double layer capacitor. Electrochem Commun. 2009;11(1):68–70. https://doi.org/10.1016/j.elecom.2008.10.039.
28. Wu Y, Sasaki T, Irie S, Sakurai K. A novel biomass-ionic liquid platform for the utilization of native chitin. Polymer. 2008;49(9):2321–7. https://doi.org/10.1016/j.polymer.2008.03.027.
29. Qin Y, Lu X, Sun N, Rogers RD. Dissolution or extraction of crustacean shells using ionic liquids to obtain high molecular weight purified chitin and direct production of chitin films and fibers. Green Chem. 2010;12(6):968–71. https://doi.org/10.1039/c003583a.
30. Setoguchi T, Kato T, Yamamoto K, Kadokawa J. Facile production of chitin from crab shells using ionic liquid and citric acid. Int J Biol Macromol. 2012;50(3):861–4. https://doi.org/10.1016/j.ijbiomac.2011.11.007.
31. Sharma M, Mukesh C, Mondal D, Prasad K. Dissolution of α-chitin in deep eutectic solvents. RSC Adv. 2013;3(39):18149–55. https://doi.org/10.1039/c3ra43404d.
32. Xu P, Zheng G-W, Zong M-H, Li N, Lou W-Y. Recent progress on deep eutectic solvents in biocatalysis. Bioresour Bioprocess. 2017;4(1):34. https://doi.org/10.1186/s40643-017-0165-5.
33. Devi R, Dhamodharan R. Pretreatment in hot glycerol for facile and green separation of chitin from prawn shell waste. ACS Sustain Chem Eng. 2017;6(1):846–53. https://doi.org/10.1021/acssuschemeng.7b03195.
34. Zhou P, Li J, Yan T, Wang X, Huang J, Kuang Z, Ye M, Pan M. Selectivity of deproteinization and demineralization using natural deep eutectic solvents for production of insect chitin (Hermetia illucens). Carbohydr Polym. 2019;225:115255. https://doi.org/10.1016/j.carbpol.2019.115255.
35. Sun X, Wei Q, Yang Y, Xiao Z, Ren X. In-depth study on the extraction and mechanism of high-purity chitin based on NADESs method. J Environ Chem Eng. 2022;10(1):106859. https://doi.org/10.1016/j.jece.2021.106859.

36. Niola F, Basora N, Chornet E, Vidal PF. A rapid method for the determination of the degree of N-acetylation of chitin-chitosan samples by acid hydrolysis and HPLC. Carbohydr Res. 1993;238:1–9. https://doi.org/10.1016/0008-6215(93)87001-9.
37. Cabrera JC, Van Cutsem P. Preparation of chitooligosaccharides with degree of polymerization higher than 6 by acid or enzymatic degradation of chitosan. Biochem Eng J. 2005;25 (2):165–72. https://doi.org/10.1016/j.bej.2005.04.025.
38. Xing R, Liu S, Yu H, Guo Z, Wang P, Li C, Li Z, Li P. Salt-assisted acid hydrolysis of chitosan to oligomers under microwave irradiation. Carbohydr Res. 2005;340(13):2150–3. https://doi.org/10.1016/j.carres.2005.06.028.
39. Wu H, Li H, Fang Z. Hydrothermal amination of biomass to nitrogenous chemicals. Green Chem. 2021;23(18):6675–97. https://doi.org/10.1039/d1gc02505h.
40. De Chavez D, Kobayashi H, Fukuoka A, Hasegawa J-Y. On the electronic structure origin of mechanochemically induced selectivity in acid-catalyzed chitin hydrolysis. Chem Eur J. 2021;125(1):187–97. https://doi.org/10.1021/acs.jpca.0c09030.
41. Kaisler M, van den Broek LAM, Boeriu CG. Chitin and chitosan as sources of bio-based building blocks and chemicals. In: Chitin and chitosan: properties and applications; 2020. p. 229–44. https://doi.org/10.1002/9781119450467.
42. Nakamura H. Application of glucosamine on human disease—osteoarthritis. Carbohydr Polym. 2011;84(2):835–9. https://doi.org/10.1016/j.carbpol.2010.08.078.
43. Jung WJ, Park RD. Bioproduction of chitooligosaccharides: present and perspectives. Mar Drugs. 2014;12(11):5328–56. https://doi.org/10.3390/md12115328.
44. Song S, Zhou H, Li X, Wang L, Li Y, Wang J. Advance of glucosamine and their derivatives as chiral ligands in asymmetric syntheses. Chin J Org Chem. 2014;34(4):706–16. https://doi.org/10.6023/cjoc201310015.
45. Zhang B, Wang D-F, Li H-Y, Xu Y, Zhang L. Preparation and properties of chitosan–soybean trypsin inhibitor blend film with anti-Aspergillus flavus activity. Ind Crop Prod. 2009;29(2–3):541–8. https://doi.org/10.1016/j.indcrop.2008.10.007.
46. Glucosamine market analysis, coherent market insights 2020.
47. Rupley JA. The hydrolysis of chitin by concentrated hydrochloric acid, and the preparation of low-molecular-weight substrates for lysozyme. Biochim Biophys Acta. 1964;83(3):245–55. https://doi.org/10.1016/0926-6526(64)90001-1.
48. Lee M-Y, Var F, Shin-ya Y, Kajiuchi T, Yang J-W. Optimum conditions for the precipitation of chitosan oligomers with DP 5–7 in concentrated hydrochloric acid at low temperature. Process Biochem. 1999;34(5):493–500. https://doi.org/10.1016/s0032-9592(98)00116-2.
49. Xia W, Liu P, Liu J. Advance in chitosan hydrolysis by non-specific cellulases. Bioresour Technol. 2008;99(15):6751–62. https://doi.org/10.1016/j.biortech.2008.01.011.
50. Einbu A, Grasdalen H, Varum KM. Kinetics of hydrolysis of chitin/chitosan oligomers in concentrated hydrochloric acid. Carbohydr Res. 2007;342(8):1055–62. https://doi.org/10.1016/j.carres.2007.02.022.
51. Novikov VY. Acid hydrolysis of chitin and chitosan. Russ J Appl Chem. 2004;77(3):484–7. https://doi.org/10.1023/B:RJAC.0000031297.24742.b9.
52. Kazami N, Sakaguchi M, Mizutani D, Masuda T, Wakita S, Oyama F, Kawakita M, Sugahara Y. A simple procedure for preparing chitin oligomers through acetone precipitation after hydrolysis in concentrated hydrochloric acid. Carbohydr Polym. 2015;132:304–10. https://doi.org/10.1016/j.carbpol.2015.05.082.
53. Zechmeister L, Tóth G. Zur Kenntnis der Hydrolyse von Chitin mit Salzsäure (II. Mitteil.). Berichte der deutschen chemischen Gesellschaft (A and B Series). 2006;65(2):161–2. https://doi.org/10.1002/cber.19320650209.
54. Mojarrad JS, Nemati M, Valizadeh H, Ansarin M, Bourbour S. Preparation of glucosamine from exoskeleton of shrimp and predicting production yield by response surface methodology. J Agric Food Chem. 2007;55(6):2246–50. https://doi.org/10.1021/jf062983a.

55. Ajavakom A, Supsvetson S, Somboot A, Sukwattanasinitt M. Products from microwave and ultrasonic wave assisted acid hydrolysis of chitin. Carbohydr Polym. 2012;90(1):73–7. https://doi.org/10.1016/j.carbpol.2012.04.064.
56. Villa-Lerma G, Gonzalez-Marquez H, Gimeno M, Lopez-Luna A, Barzana E, Shirai K. Ultrasonication and steam-explosion as chitin pretreatments for chitin oligosaccharide production by chitinases of Lecanicillium lecanii. Bioresour Technol. 2013;146:794–8. https://doi.org/10.1016/j.biortech.2013.08.003.
57. Yang W, Wang H, Zhou J, Wu S. Hydrolysis kinetics and mechanism of chitin in subcritical water. J Supercrit Fluids. 2018;135:254–62. https://doi.org/10.1016/j.supflu.2018.01.029.
58. Zhang X, Mao Y, Briber RM. Efficient production of oligomeric chitin with narrow distributions of degree of polymerization using sonication-assisted phosphoric acid hydrolysis. Carbohydr Polym. 2022;276:118736. https://doi.org/10.1016/j.carbpol.2021.118736.
59. Guo M, Wei X, Chen S, Xiao J, Huang D. Enhancing nonspecific enzymatic hydrolysis of chitin to oligosaccharides pretreated by acid and green solvents under simultaneous microwave-radiation. Int J Biol Macromol. 2022;209(Pt A):631–41. https://doi.org/10.1016/j.ijbiomac.2022.04.032.
60. Abidin MZ, Junqueira-Gonçalves MP, Khutoryanskiy VV, Niranjan K. Intensifying chitin hydrolysis by adjunct treatments - an overview. J Chem Technol Biotechnol. 2017;92(11):2787–98. https://doi.org/10.1002/jctb.5208.
61. Hou F, Ma X, Fan L, Wang D, Ding T, Ye X, Liu D. Enhancement of chitin suspension hydrolysis by a combination of ultrasound and chitinase. Carbohydr Polym. 2020;231:115669. https://doi.org/10.1016/j.carbpol.2019.115669.
62. Duan B, Zheng X, Xia Z, Fan X, Guo L, Liu J, Wang Y, Ye Q, Zhang L. Highly biocompatible nanofibrous microspheres self-assembled from chitin in NaOH/urea aqueous solution as cell carriers. Angew Chem Int Ed Engl. 2015;54(17):5152–6. https://doi.org/10.1002/anie.201412129.
63. Tamura H, Nagahama H, Tokura S. Preparation of chitin hydrogel under mild conditions. Cellulose. 2006;13(4):357–64. https://doi.org/10.1007/s10570-006-9058-z.
64. Zhang A, Wei G, Mo X, Zhou N, Chen K, Ouyang P. Enzymatic hydrolysis of chitin pretreated by bacterial fermentation to obtain pure N-acetyl-d-glucosamine. Green Chem. 2018;20(10):2320–7. https://doi.org/10.1039/c8gc00265g.
65. Ifuku S, Saimoto H. Chitin nanofibers: preparations, modifications, and applications. Nanoscale. 2012;4(11):3308–18. https://doi.org/10.1039/c2nr30383c.
66. Zhou N, Yang P, Chen J, Wei G, Wang C, Zhang A, Chen K, Ouyang P. Effect of organic solvents treatment on structure of chitin and its enzymatic hydrolysis. Polym Degrad Stab. 2022;198:109654. https://doi.org/10.1016/j.polymdegradstab.2021.109654.
67. Wang Y, Zhang A, Mo X, Zhou N, Yang S, Chen K, Ouyang P. The effect of ultrasonication on enzymatic hydrolysis of chitin to N-acetyl glucosamine via sequential and simultaneous strategies. Process Biochem. 2020;99:265–9. https://doi.org/10.1016/j.procbio.2020.09.013.
68. Berton P, Shamshina JL, Ostadjoo S, King CA, Rogers RD. Enzymatic hydrolysis of ionic liquid-extracted chitin. Carbohydr Polym. 2018;199:228–35. https://doi.org/10.1016/j.carbpol.2018.07.014.
69. Zhai M, Du J, Zhang J, Miao J, Luo J, Cao Y, Duan S. Changes in the microstructure and enzymatic hydrolysis performance of chitin treated by steam explosion, high-pressure homogenization, and γ radiation. J Appl Polym Sci. 2020;137(48):49597. https://doi.org/10.1002/app.49597.
70. Dun Y, Li Y, Xu J, Hu Y, Zhang C, Liang Y, Zhao S. Simultaneous fermentation and hydrolysis to extract chitin from crayfish shell waste. Int J Biol Macromol. 2019;123:420–6. https://doi.org/10.1016/j.ijbiomac.2018.11.088.
71. Cardozo FA, Facchinatto WM, Colnago LA, Campana-Filho SP, Pessoa A. Bioproduction of N-acetyl-glucosamine from colloidal alpha-chitin using an enzyme cocktail produced by Aeromonas caviae CHZ306. World J Microbiol Biotechnol. 2019;35(8):114. https://doi.org/10.1007/s11274-019-2694-x.

72. Xu P, Wu X-L, Guo X-X, Tang J, Zong M-H, Lou W-Y. Double-chitinase hydrolysis of crab shell chitin pretreated by ionic liquid to generate chito-oligosaccharide. ACS Sustain Chem Eng. 2018;7(1):1683–91. https://doi.org/10.1021/acssuschemeng.8b05447.
73. Poshina DN, Raik SV, Poshin AN, Skorik YA. Accessibility of chitin and chitosan in enzymatic hydrolysis: a review. Polym Degrad Stab. 2018;156:269–78. https://doi.org/10.1016/j.polymdegradstab2018.09.005.
74. Zhang A, Wang C, Chen J, Wei G, Zhou N, Li G, Chen K, Ouyang P. Efficient enzymatic hydrolysis of chitin into N-acetyl glucosamine using alkali as a recyclable pretreatment reagent. Green Chem. 2021;23(8):3081–9. https://doi.org/10.1039/d1gc00818h.
75. Sashiwa H, Fujishima S, Yamano N, Kawasaki N, Nakayama A, Muraki E, Aiba S-I. Production of N-acetyl-D-glucosamine fromβ-chitin by enzymatic hydrolysis. Chem Lett. 2001;30(4):308–9. https://doi.org/10.1246/cl.2001.308.
76. Husson E, Hadad C, Huet G, Laclef S, Lesur D, Lambertyn V, Jamali A, Gottis S, Sarazin C, Van Nhien AN. The effect of room temperature ionic liquids on the selective biocatalytic hydrolysis of chitin via sequential or simultaneous strategies. Green Chem. 2017;19(17):4122–31. https://doi.org/10.1039/c7gc01471f.
77. Carrasquillo-Flores R, Käldström M, Schüth F, Dumesic JA, Rinaldi R. Mechanocatalytic depolymerization of dry (Ligno)cellulose as an entry process for high-yield production of furfurals. ACS Catal. 2013;3(5):993–7. https://doi.org/10.1021/cs4001333.
78. Hilgert J, Meine N, Rinaldi R, Schüth F. Mechanocatalytic depolymerization of cellulose combined with hydrogenolysis as a highly efficient pathway to sugar alcohols. Energy Environ Sci. 2013;6(1):92–6. https://doi.org/10.1039/c2ee23057g.
79. Meine N, Rinaldi R, Schuth F. Solvent-free catalytic depolymerization of cellulose to water-soluble oligosaccharides. ChemSusChem. 2012;5(8):1449–54. https://doi.org/10.1002/cssc.201100770.
80. Ribeiro SL, Órfão JJM, Pereira MFR. Enhanced direct production of sorbitol by cellulose ball-milling. Green Chem. 2015;17(5):2973–80. https://doi.org/10.1039/c5gc00039d.
81. Tabasso S, Carnaroglio D, Calcio Gaudino E, Cravotto G. Microwave, ultrasound and ball mill procedures for bio-waste valorisation. Green Chem. 2015;17(2):684–93. https://doi.org/10.1039/c4gc01545b.
82. Kobayashi H, Suzuki Y, Sagawa T, Kuroki K, Hasegawa JY, Fukuoka A. Impact of tensile and compressive forces on the hydrolysis of cellulose and chitin. Phys Chem Chem Phys. 2021;23(30):15908–16. https://doi.org/10.1039/d1cp01650d.
83. Yabushita M, Kobayashi H, Kuroki K, Ito S, Fukuoka A. Catalytic depolymerization of chitin with retention of N-acetyl group. ChemSusChem. 2015;8(22):3760–3. https://doi.org/10.1002/cssc.201501224.
84. Margoutidis G, Parsons VH, Bottaro CS, Yan N, Kerton FM. Mechanochemical amorphization of α-chitin and conversion into oligomers of N-acetyl-d-glucosamine. ACS Sustain Chem Eng. 2018;6(2):1662–9. https://doi.org/10.1021/acssuschemeng.7b02870.
85. Chen X, Yang H, Zhong Z, Yan N. Base-catalysed one-step mechanochemical conversion of chitin and shrimp shells into low molecular weight chitosan. Green Chem. 2017;19(12):2783–92. https://doi.org/10.1039/C7GC00089H.
86. Khan FI, Rahman S, Queen A, Ahamad S, Ali S, Kim J, Hassan MI. Implications of molecular diversity of chitin and its derivatives. Appl Microbiol Biotechnol. 2017;101(9):3513–36. https://doi.org/10.1007/s00253-017-8229-1.
87. Pierson Y, Chen X, Bobbink FD, Zhang J, Yan N. Acid-catalyzed chitin liquefaction in ethylene glycol. ACS Sustain Chem Eng. 2014;2(8):2081–9. https://doi.org/10.1021/sc500334b.
88. Zhang J, Yan N. Formic acid-mediated liquefaction of chitin. Green Chem. 2016;18(18):5050–8. https://doi.org/10.1039/c6gc01053a.
89. Zhang Z, Li C, Wang Q, Zhao ZK. Efficient hydrolysis of chitosan in ionic liquids. Carbohydr Polym. 2009;78(4):685–9. https://doi.org/10.1016/j.carbpol.2009.06.002.

90. Luterbacher JS, Rand JM, Alonso DM, Han J, Youngquist JT, Maravelias CT, Pfleger BF, Dumesic JA. Nonenzymatic sugar production from biomass using biomass-derived gamma-valerolactone. Science. 2014;343(6168):277–80. https://doi.org/10.1126/science.1246748.
91. Chen Q, Xiao W, Zhou L, Wu T, Wu Y. Hydrolysis of chitosan under microwave irradiation in ionic liquids promoted by sulfonic acid-functionalized ionic liquids. Polym Degrad Stab. 2012;97(1):49–53. https://doi.org/10.1016/j.polymdegradstab.2011.10.014.
92. Zhang J, Yan N. Production of glucosamine from chitin by co-solvent promoted hydrolysis and deacetylation. ChemCatChem. 2017;9(14):2790–6. https://doi.org/10.1002/cctc.201601715.
93. Gözaydın G, Song S, Yan N. Chitin hydrolysis in acidified molten salt hydrates. Green Chem. 2020;22(15):5096–104. https://doi.org/10.1039/D0GC01464H.
94. Deng W, Wang Y, Yan N. Production of organic acids from biomass resources. Curr Opin Green Sustain Chem. 2016;2:54–8. https://doi.org/10.1016/j.cogsc.2016.10.002.
95. He J, Chen L, Liu S, Song K, Yang S, Riisager A. Sustainable access to renewable N-containing chemicals from reductive amination of biomass-derived platform compounds. Green Chem. 2020;22(20):6714–47. https://doi.org/10.1039/d0gc01869d.
96. Jackson RFW. Recent developments in the application of organometallic chemistry to amino acid synthesis. In: Asymmetric synthesis and application of α-amino acids. American Chemical Society; 2009. p. 2–12.
97. Ager DJ, Lefort L, de Vries JG. ChemInform abstract: catalyst screening for the synthesis of amino acids. ChemInform. 2010;41(1) https://doi.org/10.1002/chin.201001236.
98. Yan N, Wang Y. Catalyst: is the amino acid a new frontier for biorefineries? Chem. 2019;5(4):739–41. https://doi.org/10.1016/j.chempr.2019.03.016.
99. Chen X, Yang H, Yan N. Shell biorefinery: dream or reality? Chemistry. 2016;22(38):13402–21. https://doi.org/10.1002/chem.201602389.
100. Chen X, Yan N. Conversion of chitin to nitrogen-containing chemicals. In: Chemical catalysts for biomass upgrading; 2020. p. 569–90. https://doi.org/10.1002/9783527814794.ch14.
101. Zhang H, Lu Y, Wang Y, Zhang X, Wang T. D-glucosamine production from chitosan hydrolyzation over a glucose-derived solid acid catalyst. RSC Adv. 2018;8(10):5608–13. https://doi.org/10.1039/c7ra12490b.
102. Jimenez-Gomez CP, Cecilia JA. Chitosan: a natural biopolymer with a wide and varied range of applications. Molecules. 2020;25(17):3981. https://doi.org/10.3390/molecules25173981.
103. Philippe M, Ebenhan-Nappe C. Use of carbohydrates for promoting skin desquamation. OREAL; 1998. (US19980043914)
104. Park KH. A stereospecific synthesis of (+)-2-epideoxymannojirimycin and (2r,3r,4r,5r)-3,4,5-trihydroxypipecolic acid. Bull Korean Chem Soc. 1995;16(10):985–8.
105. Varela O, Nin AP, de Lederkremer RM. A short synthesis of (2S,4S,5R)-4,5,6-trihydroxynorleucine. Tetrahedron Lett. 1994;35(50):9359–62. https://doi.org/10.1016/s0040-4039(00)78542-x.
106. Kelebekli L, Çelik M, Şahin E, Kara Y, Balci M. Stereospecific synthesis of a new class of aminocyclitol with the conduramine D-2 configuration. Tetrahedron Lett. 2006;47(39):7031–5. https://doi.org/10.1016/j.tetlet.2006.07.108.
107. Ou SJ, Chen G, Lin ZH, Bai ZP, Duan CY, Mao CP. Chromium(III) complexes of D-glucosaminic acid and their effect on decreasing blood sugar in vivo. Arch Pharm (Weinheim). 2006;339(9):527–30. https://doi.org/10.1002/ardp.200600053.
108. Zhang J, Xia W, Liu P, Cheng Q, Tahirou T, Gu W, Li B. Chitosan modification and pharmaceutical/biomedical applications. Mar Drugs. 2010;8(7):1962–87. https://doi.org/10.3390/md8071962.
109. Miyazaki M, Senshu T, Tamura Z. Metal complexes of D-glucosamine and its derivatives. VII. Metal complexes of D-glucosaminic acid. Chem Pharm Bull(Tokyo). 1966;14(2):114–7. https://doi.org/10.1248/cpb.14.114.

110. Ferrari E, Grandi R, Lazzari S, Saladini M. Hg(II)-coordination by sugar-acids: role of the hydroxy groups. J Inorg Biochem. 2005;99(12):2381–6. https://doi.org/10.1016/j.jinorgbio.2005.09.005.
111. Junicke H, Kluge R, Steinborn D. Synthesis and characterization of novel platinum(IV) complexes with non-functionalized acetylated carbohydrates. J Inorg Biochem. 2000;81(1–2):43–8. https://doi.org/10.1016/s0162-0134(00)00088-x.
112. Ferrari E, Saladini M. Iron(III) complexing ability of carbohydrate derivatives. J Inorg Biochem. 2004;98(6):1002–8. https://doi.org/10.1016/j.jinorgbio.2004.02.017.
113. Abtew E, Domb AJ, Basu A. Glycopeptides derived from glucosaminic acid. Polym Chem. 2016;7(27):4447–52. https://doi.org/10.1039/c6py00858e.
114. Abtew E, Domb AJ, Basu A. Synthesis of glycopeptides from glucosaminic acid. J Polym Sci A Polym Chem. 2017;55(16):2657–62. https://doi.org/10.1002/pola.28666.
115. Zheng Y, Xu D, Zhang L, Chen X. Base-free air oxidation of chitin-derived glucosamine to glucosaminic acid by zinc oxide-supported gold nanoparticles. Chem Asian J. 2022;17(18):e202200556. https://doi.org/10.1002/asia.202200556.
116. Pringsheim H, Ruschmann G. Zur Darstellung der Glucosaminsäure. Ber Dtsch Chem Ges. 2006;48(1):680–2. https://doi.org/10.1002/cber.19150480192.
117. Wolfrom ML, Cron MJ. Acyl derivatives of D-glucosaminic acid. J Am Chem Soc. 2002;74(7):1715–6. https://doi.org/10.1021/ja01127a030.
118. Tominaga M, Nagashima M, Taniguchi I. Controlled-potential electrosynthesis of glucosaminic acid from glucosamine at a gold electrode. Electrochem Commun. 2007;9(5):911–4. https://doi.org/10.1016/j.elecom.2006.11.024.
119. Wen-xiu G, Wen-shui X. Catalytic synthesis of D-glucosaminic acid from D-glucosamine on active charcoal-supported Pd-bi catalysts. J Carbohydr Chem. 2006;25(4):297–301. https://doi.org/10.1080/07328300600723757.
120. Ohmi Y, Nishimura S, Ebitani K. Synthesis of alpha-amino acids from glucosamine-HCl and its derivatives by aerobic oxidation in water catalyzed by Au nanoparticles on basic supports. ChemSusChem. 2013;6(12):2259–62. https://doi.org/10.1002/cssc.201300303.
121. Dai J, Gözaydın G, Hu C, Yan N. Catalytic conversion of chitosan to glucosaminic acid by tandem hydrolysis and oxidation. ACS Sustain Chem Eng. 2019;7(14):12399–407. https://doi.org/10.1021/acssuschemeng.9b01912.
122. Techikawara K, Kobayashi H, Fukuoka A. Conversion of N-acetylglucosamine to protected amino acid over Ru/C catalyst. ACS Sustain Chem Eng. 2018;6(9):12411–8. https://doi.org/10.1021/acssuschemeng.8b02951.
123. Ma X, Gozaydin G, Yang H, Ning W, Han X, Poon NY, Liang H, Yan N, Zhou K. Upcycling chitin-containing waste into organonitrogen chemicals via an integrated process. Proc Natl Acad Sci U S A. 2020;117(14):7719–28. https://doi.org/10.1073/pnas.1919862117.
124. Liu B, Zhang Z. One-pot conversion of carbohydrates into furan derivatives via furfural and 5-hydroxylmethylfurfural as intermediates. ChemSusChem. 2016;9(16):2015–36. https://doi.org/10.1002/cssc.201600507.
125. Liu Y, Stähler C, Murphy JN, Furlong BJ, Kerton FM. Formation of a renewable amine and an alcohol via transformations of 3-Acetamido-5-acetylfuran. ACS Sustain Chem Eng. 2017;5(6):4916–22. https://doi.org/10.1021/acssuschemeng.7b00323.
126. Pham TT, Chen X, Yan N, Sperry J. A novel dihydrodifuropyridine scaffold derived from ketones and the chitin-derived heterocycle 3-acetamido-5-acetylfuran. Monatshefte für Chemie-Chemical Monthly. 2017;149(4):857–61. https://doi.org/10.1007/s00706-017-2112-8.
127. Pham TT, Lindsay AC, Kim SW, Persello L, Chen X, Yan N, Sperry J. Two-step preparation of diverse 3-Amidofurans from chitin. ChemistrySelect. 2019;4(34):10097–9. https://doi.org/10.1002/slct.201902765.
128. Pham TT, Gözaydın G, Söhnel T, Yan N, Sperry J. Oxidative ring-expansion of a chitin-derived platform enables access to unexplored 2-amino sugar chemical space. Eur J Org Chem. 2019;2019(6):1355–60. https://doi.org/10.1002/ejoc.201801683.

129. Pham, T.T., X. Chen, T. Söhnel, N. Yan, and J. Sperry, Haber-independent, diversity-oriented synthesis of nitrogen compounds from biorenewable chitin. Green Chem, 2020, 22(6): 1978–1984.doi:https://doi.org/10.1039/D0GC00208A.
130. Wolter FE, Schneider K, Davies BP, Socher ER, Nicholson G, Seitz O, Sussmuth RD. Total synthesis of proximicin A-C and synthesis of new furan-based DNA binding agents. Org Lett. 2009;11(13):2804–7. https://doi.org/10.1021/ol901003p.
131. Brucoli F, Natoli A, Marimuthu P, Borrello MT, Stapleton P, Gibbons S, Schatzlein A. Efficient synthesis and biological evaluation of proximicins A, B and C. Bioorg Med Chem. 2012;20(6):2019–24. https://doi.org/10.1016/j.bmc.2012.01.043.
132. Sadiq AD, Chen X, Yan N, Sperry J. Towards the shell biorefinery: sustainable synthesis of the anticancer alkaloid proximicin A from chitin. ChemSusChem. 2018;11(3):532–5. https://doi.org/10.1002/cssc.201702356.
133. Kikuchi H, Saito Y, Komiya J, Takaya Y, Honma S, Nakahata N, Ito A, Oshima Y. Furanodictine A and B: amino sugar analogues produced by cellular slime mold dictyostelium discoideum showing neuronal differentiation activity. J Org Chem. 2001;66(21):6982–7. https://doi.org/10.1021/jo015657x.
134. Noren-Muller A, Reis-Correa I Jr, Prinz H, Rosenbaum C, Saxena K, Schwalbe HJ, Vestweber D, Cagna G, Schunk S, Schwarz O, Schiewe H, Waldmann H. Discovery of protein phosphatase inhibitor classes by biology-oriented synthesis. Proc Natl Acad Sci U S A. 2006;103(28):10606–11. https://doi.org/10.1073/pnas.0601490103.
135. Chen X, Liu Y, Kerton FM, Yan N. Conversion of chitin and N-acetyl-d-glucosamine into a N-containing furan derivative in ionic liquids. RSC. Advances. 2015;5(26):20073–80. https://doi.org/10.1039/C5RA00382B.
136. Ogata M, Hattori T, Takeuchi R, Usui T. Novel and facile synthesis of furanodictines A and B based on transformation of 2-acetamido-2-deoxy-D-glucose into 3,6-anhydro hexofuranoses. Carbohydr Res. 2010;345(2):230–4. https://doi.org/10.1016/j.carres.2009.10.007.
137. Osada M, Kikuta K, Yoshida K, Totani K, Ogata M, Usui T. Non-catalytic synthesis of chromogen I and III from N-acetyl-d-glucosamine in high-temperature water. Green Chem. 2013;15(10):2960–6. https://doi.org/10.1039/c3gc41161c.
138. Osada M, Shoji S, Suenaga S, Ogata M. Conversion of N-acetyl-d-glucosamine to nitrogen-containing chemicals in high-temperature water. Fuel Process Technol. 2019;195:106154. https://doi.org/10.1016/j.fuproc.2019.106154.
139. Fu X, Dai J, Guo X, Tang J, Zhu L, Hu C. Suppression of oligomer formation in glucose dehydration by CO2 and tetrahydrofuran. Green Chem. 2017;19(14):3334–43. https://doi.org/10.1039/c7gc01115f.
140. Franich RA, Goodin SJ, Wilkins AL. Acetamidofurans, acetamidopyrones, and acetamidoacetaldehyde from pyrolysis of chitin and n-acetylglucosamine. J Anal Appl Pyrolysis. 1984;7(1–2):91–100. https://doi.org/10.1016/0165-2370(84)80043-1.
141. Chen J, Wang M, Ho C-T. Volatile compounds generated from thermal degradation of N-acetylglucosamine. J Agric Food Chem. 1998;46(8):3207–9.
142. Drover MW, Omari KW, Murphy JN, Kerton FM. Formation of a renewable amide, 3-acetamido-5-acetylfuran, via direct conversion of N-acetyl-d-glucosamine. RSC. Advances. 2012;2(11):4642–4. https://doi.org/10.1039/c2ra20578e.
143. Chen X, Gao Y, Wang L, Chen H, Yan N. Effect of treatment methods on chitin structure and its transformation into nitrogen-containing chemicals. ChemPlusChem. 2015;80(10):1565–72. https://doi.org/10.1002/cplu.201500326.
144. Wang J, Zang H, Jiao S, Wang K, Shang Z, Li H, Lou J. Efficient conversion of N-acetyl-D-glucosamine into nitrogen-containing compound 3-acetamido-5-acetylfuran using amino acid ionic liquid as the recyclable catalyst. Sci Total Environ. 2020;710:136293. https://doi.org/10.1016/j.scitotenv.2019.136293.
145. Padovan D, Kobayashi H, Fukuoka A. Facile preparation of 3-Acetamido-5-acetylfuran from N-acetyl-d-glucosamine by using commercially available aluminum salts. ChemSusChem. 2020;13(14):3594–8. https://doi.org/10.1002/cssc.202001068.

146. Zang H, Lou J, Jiao S, Li H, Du Y, Wang J. Valorization of chitin derived N-acetyl-D-glucosamine into high valuable N-containing 3-acetamido-5-acetylfuran using pyridinium-based ionic liquids. J Mol Liq. 2021;330:115667. https://doi.org/10.1016/j.molliq.2021.115667.
147. Wu C, Wang C, Zhang A, Chen K, Cao F, Ouyang P. Preparation of 3-aceta mido-5-acetylfuran from N-acetylglucosamine and chitin using biobased deep eutectic solvents as catalysts. React Chem Eng. 2022;7:1742–9. https://doi.org/10.1039/d2re00118g.
148. Chen K, Wu C, Wang C, Zhang A, Cao F, Ouyang P. Chemo-enzymatic protocol converts chitin into a nitrogen-containing furan derivative, 3-acetamido-5-acetylfuran. Mol Catal. 2021;516:112001. https://doi.org/10.1016/j.mcat.2021.112001.
149. Chen X, Chew SL, Kerton FM, Yan N. Direct conversion of chitin into a N-containing furan derivative. Green Chem. 2014;16(4):2204–12. https://doi.org/10.1039/c3gc42436g.
150. Omari KW, Besaw JE, Kerton FM. Hydrolysis of chitosan to yield levulinic acid and 5-hydroxymethylfurfural in water under microwave irradiation. Green Chem. 2012;14(5):1480–7. https://doi.org/10.1039/c2gc35048c.
151. Zhu X, Lobban LL, Mallinson RG, Resasco DE. Bifunctional transalkylation and hydrodeoxygenation of anisole over a Pt/HBeta catalyst. J Catal. 2011;281(1):21–9.
152. Osada M, Kobayashi H, Miyazawa T, Suenaga S, Ogata M. Non-catalytic conversion of chitin into Chromogen I in high-temperature water. Int J Biol Macromol. 2019;136:994–9. https://doi.org/10.1016/j.ijbiomac.2019.06.123.
153. Zang H, Li H, Jiao S, Lou J, Du Y, Huang N. Green conversion of N-Acetylglucosamine into valuable platform compound 3-acetamido-5-acetylfuran using ethanolamine ionic liquids as recyclable catalyst. ChemistrySelect. 2021;6(16):3848–57. https://doi.org/10.1002/slct.202100231.
154. Gao X, Chen X, Zhang J, Guo W, Jin F, Yan N. Transformation of chitin and waste shrimp shells into acetic acid and pyrrole. ACS Sustain Chem Eng. 2016;4(7):3912–20. https://doi.org/10.1021/acssuschemeng.6b00767.
155. Liang B, Deng Y, Liang X, Cai X, Zhang Y, Yin Y, Hu H, Huang Z, Qin Y. Enhancement of hydrothermal carbonization of chitin by combined pretreatment of mechanical activation and FeCl3. Int J Biol Macromol. 2021;189:242–50. https://doi.org/10.1016/j.ijbiomac.2021.08.125.
156. Qi M, Chen X, Zhong H, Wu J, Jin F. Base-free, vanadium-catalyzed conversion of chitin into acetic acid under low oxygen pressure. ACS Sustain Chem Eng. 2020;8(50):18661–70. https://doi.org/10.1021/acssuschemeng.0c07147.
157. Liang Q, Su H, Yan J, Leung C, Cao S, Yuan D. N-doped mesoporous carbon as a bifunctional material for oxygen reduction reaction and supercapacitors. Chin J Catal. 2014;35(7):1078–83. https://doi.org/10.1016/s1872-2067(14)60044-9.
158. Sagawa T, Kobayashi H, Murata C, Shichibu Y, Konishi K, Fukuoka A. Catalytic conversion of a chitin-derived sugar alcohol to an amide-containing isosorbide analog. ACS Sustain Chem Eng. 2019;7(17):14883–8. https://doi.org/10.1021/acssuschemeng.9b02985.
159. Sagawa T, Kobayashi H, Fukuoka A. Effect of Lewis acid on catalytic dehydration of a chitin-derived sugar alcohol. Mol Catal. 2020;498:111282. https://doi.org/10.1016/j.mcat.2020.111282.
160. Yang C, Sagawa T, Fukuoka A, Kobayashi H. Characteristic activity of phosphorous acid in the dehydration condensation of a chitin-derived nitrogen-containing sugar alcohol. Green Chem. 2021;23(18):7228–34. https://doi.org/10.1039/d1gc02193a.
161. Kobayashi H, Techikawara K, Fukuoka A. Hydrolytic hydrogenation of chitin to amino sugar alcohol. Green Chem. 2017;19(14):3350–6. https://doi.org/10.1039/c7gc01063j.
162. Osada M, Kikuta K, Yoshida K, Totani K, Ogata M, Usui T. Non-catalytic dehydration of N,N′-diacetylchitobiose in high-temperature water. RSC Adv. 2014;4(64):33651–7. https://doi.org/10.1039/c4ra06319h.
163. Fujii S, Kikuchi R, Kushida H. Formation of fructosazine. J Org Chem. 2002;31(7):2239–41. https://doi.org/10.1021/jo01345a036.

164. Rohovec J, Kotek J, Peters JA, Maschmeyer T. A clean conversion of D-glucosamine hydrochloride to a pyrazine in the presence of phenylboronate or borate. Eur J Org Chem. 2001;2001(20):3899–901. https://doi.org/10.1002/1099-0690(200110)2001:20<3899::Aid-ejoc3899>3.0.Co;2-g.
165. Jia L, Wang Y, Qiao Y, Qi Y, Hou X. Efficient one-pot synthesis of deoxyfructosazine and fructosazine fromd-glucosamine hydrochloride using a basic ionic liquid as a dual solvent-catalyst. RSC Adv. 2014;4(83):44253–60. https://doi.org/10.1039/c4ra06832g.
166. Jia L, Pedersen CM, Qiao Y, Deng T, Zuo P, Ge W, Qin Z, Hou X, Wang Y. Glucosamine condensation catalyzed by 1-ethyl-3-methylimidazolium acetate: mechanistic insight from NMR spectroscopy. Phys Chem Chem Phys. 2015;17(35):23173–82. https://doi.org/10.1039/c5cp02169c.
167. Jia L, Zhang Z, Qiao Y, Pedersen CM, Ge H, Wei Z, Deng T, Ren J, Liu X, Wang Y, Hou X. Product distribution control for glucosamine condensation: nuclear magnetic resonance (NMR) investigation substantiated by density functional calculations. Ind Eng Chem Res. 2017;56(11):2925–34. https://doi.org/10.1021/acs.iecr.6b05057.

Chapter 5
Sustainable Production of Nitriles from Biomass

Lujiang Xu, Geliang Xie, Guoqiang Zhu, Wei Chen, Chengyu Dong, Richard L. Smith Jr, and Zhen Fang

Abstract Biomass is a low cost, environmentally friendly, carbon-neutral, renewable resource that is widely distributed over the earth in diverse forms. Nitriles, which are an important class of N-containing compounds, are used as feedstocks for producing ago-chemicals, pharmaceuticals, dyes, and polymers. Transformation of biomass and biomass-derived compounds into nitriles is of prime interest and is the subject of this chapter that focuses on biomass valorization with the purpose of illustrating sustainable synthesis methods for production of bio-based nitriles. The scope of substrates used in the preparation of bio-based nitriles covers nitrogenous components, non-nitrogenous components, and raw biomass. Reaction mechanisms and typical reaction pathways are summarized and outlined. Finally, critical challenges and perspectives are provided to stimulate discussion between researchers and industry who are striving to bring production methods for bio-based nitriles to realization.

Keywords Biomass · Amination · Nitriles · Chemical conversion

5.1 Introduction

Biomass, which refers to matter derived from living organisms directly or indirectly through photosynthesis, is a widely distributed, readily available, environmental and carbon neutral resource [1–5]. Currently, biomass is the only known renewable organic carbon resource that can be sustainably converted into liquid fuels and commodity chemicals [6]. In recent years, research on the production of bio-based chemicals (furans [7], organic acids [8], N-containing compounds [8, 9], aromatics

L. Xu · G. Xie · G. Zhu · W. Chen · C. Dong · Z. Fang (✉)
Biomass Group, College of Engineering, Nanjing Agricultural University, Nanjing, China
e-mail: zhenfang@njau.edu.cn

R. L. Smith Jr
Graduate School of Environmental Studies, Tohoku University, Sendai, Japan

Z. Fang et al. (eds.), *Production of N-containing Chemicals and Materials from Biomass*, Biofuels and Biorefineries 12,
https://doi.org/10.1007/978-981-99-4580-1_5

[10]) from renewable resources has been pursued for developing continuous biomass biorefinery technologies.

Nitriles, which are a class of N-containing compounds with the cyano (–C ≡N) functional group, are important chemical products and widely used as substrates for producing resins, fibers, elastomers, pharmaceuticals, pesticides and dyestuffs and their related synthesis intermediates [11–13]. Sustainable transformation of biomass and biomass-derived compounds into nitriles (e.g. alkyl nitriles, aromatic nitriles, furonitriles) is the focus of this chapter. Valorization methods for transforming sustainable and renewable nitrogen sources and oxygenates into nitriles are summarized and discussed.

5.2 Bio-Based Nitriles Production from Renewable Nitrogen Sources

Nitrogenous biomass that has high nitrogen content, such as chitins, microalgal biomass, sewage sludge, mycelial waste, waste leather, and animal manure waste are especially useful in production schemes [14, 15]. Nitrogenous compounds can be a constituent of lignocellulosic biomass with content in some lignocellulosic biomass being about 2% [16] that is typically composed of proteins and amino acids. Use of nitrogenous biomass fractions for synthesis of N-containing chemicals not only avoids external nitrogen sources, but also simplifies chemical processes and improves process economics and energetics. Herein, advances in production of bio-based nitriles from renewable nitrogen sources are summarized.

Glutamic acid, which is one of the basic amino acids, is widely used in food, chemical and pharmaceutical industries and is one of the top 12 bio-based chemicals that can be produced from carbohydrates via fermentation processes [17]. Dai et al. [18] demonstrated bio-based synthesis of adiponitrile from glutamic acid with electrochemical methods under mild conditions. The reaction pathway from glutamic acid to adiponitrile consists of the following steps (Scheme 5.1a): (1) conversion of glutamic acid into glutamic acid 5-methyl ester (Glu-Me), (2) electro-oxidative decarboxylation to form methyl 3-cyanopropionate (CPA-Me) and (3) conversion of CPA-Me into adiponitrile via electrochemical Kolbe coupling reactions (Scheme 5.1). Le Notre et al. [19] reported that glutamic acid could be converted into acrylonitrile with yields of about 12% via a two-step oxidation decarboxylation organic synthesis process involving oxidative decarboxylation to form 3-cyanopropanoic acid followed by a decarboxylation - elimination reaction over $PdCl_2$ catalyst (Scheme 5.1). On this basis, Lammens et al. [20] developed a route to selectively produce succinonitrile from glutamic acid via a multiple-step reaction. The glutamic acid first undergoes esterification or amination reactions to form glutamic ester or glutamic amide intermediates, and then the intermediates undergo oxidative decarboxylation coupled with dehydration reactions to form succinonitrile. As reported by But et al. [21], they also form

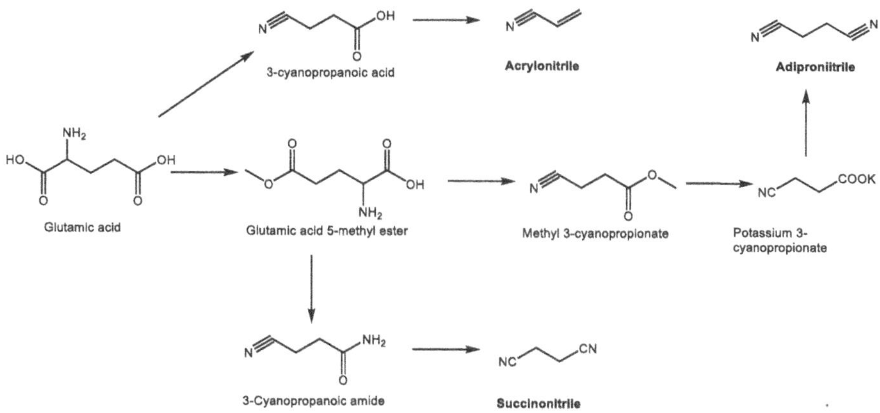

Scheme 5.1 Reaction pathways for production of nitriles from glutamic acid

acrylonitrile and succinonitrile (Scheme 5.1) through selective oxidative decarboxylation by using a vanadium chloroperoxidase enzyme (VCPO) and hydrogen peroxide. Claes et al. [22] reported that propionitrile, and phenylacetonitrile could be produced from alanine and phenylalanine via oxidative decarboxylation reactions over [Ni, Al]-LDH-WO_4 catalyst. Lammens et al. [23] suggested that bio-based succinonitrile could meet the level that allowed it to be produced competitively against petrochemical processes, because glutamic acid-based acrylonitrile processes are far superior to petrochemical process routes according to life cycle assessment.

5.3 Bio-Based Nitriles Production from Renewable Oxygenates

5.3.1 Acrylonitrile

Acrylonitrile (ACN), which is an important monomer in the polymer industry, has served as a traditional raw material for synthesis of polymer fibers, rubbers and resins [24, 25]. The annual production of ACN was over 6 million tons valued at US $13.8 billion in 2020 [25]. Polyacrylonitrile, which is a downstream product of ACN, has properties similar to actual wool and is widely used in the textile industry and moreover, it can also serve as a raw material for producing carbon nano-fibers [26–28]. Acrylonitrile butadiene styrene plastic (ABS), which is another downstream product of ACN, is an important engineering plastic with advantages of being oil-, cold-, and abrasion- resistant while having electrical insulation properties [29, 30]. Nitrile rubber or nitrile butadiene rubber (NBR), which is synthesized from butadiene, is widely used to produce parts in aviation, automotive, printing, textile and machinery manufacturing industries [31, 32]. Industrially, ACN is synthesized

Table 5.1 Reaction conditions for producing acrylonitrile (ACN) from bio-based substrates

Substrate	Reaction type	Catalyst & conditions	Yield (%)	Ref.
Glycerol	Ammoxidation	VSbNb/Al, T:400 °C, 25% O_2 and 8.6% NH_3	~48	[34]
Acrolein	Ammoxidation	$Bi_2O_3.3MoO_3$/KIT-6; T: 450 °C	82.5	[35]
Glycerol	Two-step ammoxidation	WO_3/TiO_2; T1: 280 °C, Sb-Fe-O; T2: 400 °C	36	[36]
Allyl alcohol	Ammoxidation	Sb-Fe-O T:450 °C	84	[37]
3-Hydroxypropionic acid	Amination dehydration	TiO_2; T:315 °C	>90	[25]
Ethyl acrylate	Amination dehydration	TiO_2 T:315 °C	98	[25]
Lactic acid	Amination, dehydration	NH4-ZSM-5 T1 = 230 °C acetic anhydride, T2 = 140 °C	18.8	[38]
Glutamic acid	Two-step oxidation decarboxylation	T1 = 0 °C; T2 = 110 °C	~12	[19]

from petroleum-based propylene through vapor phase ammoxidation reaction using ammonia and oxygen [33]. In recent years, many studies on the synthesis of ACN from bio-based alcohols and aldehydes have been reported. Herein, the progress of bio-based ACN production is summarized.

Substrates used for the sustainable synthesis of ACN are bio-based small molecule alcohols and aldehydes, including glycerol, acrolein, allyl alcohol, lactic acid, and 3-hydroxypropionic acid as shown in Table 5.1. Glycerol, which is a by-product generated from biodiesel production, was proposed as a feedstock for synthesizing ACN by Guerrero- Pérez et al. in 2008 [34]. Glycerol can undergo two-step ammoxidation to form ACN in which glycerol is firstly dehydrated to form acrolein and is a key intermediate for ACN production. Subsequently, acrolein reacts with ammonia to form ACN via ammoxidation. During the ammoxidation reaction, a suitable catalyst is essential for which V-Sb-Nb/Al catalyst is effective [34]. Thanh-Binh et al. [35] used $Bi_2O_3.3MoO_3$/KIT-6 to catalyze acrolein to synthesize ACN in 80% yields with 100% selectivity which demonstrates that acrolein is an important intermediate for the glycerol to ACN pathway. Liebig et al. [36] further validated the aforementioned reaction pathways by conducting tandem reactions of dehydration over WO_3/TiO_2 catalyst and an ammoxidation step over Sb-(Fe, V)-O catalyst. Furthermore, ACN in high yields (84%) and selectivities (100%) can be produced from allyl alcohol via catalytic ammoxidation over Sb-Fe-O catalysts [37].

In addition to polyols and acrolein, other bio-based compounds such as 3-hydroxypropionic acid (3-HP), ethyl acrylate, and lactic acid are good feedstocks for producing ACN via amination reactions. Karp et al. [25] presented a process for renewable ACN production from 3-HP and ethyl acrylate, which was produced from sugars via fermentation, through dehydration and nitrilation reactions with ammonia over TiO_2. The yields of ACN from 3-HP and ethyl acrylate were up to 92% and > 98%, respectively. Moreover, lactic acid could also be converted into ACN through a two-step reaction by direct amination and dehydration [38]. However, the yield of ACN from lactic acid was only about 18%, which was much lower than that from glycerol, allyl alcohol or 3-HP and might be caused by low conversion rates of lactic acid in the direct amination reaction with ammonia over zeolite catalyst.

Guerrero-Pérez et al. [33] compared manufacturing processes between petroleum-based ACN and glycerol-based ACN in terms of metrics such as materials efficiency, E factor, energy efficiency, total energy input, land use, and cost and concluded that the synthesis of ACN from glycerol via ammoxidation is a favorable and green alternative way to realize glycerol valorization. Therefore, these approaches for producing ACN from bio-derived oxygenated chemicals eliminate the risk of uncontrolled reactions and avoid the formation of hydrogen cyanide (a highly dangerous by-product) and achieve higher yields than presently used propylene ammoxidation processes. Thus, it can be concluded that industrial production of ACN from renewable substrates has great potential. Besides acrylonitrile and acetonitrile, other nitriles can be produced from renewable sources. Zhang et al. [39] showed that propionitrile (PN) could be produced from allyl alcohol and glycerol over a bimetallic catalyst, $Zn_{30}Ru_{1.0}/\gamma\text{-}Al_2O_3$, and that the yields of PN from allyl alcohol and glycerol were above 70% under optimized conditions.

5.3.2 Acetonitrile

Acetonitrile, which is used as a solvent or a key substrate in pharmaceutical, pesticide and semiconductor industries, is produced as a by-product of ACN in propylene ammoxidation or through amination of acetic acid [40, 41]. Acetic acid is a bio-based organic compound that can be produced from carbohydrates via biological fermentation or hydrothermal oxidation [42, 43]. Acetic acid can undergo a two-step reaction to form acetonitrile in which reaction with ammonia forms acetamide and subsequent dehydration forms the product acetonitrile in quantitative yields over acid catalyst (e.g., Al_2O_3, Al_2O_3 modified with H_3PO_4) [44]. Great efforts have been made on developing reaction pathways to bio-based acetonitrile from renewable substrates, such as ethanol [45], glycerol [46], cellulose [47], and actual biomass feedstocks (waste polylactic acid [48], microalgae [49] and lignocellulose [50]). Herein, research advances on producing acetonitrile from renewable resources are classified according to substrate feedstock.

1. *Synthesis of acetonitrile from ethanol*

   Ethanol, which is a large-scale bio-based liquid fuel and commodity (bulk) chemical produced via biological fermentation, can be converted into acetonitrile via catalytic ammoxidation or amination [51–53]. In the catalytic conversion of ethanol to acetonitrile, it is first transformed into acetaldehyde via oxidative dehydrogenation or via direct dihydrogen reactions. The formed acetaldehyde

Table 5.2 Summary of reactions, conditions and results for producing acetonitrile from bio-based substrates

Substrate	Reaction type	Catalyst; Reaction temperature	Yield (%)	Ref.
Ethanol	Ammoxidation	(100) Pd/TiO_2, 230 °C	93	[54]
Ethanol	Ammoxidation	Cu-Zn-Zr/calcium aluminate; 280 °C	96–99	[55]
Ethanol	Direct amination	$Co_{19.9}Ni_{3.0}/\gamma\text{-}Al_2O_3$; 420 °C	>82	[45]
Ethanol	Direct amination	$Co\text{-}Ni/Al_2O_3$, 380 °C; NH_3/alcohol = 5: 1	92.6	[56]
Glycerol	Direct amination	$Fe_{19.2}K_{0.2}/\gamma\text{-}Al_2O_3$; 525 °C; NH_3/glycerol:8:1	45.8	[46]
Alanine	Oxidative decarboxylation	[Ni, Al]-LDH-WO_4, RT[a]	99	[22]

[a]*RT* room temperature

intermediate can then be reacted with ammonia via nitrilation to form the acetonitrile product.

For producing acetonitrile via catalytic ammoxidation, Hamill et al. [54] synthesized acetonitrile with 100% selectivity by using Pd/TiO_2 catalyst at 240 °C and found that the selectivity of acetonitrile was influenced by catalyst particle size with smaller and higher density Pd particles leading to higher acetonitrile selectivities. Moreover, similar conclusions were reached regarding acetonitrile production from ethanol for a Cu-Zn-Zr/calcium aluminate catalyst in which the yield of acetonitrile was about 96% [55]. For the production of acetonitrile via direct amination, Zhang et al. [45] reported that high yields were obtained from ethanol via direct amination process over $Co\text{-}Ni/Al_2O_3$ and that the catalyst remains stable for more than 720 h. Table 5.2 summarizes advances in producing acetonitrile from ethanol via ammoxidation and direct amination reactions.

2. *Synthesis of acetonitrile from glycerol or alanine*

Glycerol can be used to synthesize acetonitrile. Zhao et al. [46] reported that 47.9% yields of acetonitrile with 100% selectivity could be obtained using $Fe_{19.2}K_{0.2}/\gamma\text{-}Al_2O_3$ as catalyst. The reaction pathway for glycerol to acetonitrile includes dehydrogenation, dehydration, decarboxylation, and amination reactions which are more complex than those for ethanol and therefore acetonitrile yields are typically low.

Alanine can be used to synthesize acetonitrile via catalytic oxidation decarboxylation over [Ni, Al]-LDH-WO_4 at room temperature (RT) as shown by Claes et al. [22], who found that near-equivalents of acetonitrile could be produced from alanine.

3. *Synthesis of acetonitrile from actual biomass feedstocks*

Acetonitrile can be produced from cellulose and from actual biomass feedstocks via thermo-catalytical conversion and ammonization (TCC-A) [47]. TCC-A is a thermal-chemical conversion process that is carried out in an ammonia atmosphere [9, 57–59]. Zhang et al. [47] showed that acetonitrile

Table 5.3 Summary of studies including actual biomass for producing acetonitrile via thermo-catalytical conversion and ammonization (TCC-A) process

Substrate	Catalyst; Reaction temperature	Yield (C%)	Ref.
Cellulose	CoO_x/HZSM-5; 600 °C	32.5	[47]
Corn cob	2%Ga/HZSM-5; 700 °C	18.4	[50]
Microalgae	2%Ga/HZSM-5; 700 °C	23.4	[49]
Polylactic acid	Co/HZSM-5(25); 650 °C	50.2	[48]
Cellulose	Bone-based HAP catalyst; 650 °C	25.1	[60]

(carbon yield, ~32%) could be directly produced from carbohydrates (cellulose, glucose, xylose) and actual biomass (bagasse, rice husk) over CoO_x/HZSM-5. Experimental data and density functional theory (DFT) calculations revealed possible reaction pathways from cellulose to acetonitrile, namely, cellulose initially decomposes to form small molecular oxygenated intermediates (acetaldehyde, glycolaldehyde), and then the intermediates react with ammonia to form acetonitrile via amination, dehydration, and dehydrogenation reactions. Formation of oxygenated intermediates from cellulose was found to be related to the presence of (–Co–O–Si–) sites in the CoOx/HZSM-5 catalyst, which in turn affects the selectivity of acetonitrile. Moreover, microalgae, corn cob and polylactic acid are also selectively converted into acetonitrile over Ga/HZSM-5, CoOx/HZSM-5, hydroxyapatite (HAP) and bone-based catalysts with high carbon yields through the TCC-A process [48–50, 60]. Table 5.3 summarizes advances in producing acetonitrile from actual biomass via the TCC-A process.

5.3.3 *Fatty Nitriles*

Fatty nitriles and their derivatives are important precursors in the pharmaceutical, polymer, and surfactant industries [61–63]. Annual production and market value of fatty nitriles and their derivatives are more than 650,000 tons and $20 billion, respectively [64]. Industrially, fatty nitriles are produced from bio-based feedstocks (fatty acids, lipids, fatty alcohols) via catalytic amination reactions rather than from petrochemical-based feedstocks [65]. Reaction mechanisms for the formation of fatty nitriles from fatty acids or lipids are very simple, namely, substrates are reacted with ammonia to form fatty amide intermediates and then the intermediates are dehydrated to form nitriles. Herein, advances for producing fatty nitriles from renewable sources are summarized and discussed.

Mekki-Berrada et al. [66] synthesized fatty nitriles from fatty acid methyl esters and revealed the effect of acid-based properties on their production, namely, strong acidity correlated with ester conversion and nitrile yield. They [67] also proposed a model to describe half-order dependence of nitrile formation on catalyst amount by analysis of kinetic data. Shirazi et al. [68] reported that near-theoretical yields of fatty nitriles could be produced from triglycerides using high acidity catalysts such as V_2O_5 or Fe_2O_3. Jamil et al. [65] synthesized fatty nitriles from triglycerides over high-silica H-beta (Hβ-75) catalyst in high yields (91%). Hu et al. [69] showed that fatty alcohols could also serve as substrates for fatty nitrile synthesis via dehydrogenation-amination reactions. Besides heterogeneous catalysts, homogeneous catalysts such as alkylidene-ruthenium-tin catalyze lipids and plant oil derivatives into fatty nitriles under mild conditions (room temperature) in near-theoretical yields [70]. Using 1,2-diol fatty ester substrates, Guicheret et al. [62] synthesized medium chain nitriles such as C7 or C8 nitriles via catalytic dehydrogenation - oxidative cleavage coupling with amination reactions.

5.3.4 Furan Nitriles

Furan-based nitriles, which belong to the class of heteroaromatic nitriles, are susceptible to electrophilic addition reactions as applied to syntheses of pharmaceuticals, fine chemicals, polymers, organic semiconductors (OSC) and other high value-added products [71, 72]. Furan-based nitriles can also be substitutes for aromatic nitriles. The platform chemicals, 5-hydroxymethylfurfural (5-HMF) and furfural are the primary substrates employed for synthesizing furan-based nitriles via ammoxidation, since they can be readily produced from carbohydrates including cellulose, hexoses, hemicellulose and xylose via dehydration reactions [7, 73–75]. Herein, research progress in synthesis routes for nitriles from bio-based furans are summarized.

Singh et al. [76] showed that 2-furonitrile could be synthesized from furfural via vapor-phase ammoxidation over Mo-Bi-V catalyst in high yields (~95%) at around 400 °C in 1978. Yao et al. reported similar results, in that furfural could be converted into 2-furonitrile over acid catalyst (HZSM-5) via vapor phase direct amination reaction [77, 78]. Nagasaki et al. [79] used an effective and reusable heterogeneous catalyst (MoO_x/SiO_2) to catalytically dehydrate 2-furamide into 2-furonitrile (96% yield). Jia et al. [80] synthesized alkali α-MnO_2/Na_xMnO_2 catalyst and synthesized 2-furonitrile from furfural via liquid phase ammoxidation-pinner tandem reactions under mild reactions (30 °C, p_{O2} = 0.5 MPa). Jia et al. [81] also synthesized a carboxylic acid-modified MnO_x catalyst and applied it to the aerobic ammoxidation of 5-HMF to form 5-hydroxymethylfuronitrile in high yields (92%) under mild reaction conditions (60 °C, p_{O2} = 0.5 MPa). Ibrahim et al. [72] synthesized 5-hydroxymethylfuronitrile from 5-HMF over iodine catalyst, and further synthesized 5-cyanofurfural, that could be used to prepare benzofuran-cyanovinyl derivatives. In addition to 2-furonitrile and 5-hydroxymethylfuronitrile, bio-based 2,5-dicyanofuran could also be produced. Xu et al. [82] showed that 2,5-dicyanofuran could be efficiently synthesized from 2,5-diformylfuran (DFF) via a two-step procedure consisting of oximation and dehydration reactions in which DFF is first reacted with hydroxylamine to form 2,5-diformylfuran dioxime as intermediate via oximation reaction, and then dehydration of the intermediate forms 2,5-dicyanofuran over Amberlyst-15 catalyst for which overall yields

Table 5.4 Substrates, products and conditions for producing bio-based furan nitriles

Substrate	Product	Catalyst	Yield (%)	Ref.
Furfural	2-furonitrile	Mo-Bi-V	~95	[76]
2-furamide	2-furonitrile	$MoOx/SiO_2$	96	[79]
Furfural	2-furonitrile	Alkali α-MnO_2/ $NaxMnO_2$	91	[80]
5-HMF	5-hydroxymethylfuronitrile	Alkali α-MnO_2/ $NaxMnO_2$	92	[81]
5-HMF	5-cyanofurfural	I_2	~100	[72]
2,5-Diformylfuran	2,5-dicyanofuran	Amberlyst-15 catalyst	82	[82]
Furfural alcohol	2-furonitrile	Co/GC@C	73	[83]
5-HMF	2,5-dicyanofuran	Co/GC@C	96	[83]
Furfural	2-furonitrile	Se,S,N-CNs-1000	91	[84]
5-HMF	5-hydroxymethylfuronitrile	Se,S,N-CNs-1000	89	[84]
2-furfuryl aldoxime	2-furonitrile	Aldoxime dehydratase	~100	[85]

approached 82%. Senthamarai et al. [83] synthesized a universal oxidation catalyst (Co/GC@C) via co-pyrolysis of cobalt-piperazine-tartaric acid complex on carbon at 800 °C and used it for conversion of furfural alcohol and 5-HMF into 2-furonitrile (73%) and 2,5-dicyanofuran (96%), respectively. Hua et al. [84] synthesized a type of non-metal tri-doped hierarchically porous carbon nanosheet catalysts (Se, S, N-CNs-1000) and used it for conversion of furfural and 5-HMF into 2-furonitrile (91%) and 5-hydroxylmethylfuronitrile (89%). Besides the above studies that use heterogeneous catalysts, furan-based nitriles have also been synthesized. Choi et al. [85] reported that 2-furonitrile can be produced from 2-furfuryl aldoxime, which is easily obtained from furfural and hydroxylamine, via an enzymatic dehydration with aldoxime dehydratase mutants as biocatalyst. Table 5.4 summarizes advances in producing furanitriles from bio-derived substrates via the TCC-A process.

5.3.5 *Aromatic Nitriles*

Aromatic nitriles are another important class of nitrogen-containing chemicals due to their motifs being present in many pharmaceuticals [86]. Industrially, aromatic nitriles are produced through cyanidation of halogenated aromatics, amination-dehydration of aromatic aldehydes (e.g. benzaldehyde, *p*-hydroxybenzaldehyde), ammoxidation of aromatic alcohols (e.g. benzyl alcohol) or ammoxidation of methyl substituted aromatics (e.g. toluene, xylenes) [87–89]. Lignin, which is one of the three main constituents of lignocellulosic biomass, is conveniently converted into aromatic oxygenated compounds including aromatic aldehydes (e.g. vanillin, syringaldehyde) or aromatic acids (e.g. vanillic acid) by oxidative degradation [90–93]. Aromatic oxygenated compounds can be readily transformed into aromatic nitriles. Herein, research progress in the preparation of bio-based aromatic nitriles is discussed according to amination-dehydration and ammoxidation pathways.

1. *Synthesis of aromatic nitriles via amination-dehydration pathway*

   Synthesis of aromatic nitriles by amination-dehydration refers to the way that aromatic substrates react with an active nitrogen source such as hydroxylamine or ammonia and undergoes amination-dehydration reactions to form the product. In sustainable synthesis methods, renewable substrates are aromatic aldehydes and aromatic esters, whereas active nitrogen sources are hydroxylamine compounds and ammonia. Nitrile formation methods can also be expressed in terms of the hydroxylamine pathway or the catalytic amination pathway.

   For the hydroxylamine pathway, hydroxylamine compounds, such as hydroxylamine hydrochloride ($NH_2OH{\cdot}HCl$) or hydroxylamine-O-sulfonic acid ($NH_2OH{\cdot}H_2SO_4$), serve as the active nitrogen source. Albarrán-Velo et al. [94] synthesized vanillonitrile (>80%) from vanillin via reaction with $NH_2OH{\cdot}HCl$ under an acetonitrile (MeCN) reflux. A similar approach was reported for vanillonitrile by Desvals et al. [95], who conducted the reaction using an acetic acid reflux. Mudshinge et al. [96] used HCl•DMPU to transform aromatic aldehydes catalytically into aromatic nitriles with acetonitrile as solvent. Khalafi-Nezhad et al. [97] synthesized a kind of green chitosan supported magnetic ionic liquid catalyst (CSMIL) and used it in the conversion of *p*-hydroxybenzaldehyde into *p*-hydroxybenzonitrile. Fe_3O_4-nanoparticles can serve as efficient catalysts for converting aromatic aldehydes (such as *p*-hydroxybenzaldehyde, vanillin) into aromatic nitriles, including *p*-hydroxybenzonitrile, and vanillonitrile [98]. Villas-Boas Hoelz et al. [99] reported that vanillin reacts with $NH_2OH{\cdot}HCl$ to form vanillonitrile (83%) in the presence of TiO_2 under microwave irradiation under solventless (i.e. reactants act as solvent) conditions. Hydroxylamine 1-sulfobutyl pyridine hydrosulfate salt ($(NH_2OH)_2{\cdot}[HSO_3$-*b*-$Py]{\cdot}HSO_4$) has been suggested as an alternative nitrogen source for $NH_2OH{\cdot}HCl$ as it has multiple roles including co-solvent, catalyst and phase separation agent [100]. Bonjour et al. [101] used hydroxylamine-O-sulfonic acid ($NH_2OH{\cdot}H_2SO_4$) to replace $NH_2OH{\cdot}HCl$ in the synthesis of aromatic nitriles such as *p*-hydroxylbenzonitrile, 3,5-dimethoxyl-4-hydroxylbenzonitrile and vanillonitrile from lignin-based aromatic aldehydes. Table 5.5 summarizes substrates and conditions used for producing bio-based aromatic nitriles via hydroxylamines.

   Bio-based aromatic nitriles including terephthalonitrile (TPN) and benzonitrile (BZN) can be produced from waste polyester plastics (polyethylene terephthalate, PET) via catalytic amination using catalytic pyrolysis with ammonia [102, 103] and Al_2O_3-based catalysts having different acidity and basicity. For example, 30.4% BZN with selectivity of 82.6% in nitriles can be produced under optimal conditions at 650 °C [102]. 52.3% TPN with selectivity of 90% in nitriles can be produced over γ-Al_2O_3-2 wt% catalyst under optimal conditions at 500 °C

Table 5.5 Substrates, products and conditions for producing bio-based aromatic nitriles via hydroxylamines

Substrate	Product	N-source	Catalyst	Yield (%)	Ref.
Vanillin	Vanillonitrile	$NH_2OH \cdot HCl$	MeCN reflux	>80	[94]
Vanillin	Vanillonitrile	$NH_2OH \cdot HCl$	Acetic acid reflux	70–91	[95]
Aromatic aldehydes	Aromatic nitriles	$NH_2OH \cdot HCl$	HCl•DMPU, MeCN, 60 °C	>80	[96]
4-Hydroxylbenzaldehye	*p*-hydroxylbenzonitrile	$NH_2OH \cdot HCl$	CSMIL	90	[97]
Aromatic aldehydes	Aromatic nitriles	$NH_2OH \cdot HCl$	Fe_3O_4-nanoparticles	>90	[98]
Aromatic aldehydes	Aromatic nitriles	$NH_2OH \cdot HCl$	TiO_2	83	[99]
Aromatic aldehydes	Aromatic nitriles	$(NH_2OH)_2 \cdot [HSO_3\text{-}b\text{-}Py] \cdot HSO_4$	$(NH_2OH)_2 \cdot [HSO_3\text{-}b\text{-}Py] \cdot HSO_4$	>90	[100]
Aromatic aldehydes	Aromatic nitriles	$NH_2OH \cdot H_2SO_4$	/	79–82	[101]

[103]. However, reaction conditions for catalytic amination methods are more critical than those for hydroxylamine methods. During catalytic pyrolysis with ammonia, PET first undergoes depolymerization reactions to form pyrolysis intermediates, such as aromatic esters or aromatic acids and then the intermediates further react via amination and dehydration reactions to form aromatic nitriles [103].

2. *Synthesis of aromatic nitriles via ammoxidation pathways*

   Synthesis of aromatic nitriles via ammoxidation is similar in method to ways used for furan nitriles, that is, by catalytic ammoxidation. Substrates for aromatic nitriles are typically aromatic aldehydes and aromatic alcohols, with an active nitrogen source (e.g. ammonia) and oxygen being necessary in the process. Herein, recent advances of catalytic systems for synthesizing aromatic nitriles are discussed. Dornan et al. [104] used Cu/TEMPO as homogeneous catalysts in the ammoxidation of aromatic aldehydes and aromatic alcohols with aqueous ammonia to produce aromatic nitriles. The ligand, 2,2-bipyridine (bipy), is necessary in the $Cu(OTf)_2$ catalytic system. Das et al. [105] used a ligand-free catalyst, $Cu(OAc)_2$, to directly catalyze ammoxidation of aromatic aldehydes into aromatic nitriles with ammonium acetate as the nitrogen source. The $Cu(OAc)_2$ used in the study was commercially available and a non-toxic homogeneous catalyst. Chen et al. [106] developed an Fe-based catalytic system ($NH_3{\cdot}H_2O/FeCl_2/NaI/Na_2S_2O_8$) that effectively converts aromatic aldehydes into aromatic nitriles for synthesis of febuxostat, which is a drug used in the treatment of hyperuricemia gout. In the catalytic system, NaI and $FeCl_2$ had the role of homogeneous catalysts, $Na_2S_2O_8$ was used as oxidant, and $NH_3{\cdot}H_2O$ was the active nitrogen source. Heterogenous catalysts have been developed for preparing aromatic nitriles via catalytic ammoxidation reactions. Zhang et al. [107] prepared a Cu/SiO_2 catalyst for vapor phase catalytic conversion of benzyl alcohol into benzonitrile (~90%) via amination-dehydrogenation. Yoon et al. [108] synthesized heterogeneous Cu@C catalyst via co-pyrolysis process, which could effectively convert aromatic aldehydes into aromatic nitriles and maintain high stability when reused. Pan et al. [109] synthesized highly disperse Co/N-doped carbon catalysts that could efficiently catalyze ammoxidation of aromatic aldehydes to form aromatic nitriles, and they also found that highly dispersed Co species could interact with the benzene ring structure, thereby promoting conversion of aromatic aldehydes and the formation of aromatic nitriles (e.g. benzonitrile, *p*-hydroxylbenzonitrile, etc.). Similarly, Wang et al. [110] prepared atomically dispersed Ru that was supported on manganese oxide nanorods (Ru/MnO_2-r) that can selectively catalyze ammoxidation of aromatic alcohols into aromatic nitriles with high efficiency and those catalysts showed good stability. Atom-dispersed Ru with a loading of 0.1 wt.% was highly active for this reaction. In addition, a non-metal catalyst (Se, S, N tri-doped hierarchically porous carbon nanosheets, Se, S, N CNs-1000) was synthesized that effectively catalyzes conversion of aromatic aldehydes into aromatic nitriles [84].

5.4 Conclusions and Future Outlook

Due to the reactivity of chemical functional groups present in biomass, many kinds of nitrogen-containing compounds can be synthesized. Much attention has been paid to the selective synthesis of nitriles from bio-based platform molecules or directly from biomass feedstocks via catalytic systems. In this chapter, recent advances and strategies have been summarized and outlined for the preparation of biobased nitriles from renewable nitrogen-containing sources and bio-based oxygenated compounds. However, research on the production of bio-based nitriles is still at an early stage and far from the goal of being on an industrial scale due to the many types of feedstocks. Therefore, multi-field researchers are needed to collaborate and devote much effort to address the following key issues.

From the above-mentioned transformations, some basic rules for the formation of the target nitriles can be understood. However, the above results are only preliminary findings, and the influencing factors and mechanisms are still not very well understood. Advanced experimental and analytical methods (e.g., in-situ NMR), electron paramagnetic resonance (EPR), photoelectron photoion coincidence spectroscopy with synchrotron radiation (iPEPICO) and in-situ FTIR) are needed to be applied to identify and track key intermediates and provide insights into the chemistry and mechanisms of N-containing compound formation. Moreover, combination of theoretical methods (quantum chemistry calculations, reactive specie simulations) with experimental studies will deepen the understanding of biomass conversion and nitrile formation mechanisms.

Approaches for producing N-containing compounds outlined in this work are mainly those for biomass-derived platform compounds including glycerol, ethanol, furan and phenol. To save much effort and cost necessary for producing specific platform compounds, it could be advantageous to obtain N-containing compounds directly from biomass. Currently, there are few studies on the preparation of nitriles from raw biomass due to the complex constituents in biomass, which pose great challenges in chemical transformation strategies for obtaining nitriles in high selectivity. In this regard, technologies for green and efficient pretreatment of biomass with sustainable solvents and/or suitable catalysts are necessary to enhance reaction efficiency. Considering that the yield and selectivity of nitriles from bio-derived platform compounds are much higher than those from raw biomass, developing technologies to produce bio-derived platform compounds more effectively could also be another fruitful approach for producing N-containing compounds.

As future outlook, many bio-based nitriles can be effectively produced from biomass feedstocks thanks to the elucidation of a number of chemical avenues. Although transformation of actual biomass into bio-based nitriles remains a challenge, step-wise approaches that are combined with pretreatment may be helpful to make progress. Development of efficient and sustainable synthesis approaches to target nitrile compounds with emphasis on high efficiency will be required for comprehensive biomass valorization.

Acknowledgments The authors are grateful to the financial supports from National Natural Science Foundation of China (No.51906112), China Postdoctoral Science Foundation (2019 M651852), Nanjing Agricultural University (77 J-0603), "Innovation & Entrepreneurship Talents" Introduction Plan of Jiangsu Province.

References

1. Kumar A, Daw P, Milstein D. Homogeneous catalysis for sustainable energy: hydrogen and methanol economies, fuels from biomass, and related topics. Chem Rev. 2022;122:385–441. https://doi.org/10.1021/acs.chemrev.1c00412.
2. Li C, Zhao X, Wang A, Huber GW, Zhang T. Catalytic transformation of lignin for the production of chemicals and fuels. Chem Rev. 2015;115:11559–624. https://doi.org/10.1021/acs.chemrev.5b00155.
3. Li J, Ding D, Xu L, Guo Q, Fu Y. The breakdown of reticent biomass to soluble components and their conversion to levulinic acid as a fuel precursor. RSC Adv. 2014;4:14985. https://doi.org/10.1039/c3ra47923d.
4. Sheldon RA. Green and sustainable manufacture of chemicals from biomass: state of the art. Green Chem. 2014;16:95–963. https://doi.org/10.1039/c3gc41935e.
5. Tuck CO, Pérez E, Horváth IT, Sheldon RA, Poliakoff M. Valorization of biomass: deriving more value from waste. Science. 2012;337:695–9. https://doi.org/10.1126/science.1218930.
6. Shuai L, Amiri MT, Questell-Santiago YM, Heroguel F, Li Y, Kim H, Meilan R, Chapple C, Ralph J, Luterbacher J. Formaldehyde stabilization facilitates lignin monomer production during biomass depolymerization. Science. 2016;354:329–33. https://doi.org/10.1126/science.aaf7810.
7. van Putten RJ, van der Waal JC, de Jong E, Rasrendra CB, Heeres HJ, de Vries JG. Hydroxymethylfurfural, A versatile platform chemical made from renewable resources. Chem Rev. 2013;113:1499–597. https://doi.org/10.1021/cr300182k.
8. Sajid M, Zhao X, Liu D. Production of 2, 5-furandicarboxylic acid (FDCA) from 5-hydroxymethylfurfural (HMF): recent progress focusing on the chemical-catalytic routes. Green Chem. 2018;20:5427–53. https://doi.org/10.1039/C8GC02680G.
9. Xu L, Shi C, He Z, Zhang H, Chen M, Fang Z, Zhang Y. Recent advances of producing biobased N-containing compounds via thermo-chemical conversion with ammonia process. Energy Fuel. 2020;34:10441–58. https://doi.org/10.1021/acs.energyfuels.0c01993.
10. Dai T, Li C, Li L, Zhao ZK, Zhang B, Cong Y, Wang A. Selective production of renewable *para*-xylene by tungsten carbide catalyzed atom-economic cascade reactions. Angew Chem Int Ed. 2018;130:1826–30. https://doi.org/10.1002/ange.201710074.
11. Ghosh P, Subba R. $FeCl_3$ mediated one-pot route to nitriles. Tetrahedron Lett. 2013;54:4885–7. https://doi.org/10.1016/j.tetlet.2013.06.134.
12. Martin A, Kalevaru VN. Heterogeneously catalyzed ammoxidation: a valuable tool for one-step synthesis of nitriles. ChemCatChem. 2010;2:1504–22. https://doi.org/10.1002/cctc.201000173.
13. Trifirò F. Some history on the new ways of synthesis of nitriles. Catal Today. 2021;363:10–4. https://doi.org/10.1016/j.cattod.2019.09.012.
14. Corma A, Iborra S, Velty A. Chemical routes for the transformation of biomass into chemicals. Chem Rev. 2007;107:2411–502. https://doi.org/10.1021/cr050989d.
15. Kumar MBA, Gao Y, Shen W, He L. Valorisation of protein waste: an enzymatic approach to make commodity chemicals. Front Chem Sci Eng. 2015;9:295–307. https://doi.org/10.1007/s11705-015-1532-4.

16. Scott E, Peter F, Sanders J. Biomass in the manufacture of industrial products—the use of proteins and amino acids. Appl Microbiol Biotechnol. 2007;75:751–62. https://doi.org/10.1007/s00253-007-0932-x.
17. Sari YW, Alting AC, Floris R, Sanders JPM, Bruins ME. Glutamic acid production from wheat by-products using enzymatic and acid hydrolysis. Biomass Bioenergy. 2014;67:451–9. https://doi.org/10.1016/j.biombioe.2014.05.018.
18. Dai J, Huang Y, Fang C, Guo Q, Fu Y. Electrochemical synthesis of adiponitrile from the renewable raw material glutamic acid. ChemSusChem. 2012;5:617–20. https://doi.org/10.1002/cssc.201100776.
19. Le Nôtre J, Scott EL, Franssen MCR, Sanders JPM. Biobased synthesis of acrylonitrile from glutamic acid. Green Chem. 2011;13:807. https://doi.org/10.1039/c0gc00805b.
20. Lammens TM, Le Nôtre J, Franssen MCR, Scott EL, Sanders JPM. Synthesis of biobased succinonitrile from glutamic acid and glutamine. ChemSusChem. 2011;4:785–91. https://doi.org/10.1002/cssc.201100030.
21. But A, Le Nôtre J, Scott EL, Wever R, Sanders JPM. Selective oxidative decarboxylation of amino acids to produce industrially relevant nitriles by vanadium chloroperoxidase. ChemSusChem. 2012;5:1199–202. https://doi.org/10.1002/cssc.201200098.
22. Claes L, Matthessen R, Rombouts I, Stassen I, De Baerdemaeker T, Depla D, Delcour J, Lagrain B, De Vos D. Bio-based nitriles from the heterogeneously catalyzed oxidative decarboxylation of amino acids. ChemSusChem. 2015;8:345–52. https://doi.org/10.1002/cssc.201402801.
23. Lammens TM, Potting J, Sanders JPM, De Boer IJM. Environmental comparison of biobased chemicals from glutamic acid with their petrochemical equivalents. Environ Sci Technol. 2011;45:8521–8. https://doi.org/10.1021/es201869e.
24. Grasselli RK, Trifirò F. Acrylonitrile from biomass: still far from being a sustainable process. Top Catal. 2016;59:1651–8. https://doi.org/10.1007/s11244-016-0679-7.
25. Karp EM, Eaton TR, Sànchez I Nogué V, Vorotnikov V, Biddy MJ, ECD T, Brandner DG, Cywar RM, Liu R, Manker LP, Michener WE, Gilhespy M, Skoufa Z, Watson MJ, Fruchey OS, Vardon DR, Gill RT, Bratis AD, Beckham GT. Renewable acrylonitrile production. Science. 2017;358:1307–10. https://doi.org/10.1126/science.aan1059.
26. Liu H, Tang Y, Lu L, Xie Y, Yuan W, Chen G, Li J, Fu T. Investigation on fiber laser cutting of polyacrylonitrile-based carbon fiber tow. J Mater Process Technol. 2019;263:151–63. https://doi.org/10.1016/j.jmatprotec.2018.08.015.
27. Nataraj SK, Yang KS, Aminabhavi TM. Polyacrylonitrile-based nanofibers—A state-of-the-art review. Prog Polym Sci. 2012;37:487–513. https://doi.org/10.1016/j.progpolymsci.2011.07.001.
28. Zhang L, Aboagye A, Kelkar A, Lai C, Fong H. A review: carbon nanofibers from electrospun polyacrylonitrile and their applications. J Mater Sci. 2014;49:463–80. https://doi.org/10.1007/s10853-013-7705-y.
29. Kumar R, Sharma SD, Deshpande S, Ghadi Y, Shaiju VS, Amols HI, Mayya YS. Acrylonitrile butadiene styrene (ABS) plastic based low cost tissue equivalent phantom for verification dosimetry in IMRT. J Appl Clin Med Phys. 2009;11:3030. https://doi.org/10.1120/jacmp.v11i1.3030.
30. Olivera S, Muralidhara HB, Venkatesh K, Gopalakrishna K, Vivek CS. Plating on acrylonitrile–butadiene–styrene (ABS) plastic: a review. J Mater Sci. 2016;51:3657–74. https://doi.org/10.1007/s10853-015-9668-7.
31. Dief Allah MM, Ali ZM, Raslan MA. Dielectric, thermal and morphological characteristics of nitrile butadiene rubber under effect filler/hybrid filler. Measurement. 2019;131:13–8. https://doi.org/10.1016/j.measurement.2018.08.036.
32. Kaiser A, Brandau S, Klimpel M, Barner-Kowollik C. Acrylonitrile-butadiene rubber (NBR) prepared via living/controlled radical polymerization (RAFT). Macromol Rapid Commun. 2010;31:1616–21. https://doi.org/10.1002/marc.201000162.

33. Guerrero-Pérez MO, Bañares MA. Metrics of acrylonitrile: from biomass vs. petrochemical route. Catal Today. 2015;239:25–30. https://doi.org/10.1016/j.cattod.2013.12.046.
34. Guerrero-Pérez MO, Bañares MA. New reaction: conversion of glycerol into acrylonitrile. ChemSusChem. 2008;1:511–3. https://doi.org/10.1002/cssc.200800023.
35. Thanh-Binh N, Dubois J, Kaliaguine S. Ammoxidation of acrolein to acrylonitrile over bismuth molybdate catalysts. Appl Catal A Gen. 2016;520:7–12. https://doi.org/10.1016/j.apcata.2016.03.030.
36. Liebig C, Paul S, Katryniok B, Guillon C, Couturier J, Dubois J, Dumeignil F, Hoelderich WF. Glycerol conversion to acrylonitrile by consecutive dehydration over WO_3/TiO_2 and ammoxidation over Sb-(Fe, V)-O. Appl Catal B Environ. 2013;132–133:170–82. https://doi.org/10.1016/j.apcatb.2012.11.035.
37. Guillon C, Liebig C, Paul S, Mamede A, Hölderich WF, Dumeignil F, Katryniok B. Ammoxidation of allyl alcohol - a sustainable route to acrylonitrile. Green Chem. 2013;15:3015. https://doi.org/10.1039/c3gc41089g.
38. Mack D, Schätzle S, Traa Y, Klemm E. Synthesis of acrylonitrile from renewable lactic acid. ChemSusChem. 2019;12:1653–63. https://doi.org/10.1002/cssc.201802914.
39. Zhang Y, Sun F, Zhang H, Ying G, Zhao J. Synthesis of nitriles from allyl alcohol derived from glycerol over a bimetallic catalyst Zn_{30} $Ru_{1.0}/\gamma$-Al_2O_3. Ind Eng Chem Res. 2018;57:4553–61. https://doi.org/10.1021/acs.iecr.8b00444.
40. Liang T, Liu X, He Y, Barbosa R, Chen G, Fan W, Xiang Y. Highly selective Sn/HZSM-5 catalyst for ethane ammoxidation to acetonitrile and ethylene. Appl Catal A Gen. 2021;610:117942. https://doi.org/10.1016/j.apcata.2020.117942.
41. Zarzycki PK, Zarzycka MB, Ślączka MM, Clifton VL. Acetonitrile, the polarity chameleon. Anal Bioanal Chem. 2010;397:905–8. https://doi.org/10.1007/s00216-010-3677-9.
42. Jiménez-Hornero JE, Santos-Dueñas IM, García-García I. Optimization of biotechnological processes. The acetic acid fermentation. Part I: the proposed model. Biochem Eng J. 2009;45:1–6. https://doi.org/10.1016/j.bej.2009.01.009.
43. La China S, Zanichelli G, De Vero L, Gullo M. Oxidative fermentations and exopolysaccharides production by acetic acid bacteria: a mini review. Biotechnol Lett. 2018;40:1289–302. https://doi.org/10.1007/s10529-018-2591-7.
44. Galanov SI, Sidorova OI, Gavrilenko MA. The process of acetonitrile synthesis over γ-Al_2O_3 promoted by phosphoric acid catalysts. Procedia Chem. 2014;10:108–13. https://doi.org/10.1016/j.proche.2014.10.020.
45. Zhang Y, Zhang Y, Feng C, Qiu C, Wen Y, Zhao J. Amination of ethanol to acetonitrile over Ni-doped Co/γ-Al_2O_3 catalyst. Catal Commun. 2009;10:1454–8. https://doi.org/10.1016/j.catcom.2009.03.019.
46. Zhang Y, Ma T, Zhao J. Study on the conversion of glycerol to nitriles over a $Fe_{19.2}K_{0.2}/\gamma$-Al_2O_3 catalyst. J Catal. 2014;313:92–103. https://doi.org/10.1016/j.jcat.2014.02.014.
47. Zhang Y, Yuan Z, Hu B, Deng J, Yao Q, Zhang X, Liu X, Fu Y, Lu Q. Direct conversion of cellulose and raw biomass to acetonitrile by catalytic fast pyrolysis in ammonia. Green Chem. 2019;21:812–20. https://doi.org/10.1039/C8GC03354D.
48. Yuan Z, Zhang X, Yao Q, Zhang Y, Fu Y. Production of acetonitrile via catalytic fast pyrolysis of biomass derived polylactic acid under ammonia atmosphere. J Anal Appl Pyrolysis. 2019;140:376–84. https://doi.org/10.1016/j.jaap.2019.04.017.
49. Yao Q, Zhang Y, Fu Y. Sustainable production of acetonitrile from microalgae via catalytic fast pyrolysis with ammonia over Ga/HZSM-5 catalysts. ACS Sustain Chem Eng. 2019;7:16173–81. https://doi.org/10.1021/acssuschemeng.9b03102.
50. Zhang X, Yuan Z, Yao Q, Zhang Y, Fu Y. Catalytic fast pyrolysis of corn cob in ammonia with Ga/HZSM-5 catalyst for selective production of acetonitrile. Bioresour Technol. 2019;290:121800. https://doi.org/10.1016/j.biortech.2019.121800.

51. Jambo SA, Abdulla R, Mohd Azhar SH, Marbawi H, Gansau JA, Ravindra P. A review on third generation bioethanol feedstock. Renew Sust Energ Rev. 2016;65:756–69. https://doi.org/10.1016/j.rser.2016.07.064.
52. Manochio C, Andrade BR, Rodriguez RP, Moraes BS. Ethanol from biomass: a comparative overview. Renew Sust Energ Rev. 2017;80:743–55. https://doi.org/10.1016/j.rser.2017.05.063.
53. Puligundla P, Smogrovicova D, Mok C, Obulam VSR. A review of recent advances in high gravity ethanol fermentation. Renew Energy. 2019;133:1366–79. https://doi.org/10.1016/j.renene.2018.06.062.
54. Hamill C, Driss H, Goguet A, Burch R, Petrov L, Daous M, Rooney D. Mild temperature palladium-catalyzed ammoxidation of ethanol to acetonitrile. Appl Catal A Gen. 2015;506: 261–7. https://doi.org/10.1016/j.apcata.2015.09.030.
55. Belov VV, Markov VI, Sova SB, Golosman EZ, Nechugovskii AI. New composite catalysts for the synthesis of acetonitrile from ethanol and ammonia, prepared using calcium aluminates (Talyum). Russ J Appl Chem. 2016;89:414–20. https://doi.org/10.1134/S1070427216030113.
56. Feng C, Zhang Y, Zhang Y, Wen Y, Zhao J. Study on alumina-supported cobalt–nickel oxide catalyst for synthesis of acetonitrile from ethanol. Catal Lett. 2011;141:168–77. https://doi.org/10.1007/s10562-010-0468-z.
57. Xu L, Han Z, Yao Q, Deng J, Zhang Y, Fu Y, Guo Q. Towards the sustainable production of pyridines via thermo-catalytic conversion of glycerol with ammonia over zeolite catalysts. Green Chem. 2015;17:2426–35. https://doi.org/10.1039/C4GC02235A.
58. Xu L, Jiang Y, Yao Q, Han Z, Zhang Y, Fu Y, Guo Q, Huber GW. Direct production of indoles via thermo-catalytic conversion of bio-derived furans with ammonia over zeolites. Green Chem. 2015;17:1281–90. https://doi.org/10.1039/C4GC02250E.
59. Xu L, Yao Q, Deng J, Han Z, Zhang Y, Fu Y, Huber GW, Guo Q. Renewable N-heterocycles production by thermocatalytic conversion and ammonization of biomass over ZSM-5. ACS Sustain Chem Eng. 2015;3:2890–9. https://doi.org/10.1021/acssuschemeng.5b00841.
60. Jia T, Zhou F, Ma H, Zhang Y. A highly stable waste animal bone-based catalyst for selective nitriles production from biomass via catalytic fast pyrolysis in NH_3. J Anal Appl Pyrolysis. 2021;157:105217. https://doi.org/10.1016/j.jaap.2021.105217.
61. Gomes Ferreira De Paula F, Gonçalves Pereira RC, Pereira Paulo BF, Ardisson JD, de Lourdes Cardeal Z, Costa Menezes H, Paula de Carvalho Teixeira A, Montero Lago R. A new pyrolytic process with potential to convert free fatty acids into long chain nitriles and H_2 intermediated by Fe nitrate. J Anal Appl Pyrolysis. 2020;145:104726. https://doi.org/10.1016/j.jaap.2019.104726.
62. Guicheret B, Bertholo Y, Blach P, Raoul Y, Métay E, Lemaire M. A two-step oxidative cleavage of 1, 2-diol fatty esters into acids or nitriles by a dehydrogenation-oxidative cleavage sequence. ChemSusChem. 2018;11:3431–7. https://doi.org/10.1002/cssc.201801640.
63. Legras JL, Chuzel G, Arnaud A, Galzy P. Natural nitriles and their metabolism. World J Microbiol Biotechnol. 1990;6:83–108. https://doi.org/10.1007/BF01200927.
64. Hayes DG. Fatty acids based surfactants and their uses. Fatty Acids. Elsevier Inc. 2017:355–84. https://doi.org/10.1016/B978-0-12-809521-8.00013-1.
65. Jamil MAR, Siddiki SMAH, Touchy AS, Rashed MN, Poly SS, Jing Y, Ting KW, Toyao T, Maeno Z, Shimizu K. Selective transformations of triglycerides into fatty amines, amides, and nitriles by using heterogeneous catalysis. ChemSusChem. 2019;12:3115–25. https://doi.org/10.1002/cssc.201900365.
66. Mekki-Berrada A, Bennici S, Gillet JP, Couturier JL, Dubois JL, Auroux A. Fatty acid methyl esters into nitriles: acid–base properties for enhanced catalysts. J Catal. 2013;306:30–7. https://doi.org/10.1016/j.jcat.2013.05.032.
67. Mekki-Berrada A, Bennici S, Gillet J, Couturier J, Dubois J, Auroux A. Ammoniation-dehydration of fatty acids into nitriles: heterogeneous or homogeneous catalysis? ChemSusChem. 2013;6:1478–89. https://doi.org/10.1002/cssc.201300210.

68. Shirazi Y, Tafazolian H, Viamajala S, Varanasi S, Song Z, Heben MJ. High-yield production of fatty nitriles by one-step vapor-phase thermocatalysis of triglycerides. ACS Omega. 2017;2: 9013–20. https://doi.org/10.1021/acsomega.7b01502.
69. Hu Y, Jin S, Zhang Z, Zhang L, Deng J, Zhang H. One-step synthesis of nitriles by the dehydrogenation–amination of fatty primary alcohols over Cu/m-ZrO_2. Catal Commun. 2014;54:45–9. https://doi.org/10.1016/j.catcom.2014.05.010.
70. Miao X, Blokhin A, Pasynskii A, Nefedov S, Osipov SN, Roisnel T, Bruneau C, Dixneufet P. Alkylidene-ruthenium-tin catalysts for the formation of fatty nitriles and esters via cross-metathesis of plant oil derivatives. Organometallics. 2010;29:5257–62. https://doi.org/10.1021/om100372b.
71. Baliani A, Bueno GJ, Stewart ML, Yardley V, Brun R, Barrett MP, Gilbert IH. Design and synthesis of a series of melamine-based nitroheterocycles with activity against Trypanosomatid parasites. J Med Chem. 2005;48:5570–9. https://doi.org/10.1021/jm050177+.
72. Ibrahim N, Moussallem C, Allain M, Segut O, Gohier F, Frère P. Exploring the electronic properties of extended benzofuran-cyanovinyl derivatives obtained from lignocellulosic and carbohydrate platforms raw materials. ChemPlusChem. 2021;86:475–82. https://doi.org/10.1002/cplu.202100062.
73. Climent MJ, Corma A, Iborra S. Heterogeneous catalysts for the one-pot synthesis of chemicals and fine chemicals. Chem Rev. 2011;111:1072–133. https://doi.org/10.1021/cr1002084.
74. Huber GW, Iborra S, Corma A. Synthesis of transportation fuels from biomass: chemistry, catalysts, and engineering. Chem Rev. 2006;106:4044–98. https://doi.org/10.1021/cr068360d.
75. Mariscal R, Maireles-Torres P, Ojeda M, Sádaba I, López GM. Furfural: a renewable and versatile platform molecule for the synthesis of chemicals and fuels. Energy Environ Sci. 2016;9:1144–89. https://doi.org/10.1039/c5ee02666k.
76. Singh P, Miwa Y, Okada J. Vapor-phase ammoxidation of 5-membered heterocyclic compounds. Ammoxidation of furfural and 2-thiophenealdehyde. Chem Pharm Bull. 1978;26: 2838–43.
77. Yao Q, Xu L, Han Z, Zhang Y. Production of indoles via thermo-catalytic conversion and ammonization of bio-derived furfural. Chem Eng J. 2015;280:74–81. https://doi.org/10.1016/j.jaap.2016.08.005.
78. Yao Q, Xu L, Zhang Y, Fu Y. Enhancement of indoles production and catalyst stability in thermo-catalytic conversion and ammonization of furfural with NH_3 and N_2 environments. J Anal Appl Pyrolysis. 2016;121:258–66. https://doi.org/10.1016/j.jaap.2016.08.005.
79. Nagasaki YH, Tamura M, Yabushita M, Nakagawa Y, Tomishige K. Dehydration of amides to nitriles over heterogeneous silica-supported molybdenum oxide catalyst. ChemCatChem. 2022;14:e202101846. https://doi.org/10.1002/cctc.202101846.
80. Jia X, Ma J, Wang M, Li X, Gao J, Xu J. Alkali α-MnO_2/NaxMnO_2 collaboratively catalyzed ammoxidation-pinner tandem reaction of aldehydes. Cat Sci Technol. 2016;6:7429–36. https://doi.org/10.1039/C6CY00874G.
81. Jia X, Ma J, Xia F, Xu Y, Gao J, Xu J. Carboxylic acid-modified metal oxide catalyst for selectivity-tunable aerobic ammoxidation. Nat Commun. 2018;9:933–40. https://doi.org/10.1038/s41467-018-03358-x.
82. Xu Y, Jia X, Ma J, Gao J, Xia F, Li X, Xu J. Efficient synthesis of 2, 5-dicyanofuran from biomass-derived 2, 5-diformylfuran via an oximation–dehydration strategy. ACS Sustain Chem Eng. 2018;6:2888–92. https://doi.org/10.1021/acssuschemeng.7b03913.
83. Senthamarai T, Chandrashekhar VG, Rockstroh N, Rabeah J, Bartling S, Jagadeesh RV, Beller M. A "universal" catalyst for aerobic oxidations to synthesize (hetero)aromatic aldehydes, ketones, esters, acids, nitriles, and amides. Chem. 2022;8:508–31. https://doi.org/10.1016/j.chempr.2021.12.001.

84. Hua M, Song J, Huang X, Liu H, Fan H, Wang W, He Z, Liu Z, Han B. Highly efficient oxidative cyanation of aldehydes to nitriles over se, S, N-tri-doped hierarchically porous carbon nanosheets. Angew Chem Int Ed. 2021;60:21479–85. https://doi.org/10.1002/anie.202107996.
85. Choi JE, Shinoda S, Asano Y, Gröger H. Aldoxime dehydratase mutants as improved biocatalysts for a sustainable synthesis of biorenewables-based 2-furonitrile. Catalysts. 2020;10:362. https://doi.org/10.3390/catal10040362.
86. Zhan W, Tong M, Ji L, Zhang H, Ge Z, Wang X, Liu R. Continuous-flow synthesis of nitriles from aldehydes via Schmidt reaction. Chin Chem Lett. 2019;30:973–6. https://doi.org/10.1016/j.cclet.2019.01.006.
87. Bornschein C, Werkmeister S, Wendt B, Jiao H, Alberico E, Baumann W, Junge H, Junge K, Beller M. Mild and selective hydrogenation of aromatic and aliphatic (di)nitriles with a well-defined iron pincer complex. Nat Commun. 2014;5:4111. https://doi.org/10.1038/ncomms5111.
88. Hu H, Wu S, Yan F, Makha M, Sun Y, Du C, Li Y. Recent developments in electrosynthesis of nitriles and electrocatalytic cyanations. J Energy Chem. 2022;70:542–75. https://doi.org/10.1016/j.jechem.2022.02.054.
89. Yan G, Zhang Y, Wang J. Recent advances in the synthesis of aryl nitrile compounds. Adv Synth Catal. 2017;359:4068–105. https://doi.org/10.1002/adsc.201700875.
90. Lin F, Liu C, Wang X, Hu C, Wu S, Xiao R. Catalytic oxidation of biorefinery corncob lignin via zirconium(IV) chloride and sodium hydroxide in acetonitrile/water: a functionality study. Sci Total Environ. 2019;675:203–12. https://doi.org/10.1016/j.scitotenv.2019.04.224.
91. Liu C, Wu S, Zhang H, Xiao R. Catalytic oxidation of lignin to valuable biomass-based platform chemicals: a review. Fuel Process Technol. 2019;191:181–201. https://doi.org/10.1016/j.fuproc.2019.04.007.
92. Song Y, Motagamwala AH, Karlen SD, Dumesic JA, Ralph J, Mobley JK, Crocker M. A comparative study of secondary depolymerization methods on oxidized lignins. Green Chem. 2019;21:3940–7. https://doi.org/10.1039/C9GC01663E.
93. Wang Y, Sun S, Li F, Cao X, Sun R. Production of vanillin from lignin: the relationship between β-O-4 linkages and vanillin yield. Ind Crop Prod. 2018;116:116–21. https://doi.org/10.1016/j.indcrop.2018.02.043.
94. Albarrán-Velo J, López-Iglesias M, Gotor V, Gotor-Fernández V, Lavandera I. Synthesis of nitrogenated lignin-derived compounds and reactivity with laccases. Study of their application in mild chemoenzymatic oxidative processes. RSC Adv. 2017;7:50459–71. https://doi.org/10.1039/C7RA10497A.
95. Desvals A, Baudron SA, Bulach V, Hoffmann N. Photocycloadditions of arenes derived from lignin. J Org Chem. 2021;86:13310–21. https://doi.org/10.1021/acs.joc.1c01361.
96. Mudshinge SR, Potnis CS, Xu B, Hammond GB. HCl·DMPU-assisted one-pot and metal-free conversion of aldehydes to nitriles. Green Chem. 2020;22:4161–4. https://doi.org/10.1039/D0GC00757A.
97. Khalafi-Nezhad A, Mohammadi S. Chitosan supported ionic liquid: a recyclable wet and dry catalyst for the direct conversion of aldehydes into nitriles and amides under mild conditions. RSC Adv. 2014;4:13782. https://doi.org/10.1039/c3ra43440k.
98. Ghosh P, Saha B, Pariyar GC, Tamang A, Subba R. Fe_3O_4-nanoparticles catalyzed an efficient synthesis of nitriles from aldehydes. Tetrahedron Lett. 2016;57:3618–21. https://doi.org/10.1016/j.tetlet.2016.06.125.
99. Villas-Boas Hoelz L, Gonçalves BT, Barros JC, Mendes Da Silva JF. Solvent free, microwave assisted conversion of aldehydes into nitriles and oximes in the presence of $NH_2OH{\cdot}HCl$ and TiO_2. Molecules. 2010;15:94–9. https://doi.org/10.3390/molecules15010094.
100. Gatera T, Zhuang D, Yan R. One-pot, catalyst-free synthesis of nitriles from aldehydes using aminating reagent $MsONH_3OTf$. Synlett. 2022;33:182–6. https://doi.org/10.1055/s-0041-1737293.

101. Bonjour O, Nederstedt H, Arcos-Hernandez MV, Laanesoo S, Vares L, Jannasch P. Lignin-inspired polymers with high glass transition temperature and solvent resistance from 4-hydroxybenzonitrile, vanillonitrile, and syringonitrile methacrylates. ACS Sustain Chem Eng. 2021;9:16874–80. https://doi.org/10.1021/acssuschemeng.1c07048.
102. Xu L, Na X, Zhang L, Dong Q, Dong G, Wang Y, Fang Z. Selective production of terephthalonitrile and benzonitrile via pyrolysis of polyethylene terephthalate (PET) with ammonia over ca$(OH)_2/Al_2O_3$ catalysts. Catalysts. 2019;9:436. https://doi.org/10.3390/catal9050436.
103. Xu L, Zhang L, Song H, Dong Q, Dong G, Kong X, Fang Z. Catalytic fast pyrolysis of polyethylene terephthalate plastic for the selective production of terephthalonitrile under ammonia atmosphere. Waste Manag. 2019;92:97–106. https://doi.org/10.1016/j.wasman.2019.05.011.
104. Dornan LM, Cao Q, Flanagan JCA, Crawford JJ, Cook MJ, Muldoon MJ. Copper/TEMPO catalysed synthesis of nitriles from aldehydes or alcohols using aqueous ammonia and with air as the oxidant. Chem Commun. 2013;49:6030. https://doi.org/10.1039/c3cc42231c.
105. Das AK, Nandy S, Bhar S. $Cu(OAc)_2$ catalysed aerobic oxidation of aldehydes to nitriles under ligand-free conditions. RSC Adv. 2022;12:4605–14. https://doi.org/10.1039/D1RA07701E.
106. Chen H, Sun S, Xi H, Hu K, Zhang N, Qu J, Zhou Y. Catalytic oxidative conversion of aldehydes into nitriles using $NH_3{\cdot}H_2O/FeCl_2/NaI/Na_2S_2O_8$: a practical approach to febuxostat. Tetrahedron Lett. 2019;60:1434–6. https://doi.org/10.1016/j.tetlet.2019.04.043.
107. Zhang Y, Zhao X, Zhang H, Yan X, Zhao J. Conversion of benzyl alcohol to benzonitrile over a $Cu_{10.3}/SiO_2$ catalyst. Appl Catal A Gen. 2016;522:45–53. https://doi.org/10.1016/j.apcata.2016.04.031.
108. Yoon Y, Kim BR, Lee CY, Kim J. Heterogeneous copper-catalyzed aerobic oxidative conversions of benzaldehydes with aqueous ammonia to give benzonitriles. Asian J Org Chem. 2016;5:746–9. https://doi.org/10.1002/ajoc.201600121.
109. Pan L, Fu W, Zhang L, Wang S, Tang T. Highly dispersed Co species in N-doped carbon enhanced the aldehydes ammoxidation reaction activity. Mol Catal. 2022;518:112087. https://doi.org/10.1016/j.mcat.2021.112087.
110. Wang H, Xu D, Guan E, Wang L, Zhang J, Wang C, Wang S, Xu H, Meng X, Yang B, Gates B, Xiao F. Atomically dispersed ru on manganese oxide catalyst boosts oxidative cyanation. ACS Catal. 2020;10:6299–308. https://doi.org/10.1021/acscatal.0c00485.

Chapter 6
Catalytic Upgrading of Bio-Based Ketonic Acids to Pyrrolidones with Hydrogen Donor Sources

Yiyuan Jiang, Yixuan Liu, Jinshu Huang, and Hu Li

Abstract Comprehensive utilization of biomass resources is necessary for the sustainable production of value-added chemicals and biofuels. For hydrogenative conversion processes, the type of hydrogen donor source among the many catalytic parameters plays an important role in upgrading of bio-based ketonic acids (mainly including biomass-derived levulinic acid (LA) and its derivatives) to pyrrolidones. In this chapter, one-pot processes for the reductive amination of bio-based ketonic acids and their advantages are introduced based on the classification of hydrogen donor sources, and the involved mechanisms and reaction parameters (e.g., reaction time, temperature, pressure, and reactor type) are discussed. The functional requirements of the catalyst are highlighted, and the effect of catalyst composition and properties on its activity and reaction path is described in detail. An outlook on the development of promising reduction systems for biomass conversion is given.

Keywords Biomass · Levulinic acid · Amination · Hydrogen donor source · Pyrrolidone

6.1 Introduction to Reductive Amination of Levulinic Acid (LA)

Selective conversion of biomass substrates to value-added organic compounds is necessary for reducing the use of fossil energy, protecting the environment, and promoting sustainable development of society [1, 2]. Levulinic acid (LA) is ranked by the U.S. Department of Energy as one of the top ten most promising and value-added biomass derivatives [3]. LA can be derived from cellulose and hemicellulose

Y. Jiang · Y. Liu · J. Huang · H. Li (✉)
National Key Laboratory of Green Pesticide, Key Laboratory of Green Pesticide & Agricultural Bioengineering, Ministry of Education, State-Local Joint Laboratory for Comprehensive Utilization of Biomass, Center for R&D of Fine Chemicals, Guizhou University, Guiyang, Guizhou, China
e-mail: hli13@gzu.edu.cn

Z. Fang et al. (eds.), *Production of N-containing Chemicals and Materials from Biomass*, Biofuels and Biorefineries 12,
https://doi.org/10.1007/978-981-99-4580-1_6

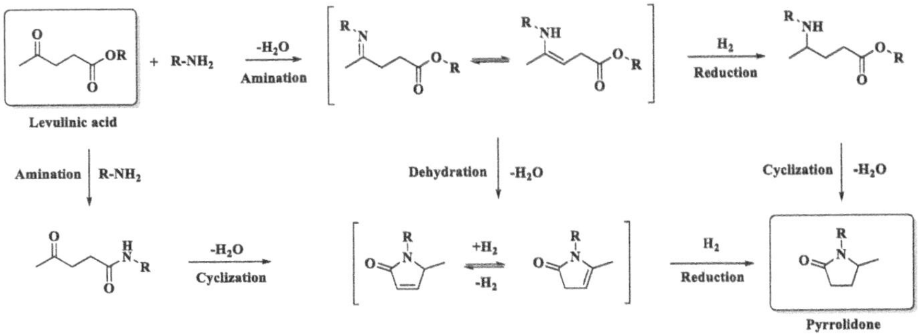

Fig. 6.1 Mechanism of catalytic LA reductive amination to pyrrolidone

under acidic conditions [4–6] and can undergo hydrogenation, oxidation, reductive amination, and other reactions to produce 1,4-propanediol, 2-methyltetrahydrofuran, succinic acid, γ-valerolactone, pyrrolidones, etc. [7–11]. Among these value-added small organic molecules, N-substituted pyrrolidines are widely used in cosmetics, food, medicine, printing, dyeing, and fuel industries [12–14]. Therefore, the efficient conversion of LA into N-substituted pyrrolidone and pyrrolidine compounds has become one of the potential ways for biomass valorization [15].

In particular, the use of biomass-derived LA has a series of advantages over the conventional two-stage liquid phase production process, in which fossil lactones are used as raw materials to react with alkylamines, followed by hydrogenation [16, 17]. On the one hand, LA has the unique molecular structure of two functional groups with carbonyl and carboxyl groups [18]. Based on this, LA can be expected to be able to directly convert aldehydes and ketones into secondary amines by one-pot method (single reactor) or primary reaction through reductive amination, in which all substrates and catalysts are mixed in the same operating unit [19–21]. Hydrogen and metal catalysts facilitate the catalytic conversion process without intermediate reaction, separation, and purification steps [22]. When functionalized carbonyl substrates (e.g., keto acids) participate in a reaction, the generated amine will immediately undergo an intramolecular cyclization reaction to obtain the corresponding lactam [23, 24]. On the other hand, it is necessary to reduce the dependence of human society on fossil fuels, energy consumption, and carbon dioxide emissions, comprehensively considering the environmental impact of chemical production processes [25–28]. All of these factors have led researchers to propose conversion schemes for replacing non-renewable fossil sources.

In the process of cascade amination and amidation of LA and its esters to generate pyrrolidone, the substrate successively undergoes the formation of imine by condensation of LA and amine, hydrogenation, and intramolecular amidation to obtain the final product [24, 29]. Fig. 6.1 shows the reduction, amination, hydrogenation, and other processes involved in the process. In the currently reported work, H_2 and formic acid (FA) are mainly used as hydrogen donor sources [30–32]. Ammonium formate (AF), hydrosilane, borohydric compounds, and other substances have been gradually used in organic synthesis as hydrogen donor sources [33, 34]. Catalytic systems of reductive amination of LA to form pyrrolidone from these hydrogen donor sources are introduced in the following sections.

6.2 Hydrogen Gas

H_2 is a clean and economical energy source, and its environmental impact as a reducing agent is relatively low [35], but most of the cascade reactions require high temperature or pressurized conditions. Even though H_2 production technology is relatively mature, there is limited H_2 capacity in high-pressure tanks or high-specific surface area materials for storage are needed [36, 37]. Safety risks existing in transportation and the use of H_2 in biomass conversion processes are still being assessed [37, 38].

6.2.1 Noble Metal-Based Catalysts

In developed systems, noble metal catalysts with hydrogenation functions usually achieve good reaction rates [39, 40]. As one of the most likely side reactions during the reaction process, C=O is easily reduced to C-OH and further esterified to form the by-product γ-valerolactone (GVL). The use of acidic carriers can prevent the side occurrence of reactions to some extent. ZrO_2 has stronger Lewis acid sites than Al_2O_3 and TiO_2. The strong Lewis acid-base effect in the Pd-based catalyst Pd/ZrO_2 prepared with ZrO_2 as the carrier greatly enhances the electrophilicity of the C=O group, promotes the amination reaction to generate amine intermediates with high selectivity, and avoids direct hydrogenation of LA to form GVL (Table 6.1) [41].

Ir-based catalysts mainly rely on the auxiliary effect of ligands to improve or increase the hydrogenation activity of the catalyst. For example, the Cp*Ir catalyst of bipyridine ligand had both dimethylamino (NMe_2) and o-hydroxyl (o-OH) group [42], in which The o-OH could promote H_2 dissociation, and the electron-giving ability of NMe_2 and o-OH increased the hydrogenation activity of the catalyst. A range of aromatic amines with electron-donating groups were obtained in 95% yield after 7 h of reaction at 20 bar hydrogen partial pressure in an aqueous solution (Fig. 6.2). For amines with electron-absorbing groups, it would affect the production rate of imine and require a long reaction time to achieve a favorable pyrrolidone yield (16 h, 89%).

The Ir1 catalyst of the zwitterionic iridium complex prepared using 2,2-dipyridine ligand was not sensitive to electronic effects [43]. Both the electron donor and the electron-absorbing substituent aniline could be converted to pyrrolidones with good yields (74–93%) (Fig. 6.2). Increasing the amount of catalyst showed a promoting effect on the tolerance of large-volume amines. Glucose was directly used as the

Table 6.1 Synthesis of pyrrolidone by amination over noble metal catalysts with H2 as a hydrogen donor source

Entry	Catalyst	Substrate	Amine	Pressure (bar)	Solvent	Tem (°C)	Time (h)	Yield [a] (%)	Ref.
1	Pd/ZrO_2	LA	*n*-octylamine	5	–	90	12	98.7	[41]
2	Cp*Ir	LA	Aniline	20	Water	80	10	96	[42]
3	Ir1	LA	4-methoxy-aniline	5	–	110	16	98	[43]
4	Ir-PVP	LA	Benzylamine	5	–	30	24	95	[44]
5	Ir/SiO_2-SO_3H	LA	Amine	34	–	30	24	95	[45]
6	Ru/C	Ita [a]	NH_3	150	Water	200	6	90	[46]
7	Ru-PP/CNTs	EL	Propylamine	30	–	120	24	98.7	[47]
8	Pt/TiO_{2D}	EL	Aniline	5	–	150	18	100	[49]
9	Pt/TiO_2-NT	EL	2-phenylethyl-amine	10	–	120	8	97	[52]
10	Pt/P-TiO_2	LA	N-octylamine	r.t.	Methanol	r.t.	3	97	[39]
11	Pt/c-C	LA	Aniline	20	Methanol	30	5	96	[54]
12	Pt@PW65S	LA	N-heptylamine	4	–	120	18	100	[55]
13	Pt-MoOx	LA	n-octylamine	3	–	100	20	95	[58]

[a]Ita: Itaconic acid
[b]Yield of pyrrolidone

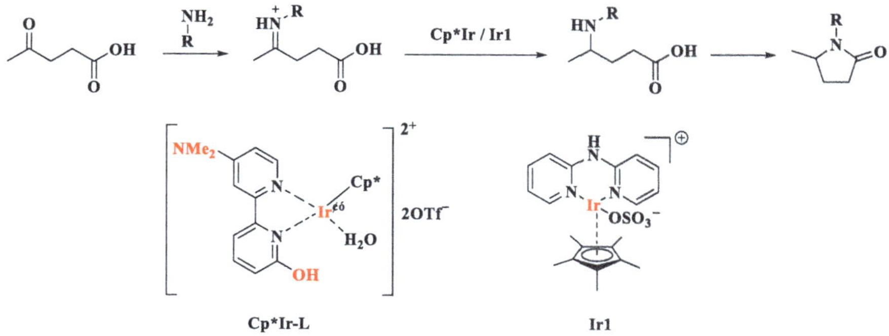

Fig. 6.2 Mechanism of pyrrolidone formation over an Ir-based catalyst

substrate, and pH was adjusted with sodium bicarbonate. The separation yield of pyrrolidone was 31% at 110 °C for 16 h under 20 bar H_2.

Amine compounds are generally obtained by hydrogenation of nitroaromatic hydrocarbons or nitriles, so the direct reaction of nitroaromatic hydrocarbons or nitriles with LA is an atomically economical strategy. Ir-PVP catalysts with small-size polyvinylpyrrolidone (PVP) stabilized Ir nanoparticles showed a good catalytic performance in the two-stage reaction of nitro compounds and nitrile hydrogenation to primary amines and hydrogenated secondary amines [44]. The Ir/SiO_2-SO_3H catalyst was prepared from SiO_2, 3-mercaptopropyltrimethoxysilane, and Ir-PVP [45]. It proved by thermodynamic and density functional theory (DFT) calculations that the sulfonic acid group (SO_3H) interacts with the carboxyl group of the substrate, and only the C2 of the carbonyl group was allowed to be used for nucleophilic addition to form imine and promote the cyclization of primary amines, whereas pyrrolidone was obtained as the final product. Due to the HOMO-LUMO interaction between nitrogen atoms and carboxyl groups, the cyclization step was determined by the electronegativity of the amine intermediate.

Ru used as a catalyst can not only activate H_2 but also activates C-O bonds. Activated carbon with good stability in an aqueous solution was selected as the carrier, and a Ru/C catalyst was prepared with itaconic acid as the substrate to explore the influence of conditions on the reaction [46]. It was found that the heat-driven reaction activation energy was high under high-temperature conditions and the solubility of H_2 was large under high-pressure conditions. Thus, the catalyst has enough H_2 activation C=O to increase the reaction rate of the reduction step. When no solvent was added, the system was too viscous, and the methylene group had adverse side reactions, leading to the yield being low. Accordingly, the substrate was diluted with water, and the methylene could be quickly reduced and the yield was increased to 65%. Another more efficient example was the Ru-PP/CNTs catalyst prepared using polymeric porphyrin-functionalized carbon nanotubes (PP/CNTs) [47]. Due to the strong π-π packing between the highly crosslinked PP porphyrin unit and the outer wall of the carbon nanotube, a layer of high-linked polymer metal porphyrin coating was formed on the surface of the catalyst, showing a bilayer structure of the amorphous polymer with a rough outer surface and the inner carbon nanotube core. The active site of catalyst hydrogenation was the central porphyrin unit of Ru. However, due to the photosensitivity of Ru porphyrin, transmission electron microscopy will damage it, so it was necessary to choose a suitable characterization method. Large H_2 storage systems directly connected to the reactor require an H_2 pressure of 30 bar, which poses a certain safety risk.

The reaction conditions of Pt-based catalysts are relatively mild. Studies have reported that suitable Brønsted acid or Lewis acid could activate imine reduction. Besides, transition metals could also play a certain role in promoting hydrogenation reduction systems of H_2 [48]. Therefore, Pt-based catalysts often used TiO_2-containing acid sites and transition metal elements as support.

The smaller the grain size of the supported metal, the higher the adsorption ratio of imine at the Pt site with low surface coordination, so the small size of the 0.2% Pt/TiO_{2D} catalyst showed excellent activity [49]. At the same time, the electronic effect had a great influence on the hydrogenation reaction [50], and the formation rate of imine in the H_2 atmosphere was about 6.5 times higher than that of the nitrogen atmosphere. In addition, the acidity of the catalyst was also conducive to the formation of iminomines. Hattori proposed that molecular H_2 was freely adsorbed in metals to form hydrogen atoms, which are transferred to the Lewis acid site and release an electron to become a proton. This proton stabilizes on the oxygen atom to form a proton acid center, which acted as the active site catalyzed by the hydrogenation reaction [51]. The H_2 on the Pt-based catalyst dissociated to form a proton acid center, which further promoted the formation of imine as a new acidic site, thereby improving reactivity. The steric hindrance of long chains and cycloalkanes hindered the adsorption and hydrogenation of iminomines. In addition to chain length and ring size, whether to extend the reaction time was also a factor affecting the yield and selectivity. The fixed-bed continuous reactors had advantages in the industrial processing of large-volume amines with high selectivity, but the obtained 80% yield was still lower than in batch reactors. After five consecutive cycles, the yield of the catalyst decreased by about 6%, mainly due to the adsorption of organic compounds on the surface of the catalyst, and the activity of the catalyst was restored after calcination. The same research group fine-tuned hydrogenation and acid functions to prepare Pt/TiO_2-NT catalyst with TiO_2 as the support [52]. Because Lewis acid content was only half of 0.2% Pt/TiO_{2D}, its reaction rate for aniline and ethyl levulinate (EL) was greatly reduced. However, for the one-pot reaction of nitrobenzene and EL, 0.2% Pt/TiO_{2D} could not continue the cascade reaction due to the toxic effect of organic matter on the Lewis site, while Pt/TiO_2-NT produced pyrrolidone with a high yield (>90%). It could be attributed to the fact that the interaction between the nitro grope and Pt-Ti interface sites was stronger than that with support sites. Especially in the hydrogenation step, the position of the Pt-Ti interface was the active site, which effectively avoids rapid inactivation of the catalyst caused by the strong adsorption of nitrobenzene.

The pt./P-TiO_2 catalyst prepared by porous TiO_2 nanosheets can realize the conversion of H_2 as a hydrogen donor source at room temperature and pressure, and the main factors can be classified according to the following three points

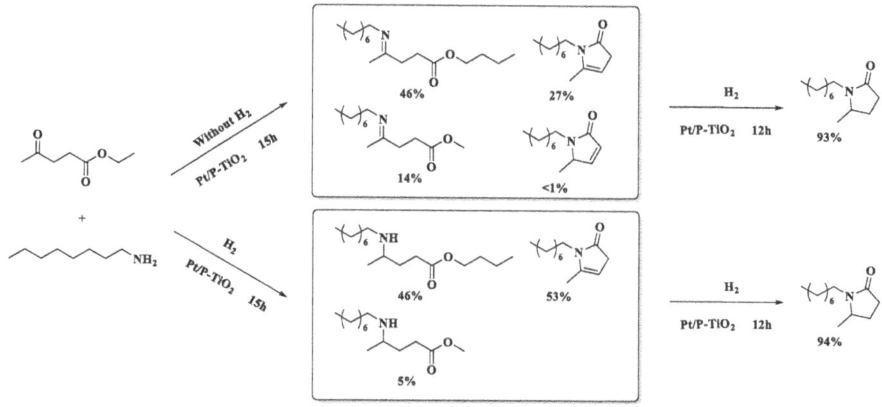

Fig. 6.3 Two catalytic reaction paths and product distribution of Pt/P-TiO_2 catalyst with or without hydrogenation

[39]. (1) compared with conventional TiO_2 support, the Ti atoms of the prepared catalyst are rich in positive charge and have stronger acidity, which effectively promoted the condensation formation and cyclization of intermediates; (2) the unique structure of porous nanosheets exposed more acidic sites, which enhanced the interaction between the catalyst and the carbonyl group; and (3) the lower electron density at the Pt catalytic site facilitated the desorption of products and reactants, resulting in higher catalytic activity [43, 53].

However, it should be noted that solvent-free systems would lead to high viscosity, especially at room temperature. More solvent dilution buffers are required, and the use of polar solvents such as methanol promotes the condensation of carbonyls and amino groups well. When methyl levulinate is used as a substrate, the solvent cannot be used due to the weak interaction between the ester group and the amino group and the low viscosity. A wide range of substrates is observed under pressurized H_2, and certainly, substrate selectivity is better under pressurized hydrogen (Fig. 6.3).

Another example of effective catalysis at room temperature and pressure is through Pt/c-C catalysts [54]. The C-C carrier is rich in oxygen-containing groups, which easily form hydrogen bonds with organic molecules to provide strong acidic sites for the condensation reaction. At the same time, the high degree of graphitization of the carrier easily affects the electronic structure of Pt nanoparticles, and the synergistic effect of the two gives the catalyst its high performance. By increasing the H_2 pressure from atmospheric pressure to 20 bar, the reaction time can be shortened from 3 h to 0.5 h and the yield can also reach 90%. Compared with Pt/P-TiO_2, the c-C carrier is cheaper and easier to obtain and has less influence on the environment.

Bifunctional catalysts are mostly used in one-pot synthesis methods to avoid a large amount of different catalysts and lower the number of reaction steps. It has been reported that solid acid catalysts containing Brønsted acid sites (e.g., PW65S,

PW79S-02%X-link) have a higher catalytic activity than solid acid catalysts containing Lewis acid sites (e.g., TiMNL, 2%Nb@TiMNL) [55]. Considering the C-N bond affinity, the supported Pt embodies higher hydrogenation reactivity than Pd metal nanoparticles. Due to the high acidity of the solid acid and the excellent heat resistance of the perfluorinated backbone, the prepared Pt@PW65S bifunctional catalyst could obtain 99% yield even under solvent-free conditions.

MoOx, TiOx, VOx, WOx, and other transition metal oxides with Lewis acidity and reducibility had a catalytic effect on promoting C=O and C-O bond reduction [56, 57]. Therefore, supported binary metal catalysts (M^1-M^2O_X/M^3O_X) showed a more prominent catalytic effect than single metal catalysts. Pt-MoO_X/TiO_2 stood out among catalysts of different metals and supports with a high yield of 99% [58]. In addition, the yield of pyrrolidone catalyzed by Pd-MoO_X/TiO_2 could reach 76%, and the yield of the remaining 22 catalysts was less than 20%. The high activity can be attributed to the acid-base interaction between the acid site of Pt-MoO_X/TiO_2 and the carboxyl groups and intermediates in LA, and the effective polarization of carbonyl groups promoted the formation and cyclization of iminomines.

6.2.2 Non-noble Metal-Based Catalysts

Replacing expensive noble metal catalysts with non-noble metal catalysts to promote conversion reactions is highly desirable. However, to achieve good catalytic performance with non-noble metal catalysts, it is necessary to design bimetallic or bifunctional catalysts while considering reaction conditions [59].

In the preparation of the bimetallic catalyst Co-Zr@Chitosan [60], Zr has a strong electronic state effect after adding Co, and the active site change from the spherical agglomeration into smaller and more dispersed particles. Moreover, the Co-Zr alloy increases the number of active sites and enhances the adsorption effect of the catalyst. In terms of the reaction process, the n-valence state of imine forms bonding and counterbonding with the metal d-valence state on the catalyst. The center of the d-state can regulate the bonding and counterbonding states, which can indicate the binding state of imine and catalyst [61, 62]. The influence of the alloy electronic effect could be understood as the center of the d band of Co-Zr was shifted in the direction of the Fermi level. Therefore, the high-charge anti-bond orbital of the reactant is more easily combined with the Co-Zr catalyst, resulting in higher product selectivity, which is the essence of bimetallic synergy. Due to the excess electron pairs, NH_3 is adsorbed to the center of Lewis acid, resulting in a lower yield of the magnetically charged catalyst in a recovery cycle. However, the yield can be increased to 86.7% by annealing H_2 reduction.

Ni-based catalysts have been widely used in catalytic systems for transforming biomass-derived compounds. The (CNFx@Ni@CNTs) catalyst prepared by the atomic deposition method encapsulates Ni nanoparticles in porous carbon nanotubes (CNTs) to effectively prevent the leaching and sintering of active metal [63]. The thickness of the carbon nanofilm is adjusted by controlling the number of

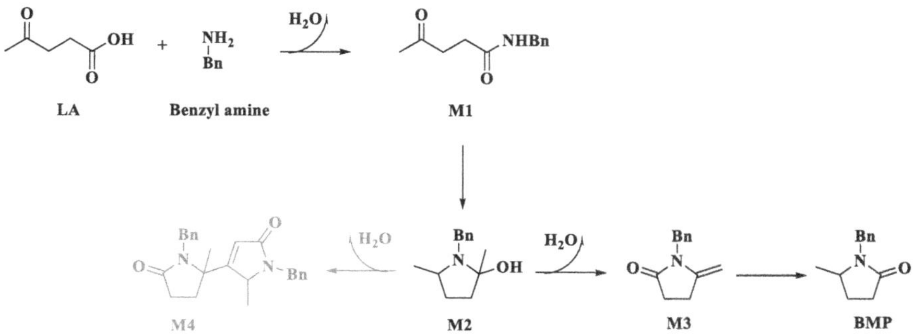

Fig. 6.4 Reductive amination of LA with benzyl aniline catalyzed by CNFx@Ni@CNTs

depositions, and its defective porous structure allowed the encapsulated metal content in good contact with the substrate. The best CNF30@Ni@CNTs could be reused 20 times while maintaining well activity. More interestingly, the amide intermediates first form during the reaction, rather than the imine. The amide intermediates then cyclize in tandem and undergo intramolecular dehydration to generate M3 intermediates. The catalyst effectively hydrogenates the C=C bond in M3 to produce the 1-benzyl-5-methylpyrrolidin-2-one (BMP) product instead of dehydration between M2 molecules to form the byproduct M4 (Fig. 6.4).

The use of solid carriers and continuous flow reactors can improve catalytic performance. NiP/SiO_2-A600 was used as a bifunctional catalyst to effectively convert EL and *n*-hexylamine in a continuous flow reactor for 6 h [64]. Parameters such as phosphorus precursor, reduction temperature, carrier type, reactant ratio, and acid diluent have been proven to affect catalytic performance. These catalyst parameters mainly depend on the hydrogenation capacity (responsible for the hydrogenation reaction) and acid amount (responsible for imine formation and intramolecular amidation) of different catalysts. Among them, SiO_2 support was superior to Al_2O_3 support. When pyrrolidone was synthesized by Al_2O_3 support, EL was adsorbed through the carbonyl group of the Lewis site, which was directly hydrogenated to generate GVL and further hydrogenated to generate 1,4-pentanediol, thereby reducing the selectivity of the final product. Hydrogen atoms were considered to be almost impossible to spill into non-reducing supports, and hydrogenation could occur at the boundary between metal catalysts and supports.

The prepared FeNi alloy catalyst increased the metal loading (21 wt% Fe, 31 wt% Ni) to prevent the graphite shell from being overwrapped [65]. Using a benchtop continuous flow reactor with an internal hydrogen donor source, the yield was basically stable in the flow state for 52.5 h (Table 6.2). When excess LA (2 equivalents) was added and the H_2 pressure was raised (25 bar to 50 bar), the selectivity could be improved by using a mixture of ethanol and 2-methylfuran as the solvent. The metal cooperated with the unsaturated matrix to promote the dissociation of H_2, and then transferred the dissociated hydrogenated species to the matrix for catalytic hydrogenation to complete the reaction. Fe was the preferential adsorption site for

Table 6.2 Synthesis of pyrrolidone by amination over non-noble metal catalysts with H2 as a hydrogen donor source

Entry	Catalyst	Substrate	Amine	Pressure (bar)	Solvent	Tem (°C)	Time (h)	Yield[a] (%)	Reusabi-lity [b]	Ref.
1	Co-Zr@Chitosan-20	LA	NH_3	30	Water	130	24	92.8	1	[60]
2	CNF@Ni@CNTs	LA	Benzylamine	30	GVL	130	6	99	–	[63]
3	Ni_2P/SiO_2-A600	EL	N-hexylamine	10	–	170	–	98	6 h	[64]
4	FeNi	LA	Phenethyl-amine	10	EtOH/2-MTHF	150	–	87	52.5 h	[65]

[a] The number of the catalyst recycle/continuous flow time
[b] Yield of 5MPs

hydrogen on the surface of the prepared alloy catalyst. The reaction began with direct acid hydrolysis of glucose and yielded about 40% of pyrrolidone.

In the catalytic system of hydrogen as a hydrogen donor, catalysts such as noble metals and non-noble metal bases show good catalytic activity in hydroreduction. Too strong or too weak an interaction between the catalyst surface and the adsorbate results in poor catalytic activity [63]. Weak chemical adsorption of intermediates makes imine difficult to activate, and strong adsorption of H_2 will lead to excessive blockage of metal surfaces. Therefore, it is necessary to prepare suitable catalysts for the overall reaction.

6.3 Formic Acid

Compared with LA reductive amination using gaseous H_2 as the hydrogen donor source, FA as a hydrogen donor source does not require external ventilation and can lower the generation of chemical waste in a biorefinery process. The mixture of LA/FA was obtained directly from the hydrolysis of biomass such as cellulose, and the FA could be selectively decomposed in situ to generate H_2 using a metal catalyst or be hydrogenated by transfer hydrogenation. However, the CO that may be produced during the FA dehydration process is toxic to the metal, and the acidic conditions may corrode the equipment, which makes the preparation of the catalyst and the selection of the reactor require certain corrosion resistance [15, 23].

Au nanoparticles (Au/ZrO_2-VS) deposited on acid-resistant ZrO_2 support could be used for a one-pot reaction of LA and FA with an equal molar amount [66]. The Pt, Pd-based catalysts supported on the ZrO_2 support were basically inactive due to the almost complete inhibition of the key LA hydrogenation step in the case of FA decomposition to form CO. Conventional SiO_2, activated carbon, Al_3O_2, CeO_2, and other carrier-supported Au-based catalysts would dissolve and leach under highly acidic conditions, so the reaction cannot be carried out. The synergistic effect of the carrier and metal in the Au/ZrO_2-VS catalyst could excite negatively charged gold clusters, which promote electron back-transport from Au to HCOO* intermediates [67]. The C-H bonds in the FA are more easily cleaved and dehydrogenated for subsequent cascade reactions. It was worth mentioning that ultra-small particle size catalysts prepared using an improved deposition method were essential for fast and efficient catalytic reactions [68].

Another catalytic protocol proposed by Wei's team was to use Ir-based cyclization Ir *1a* with OMe substituents as a catalyst [69]. HCOONa was added to adjust the pH in the system with FA as the hydrogen donor source, and it was found that the Ir *1a* catalyst only participated in the in-situ reduction of imine intermediates and did not participate in the subsequent cyclization step (Fig. 6.5a). Ru-NPs is a ruthenium nanoparticle catalyst prepared by the thermal decomposition of $Ru_3(CO)_{12}$ [70]. There are two reaction pathways, namely reduction of imine by the decomposition of FA, and formation of hydrogenated intermediates, respectively (Fig. 6.5b).

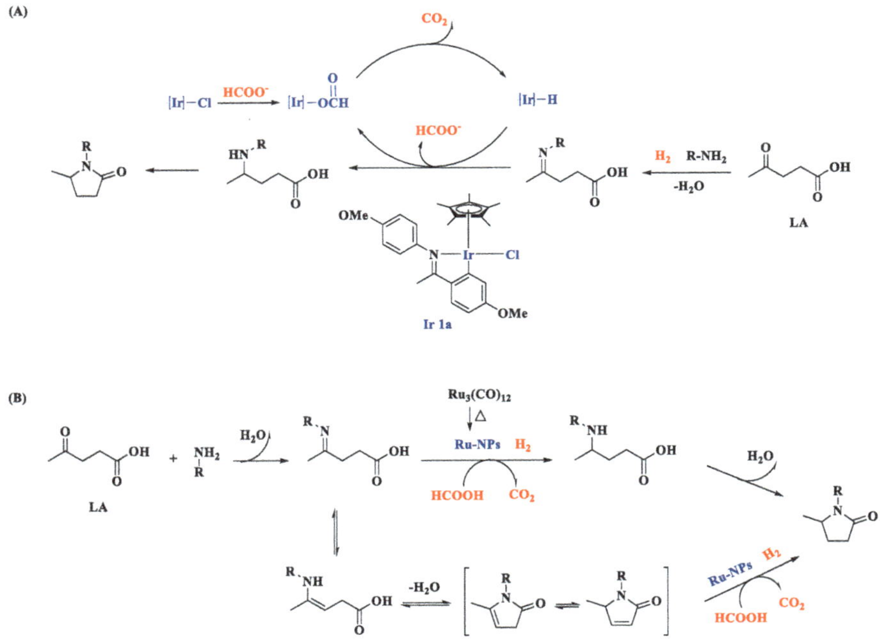

Fig. 6.5 Catalytic pathways of Ir 1a (**a**) and Ru-NPs (**b**) with FA as hydrogen donor source

The addition of alkali could promote the decomposition of FA into H_2 with CO_2 as the hydrogen donor source. On this basis, the $Fe_3(CO)_{12}$ cluster was used as a catalyst to try two new systems by Metzker et al. Under the action of the catalyst, FA, imidazole (InM), ethylamine and LA were added, and the pyrrolidone yield reached 87% [71]. The use of ethylamine only as a base actually improved the conversion of FA, and could also participate in reductive amination as a substrate, thereby avoiding the addition of other bases. In addition, $Fe_3(CO)_{12}$ was used directly in the hydrolysis reaction from bagasse waste, resulting in a crude pyrrolidone yield of approximately 73%.

In addition to noble metal and non-noble metal catalytic systems, metal-free reaction systems were also hot research topic in recent years. Based on the Leuckart Wallach reaction [72], Xiao et al. [73] developed a catalyst-free reaction system for the catalytic conversion of LA and amide with FA as a hydrogen donor source and DMSO as a solvent. With the addition of 5 equivalents of FA and 1 equivalent of triethylamine relative to LA, the conversion rate of LA was greatly increased to 89% and the BMP yield was 87% after 4 h of reaction at 100 °C. The sulfinyl group was essential for the conduct of the reaction, and its high solvent alkalinity was expected to promote the nucleophilic attack of amines on ketones and the nucleophilic attack of formate on imide. In addition, the acidity of the equilibrium system could not be ignored to obtain favorable reaction progress. However, DMSO with a boiling point of 189 °C was difficult to handle, which brings certain difficulties to the separation and purification of products.

The use of continuous flow reactors has a much larger specific surface area than that of batch reactors, which significantly improves the heat exchange and mixing rates of catalyst-free systems. Building on the former, Zhang's team used FA as a hydrogen donor source with LA and *n*-butylamine as model reaction to screen a series of solvents [74]. It was found that the use of acetonitrile instead of DMSO could reduce the proportion of high boiling impurities and improve selectivity. A slight excess of FA and *n*-butylamine provided enough hydrogen to guarantee good yields. When the ratio of LA:FA:*n*-butylamine was 1:2:2, acidity imbalance and complex separation could be avoided. At the same time, the optimal residence time was only 10 min, which greatly reduced the reaction time.

In heat treatment under harsh conditions without the use of catalysts and solvents, FA reacted directly with LA and amine to obtain a series of 5-methylpyrrolidone derivatives [15]. Increased the temperature and pressure, so that FA could only produce pyrrolidone and formamide (FAM), under the condition that water and amine were present at the same time, and the reaction time was shortened. The system could also have a yield of about 80% for most aliphatic amines with low boiling points and low viscosity. At the same time, the e-factor of the system was as low as 0.2, which had broad industrial application potential.

Li et al. [75] effectively completed a series of bio-based ketoic acids to undergo N-lactam conversion with LA for 3 h using FA as a hydrogen donor source and FAM as a nitrogen source. When the LA/FA/FAM ratio was 1:10:3, the reaction was converted faster. Among them, FA not only acts as a hydrogen donor, but also manifests itself as an acid promoting the formation of C-N bonds in the hydro-amination process. It was critical that there were sufficient sources of nitrogen and hydrogen for amination and transfer hydrogenation, respectively. DFT calculations and kinetic experiments showed that the initial rapid transfer hydroamination helped to increase the reaction rate, while the amidation process facilitated subsequent cyclization reactions to obtain the required lactam. In addition, the design of suitable reactors had some potential for large-scale synthesis.

6.4 Ammonium Formate

The reductive amination reaction system of AF added to LA can decompose H_2 and NH_3, and act as both hydrogen and nitrogen sources. Under the action of a catalyst, aldehydes or ketones are reacted with AF, ammonia is nucleophilically added to carbonyls, and then imine is generated after dehydration, which is subsequently reduced for subsequent cyclization steps [76–78].

Using N-heterocyclic carbon compounds (NHCs) as carriers for Ru metal particles, NHC-Ru catalysts were prepared to directly catalyze the reduction of LA and AF without solvents [79]. This method of tandem amination directly from ketones or aldehydes without adding additional amines, and demonstrated a high degree of

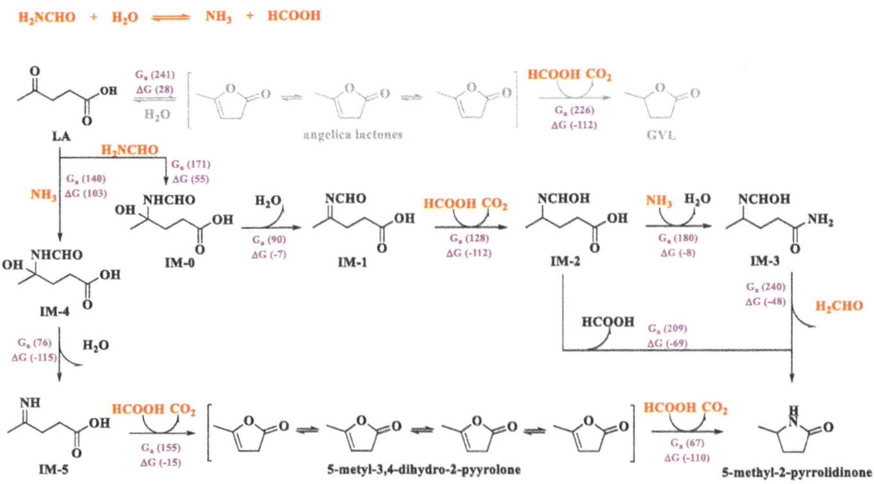

Fig. 6.6 Cycloamination path of LA under H_2NCHO and water. Calculated free energies and enthalpies given in parentheses (kJ/mol). IM: intermediate

environmental friendliness and relatively low synthesis costs. In continuous large-volume conversions, the robust and thermally stable carrier allowed the catalyst to be reused 37 times.

Another example was the reaction with water using a Raney nickel catalyst in a Polytetrafluoroethylene reactor [80]. Quadruple equivalent FAM was rapidly decomposed into NH_3, H_2 and CO_2 at 180 °C. The complete conversion of LA avoided the formation of by-product GVL, and 5-methyl-2-pyrrolidone was obtained in 94% yield. The reaction using a mixture of FA, LA, and primary amines generated a series of BMPs with yields of 90–95%.

A typical catalyst-free strategy was developed by Li et al. [81, 82]. The method was achieved by water-induced in situ release of FA from AF and FAM to provide 5-methylpyrrolidone (MPL) and a range of N-(un)substituted lactams. Deionized water was involved in the in situ release of FA from N-formyl compounds (e.g., AF or FAM) and inhibited the production of chemically stable by-products (e.g., N-formyl-5-methyl-2-pyrrolidone) under anhydrous conditions. DFT calculations showed that the water-assisted release of FA was a rate-determining step, highlighting the importance of hydrogen transfer reaction. At the same time, 5-methyl-3,4-dihydro-2-pyrrolidone and its tautomeric structure were confirmed as key intermediates in the reaction (Fig. 6.6). Even if EL with different acidity were used as substrates, about 90% yield could be obtained. Comparative experiments showed that the flow reactor only took 20 min to achieve a large-scale yield, which greatly shortened the time compared with the batch reactor which takes 4 h.

6.5 Hydrosilane

The use of air-stable and safe hydrosilane as a mild reducing agent in the production of pyrrolidone has also been reported [83, 84]. The catalyst focuses on the activation of hydrosilane and promotes cyclization. And the preparation of catalysts with appropriate acidity plays a positive role in avoiding excessive reduction. At the same time, when using hydrosilane as a hydrogen donor, it is important to note that by-product siloxanes will be produced.

A series of In-based catalysts have been reported in a one-pot catalytic system using $PhSiH_3$ as a reducing agent. Ketamine was first formed with a primary amine over the catalysis of $In(OAc)_3$ at 120 °C [83]. Then, the silicon-based esters were formed by dehydrosilylation of carboxylic acids [85], and accompanied by the formation of H_2 and iminosilylation. Finally, lactam products were obtained by cyclization (Fig. 6.7a). The generated siloxane waste could be removed by adding methanol precipitate. Lewis acidity of InI_3 was stronger than that of $In(OAc)_3$, resulting in the overreduction of lactams to N-substituted pyrrolidine and piperidine [86, 87].

Another set of reactions that occurred under relatively mild conditions were catalyzed by $AlCl_3$ and $RuCl_3$. Taking LA and aniline as the reaction model, 3 equivalents of $PhSiH_3$ were added [88]. At room temperature, 95% of 5-methyl-1-phenylpyrrolidine 2-ketone was catalyzed by $AlCl_3 \cdot 6H_2O$, and 93% of 2-methyl-1-phenylpyrrolidine and a small amount of 2-methyl-1-phenylpyrrole were catalyzed at 45 °C. Possible reaction pathways are shown in Fig. 6.7b. Both $AlCl_3$ and $RuCl_3$ could catalyze the dehydrating cyclization of imines into 5-methyl-1-phenylpyrrolidin-2-one [89], but under the catalysis of $RuCl_3$, further conversion is observed. The product 2-methyl-1-phenylpyrrole, was obtained by hydrodehydration of intermediates.

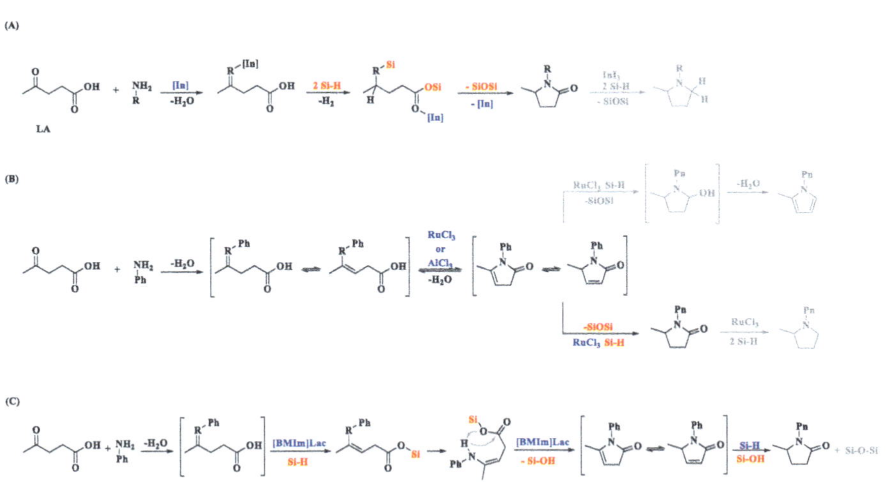

Fig. 6.7 Three catalytic examples using (a & b) $PhSiH_3$ and (c) $(EtO)_3SiH$ as a hydrogen donor

The ionic liquid, 1-Butyl-3-methylimidazolium lactate ([BMIm][Lac]) could be applied to metal-free catalyzed systems as a mission-specific ionic liquid [90]. Using triethoxysilane ($(EtO)_3SiH$) which has a lower price than $PhSiH_3$ as the hydrogen donor source [91], [BMIm][Lac] could obtain a range of pyrrolidone products at 80 ° C without solvent due to its good solubility. Lactate anion-based catalysts with both acidic (COO) and basic (OH) sites played a key role in activating $(EtO)_3SiH$ and promoting cyclization. The reaction process mainly includes the formation of ketimine, while the reaction with hydrogensilane to silicon ether, cyclization and reduction to pyrrolidone, is accompanied by a small amount of siloxane (Fig. 6.7c).

6.6 Boron Hydride

As a borohydride, ammonia borane (AB) is regarded as one of the promising hydrogen donor sources to replace H_2 due to its safe transport and controlled-release hydrogen characteristics [34, 92, 93]. Besides, AB as a water-soluble compound can decompose H_2 under specific conditions, with its high hydrogen content of 19.8 wt% and relatively low H_2 release temperature (<350 °C). In the process of transformation of LA, boron hydride has attracted attention as a hydrogen donor [33, 94, 95]. In general, both metal catalyst systems and catalyst-free systems can react, mainly relying on the formation of imines and the dihydrogen transfer of AB.

AuPd nanoparticles were loaded on carbon support to prepare a C-$Au_{66}Pd_{34}$ catalyst [96]. The catalyst was effectively used to generate atmospheric pressure H_2 and reduce amination reactions. First, the catalyst co-decomposes AB with methanol to produce H_2, which was collected and used in a balloon for subsequent reactions. Biological keto acids such as EL and LA were then proved to be able to obtain a series of lactam products by reducing amination with H_2 and catalysts. During the reaction, the metal composition and proportion of the catalyst played an important role. Too much Pd slowed down the reaction kinetics, and too much Au sharply reduced the catalytic activity, both of which worked best at a ratio of 66:34. It was worth mentioning that the catalyst showed good reusability. After the catalyst was reused 10 times, the morphology did not change significantly, and the catalyst still showed good activity after being placed in air for 1 year for C-$Au_{66}Pd_{34}$.

Another example of AB splitting H_2 as a hydrogen donor source was the Cu/H-ZSM-5 catalyst [97]. Water was added to the reaction system, and hydrogen bonds were formed at the interface, promoting the formation of imine and subsequent hydrogenation [98]. In addition, water was generated during the production of LA from biomass, so water not only did not need to be separated, but also speeded up the reaction. There was a strong polarization between the Cu-based catalyst and the substrate, mainly due to the interaction between the C=O of the substrate and the Lewis/Brønsted acid sites on the catalyst surface.

The catalytic reactions of AB in different solvents with a series of biomass-based keto acids and amines were explored [99]. It was confirmed that the reductive amination reaction occurred through imine production and dihydrotransfer of

AB. It was worth mentioning that AB could achieve a high yield of 92% with only 0.5 equivalent to the atomic economy. The choice of solvent had a great influence on the mode of action of AB. In proton solvent systems of methanol and water, AB was decomposed into borane and ammonia. In aprotic solvents (e.g., DMSO), AB was converted to polyiminoborane (PIB), polyaminoborane (PB), cyclotriborazane (CTB), and polyborazylene (PBZ). In this pathway, AB produced two-fold equivalent hydrogen by dihydrogen transfer. In the substrate expansion, alkyl amine, benzylamine, long-chain aliphatic amine, etc. could achieve an excellent yield (>90%). However, the aromatic substituent amine ketone group was easily reduced to hydroxyl compounds to obtain moderate yields and the use of EL produced a range of ester by-products.

6.7 Conclusions and Outlook

Hydrogen donor sources such as H_2, FA, AF, hydrosilane, and AB are effective in the conversion of biomass ketoacid derivatives (e.g., LA to pyrrolidone). Suitable catalysts have a significant role in promoting the activation of H_2 and the succeeding hydrogenation of substrates and intermediates. In particular, when the catalyst has suitable acid-base sites and solid support, the interaction of bifunctional metals or alloys can make the corresponding catalytic systems achieve good yields. The type of substrate, the choice of solvent, the setting of temperature and pressure, and the design of the reactor are also influencing factors in the choice of the hydrogen donor source.

Some issues in establishing clear reaction systems and catalytic materials can provide a reference for further development of suitable hydrogen donor types:

1. It is necessary to reduce the pressure and temperature of using H_2, and try to catalyze the reaction under mild conditions to improve safety.
2. Acid-resistant and corrosion-resistant catalysts and reactor materials are the key support for acidic FA systems, which minimize the toxic effect of CO on metal catalysts.
3. The AF system reduces the use of exogenous amines, and higher temperature and water addition can inhibit the production of by-products.
4. The selection of cheap and large quantities of hydrosilane is still a potential direction, and by-products (e.g., siloxanes and silyl ethers) need to be removed and purified efficiently.
5. Appropriate catalysts and solvents need to be selected to accelerate the green and efficient hydrolysis of AB, expand other systems of borohydride, and study other borohydride systems.
6. Other hydrogen donor sources can be developed to efficiently convert and upgrade a wide range of biomass substrates under mild conditions.

In both laboratory and industrial synthesis, there are numerous opportunities to develop biomass conversion systems mediated by various materials as hydrogen donors. This chapter has mainly introduced some realistic methods and processes in the catalytic production of lactams like pyrrolidones from biomass feedstocks.

References

1. Li H, Guo H, Su Y, Hiraga Y, Fang Z, Hensen EJ, Watanabe M, Smith RL. N-formyl-stabilizing quasi-catalytic species afford rapid and selective solvent-free amination of biomass-derived feedstocks. Nat Commun. 2019;10:699. https://doi.org/10.1038/s41467-019-08577-4.
2. Meng Y, Jian Y, Li J, Wu H, Zhang H, Saravanamurugan S, Yang S, Li H. Surface-active site engineering: synergy of photo-and supermolecular catalysis in hydrogen transfer enables biomass upgrading and H_2 evolution. Chem Eng J. 2023;452:139477. https://doi.org/10.1016/j.cej.2022.139477.
3. Arias PL, Cecilia JA, Gandarias I, Iglesias J, López Granados M, Mariscal R, Morales G, Moreno-Tost R, Maireles-Torres P. Oxidation of lignocellulosic platform molecules to value-added chemicals using heterogeneous catalytic technologies. Cat Sci Technol. 2020;10:2721–57. https://doi.org/10.1039/D0CY00240B.
4. Lin L, Han X, Han B, Yang S. Emerging heterogeneous catalysts for biomass conversion: studies of the reaction mechanism. Chem Soc Rev. 2021;50:11270–92. https://doi.org/10.1039/D1CS00039J.
5. Tiong YW, Yap CL, Gan S, Yap WSP. Conversion of biomass and its derivatives to levulinic acid and levulinate esters via ionic liquids. Ind Eng Chem Res. 2018;57:4749–66. https://doi.org/10.1039/D1CS00039J.
6. Nordin N, Md Illias R, Manas NHA, Ramli ANM, Selvasembian R, Azelee NIW, Rajagopal R, Thirupathi A, Chang SW, Ravindran B. Highly sustainable cascade pretreatment of low-pressure steam heating and organic acid on pineapple waste biomass for efficient delignification. Fuel. 2022;321:124061. https://doi.org/10.1016/j.fuel.2022.124061.
7. Tan J, Li Y, Tan X, Wu H, Li H, Yang S. Advances in pretreatment of straw biomass for sugar production. Front Chem. 2021;9:696030. https://doi.org/10.3389/fchem.2021.696030.
8. Li T, Chen C, Brozena AH, Zhu JY, Xu L, Driemeier C, Dai J, Rojas OJ, Isogai A, Wagberg L, Hu L. Developing fibrillated cellulose as a sustainable technological material. Nature. 2021;590:47–56. https://doi.org/10.1038/s41586-020-03167-7.
9. Liu Y, Chen D, Li M, Zhang H, Li H. Catalytic stereoselective conversion of biomass-derived 4′-methoxypropiophenone to trans-anethole with a bifunctional and recyclable hf-based polymeric nanocatalyst. Polymers. 2021;13:2808. https://doi.org/10.3390/polym13162808.
10. Pileidis FD, Titirici MM. Levulinic acid biorefineries: new challenges for efficient utilization of biomass. ChemSusChem. 2016;9:562–82. https://doi.org/10.1002/cssc.201501405.
11. Li Z, Jiang Y, Li Y, Zhang H, Li H, Yang S. Advances in diels–alder/aromatization of biomass furan derivatives towards renewable aromatic hydrocarbons. Cat Sci Technol. 2022;12:1902–21. https://doi.org/10.1039/D1CY02122B.
12. Bhardwaj V, Gumber D, Abbot V, Dhiman S, Sharma P. Pyrrole: a resourceful small molecule in key medicinal hetero-aromatics. RSC Adv. 2015;5:15233–66. https://doi.org/10.1039/C4RA15710A.
13. Roy S, Reiser O. A catalytic multicomponent approach for the stereoselective synthesis of cis-4,5-disubstituted pyrrolidinones and tetrahydro-3H-pyrrolo[3,2-c]quinolines. Angew Chem Int Ed. 2012;51:4722–5. https://doi.org/10.1002/anie.201107831.
14. Hao L, Zhu K, Zhang S, Yu D. The green preparation of poly n-vinylpyrrole nanoparticles. RSC Adv. 2016;6:90354–9. https://doi.org/10.1039/C6RA18234H.
15. Ledoux A, Sandjong Kuigwa L, Framery E, Andrioletti B. A highly sustainable route to pyrrolidone derivatives-direct access to biosourced solvents. Green Chem. 2015;17:3251–4. https://doi.org/10.1039/C5GC00417A.

16. Davies J, Svejstrup TD, Fernandez Reina D, Sheikh NS, Leonori D. Visible-light-mediated synthesis of amidyl radicals: transition-metal-free hydroamination and N-arylation reactions. J Am Chem Soc. 2016;138:8092–5. https://doi.org/10.1021/jacs.6b04920.
17. He J, Chen L, Liu S, Song K, Yang S, Riisager A. Sustainable access to renewable n-containing chemicals from reductive amination of biomass-derived platform compounds. Green Chem. 2020;22:6714–47. https://doi.org/10.1039/D0GC01869D.
18. Yan L, Yao Q, Fu Y. Conversion of levulinic acid and alkyl levulinates into biofuels and high-value chemicals. Green Chem. 2017;19:5527–47. https://doi.org/10.1039/C7GC02503C.
19. Gomez S, Peters JA, Maschmeyer T. The reductive amination of aldehydes and ketones and the hydrogenation of nitriles: mechanistic aspects and selectivity control. Adv Syn Catal. 2002;344: 1037–57. https://doi.org/10.1002/1615-4169(200212)344:10<1037::AID-ADSC1037>3.0.CO;2-3.
20. Di Menno Di Bucchianico D, Wang Y, Buvat JC, Pan Y, Casson Moreno V, Leveneur S. Production of levulinic acid and alkyl levulinates: a process insight. Green Chem. 2022;24: 614–46. https://doi.org/10.1039/D1GC02457D.
21. Xue Z, Yu D, Zhao X, Mu T. Upgrading of levulinic acid into diverse n-containing functional chemicals. Green Chem. 2019;21:5449–68. https://doi.org/10.1039/C9GC02415H.
22. Besson M, Gallezot P, Pinel C. Conversion of biomass into chemicals over metal catalysts. Chem Rev. 2014;114:1827–70. https://doi.org/10.1021/cr4002269.
23. Li H, Guo H, Fang Z, Aida TM, Smith RL. Cycloamination strategies for renewable n-heterocycles. Green Chem. 2020;22:582–611. https://10.1039/C9GC03655E
24. Moreno-Marrodan C, Liguori F, Barbaro P. Sustainable processes for the catalytic synthesis of safer chemical substitutes of N-methyl-2-pyrrolidone. Mol Catal. 2019;466:60–9. https://doi.org/10.1016/j.mcat.2019.01.014.
25. El-Sayed TH, Aboelnaga A, El-Atawy MA, Hagar M. Ball milling promoted n-heterocycles synthesis. Molecules. 2018;23:1348. https://doi.org/10.3390/molecules23061348.
26. Zhang S, Pan H, Huang J, Li Y, Zhang H. A highly effective biomass-derived solid acid catalyst for biodiesel synthesis through esterification. Front Chem. 2022;10:882235. https://doi.org/10.3389/fchem.2022.882235.
27. Li Y, Zhu K, Jiang Y, Chen L, Zhang H, Li H, Yang S. Biomass-derived hydrophobic metal-organic frameworks solid acid for green efficient catalytic esterification of oleic acid at low temperatures. Fuel Process Technol. 2023;239:107558. https://doi.org/10.1016/j.fuproc.2022.107558.
28. Wang H, Peng X, Zhang H, Yang S, Li H. Microorganisms-promoted biodiesel production from biomass: a review, Energ. Conv Manage. 2021;X 12:100137. https://doi.org/10.1016/j.ecmx.2021.100137.
29. Irrgang T, Kempe R. Transition-metal-catalyzed reductive amination employing hydrogen. Chem Rev. 2020;120:9583–674. https://doi.org/10.1021/acs.chemrev.0c00248.
30. Chaparro-Garnica J, Navlani-García M, Salinas-Torres D, Berenguer-Murcia Á, Morallón E, Cazorla-Amorós D. Efficient production of hydrogen from a valuable CO_2-derived molecule: formic acid dehydrogenation boosted by biomass waste-derived catalysts. Fuel. 2022;320: 123900. https://doi.org/10.1016/j.fuel.2022.123900.
31. Li Z, Cheng J, Yan H, Lei Z, Yan J, Ren S, Wang Z, Kang S, Shui H. Ethanol and formic acid as synergistic solvents for converting lignite to platform chemicals. Fuel. 2021;293:120491. https://doi.org/10.1016/j.fuel.2021.120491.
32. Xu H, Ju J, Li H. Toward efficient heterogeneous catalysts for in-situ hydrodeoxygenation of biomass. Fuel. 2022;320:123891. https://doi.org/10.1016/j.fuel.2022.123891.
33. Zhao W, Li H, Zhang H, Yang S, Riisager A. Ammonia borane-enabled hydrogen transfer processes: insights into catalytic strategies and mechanisms. Green Energ Environ. 2023; https://doi.org/10.1016/j.gee.2022.03.011.
34. Ye M, Wu J, Fan G. Hierarchical porous cobalt/carbon hybrid anchored Ru-catalyzed ammonia-borane hydrolysis for efficient H_2 release. Fuel. 2022;321:123982. https://doi.org/10.1016/j.fuel.2022.123982.

35. Guo S, Li X, Li J, Wei B. Boosting photocatalytic hydrogen production from water by photothermally induced biphase systems. Nat Commun. 2021;12:1343. https://doi.org/10.1038/s41467-021-21526-4.
36. Ahmed SF, Mofijur M, Chowdhury SN, Nahrin M, Rafa N, Chowdhury AT, Nuzhat S, Ong HC. Pathways of lignocellulosic biomass deconstruction for biofuel and value-added products production. Fuel. 2022;318:123618. https://doi.org/10.1016/j.fuel.2022.123618.
37. Ahmed A, Seth S, Purewal J, Wong-Foy AG, Veenstra M, Matzger AJ, Siegel DJ. Exceptional hydrogen storage achieved by screening nearly half a million metal-organic frameworks. Nat Commun. 2019;10:1568. https://doi.org/10.1038/s41467-019-09365-w.
38. Zhang P, Guo Y, Chen J, Zhao Y, Chang J, Junge H, Beller M, Li Y. Streamlined hydrogen production from biomass. Nat Catal. 2018;1:332–8. https://doi.org/10.1038/s41929-018-0062-0.
39. Xie C, Song J, Wu H, Hu Y, Liu H, Zhang Z, Zhang P, Chen B, Han B. Ambient reductive amination of levulinic acid to pyrrolidones over pt nanocatalysts on porous TiO_2 nanosheets. J Am Chem Soc. 2019;141:4002–9. https://doi.org/10.1021/jacs.8b13024.
40. Bukhtiyarova MV, Bukhtiyarova GA. Reductive amination of levulinic acid or its derivatives to pyrrolidones over heterogeneous catalysts in the batch and continuous flow reactors: a review. Renew Sust Energ Rev. 2021;143:110876. https://doi.org/10.1016/j.rser.2021.110876.
41. Zhang J, Xie B, Wang L, Yi X, Wang C, Wang G, Dai Z, Zheng A, Xiao F-S. Zirconium oxide supported palladium nanoparticles as a highly efficient catalyst in the hydrogenation-amination of levulinic acid to pyrrolidones. ChemCatChem. 2017;9:2661–7. https://doi.org/10.1002/cctc.201600739.
42. Xu Z, Yan P, Jiang H, Liu K, Zhang ZC. Iridium-catalyzed reductive amination of levulinic acid to pyrrolidinones under H_2 in water. Chin J Chem. 2017;35:581–5. https://doi.org/10.1002/cjoc.201600726.
43. Wang S, Huang H, Bruneau C, Fischmeister C. Selective and efficient iridium catalyst for the reductive amination of levulinic acid into pyrrolidones. ChemSusChem. 2017;10:4150–4. https://doi.org/10.1002/cssc.201701299.
44. Chaudhari C, Shiraishi M, Nishida Y, Sato K, Nagaoka K. One-pot synthesis of pyrrolidones from levulinic acid and amines/nitroarenes/nitriles over the Ir-PVP catalyst. Green Chem. 2020;22:7760–4. https://doi.org/10.1039/D0GC01725F.
45. Martínez JJ, Silva L, Rojas HA, Romanelli GP, Santos LA, Ramalho TC, Brijaldo MH, Passos FB. Reductive amination of levulinic acid to different pyrrolidones on Ir/SiO_2-SO_3H: elucidation of reaction mechanism. Catal Today. 2017;296:118–26. https://doi.org/10.1016/j.cattod.2017.08.038.
46. Belle A, Tabanelli T, Fiorani G, Perosa A, Cavani F, Selva M. A multiphase protocol for selective hydrogenation and reductive amination of levulinic acid with integrated catalyst recovery. ChemSusChem. 2019;12:3343–54. https://doi.org/10.1002/cssc.201900925.
47. Zhang T, Ge Y, Wang X, Chen J, Huang X, Liao Y. Polymeric ruthenium porphyrin-functionalized carbon nanotubes and graphene for levulinic ester transformations into gamma-valerolactone and pyrrolidone derivatives. ACS Omega. 2017;2:3228–40. https://doi.org/10.1021/acsomega.7b00427.
48. Yang J, Xue J, Cheng J. Understanding the role of thermodynamics in catalytic imine reductions. Chem Soc Rev. 2019;48:2913–26. https://doi.org/10.1039/C9CS00036D.
49. Vidal JD, Climent MJ, Concepcion P, Corma A, Iborra S, Sabater MJ. Chemicals from biomass: Chemoselective reductive amination of ethyl levulinate with amines. ACS Catal. 2015;5:5812–21. https://doi.org/10.1021/acscatal.5b01113.
50. Zaccheria F, Santoro F, Iftitah ED, Ravasio N. Brønsted and Lewis solid acid catalysts in the valorization of citronellal. Catalysts. 2018;8:410. https://doi.org/10.3390/catal8100410.
51. Hattori H, Shishido T. Molecular hydrogen-originated protonic acid site as active site on solid acid catalyst. Catal Surv Jpn. 1997;1:205–13. https://doi.org/10.1023/A:1019081031021.

52. Vidal JD, Climent MJ, Corma A, Concepcion DP, Iborra S. One-pot selective catalytic synthesis of pyrrolidone derivatives from ethyl levulinate and nitro compounds. ChemSusChem. 2017;10:119–28. https://doi.org/10.1002/cssc.201601333.
53. Chen Z, Ibrahim I, Hao D, Liu X, Wu L, Wei W, Su D, Ni B. Controllable design of nanoworm-like nickel sulfides for efficient electrochemical water splitting in alkaline media. Materials Today Energy. 2020;18:100573. https://doi.org/10.1016/j.mtener.2020.100573.
54. Wu Y, Zhao Y, Wang H, Zhang F, Li R, Xiang J, Wang Z, Han B, Liu Z. Ambient reductive synthesis of n-heterocyclic compounds over cellulose-derived carbon supported pt nanocatalyst under H_2 atmosphere. Green Chem. 2020;22:3820–6. https://doi.org/10.1039/D0GC01177K.
55. Barbaro P, Liguori F, Oldani C, Moreno-Marrodán C. Sustainable catalytic synthesis for a bio-based alternative to the reach-restricted N-methyl-2-pyrrolidone. Adv Sust Syst. 2020;4: 1900117. https://doi.org/10.1002/adsu.201900117.
56. Boffa A, Lin C, Bell A, Somorjai GA. Promotion of CO and CO_2 hydrogenation over rh by metal oxides: the influence of oxide Lewis acidity and reducibility. J Catal. 1994;149:149–58. https://doi.org/10.1006/jcat.1994.1280.
57. Nakagawa Y, Tamura M, Tomishige K. Catalytic reduction of biomass-derived furanic compounds with hydrogen. ACS Catal. 2013;3:2655–68. https://doi.org/10.1021/cs400616p.
58. Touchy AS, Hakim Siddiki SMA, Kon K, Shimizu K. Heterogeneous pt catalysts for reductive amination of levulinic acid to pyrrolidones. ACS Catal. 2014;4:3045–50. https://doi.org/10.1021/cs500757k.
59. Xue Z, Liu Q, Wang J, Mu T. Valorization of levulinic acid over non-noble metal catalysts: challenges and opportunities. Green Chem. 2018;20:4391–408. https://doi.org/10.1039/C8GC02001A.
60. Wu P, Li H, Fang Z. Synergistic catalysis of Co-Zr/CNx bimetallic nanoparticles enables reductive amination of biobased levulinic acid. Adv. Sust. Syst. 2022;6:2100321. https://doi.org/10.1002/adsu.202100321.
61. Bera B, Chakraborty A, Kar T, Leuaa P, Neergat M. Density of states, carrier concentration, and flat band potential derived from electrochemical impedance measurements of N-doped carbon and their influence on electrocatalysis of oxygen reduction reaction. J Phys Chem C. 2017;121: 20850–6. https://doi.org/10.1021/acs.jpcc.7b06735.
62. Mao S, Wang C, Wang Y. The chemical nature of N-doping on N-doped carbon supported noble metal catalysts. J Catal. 2019;375:456–65. https://doi.org/10.1016/j.jcat.2019.06.039.
63. Gao G, Sun P, Li Y, Wang F, Zhao Z, Qin Y, Li F. Highly stable porous-carbon-coated ni catalysts for the reductive amination of levulinic acid via an unconventional pathway. ACS Catal. 2017;7:4927–35. https://doi.org/10.1021/acscatal.7b01786.
64. Wang Y, Nuzhdin AL, Shamanaev IV, Kodenev EG, Gerasimov EY, Bukhtiyarova MV, Bukhtiyarova GA. Effect of phosphorus precursor, reduction temperature, and support on the catalytic properties of nickel phosphide catalysts in continuous-flow reductive amination of ethyl levulinate. Int J Mol Sci. 2022;23:1106. https://doi.org/10.3390/ijms23031106.
65. Chieffi G, Braun M, Esposito D. Continuous reductive amination of biomass-derived molecules over carbonized filter paper-supported FeNi alloy. ChemSusChem. 2015;8:3590–4. https://doi.org/10.1002/cssc.201500804.
66. Liu X, He L, Liu Y, Cao Y. Supported gold catalysis: from small molecule activation to green chemical synthesis. Acc Chem Res. 2014;47:793–804. https://doi.org/10.1021/ar400165j.
67. Tedsree K, Chan CWA, Jones S, Cuan Q, Li WK, Gong XQ, Tsang SCE. ^{13}C nmr guides rational design of nanocatalysts via chemisorption evaluation in liquid phase. Science. 2011;332:224–8. https://doi.org/10.1126/science.1202364.
68. Oliver-Meseguer J, Cabrero-Antonino JR, Domínguez I, Leyva-Pérez A, Corma A. Small gold clusters formed in solution give reaction turnover numbers of 10^7 at room temperature. Science. 2012;338:1452–5. https://doi.org/10.1126/science.1227813.
69. Wei Y, Wang C, Jiang X, Xue D, Li J, Xiao J. Highly efficient transformation of levulinic acid into pyrrolidinones by iridium catalysed transfer hydrogenation. Chem Commun. 2013;49: 5408–10. https://doi.org/10.1039/C3CC41661E.

70. Ortiz-Cervantes C, Flores-Alamo M, García JJ. Synthesis of pyrrolidones and quinolines from the known biomass feedstock levulinic acid and amines. Tetrahedron Lett. 2016;57:766–71. https://doi.org/10.1016/j.tetlet.2016.01.018.
71. Metzker G, Dias RMP, Burtoloso ACB. Iron-catalyzed reductive amination from levulinic and formic acid aqueous solutions: an approach for the selective production of pyrrolidones in biorefinery facilities. Chemistry Select. 2018;3:368–72. https://doi.org/10.1002/slct.201702580.
72. Kitamura M, Lee D, Hayashi S, Tanaka S, Yoshimura M. Catalytic leuckart-wallach-type reductive amination of ketones. J Org Chem. 2002;67:8685–7. https://doi.org/10.1021/jo0203701.
73. Wei Y, Wang C, Jiang X, Xue D, Liu Z, Xiao J. Catalyst-free transformation of levulinic acid into pyrrolidinones with formic acid. Green Chem. 2014;16:1093–6. https://doi.org/10.1039/C3GC42125B.
74. Ma T, Zhang H, Yin G, Zhao J, Zhang Y. Catalyst-free reductive amination of levulinic acid to n-substituted pyrrolidinones with formic acid in continuous-flow microreactor. J Flow Chem. 2018;8:35–43. https://doi.org/10.1007/s41981-018-0005-6.
75. Wu H, Dai W, Saravanamurugan S, Li H, Yang S. Quasi-catalytic approach to n-unprotected lactams via transfer hydro-amination/cyclization of biobased keto acids, ACS sustainable chem. Eng. 2019;7:10207–13. https://doi.org/10.1021/acssuschemeng.9b00412.
76. Skachilova SY, Zheltukhin NK, Sergeev VN, Davydova NK. Reductive amination of sterically hindered arylaminoketones using a modified leuckart reaction. Pharmaceut Chem J. 2018;52: 545–9. https://doi.org/10.1007/s11094-018-1857-7.
77. Leuckart R. Ueber eine neue bildungsweise von tribenzylamin. Ber Dtsch Chem Ges. 1885;18: 2341–4. https://doi.org/10.1002/cber.188501802113.
78. Chubb F, Hay AS, Sandin RB. The leuckart reaction of some 1, 5-diketones. J Am Chem Soc. 1953;75:6042–4. https://doi.org/10.1021/ja01119a508.
79. Sun Z, Chen J, Tu T. NHC-based coordination polymers as solid molecular catalysts for reductive amination of biomass levulinic acid. Green Chem. 2017;19:789–94. https://doi.org/10.1039/C6GC02591A.
80. Amarasekara AS, Lawrence YM. Raney-ni catalyzed conversion of levulinic acid to 5-methyl-2-pyrrolidone using ammonium formate as the H and N source. Tetrahedron Lett. 2018;59: 1832–5. https://doi.org/10.1016/j.tetlet.2018.03.087.
81. Li H, Wu H, Zhang H, Su Y, Yang S, Hensen EJM. A facile direct route to n-(un)substituted lactams by cycloamination of oxocarboxylic acids without external hydrogen. ChemSusChem. 2019;12:3778–84. https://doi.org/10.1002/cssc.201901780.
82. Wu H, Yu Z, Li Y, Xu Y, Li H, Yang S. Hot water-promoted catalyst-free reductive cycloamination of (bio-)keto acids with $HCOONH_4$ toward cyclic amides. J Supercrit Fluid. 2020;157:104698. https://doi.org/10.1016/j.supflu.2019.104698.
83. Ogiwara Y, Uchiyama T, Sakai N. Reductive amination/cyclization of keto acids using a hydrosilane for selective production of lactams versus cyclic amines by switching of the indium catalyst. Angew Chem Int Ed. 2016;55:1864–7. https://doi.org/10.1002/ange.201509465.
84. Ganem B. Inhibitors of carbohydrate-processing enzymes: design and synthesis of sugar-shaped heterocycles. Acc Chem Res. 1996;29:340–7. https://doi.org/10.1021/ar9502184.
85. Miyamoto K, Motoyama Y, Nagashima H. Selective reduction of carboxylic acids to aldehydes by a ruthenium-catalyzed reaction with 1,2-bis(dimethylsilyl) benzene. Chem Lett. 2012;41: 229–31. https://doi.org/10.1246/cl.2012.229.
86. Cheng C, Brookhart M. Iridium-catalyzed reduction of secondary amides to secondary amines and imines by diethylsilane. J Am Chem Soc. 2012;134:11304–7. https://doi.org/10.1021/ja304547s.
87. Li Y, Molina de La Torre JA, Grabow K, Bentrup U, Junge K, Zhou S, Brückner A, Beller M. Selective reduction of amides to amines by boronic acid catalyzed hydrosilylation. Angew Chem Int Ed. 2013;52:11577–80. https://doi.org/10.1002/anie.201304495.

88. Wu C, Luo X, Zhang H, Liu X, Ji G, Liu Z, Liu Z. Reductive amination/cyclization of levulinic acid to pyrrolidones versus pyrrolidines by switching the catalyst from $AlCl_3$ to $RuCl_3$ under mild conditions. Green Chem. 2017;19:3525–9. https://doi.org/10.1039/C7GC00999B.
89. Azizi N, Edrisi M. Deep eutectic solvent catalyzed eco-friendly synthesis of imines and hydrobenzamides. Monatshefte für Chemie-Chemical Monthly. 2015;146:1695–8. https://doi.org/10.1007/s00706-015-1447-2.
90. Wu C, Zhang H, Yu B, Chen Y, Ke Z, Guo S, Liu Z. Lactate-based ionic liquid catalyzed reductive amination/cyclization of keto acids under mild conditions: a metal-free route to synthesize lactams. ACS Catal. 2017;7:7772–6. https://doi.org/10.1021/acscatal.7b02231.
91. Addis D, Das S, Junge K, Beller M. Selective reduction of carboxylic acid derivatives by catalytic hydrosilylation. Angew Chem Int Ed. 2011;50:6004–11. https://doi.org/10.1002/anie.201100145.
92. Asim M, Zhang S, Wang Y, Maryam B, Sajid M, Shi C, Pan L, Zhang X, Zou J. Self-supporting NiCop for hydrogen generation via hydrolysis of ammonia borane. Fuel. 2022;318:123544. https://doi.org/10.1016/j.fuel.2022.123544.
93. Feng Y, Liao J, Chen X, Wang H, Guo B, Li H, Zhou L, Huang J, Li H. Synthesis of rattle-structured $CuCo_2O_4$ nanospheres with tunable sizes based on heterogeneous contraction and their ultrahigh performance toward ammonia borane hydrolysis. J Alloy Compd. 2021;863: 158089. https://doi.org/10.1016/j.jallcom.2020.158089.
94. Hasenbeck M, Becker J, Gellrich U. Efficient organocatalytic dehydrogenation of ammonia borane. Angew Chem Int Ed. 2020;59:1590–4. https://doi.org/10.1002/anie.201910636.
95. Demirci UB. Ammonia borane, a material with exceptional properties for chemical hydrogen storage. Int J Hydrogen Energ. 2017;42:9978–10013. https://doi.org/10.1016/j.ijhydene.2017.01.154.
96. Muzzio M, Yu C, Lin H, Yom T, Boga DA, Xi Z, Li N, Yin Z, Li J, Dunn JA, Sun S. Reductive amination of ethyl levulinate to pyrrolidones over AuPd nanoparticles at ambient hydrogen pressure. Green Chem. 2019;21:1895–9. https://doi.org/10.1039/C9GC00396G.
97. Boosa V, Varimalla S, Dumpalapally M, Gutta N, Velisoju VK, Nama N, Akula V. Influence of brønsted acid sites on chemoselective synthesis of pyrrolidones over H-ZSM-5 supported copper catalyst. Appl Catal B Environ. 2021;292:120177. https://doi.org/10.1016/j.apcatb.2021.120177.
98. Butler RN, Coyne AG. Water: Nature's reaction enforcer-comparative effects for organic synthesis "in-water" and "on-water". Chem Rev. 2010;110:6302–37. https://doi.org/10.1021/cr100162c.
99. Zhao W, Meier S, Yang S, Riisager A. Insights into ammonia borane-enabled green synthesis of n-substituted lactams from biomass-derived keto acids and amines, ACS sustainable chem. Eng. 2021;9:4377–82. https://doi.org/10.1021/acssuschemeng.1c00211.

Part II
Production of N-Containing Compounds via Biological Processes

Chapter 7
Microbial Production of Amine Chemicals from Sustainable Substrates

Fernando Pérez-García, Luciana F. Brito, and Volker F. Wendisch

Abstract With global population projected to be almost 10 billion people by 2050, the demand for food, chemicals and energy is expected to rise continuously. Nitrogen-containing (N-containing) compounds play a key role in many aspects of life and commercial processes, from the industrial production of fertilizers to the building blocks of life. Bio-based production of these N-containing compounds is gaining importance and replacing chemical synthesis approaches because of their economic and environmental advantages. Microbial synthesis often offers one-step fermentations and formation of less hazardous wastes as advantages compared to chemical synthesis. Moreover, microbial synthesis has additional benefits that include the use of renewable sources like side stream biomasses, adding environmental-friendly value to the bioprocesses. Many genetic engineering technologies have important impact in the field of metabolic engineering with microbial workhorses like the bacteria *Escherichia coli* or *Corynebacterium glutamicum*. The product spectrum of these production hosts has been expanded extensively to comprise valuable N-containing chemicals like amino acids, diamines, or lactams among others, which have great interest, for instance, as feed additives, as bioplastics precursors, or as molecules with relevant pharmacological activity.

This chapter summarizes the most recent advances in metabolic engineering of biotechnologically relevant microbial hosts to produce N-containing chemicals.

Keywords N-containing compounds · Amino acids derivatives · Diamines · Metabolic engineering · Synthetic biology · Microbial bioprocesses · Renewable substrates · *Escherichia coli* · *Corynebacterium glutamicum*

F. Pérez-García (✉) · L. F. Brito
Department of Biotechnology and Food Science, Norwegian University of Science and Technology, Trondheim, Norway
e-mail: fernando.perez-garcia@ntnu.no

V. F. Wendisch
Genetics of Prokaryotes, Faculty of Biology & Center for Biotechnology (CeBiTec), Bielefeld University, Bielefeld, Germany

Z. Fang et al. (eds.), *Production of N-containing Chemicals and Materials from Biomass*, Biofuels and Biorefineries 12,
https://doi.org/10.1007/978-981-99-4580-1_7

7.1 Metabolic Engineering for the Production of Amino Acids as N-Containing Building Blocks

Green chemical production by microbial processes is essential for the development of a sustainable society. To date, several N-containing compounds, including proteinogenic and non-proteinogenic amino acids, diamines, and their N-functionalized forms have been produced via bioprocesses with microbial cell factories (Figs. 7.1, 7.2, 7.3 and 7.4). Recent developments in metabolic engineering and synthetic biology technologies have enabled the extension, creation and fine-tuning of metabolic pathways to produce such compounds in a competent manner. Amino acids can be obtained from diverse processes as, for instance, L-aspartic acid via enzyme catalysis [1], DL-methionine via chemical synthesis [2], or L-cysteine by extraction of natural sources like human hair [3]. However, amino acid production by fermentation is one of the most important fields within industrial biotechnology

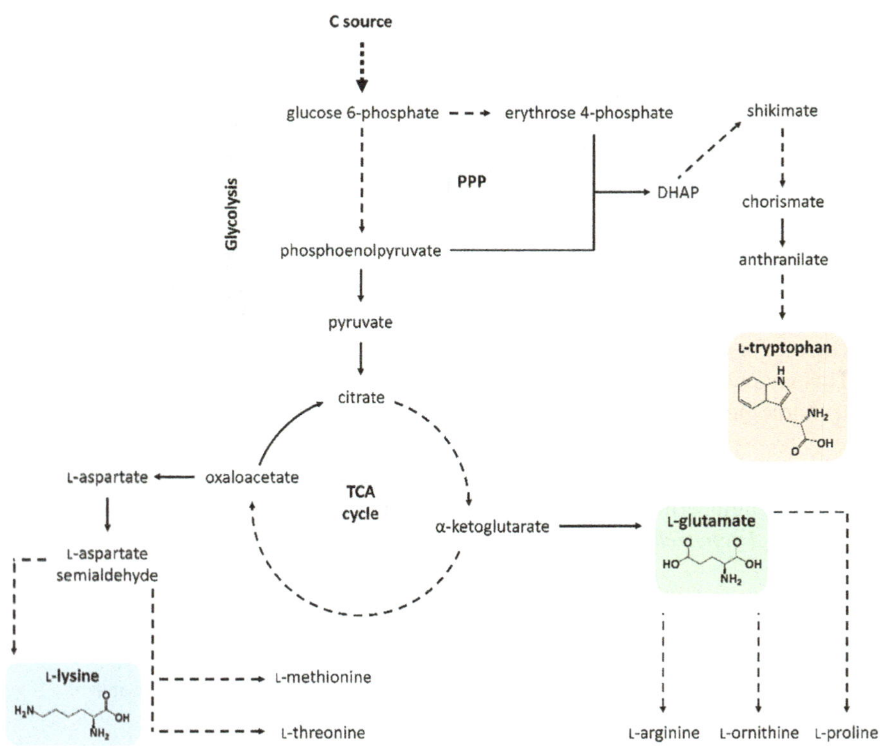

Fig. 7.1 *Biosynthetic pathways of amino acids connected with the central carbon metabolism.* Central metabolic pathways depicted: glycolysis, pentose phosphate pathway (PPP), and tricarboxylic acid cycle (TCA cycle). The biosynthesis of L-Glutamate family amino acids and L-aspartate family amino acids is directly linked to the TCA cycle. While L-tryptophan biosynthesis is linked to the PPP

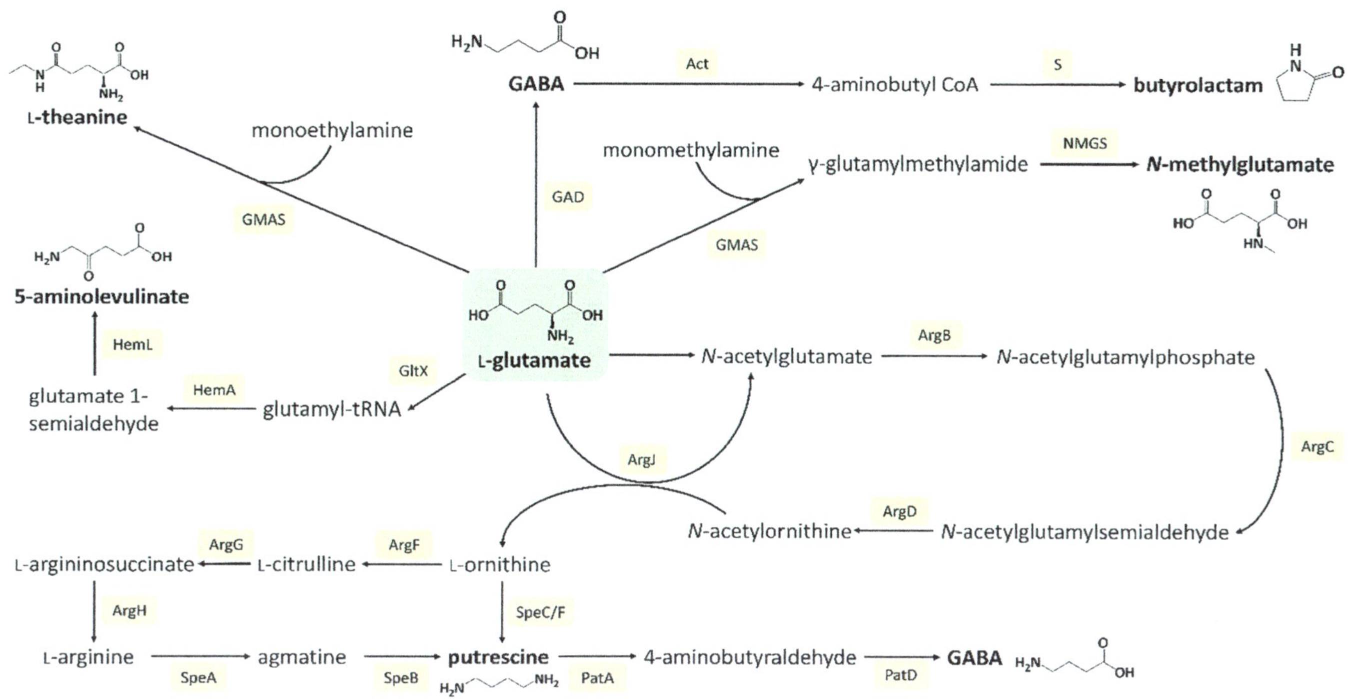

Fig. 7.2 *Biosynthetic pathways of* L*-glutamate derived compounds.* Act β-alanine CoA transferase, ArgB *N*-acetylglutamate kinase, ArgC *N*-acetylglutamylphosphate reductase, ArgD *N*-acetylornithine aminotransferase, ArgF ornithine transcarbamoylase, ArgG argininosuccinate synthase, ArgH argininosuccinate lyase, ArgJ ornithine acetyltransferase, GABA γ-aminobutyric acid, GAD glutamate decarboxylase, GltX glutamyl-tRNA synthetase, GMAS γ-glutamylmethylamide synthetase, HemA glutamyl-tRNA reductase, HemL glutamate-1-semialdehyde aminotransferase, NMGS *N*-methylglutamate synthase, PatA putrescine transaminase, PatD γ-aminobutyraldehyde dehydrogenase, S spontaneous, SpeA biosynthetic arginine decarboxylase, SpeB agmatinase, SpeC biosynthetic ornithine decarboxylase, SpeF degradative ornithine decarboxylase

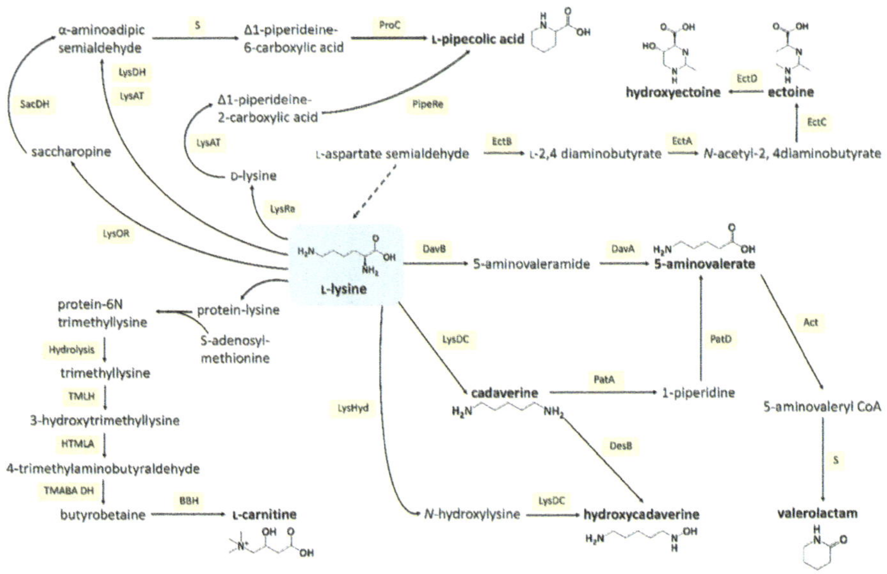

Fig. 7.3 *Biosynthetic pathways of* L*-lysine derived compounds*. Act β-alanine CoA transferase, BBH butyrobetaine hydroxylase, DavA 5-aminovaleramidase, DavB lysine monooxygenase, EctA diaminobutyrate acetyl transferase, EctB diaminobutyrate transaminase, EctC ectoine synthase, EctD ectoine hydroxylase, HTMLA hydroxytrimethyllysine aldolase, LysAT lysine aminotransferase, LysDC lysine decarboxylase, LysDH lysine dehydrogenase, LysHyd lysine hydroxylase, LysOR lysine oxidoreductase, LysRa lysine racemase, PatA putrescine transaminase, PatD γ-aminobutyraldehyde dehydrogenase, PipeRe piperidine carboxylate reductase, ProC pyrroline 5-carboxylate reductase, S spontaneous, SacDH saccharopine dehydrogenase, TMABA DH trimethylaminobutyraldehyde dehydrogenase, TMLH trimethyllysine hydroxylase

[4]. According to Industry-Experts (Industry-Experts 2022), the global demand for amino acids is projected to register a robust compound annual growth rate of 5%, mainly driven by above average growth in animal feed sector and sustained growth in nutraceutical and pharmaceutical sectors. Animal feed constitutes the largest, and the fastest growing application for amino acids globally for which the global value market is projected to reach US $15.8 billion by 2026. The global volume consumption of amino acids in animal feed and nutraceutical applications is projected to grow with corresponding compound annual growth rate of 6.1% and 5.5% respectively (Industry-Experts, 2022). L-Glutamic acid emerged as the leading amino acid accounting for over 40% of total amino acid market volume. L-Lysine, L-threonine, L-methionine and L-tryptophan are majorly consumed feed additives. L-Tryptophan is expected to be the fastest growing amino acid with an estimated compound annual growth rate of 18.0% from 2015 to 2022 (Foods & Beverages Research Report 2022).

To cope with this fast-increasing demand, two challenges have to be tackled. First, white biotechnology needs to be at technological forefront. From the classic mutagenesis and screening to the most advance approaches like model-driven

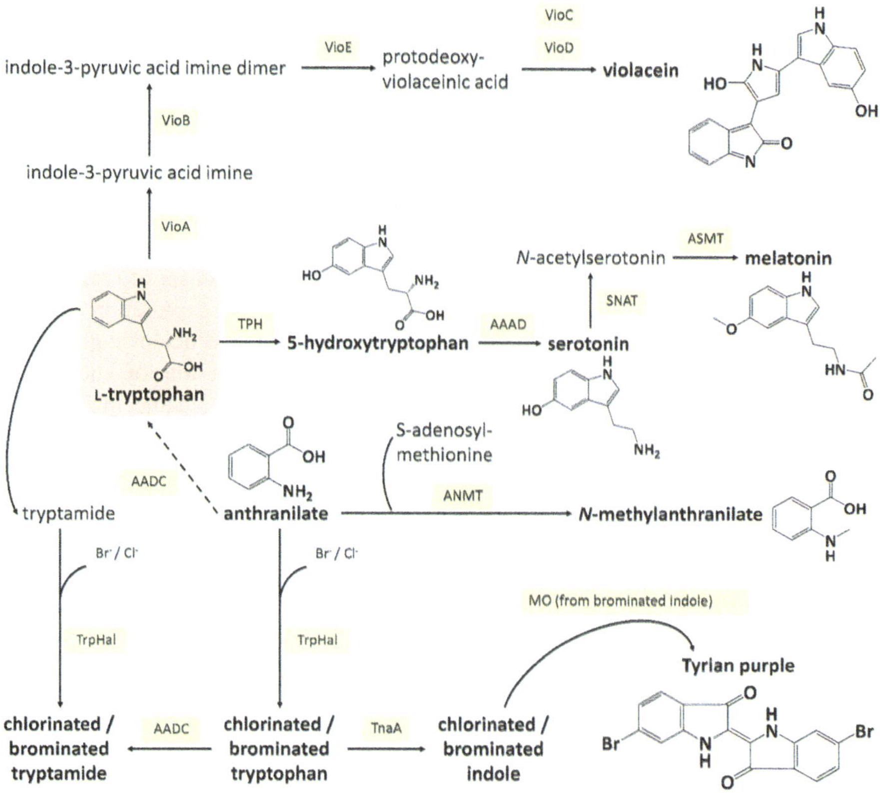

Fig. 7.4 *Biosynthetic pathways of* L*-tryptophan-derived compounds*. AAAD hydroxytryptophan decarboxylase, AADC aromatic amino acid decarboxylase, ANMT *N*-methylanthranilate transferase, ASMT hydroxyindole-o-methyltransferase, MO monooxygenase, SNAT serotonin *N*-acetyltransferase, TnaA tryptophanase, TPH tryptophan hydroxylase, TrpHal tryptophan halogenase, VioA flavoenzyme L-tryptophan oxidase, VioB iminophenyl-pyruvate dimer synthase, VioC Violacein synthase, VioD Protodeoxyviolaceinate monooxygenase, VioE Violacein biosynthesis protein VioE

metabolic engineering, systems and synthetic biology, or high-throughput sequencing methods, microbial cell factories have been on-demand shaped. Second, side streams from agri- and aqua-culture and other feedstocks without competing uses as food and feed have to be accessed.

Typically, to overcome metabolic bottlenecks in the microbial production of N-containing compounds, modification of terminal pathways, engineering of central carbon metabolism for increased supply of precursors, and engineering of product secretion are common targets. These strategies have been widely applied to establish competent cell-factories for the production of L-glutamate, L-aspartate family amino acids, and L-tryptophan.

7.1.1 L-Glutamate

The microbial production of amino acids is dominated by two main important producing microorganisms: *Corynebacterium glutamicum* and *Escherichia coli* [4–6] (Table 7.1). Since the first discovery of *C. glutamicum* as natural L-glutamate producer in Japan in 1957, strain development has become an intense competing spot of research. The amino acid L-glutamate covers nearly two thirds of the amino acid market being the major bulk amino acid [5] and it is mainly produced by fermentation of *C. glutamicum*. For several decades, the mechanism of L-glutamate overproduction in *C. glutamicum* has been the point of exploration by the researchers. Certain treatments can trigger *C. glutamicum* as efficient L-glutamate producer such as suboptimal supply of biotin or the addition of membrane destabilizers like β-lactam antibiotics or detergents [29]. A valuable insight into the secretion mechanism was the identification of the NCgl1221 gene product as a L-glutamate exporter [29]. This gene encodes the YggB protein, which is described as a putative mechanosensitive channel [30]. Overexpression of the wild-type NCgl1221 gene increases L-glutamate secretion while its disruption greatly suppresses L-glutamate secretion accompanied by an increase in the intracellular L-glutamate pool. Based on these findings, the initial proposed secretion mechanism involved the alteration of the membrane fluidity under the induction conditions by inhibiting lipid or peptidoglycan synthesis, which would trigger conformational changes in YggB enabling L-glutamate secretion. Currently, findings support the correlation between the structural and functional alteration of cell envelope and the variation of intracellular metabolism. The connection lays in the 2-oxoglutarate dehydrogenase complex (ODHC) and glutamate dehydrogenase (GDH) activities levels which regulate the flux towards L-glutamate from the tricarboxylic acid (TCA) cycle. The decrease in the ODHC activity and the increase of the GDH activity is crucial for the L-glutamate production [7]. The OdhI protein was identified as a regulator of ODHC [31]. The unphosphorylated form of OdhI binds to the ODHC subunit OdhA inhibiting the ODHC activity. Besides, the phosphoserine/threonine protein phosphatase PknG catalyzes the phosphorylation of OdhI which prevents the inhibition of ODHC [32]. Additionally, disruption of the *odhI* gene abolished L-glutamate production even under the induction conditions [32]. In the proposed mechanism, the induction treatments enhance the synthesis of the unphosphorylated protein OdhI and thereby inhibit ODHC activity. Besides, the enhancement of the anaplerotic pathways has also been proven to be an effective approach to increase the carbon flux for L-glutamate [8, 33].

7.1.2 L-Lysine

The market demand of L-lysine ranks just next to L-glutamate, with a current annual production of over 2.3 million tons [34]. L-Lysine is an essential amino acid which

Table 7.1 Relevant microbial amino acid biofactories developed via rational genetic engineering

Strain	Product	Bioprocess	Production values*	Reference
C. glutamicum AJ13133	L-glutamate	Batch	T: 40.1, Y: 0.63, V: 1.4	[7]
C. glutamicum DR1	L-glutamate	Batch	T: 36.9, Y: 0.33, V: 1.5	[8]
C. glutamicum GDK-9*Δldh*	L-glutamate	Fed-batch	T: 120, V: 5	[9]
C. glutamicum DM1933	L-lysine	Shake flasks	T: 4.8, Y: 0.12, V: 0.1	[10]
C. glutamicum LYS-12	L-lysine	Fed-batch	T: 120, Y: 0.55, V: 4.0	[11]
C. glutamicum JL-69Ptac-M *gdh*	L-lysine	Fed-batch	T: 181.5, Y: 0.65, V: 3.8	[12]
E. coli NT1003	L-lysine	Fed-batch	T: 134.9, Y: 0.45, V: 1.9	[13]
C. glutamicum DM1800-96-T (pEKEx2-$rhtC_{Ec}$)	L-threonine	Shake flasks	T: 6.4, Y: 0.16, V: 0.21	[14]
E. coli MG422 *rhtA23*(pAYC32-*thrA*^{r}*BC*)	L-threonine	Batch	T: 36.3	[15]
E. coli TH27C (pBRThrABCR3)	L-threonine	Fed-batch	T: 82.4, Y: 0.39, V: 1.5	[16]
E. coli MDS-205	L-threonine	Batch	T: 40.1, Y: 0.4, V: 1.3	[17]
C. glutamicum JHI3-156 (pDXW-8-*lrp-brnFE*)	L-isoleucine	Fed-batch	T: 26.9, Y: 0.12, V: 0.37	[18]
C. glutamicum K2P55	L-isoleucine	Fed-batch	T: 14.3, Y: 0.14, V: 0.18	[19]
C. glutamicum IWJ001(pDXW-8-*gnd-fbp-pgl*)	L-isoleucine	Fed-batch	T: 29.0, Y: 0.14, V: 0.36	[20]
E. coli ILE03	L-isoleucine	Fed-batch	T: 9.5, Y: 0.14, V:1.6	[21]
E. coli ΔmetJΔmetIΔlysA(pTrcA*H)	L-methionine	Fed-batch	T: 9.7, Y: 0.14, V:0.2	[22]
C. glutamicum QW102(pJYW-4-*hom*^{m}-*lysC*^{m}-*brnFE)*	L-methionine	Fed-batch	T: 6.3, V:0.1	[23]
C. glutamicum LY-5	L-methionine	Fed-batch	T: 6.8, Y: 0.07, V:0.09	[24]
C. glutamicum KY9218(pIK9960)	L-tryptophan	Fed-batch	T: 58, V:0.73	[25]
E. coli FB-04(pta1)	L-tryptophan	Fed-batch	T: 44, Y: 0.13, V:0.8	[26]
E. coli S028	L-tryptophan	Fed-batch	T: 40, Y: 0.15, V:0.6	[27]
E. coli KW023	L-tryptophan	Fed-batch	T: 39.7, Y: 0.17, V:1.6	[28]

**T* titer in g/L, *Y* product yield per substrate in g/g, *V* volumetric productivity in g/L/h has significant value as feed additive to promote growth of animals like poultry and swine. In-depth research has been undertaken for decades to engineer the metabolism of *C. glutamicum* for L-lysine overproduction. Pyruvate from glycolysis and oxaloacetate from the TCA cycle are direct precursors of L-lysine. A favorable anaplerotic net flux for L-lysine production was achieved by deleting the decarboxylating phosphoenolpyruvate (PEP) carboxykinase gene (*pck*) and by overexpression the pyruvate carboxylase gene (*pyc*) or by introducing a point mutation (pyc^{P458S}) [33, 35, 36]. Lowering the activities of citrate synthase and isocitrate dehydrogenase genes of the TCA cycle led to an increase in the pyruvate availability for L-lysine biosynthesis [35, 37]. Additionally, the deletion of the succinyl-CoA-synthetase genes *sucCD* of the TCA cycle also increased L-lysine production by 60% [38]. At the beginning of L-lysine biosynthesis, oxaloacetate is transaminated to L-aspartate, which is later phosphorylated by the enzyme aspartokinase LysC. The aspartokinase LysC of *C. glutamicum* is feedback-regulated and a key enzyme of L-lysine production. L-Lysine or L-threonine feedback-resistant LysC variants are caused by mutations in the β-subunit of the *lysC* gene [39]. The point mutated $LysC^{T311I}$ variant also showed feedback-resistance which led to L-lysine production [40]. An additional and common strategy to de-bottleneck the L-lysine biosynthetic pathway is the overexpression and/or duplication of genes such as dihydrodipicolinate synthase (*dapA*), diaminopimelate epimerase (*dapF*), *N*-succinyldiaminopimelate aminotransferase (*dapC*), dihydrodipicolinate reductase (*dapB*), meso-diaminopimelate dehydrogenase (*ddh*), and *lysA* [10, 11]. L-Lysine biosynthesis requires NADPH as cofactor for several reactions, hence a proper supply of such cofactor is of great importance here. Under aerobic conditions NADPH is mainly generated through the pentose phosphate pathway (PPP). As shown by ^{13}C-labeling experiments, NADPH supply by PPP correlates with the L-lysine yield [41]. The enhancement of the PPP flux was achieved through diverse approaches. For instance, by altering the glycolysis pathway with the deletion of the phosphoglucose isomerase gene (*pgi*) or with the overexpression of the fructose 1,6-bisphosphatase gene (*fbp*) the carbon flux is forced through the PPP [42, 43]. Another approach was the overexpression of the glucose 6-phosphate dehydrogenase gene or a point mutation variant of such gene that re-route glucose-6-phosphate from the glycolysis to the PPP [11, 40]. NADPH supply was also increased via expression in *C. glutamicum* of the membrane-integral nicotinamide nucleotide transhydrogenase PntAB from *E. coli*, which uses NADH and the membrane potential to reduce $NADP^+$ to NADPH [44]. By-product formation was also optimized, the metabolic intermediate of L-lysine biosynthesis, L-aspartyl-semialdehyde, is also a precursor of L-threonine, L-isoleucine, and L-methionine biosynthesis. Homoserine dehydrogenase (*hom*) catalyses the reduction of L-aspartyl-semialdehyde to L-homoserine and this activity was reduced by introducing alleles for less active enzyme variants [45]. Finally, L-lysine export could be improved by overexpression of *lysE* gene, which code for the L-lysine export system in *C. glutamicum* [46]. Systems metabolic engineering has been successfully applied in the development of L-lysine high-producing strains as, for example, Becker et al. obtained the genetically defined LYS-12 strain that produced 120 g/L of L-lysine with a glucose yield and volumetric productivities of 0.55 g/g and 4.0 g/L/h respectively [11]. More recently, the strain *C. glutamicum* JL-$69_{Ptac\text{-}M}$ *gdh* was obtained via rational engineering which produced 181.5 g/L of L-lysine with a glucose yield of 0.65 g/g [12] (Table 7.1).

Similar strategies have been applied to *E. coli* for the developed for L-lysine producing strains. For instance, the strain *E. coli* NT1003 is a threonine- and methionine-deficient *E. coli* that overexpresses *ppc* gene encoding phosphoenolpyruvate (PEP) carboxylase, *pntB* gene encoding pyridine nucleotide transhydrogenase, and *aspA* gene encoding aspartate ammonia-lyase. The NT1003 strain produced 134.9 g/L of L-lysine with a glucose yield of 0.45 g/g [13] (Table 7.1).

7.1.3 Production of Other L-Aspartate Family Amino Acids

The amino acids L-lysine, L-threonine, L-isoleucine and L-methionine share L-aspartate as common precursor and, therefore, certain engineering approaches from the development of L-lysine producers also apply to these amino acids. Those approaches are strengthening anaplerotic reactions to provide more oxaloacetate, engineering NADPH synthesis pathways, and elimination of by-products formation. L-Threonine is synthesized through five enzymatic steps from L-aspartate of which the phosphorylation of L-aspartate is the first and a key step. While *C. glutamicum* has only one aspartokinase feedback-regulated by L-lysine and L-threonine, *E. coli* has three aspartokinase isoenzymes encoded by the *thrA*, *metL*, and *lysC* genes, which are subjected to different regulations [47]. The second key enzyme of the pathway is the homoserine dehydrogenase that controls the carbon flux towards homoserine synthesis. *C. glutamicum* possesses one homoserine dehydrogenase encoded by *hom* gene, however in *E. coli* the aspartokinases encoded by *thrA* and *metL* have also homoserine dehydrogenase activity. The conversion of homoserine to L-threonine is carried out by the homoserine kinase encoded by *thrB* and threonine synthase encoded by *thrC*. The homoserine kinase is the third key enzyme of the pathway, which competes with homoserine *O*-succinyltransferase, the first enzyme of the methionine branch [47]. In *E. coli,* the most important modification is the overexpression of *thrABC* operon and especially the overexpression of the deregulated mutant of *thrA*. Additionally, the removal of negative regulation of *lysC* by L-lysine also enhanced threonine production [16]. A similar approach was used in *C. glutamicum* by heterologous expression of the feedback resistant *hom* variant, *thrB* and *thrE* in the L-lysine producer DM1800 [14]. To increase precursor supply in L-threonine production, it is necessary to disrupt the pathways of L-lysine and L-methionine. In this regard, the deletion of the genes encoding DAP decarboxylase (*lysA*) and homoserine succinyltransferase (*metA*) in *E. coli*, and the DAP synthase (*dapA*) and homoserine acetyltransferase (*metX*) in *C. glutamicum* enhanced L-threonine production [16]. On the other hand, the breakdown of L-threonine was reduced by mutating the threonine dehydratase genes *ilvA* in *E. coli* and *C. glutamicum* [14, 16]. Transport engineering systems can improve L-threonine secretion while reducing extracellular accumulation of threonine. In *E. coli,* the increased expression of the threonine exporter gene *rhtA* improved threonine production [15]. The gene *rhtA* from *E. coli* was also overexpressed in *C. glutamicum* L-threonine producing strains [14]. The combination of several of these strategies led to the most successful L-threonine overproducing from the parental strain *E. coli* W3110. The resulting strain reached a titer and yield of 82.3 g/L and 0.39 g/g, respectively [16] (Table 7.1).

L-Isoleucine is synthesized through five enzymatic reactions from L-threonine as a precursor. In the isoleucine synthesis branch after threonine, threonine dehydratase (TD) encoded by *ilvA* and acetohydroxy acid synthase encoded by *ilvIH* (AHAS) are key enzymes subjected to feedback inhibition by L-isoleucine in *C. glutamicum.* In *E. coli* TD is encoded by *ilvA* or *tdcB* [47]. A previously constructed L-threonine producer *E. coli* TH20 was used as platform for further metabolic engineering to produce isoleucine [21]. The *ilvA*$^{L447F,\ L451A}$ mutant variant, which encoded an active and feedback resistant TD was overexpressed in the strain TH20 together with a feedback resistant *ilvIH* mutant variant yielding the strain ILE01. This strain was able to produce 0.32 g/L L-isoleucine from 20 g/L of glucose as carbon source. The acetohydroxy acid isomeroreductase (*ilvC*), branched chain amino acids (BCAA) aminotransferase (*ilvE*) and dihydroxy acid dehydratase (*ilvD*) genes are involved in the L-isoleucine biosynthetic pathway. The native promoters of the *ilvEDA* operon and the *ilvC* gene were replaced with the strong *trc* promoter by homologous recombination in the chromosome; resulting in the strain ILE02 which produced 2.1 g/L from 20 g/L of glucose. Finally, further overexpression of the *ygaZH* genes encoding an L-isoleucine exporter yielded the strain ILE03, which produced 2.8 g/L from 20 g/L of glucose. In a glucose-based fed-batch fermentation the strain ILE03 reached 9.46 g/L after 60 h [21] (Table 7.1).

Methionine synthesis competes with the formation of threonine and isoleucine at the homoserine branch point. In *C. glutamicum,* homoserine is condensed with acetyl-CoA to produce acetyl homoserine by the action of the enzyme homoserine acetyltransferase (HAT) encoded by *metX.* However, in *E. coli*, homoserine is condensed with succinyl-CoA to produce succinyl homoserine by the enzyme homoserine succinyltransferase (HST) encoded by *metA.* In addition, methionine synthesis in *C. glutamicum* is performed by two parallel pathways. The acetyl homoserine is converted into homocysteine either by transsulfuration pathway or by direct sulfhydrylation utilizing inorganic sulfur instead of cysteine. Regarding transport systems, *C. glutamicum* possesses a high affinity ABC-type importer and a low affinity importer for methionine uptake and the BCAA exporter BrnFE controls the methionine export. In *E. coli*, the *metD* and *metP* genes encode a high affinity and a low affinity importer system of methionine, respectively, while the exporter YjeH is responsible for methionine secretion [47]. *E. coli* W3110 was engineered to produce L-methionine. Deletion of the transcriptional repressor gene *metJ* and overexpression of homoserine O-succinyltransferase and the efflux transporter genes *metA* and *yjeH*, respectively, resulted in a L-methionine production of 0.4 g/L. The partial inactivation of the L-methionine import system MetD resulted in a 43.6% higher accumulation of L-methionine. In addition, blocking the L-lysine biosynthesis pathway by deleting the *lysA* gene increased 8.5-fold the L-methionine titer. Besides, addition of sodium thiosulfate to the media led to an increase of fermentation titer of 11.5%. Finally, after optimization constructed the final *E. coli* strain was able to produce 9.7 g/L of L-methionine with a volumetric productivity of 0.2 g/L/h in a 5 L bioreactor [22]. The *C. glutamicum* wild type strain ATCC 13032 was also metabolically engineered for methionine production. The methionine uptake system gene *metD* was deleted to avoid re-uptake of methionine. Afterwards, random mutagenesis was performed to remove feedback inhibition by metabolic end-products resulting in the strain *C. glutamicum* ENM-16. This strain was further engineered to decrease competitive branch pathways by changing the start codon of the *dapA* gene and deleting the *thrB* gene. Methionine precursor supply was improved by the point mutations in $lysC^{C932T}$ and $pyc^{G1A,\ C1372T}$. Furthermore, glucose-6-phosphate dehydrogenase and 6-phosphogluconate dehydrogenase in the PPP were mutated to enrich the NADPH pool. The final strain *C. glutamicum* LY-5 reached 6.85 ± 0.23 g/L of methionine after 72 h in a glucose-based fed-batch fermentation [24] (Table 7.1).

7.1.4 L-Tryptophan

Tryptophan is an aromatic amino acid that has been widely used in food industry since it participates in the regulation of protein synthesis and metabolic network *in vivo*. It is also a key precursor of industrial and pharmaceutical compounds such as melatonin and serotonin [48] or the flavouring compound indole [49]. To date, the industrial scale production of L-tryptophan is carried out via microbial fermentations of *C. glutamicum* and *E. coli*. Biosynthesis of the aromatic amino acids begins with the condensation of PEP and erythrose 4-phosphate (E4P) by the key enzyme 3-deoxy-D-rabino-heptulosonate 7-phosphate synthase (DS), which then proceeds to chorismate. From chorismate the pathways branches to L-tryptophan, L-tyrosine, and L-phenylalanine. *C. glutamicum* has two DS enzymes encoded by *aro* and *aroII*. While the first DS (*aro*) is feedback regulated by L-phenylalanine and L-tyrosine, the second DS (*aroII*) is only sensitive to L-tyrosine [50]. *E. coli* has 3 DS enzymes encoded by *aroF*, *aroG* and *aroH* which are feedback regulated by L-tyrosine, L-phenylalanine and L-tryptophan, respectively [51]. Also in *E. coli*, L-tryptophan feedback regulates the first specific enzymes of the L-tryptophan biosynthetic pathway encoded by *trpED*, while in *C. glutamicum* L-tryptophan also feedback regulates the next enzymatic step encoded by *trpD*. Another improvement is achieved when removing the feedback regulation from the primary competing branches, meaning the biosynthetic pathways of L-tyrosine and L-phenylalanine are the primary competing branches. To address this problem, disrupting the competing pathways was performed by deleting the genes *pheA* and *tyrA* [52]. Additionally, L-tryptophan also controls at transcriptional level all the L-tryptophan biosynthetic steps in *E. coli* and *C. glutamicum* [50, 51]. In the process of developing L-tryptophan high-producing strain, increasing the accessible amounts of precursors is an important strategy. Increasing the PEP and E4P pool is this regard. PEP can be increased by modifying the flux through the PEP-Pyruvate metabolic node, which can be achieved, for instance, by deleting the pyruvate kinase or the PEP carboxylase genes [53]. Besides, the phosphotransferase (PTS) systems naturally utilize PEP as phosphoryl group donor in the process of glucose uptake, hence replacement with PTS-independent transport systems can improve PEP availability [28]. L-Serine is an important precursor for L-tryptophan. Engineering the L-serine biosynthetic pathway, for instance by alleviating the feedback regulation, increased the L-serine pool [51]. Relieving the feedback inhibition, repression and attenuation of the L-tryptophan biosynthesis is a prerequisite for increasing the productivity of desired biochemicals. Desensitizing the DS enzymes has proven to increase the carbon flux to the common L-tryptophan, L-tyrosine, and L-phenylalanine biosynthetic pathways. For instance, mutating the binding region of AroG or by specific amino acids substitutions ($AroG^{L76V}$ and $AroF^{P148L,\ Q152I}$) deregulate the feedback inhibition of DS [54, 55]. Responding to intracellular L-tyrosine and L-tryptophan, transcription factors TyrR and TrpR acted as repressor or activator in L-tryptophan metabolism [56]. Deletion of the gene *trpR* could eliminate *trp* operon repression [28]. Depending on the leader peptide of *trpL*, attenuation mechanism regulates transcription and translation of the *trp* operon at the same time. Deletion of *trpL* sequence can overcome this attenuation control [57]. Finally, Mtr and TnaB are L-tryptophan-specific permeases [58] while AroP imports all aromatic amino acids [59].

Several of these strategies have been applied to develop the *E. coli* strain KW023. In this strain the *trp* operon (*trpEDCBA*) and a feedback resistant variant of AroG are overexpressed. The pyruvate kinase (*pykF*) and the PTS system HPr (*ptsH*). Repression by *trpR* and attenuation by *trpL* were inactivated. The glucokinase (*glk*) and galactose permease (*galP*) were overexpressed to overcome slow growth. Acetate production as by-product was reduced by repressive regulation of phosphate acetyltransferase (*pta*) expression. In the end, the strain KW023 produced 39.7 g/L of L-tryptophan with a yield of 0.17 g/g and a volumetric productivity of 1.6 g/L/h in a fed-batch fermentation [28]. With regard to *C. glutamicum*, a classically derived L-tryptophan producing strain was improved via heterologous expression of a feedback resistant variant of DS, the 6-phosphogluconate dehydrogenase gene, and the tryptophan-biosynthetic gene cluster in from *C. glutamicum* ATCC31833, and transketolase gene [25]. The new strain produced 58 g/L of L-tryptophan after 80 h in a sucrose-based fed-batch fermentation (Table 7.1).

7.2 Extending the Metabolic Pathways of Amino Acid Biosynthesis

The capability to produce substantial amounts of amino acids has gained great attention since the amino acids can be used as precursors to produce other high value-added chemicals (Table 7.2). Thanks to the recent developments in the fields

Table 7.2 Relevant microbial biofactories for the production of amino acid derivatives

Strain	Product	Bioprocess	Production values[a]	Reference
	From L-glutamate:			
C. glutamicum NA6	Putrescine	Shake flasks	T: 5.1, Y:0.26, V: 0.21	[60]
C. glutamicum PUT21	Putrescine	Fed-batch	T: 19, Y: 0.16, V:0.55	[61]
E. coli XQ52(p15SpeC)	Putrescine	Fed-batch	T: 24.2, Y: 0.17, V: 0.75	[62]
C. glutamicum GABA6F	GABA	Fed-batch	T: 63.2, Y: 0.24, V: 1.13	[63]
C. glutamicum R4a-B2B1^{mut}	GABA	Shake flasks	T: 26.5, Y: 0.27, V: 0.44	[64]
E. coli BL21 (pET32a-*gadABC*)	GABA	Fed-batch	T: 31.3, V: 0.55	[65]
E. coli BLM01	Butyrolactam	Fed-batch	T: 54.1, Y: 0.12, V: 0.58	[66]
C. glutamicum CA1P4	5AVA	Fed-batch	T: 16.3, V: 0.42	[67]
E. coli RrGI	5AVA	Shake flasks	T: 7.5, V: 0.6	[68]
E. coli TEA6Δggt	L-theanine	Fed-batch	T: 16.1, Y: 0.13, V: 0.33	[69]
P. putida Thea1	L-theanine	Fed-batch	T: 10, Y: 0.03, V: 0.42	[70]
P. putida NMG3	*N*-methylglutamate	Fed-batch	T: 17.9, Y: 0.11, V: 0.12	[71]
	From L-lysine:			
E. coli XQ56 (p15CadA)	Cadaverine	Fed-batch	T: 9.6, Y: 0.12, V: 0.32	[72]
C. glutamicum DAP-16	Cadaverine	Fed-batch	T: 88, Y: 0.57, V: 2.2	[73]
C. glutamicum HCad1	Hydroxycadaverine	Fed-batch	T: 8.8, Y: 0.31, V: 1.5	[74]
C. glutamicum 5AVA3	5AVA	Shake flasks	T: 5.1, Y: 0.13, V: 0.12	[75]
C. glutamicum AVA-3	5AVA	Fed-batch	T: 28, Y: 0.13, V: 0.56	[76]
C. glutamicum AVA2 (p36davAB3)	5AVA	Fed-batch	T: 33.1, Y: 0.1, V: 0.22	[77]
E. coli VLM01	Valerolactam	Fed-batch	T: 1.2, Y: 0.005, V: 0.02	[66]
C. glutamicum PIPE4	L-PA	Fed-batch	T: 14, Y: 0.2, V: 0.3	[78]
E. coli XQ-11-4	L-PA	Fed-batch	T: 61, Y: 0.3, V: 1.02	[79]
C. glutamicum Ecto5	Ectoine	Fed-batch	T: 22, Y: 0.16, V: 0.32	[80]

(continued)

Table 7.2 (continued)

Strain	Product	Bioprocess	Production values[a]	Reference
C. glutamicum ectABC^{opt}	Ectoine	Fed-batch	T: 65.3, Y: 0.19, V: 1.2	[81]
E. coli ECT05	Ectoine	Fed-batch	T: 25.1, Y: 0.11, V: 0.84	[82]
E. coli HECT31	Hydroxyectoine	Shake flasks	T: 14.9, Y: 1.9, V: 0.42	[83]
E. coli BW25113*Δcai-fix* (pTrc99A-TML2Car)	L-carnitine	Shake flasks	T: 0.003, V: 5.00E-05	[84]
	From L-tryptophan:			
C. glutamicum 21,850 (pEC-C-vio1)	Violacein	Fed-batch	T: 5.4, Y: 0.05, V: 0.05	[85]
E. coli B8(pTRPH1)(pVio-VioE)	Violacein	Fed-batch	T: 4.5, V: 0.1	[86]
C. freundii (pCom10vio)	Violacein	Fed-batch	T: 4.13, V: 0.08	[87]
E. coli HTP101-LMT	Hydroxytryptophan	Fed-batch	T: 5.1, V: 0.13	[88]
E. coli TrpD-Pl (pCtAAAH-LC) / *E. coli ΔtnaA(*pCOLAJ23-TDC)	Serotonin	Two-step fermentation	T: 0.15, V: 1.38E-03	[89]
E. coli EcMEL8	Melatonin	Fed-batch	T: 0.65, V: 3.61E-03	[90]
E. coli trpD9923 (pJLB*aroG*^{fbr}*tktA)*	Anthranilate	Fed-batch	T: 14, Y: 0.2, V: 0.39	[91]
C. glutamicum YTM8 (pSH36HTc)(pEKGH)	*N*-methylanthranilate	Two-phase fed-batch	T: 5.74, Y: 0.02, V: 0.05	[92]
E. coli ZWA4 (pBBR1G^{fbr}A^{fbr}E^{fbr}) (pTacT)	*N*-methylanthranilate	Two-phase fed-batch	T: 4.47, Y: 0.045, V: 0.06	[92]
C. glutamicum HalT1	Chlorinated tryptophan	Shake flasks	T: 1.41E-03, Y: 3.53E-05, V: 4.70E-04	[93]
C. glutamicum HalT2	Brominated tryptophan	Fed-batch	T: 1.2, V: 0.02	[94]
E. coli ΔtnaA TnaA + *E. coli ΔtnaA*MaFMO	Tyrian purple	Two-cell reaction system	T: 0.32, V: 0.04	[95]
C. glutamicum HaloInd Ec	7-Cl-indole	Shake flasks	T: 0.016, V: 0.00022	[96]
C. glutamicum HaloInd Ec	7-Br-indole	Shake flasks	T: 0.026, V: 0.00032	[96]
C. glutamicum HaloTra Cs	7-Br-tryptamine	Fed-batch	T: 0.36, V: 0.0083	[96]
	From L-isoleucine:			
C. glutamicum HIL18	4-hydroxyisoleucine	Fed-batch	T: 34.2, Y: 0.15, V: 0.53	[97]

(continued)

Table 7.2 (continued)

Strain	Product	Bioprocess	Production values[a]	Reference
C. glutamicum ST17	4-hydroxyisoleucine	Shake flasks	T: 19.9, Y: 0.16, V: 0.14	[98]

[a]*T* titer in g/L, *Y* product yield per substrate in g/g, *V* volumetric productivity in g/L/h

of metabolic engineering, synthetic biology and sequence analysis, the extension of metabolic pathways from amino acids have been enabled towards bio-based monomers as well as building blocks for healthcare products and pharmaceuticals. Here, proper supply of amino acids as initial precursors for the extended pathways is needed, which can be tackled by the metabolic strategies described previously in this chapter. Additionally, other factors that have to be considered when producing not native compound with cell-factories. For instance, toxicity of the new compound that could disturb the cells growth, an adequate expression balance of non-native genes to avoid metabolic drawbacks or cofactors imbalances, and the present or absent of appropriate cell secretion systems for the compound to be produced.

7.2.1 Value-Added N-Containing Chemicals Derived from L-Glutamate

The amino acid L-glutamate can serve as precursor for the biosynthesis of the diamine putrescine and the non-proteinogenic amino acids γ-aminobutyric acid (GABA), 5-Aminolevulinic acid (5-ALA), L-theanine, and *N*-methylglutamate as depicted in Fig. 7.2.

7.2.1.1 Putrescine

The diamine putrescine (1,4-diaminobutane) can be produced by chemical synthesis via the addition of hydrogen cyanide to acrylonitrile, afterwards the resulting succinonitrile is hydrogenated [99]. Putrescine is used for synthesizing the polyamine nylon 4,6 together with adipic acid. Nylon 4,6 possesses interesting mechanical and physical properties such as tensile strength, solvent resistance, and crystallization rate that are comparable to those of nylon 6,6, being one of the most common polyamides used in the textile and plastic industries [100]. Putrescine can be also biologically synthesized from L-glutamate through L-ornithine or L-arginine as intermediates [101] (Fig. 7.2). The ornithine decarboxylase gene *speC* from *E. coli* was introduced in *C. glutamicum* enabling the synthesis of putrescine [101]. In order to increase the supply of L-ornithine from L-glutamate, the *argF* gene encoding ornithine transcarbamylase was deleted resulting in a higher titer of 19 g/L in fed-batch fermentation [61]. In addition, the gene responsible for putrescine acetylation in *C. glutamicum snaA* was identified and deleted to minimize the generation of acetylputrescine as a by-product [60]. The gene *snaA* is also responsible for the acetylation of the diamine cadaverine. Strategies associated with L-glutamate overproduction were also used here. For instance, by decreasing fivefold the activity of ODHC in *C. glutamicum* putrescine production increased 28% [102]. The strain *C. glutamicum* NA6 produced 5.1 g/L of putrescine with a yield of 0.26 g/L and a volumetric productivity of 0.21 g/L/h [60]. Other metabolic strategies applied to *C. glutamicum* to increase putrescine production were increase in the ATP, pyruvate, and the NADPH supply yielding the strain *C. glutamicum* PUT21 [103]. This strain produced 19 g/L of putrescine in fed-batch fermentation. Although *C. glutamicum* has a higher tolerance to putrescine than *E. coli* [101], the latest has been also used as platform for the production of putrescine. To date, the best *E. coli* putrescine producer obtained via rational strain engineering in the strain XQ52(p15SpeC) [62]. To construct such strain the following modifications were carried out: deletion of the spermidine synthase gene (*speE*) to avoid putrescine consumption; deletion of the spermidine acetyltransferase gene (*speG*) to avoid acetylation of putrescine; deletion of the ornithine carbamoyltransferase gene (*argI*) to reduce the flux from L-ornithine toward L-arginine; deletion of the putrescine importer gene (*puuP*); deletion of the putrescine ligase gene (*puuA*), plasmid-based and chromosome-based overexpression of the ornithine decarboxylase gene (*speC*); replacing the native promoter of the L-arginine biosynthesis operon by pTrc to increase provision of L-ornithine as precursor; and disrupting global stress regulator gene *rpoS* to reduce the stress response under putrescine overproduction condition. A fed-batch fermentation with the resulting strain XQ52(p15SpeC) yielded 24.2 g/L of putrescine with a volumetric productivity of 0.75 g/L/h [62] (Table 7.2).

7.2.1.2 GABA

The non-proteinogenic amino acid GABA is bioactive component of analgetic drugs, anti-anxiety drugs, and diuretics in the pharmaceutical and food industries [104, 105]. Furthermore, GABA can be cyclized to form the highly stable lactam 2-pyrrolidone, or butyrolactam, which can be chemically converted to bio-based polyamide 4 [99]. GABA can be synthesized via decarboxylation of L-glutamate by the glutamate decarboxylase encoded by the *gad* gene (Fig. 7.2). The heterologous expression of the *gad* gene from *Lactobacillus brevis*, *Lactobacillus plantarum*, or *E. coli* enabled GABA production in *C. glutamicum* [106]. To efficiently improve GABA production, the *pck* encoding the PEP carboxylase gene was overexpressed, the malate dehydrogenase gene was deleted, and promoter and ribosomal binding site sequences of the *gad* gene were optimized [64]. Production of 25 g/L of putrescine was achieved. Additionally, a new two-step reactions pathway was constructed to produce GABA from putrescine [107] (Fig. 7.2). In this pathway, putrescine is first converted to 4-amino-1-butanal by the putrescine aminotransferase encoded by the *patA* gene from *E. coli* and afterwards GABA is produced from

4-amino-1-butanal by the γ-aminobutyraldehyde dehydrogenase encoded by the *patD* gene also from *E. coli* [107]. The *speC*, *patA*, and *padD* genes from *E. coli* were heterologously expressed in a *C. glutamicum* strain with decreased ODHC activity, deletion of L-glutamic acid and putrescine exporter genes, deletion of GABA importer gene, deletion of the pathway that converts GABA to succinic acid, and deletion of the gene responsible for putrescine acetylation The final strain was tested in fed-batch fermentation peaking 63.2 g/L of GABA [63]. Another *C. glutamicum* GABA producing strain with deletion of the pathways towards by-product pools of L-arginine, L-proline, and L-lysine and the introduction of the pyridoxal kinase gene from *Lactobacillus plantarum* to increase pyridoxal 5′--phosphate formation produced 70.6 g/L of GABA in fed-batch fermentation [108] (Table 7.2).

In *E. coli* the overexpression of the two endogenous glutamate decarboxylase genes *gadA* and *gadB* and the deletion of the aminobutyrate aminotransferase gene *gabT* led to 5.46 g/L of GABA production [109]. In another study, the enzymes for the regeneration of the coenzyme factor pyridoxal 5′-phosphate from *Streptomyces cerevisiae* SC288 were targeted and engineered to improve GABA production [110]. The strain *E. coli* BL21 (pET32a-*gadABC*) overexpressed 3 different glutamate decarboxylase genes enabling the production of 31.3 g/L GABA after 57 h in fed-batch fermentation [65].

In lactic bacteria, GABA production does not proceed from glucose, but from L-glutamate to the culture medium. *Lactobacillus brevis* cells with high glutamate decarboxylase activity allowed biotransformation of L-monosodium glutamate to GABA with an overall yield > 90% [111]. After medium optimization and operation in fed-batch mode at 32 °C and pH 5.0 led to a final GABA titer of around 113 g/L [112] (Table 7.2).

7.2.1.3 5-Aminolevulinic acid

5-ALA is a non-proteinogenic amino acid involved in tetrapyrrole synthesis and that has been in increasing demand in cosmetics, pharmaceutical, and agriculture industries to produce, for instance, heme, porphyrin, chlorophyll, and vitamin B12 [113]. 5-ALA can be synthetized in three enzyme steps from L-glutamate (Fig. 7.2). In the first step, L-glutamate is converted to glutamyl-tRNA by the glutamyl-tRNA ligase encoded by the *gltX* gene. Glutamyl-tRNA is then converted to glutamate 1-semialdehyde by the glutamyl-tRNA reductase encoded by the *hemA* gene. Finally, glutamate 1-semialdehyde is transaminated by glutamate 1-semialdehyde aminotransferase encoded by the *hemL* gene generating 5-ALA [114].

C. glutamicum possesses the *gltX*, *hemA*, and *hemL* genes, however production of 5-ALA has been achieved by heterologous expression of the *hemL* gene from *E. coli* as well as by heterologous expression of *hemA* from different microorganisms such as *Salmonella typhimurium*, *Salmonella arizonae*, or *Rhodobacter sphaeroides* [113, 115, 116]. To date, the highest titer of 5-ALA reached using *C. glutamicum* strains was 16.3 g/L which was achieved by overexpression of the *hemA* gene from

Rhodopseudomonas palustris and the endogenous PEP carboxylase *ppc* gene together with the addition of glycine to the medium [67] (Table 7.2).

Recently, the clustered regularly interspaced short palindromic repeats interference (CRISPRi) tool was applied in *E. coli* to fine-tune the degradation of 5-ALA by HemB. This approach led to a production improvement of 493.1% compared to the original strain. In fed-batch fermentation the HemB-modified strain produced 2 g/L of 5-ALA in 42 h [117]. Downregulating of *hemB* was also achieved via synthetic antisense RNAs [118]. The plasmid-free *E. coli* strain RrGI expressing ALA synthase from *Rhodobacter sphaeroides*, the non-specific ALA exporter RhtA, and the chaperones DnaK and GroELS led to 7.47 g/L of 5-ALA [68] (Table 7.2).

7.2.1.4 L-Theanine

L-Theanine is a non-proteinogenic amino acid normally found in tea leaves [119]. L-Theanine exerts mitigative effects in the case of cancer, cardiovascular diseases, and obesity [120]. The synthesis of L-theanine takes place from L-glutamate via a one-step enzymatic reaction catalyzed by the γ-glutamylmethylamide synthetase encoded by the *gmas* gene [121] (Fig. 7.2). The *gmas* gene from *Methylovorus mays* was used to enable L-theanine production in *C. glutamicum*. Together with the deletion of the L-glutamate exporter gene *yggB* the new strains produced 42 g/L of L-theanine in fed-batch fermentation [121]. Production of L-theanine uses the toxic and highly flammable precursor ethylamine, however, a *E. coli* strain was engineered for the synthesis of L-theanine in the absence of supplemental ethylamine [69]. In this strain, the transaminase PpTA from *Pseudomonas putida* KT2440 converts acetaldehyde to ethylamine and the γ-glutamylmethylamide synthetase from *Pseudomonas syringae* pv. syringae B728a catalyzes the condensation of L-glutamate and ethylamine to produce L-theanine. The final strain TEA6*Δggt* produced 16.1 g/L of L-theanine after 48 h in fed-batch fermentation [69]. In *P. putida*, expression of the *gmas* gene from *Methylobacterium extorquens* DM4 enabled production of L-theanine. A titer of 2.7 g/L of L-theanine reached in shake flasks with minimal medium containing monoethylamine and glucose. Heterologous expression of the *xylXABCD* operon from *Caulobacter crescentus* in *P. putida* enabled xylose utilization. Fed-batch fermentation with glucose plus xylose as carbon sources yielded 10 g/L of L-theanine [70] (Table 7.2).

7.2.1.5 *N*-Methylglutamate

N-Alkylated amino acids can be used for the synthesis of peptide-based drugs named peptidomimetics with enhanced half-lives and stability against proteolysis or their membrane permeability [122]. *N*-Methylglutamate production was established in *P. putida* KT2440 by heterologous expression of the γ-glutamylmethylamide synthetase gene *gmaS* and the *N*-methylglutamate synthase genes *mgsABC* from

M. extorquens (Fig. 7.2). A fed-batch cultivation process yielded 17.9 g/L of *N*-methylglutamate after 107 h from glycerol and monomethylamine [71] (Table 7.2).

7.2.2 *Value-Added N-Containing Chemicals Derived from L-Lysine*

The amino acid L-lysine is an important feed additive that promotes growth of animals like poultry and swine. Additionally, L-lysine is the precursor to diamine cadaverine, quaternary ammonium compound L-carnitine, and the non-proteinogenic amino acids ectoine, L-pipecolic acid (LPA) and 5 aminovalerate (5AVA) as depicted in the Fig. 7.3.

7.2.2.1 Cadaverine

Cadaverine is a five-carbon (C5) diamine [72]. Cadaverine is found mainly in dead organic matter, being the reason in part for the strong smell of putrefaction [123]. The L-lysine decarboxylase is responsible for the formation of cadaverine [124] (Fig. 7.3), and it has been identified in bacteria, cyanobacteria and plants [125]. Cadaverine plays a structural role in bacteria since it can form covalent bonds with the peptidoglycan [126]. Additionally, in *E. coli* the cadaverine formed by the L-lysine decarboxylase CadA is part of the acid stress response system [127]. With regard to industry, cadaverine is a platform chemical to produce chelating agents, polyurethanes and polyamides. Polymerization of cadaverine with the dicarboxylate succinate or with sebacic acid yielded the polyamides PA-5,4 and PA-5,10, respectively [128]. Cadaverine can be produced by chemical synthesis or biologically via whole-cell biocatalysis, and *de novo* biosynthesis [129]. Lysine decarboxylation is the key step for cadaverine synthesis and, hence, genes encoding lysine decarboxylase are central to engineer cadaverine producers from lysine overproducing strains [124]. By overexpressing the endogenous lysine decarboxylase gene *cadA* in a *E. coli* lysine producer enabled overproduction of cadaverine to a final titer of 69 g/L (Patent US 7189543 B2). By plasmid-based overexpression of the *cadA*, avoiding side reactions with other diamines as putrescine as well as by increasing the flux toward L-lysine, the *E. coli* XQ56 strain produced 9.6 g/L cadaverine with a volumetric productivity of 0.32 g/L/h in fed-batch fermentation [72] (Table 7.2). On the other hand, cadaverine is not a native metabolite in *C. glutamicum* [124]. Cadaverine production in *C. glutamicum* was enabled by heterologous expression of either the lysine decarboxylase genes from *E. coli cadA* [124] or *ldcC* [130]. Since the pH optimum of LdcC is close to the neutral intracellular pH of *C. glutamicum,* LdcC is more suitable than CadA to establish cadaverine production in *C. glutamicum* as CadA is most active at acidic pH of 5.5. In a *C. glutamicum* strain with 12 genomic optimizations for the overproduction of L-lysine [11], Kind et al. [130] constructed genome-based expression of *ldcC* under the control of the strong constitutive *C. glutamicum tuf* promoter [130]. *N*-acetylcadaverine formation was avoided by deleting the *N*-acyltransferase gene *snaA* [102]. Furthermore, to prevent loss of L-lysine the L-lysine exporter gene *lysE* can be deleted [46]. The L-arginine and diamines export system CgmA was overexpressed [131]. The final *C. glutamicum* DAP-16 produced 88 g/L of cadaverine with a volumetric productivity of 2.2 g/L/h in fed-batch fermentation [73] (Table 7.2). Alternatively, cadaverine export via the cadaverine-lysine antiporter gene *cadB* from *E. coli* was shown [132]. Recently, the overexpression of the gene *dr1558* encoding a response regulator from *Deinococcus radiodurans* improved cadaverine production in *C. glutamicum*. Differential gene expression analysis revealed that the PEP carboxykinase (*pck*) was downregulated, and pyruvate kinase (*pyk*) and other lysine biosynthesis genes were upregulated [133]. *De novo* production of 3-hydroxycadaverine has been also demonstrated in *C. glutamicum*. The lysine hydroxylase from *Flavobacterium johnsoniae* was used to generate 4-hydroxylysine from L-lysine, which was further decarboxylated to 3-hydroxycadaverine via the LdcC from *E. coli* [74].

7.2.2.2 5-Aminovalerate

5-aminovalerate or 5AVA is a non-proteinogenic δ-amino acid that comprising pentanoic acid with an amino substituent at C5 [134], known to be a natural L-lysine degradation product. Some *Pseudomonas* sp. can use 5AVA as a sole carbon and nitrogen source [135]. 5AVA is a precursor of valerolactam, a C5 platform for synthesizing 5-hydroxyvalerate, glutarate and 1,5-pentanediol, and a building block for the production of polyamide-5 (PA5) [134]. 5AVA is an intermediate of the aminovalerate pathway of L-lysine degradation existing, for instance in *P. putida* and many *Pseudomonas* sp. In this pathway, L-lysine is converted to 5AVA by the sequential catalysis of lysine 2-monooxygenase encoded by *davB* and δ-aminovaleramidase encoded by *davA* [134] (Fig. 7.3).

An *E. coli* strain expressing *davAB* genes from the *P. putida* grown in a medium supplemented with 20 g/L of glucose and 10 g/L of L-lysine produced 3.6 g/L of 5AVA, although 3 g/L of L-lysine remained no consumed [136]. In a similar approach, an *E. coli* L-lysine producer with the *cadA* and *ldcC* genes deleted was used as platform to heterologous express *davAB* from *P. putida*. The deletion of *cadA* and *ldcC* genes prevented the decarboxylation of L-lysine to cadaverine as undesired product. The final strain produced 0.86 g/L of 5AVA from 25 g/L of glucose in 48 h [134] (Table 7.2). The overexpression of the *davAB* genes together with the 4-aminobutyrate transporter PP2911 from *P. putida* in *E. coli* improved the 5AVA titer and volumetric productivity [137].

In *C. glutamicum*, the *davBA* genes from *P. putida* were integrated in the locus *bioD* under the control of the constitutive promoter pTuf of the L-lysine overproducing strain LYS-12 [11]. Reduction of L-lysine and glutarate as by-products was achieved by disrupting the genes *lysE* and *gabT*. The resulting AVA-3 strain produced 28 g/L of 5AVA with a volumetric productivity of 0.9 g/L/h in fed-batch fermentation experiments [76]. Similarly, expressing a codon-optimized *davA* gene fused with His6-Tag and the *davB* gene under the control of the strong synthetic H36 promoter as well as deleting *gabT* in a L-lysine producing *C. glutamicum* led to a titer of 33.1 g/L/1 of 5AVA after 153 h of cultivation in bioreactors [77] (Table 7.2).

A different pathway to 5AVA production that does not depend on molecular oxygen was set up as an alternative [75]. A three-step route comprising L-lysine decarboxylase (LdcC), putrescine transaminase (PatA), and γ-aminobutyraldehyde dehydrogenase (PatD) from *E. coli* was established and tested to produce 5AVA [75] (Fig. 7.3). To improve glucose consumption the transcriptional repressor gene *sugR* was disrupted. Moreover, to reduce the formation of the by-products L-lactate, *N*-acetylcadaverine, glutarate, and cadaverine, the genes *ldhA*, *snaA*, *cgmA*, and *gabTDP* were also disrupted [75, 138]. The resulting strain 5AVA3 produced 5.1 g/L of 5AVA during shake flask cultivations [75] (Table 7.2).

7.2.2.3 L-Pipecolic Acid

Pipecolic acid or L-PA is a non-proteinogenic cyclic α-amino acid derived from L-lysine [139]. In nature, L-PA is widely distributed in plants, animals, and microorganisms and it has different functions including interactions between organisms [140, 141]. In bacteria such as *P. putida* and *Agrobacterium tumefaciens* L-PA is an intermediate of the catabolic pathways of L-lysine [142]. L-PA and its *N*-hydroxy forms are critical elements of plant systemic immunity [141]. Besides, L-PA can serve as an osmoprotectant for *E. coli* [143] and *C. glutamicum* [144]. L-PA is an important precursor of secondary metabolites with pharmacological activities such as immunosuppressants, peptide antibiotics or piperidine alkaloids [138]. Currently, pure L-PA enantiomer is mainly prepared by chemical enantioselective synthesis, stereoselective transformation, and biosynthesis [139, 145, 146]. There are two main routes for converting L-lysine to L-PA, one pathway proceeds via ε-amino-α-ketocaproic acid and Δ1-piperideine-2-carboxylic acid, which involves loss of the α-amino nitrogen of L-lysine and would lead to incorporation of the ε-nitrogen into pipecolic acid. The other is a pathway through α-aminoadipic-semialdehyde and Δ1-piperideine-6-carboxylic acid, leading to loss of the ε-amino group of lysine and entry of the α-nitrogen into pipecolic acid [147, 148] (Fig. 7.3).

The *C. glutamicum* L-lysine producer strain GRLys1 was further engineered for a more efficient glucose metabolism in order to increase the productivity of L-lysine. The heterologous L-lysine 6-dehydrogenase gene (*lysDH*) from *Silicibacter pomeroyi* was expressed to oxidize L-lysine, such reaction was followed by spontaneous cyclization and reduction by endogenous pyrroline-5-carboxylate reductase yielding L-PA formation [138]. Further transport engineering and gene expression optimization yielded the strain PIPE4, which was tested in fed-batch fermentation. The L-PA titer and yield values achieved were 14 g/L and 0.2 g/g, respectively [78] (Table 7.2). In *E. coli*, the *pipA* gene from *Streptomyces pristinaespiralis* ATCC25486 encoding lysine cyclodeaminase was amplified enabling 5 g/L of L-

PA with a yield of 0.13 g/g from glucose in fed-batch cultivation [149]. Several copies of the *pipA* gene were introduced in the strain XQ-11, which is a *E. coli* L-lysine producer obtained after several rounds of random mutagenesis [79]. L-Lysine riboswitch to dynamically regulate the expression of *lysP* and *lysO* genes encoding a L-lysine import and export system respectively. The final strain XQ-11-4 produced 61 ± 3.4 g/L of L-PA with a volumetric productivity of 1.02 g/L/h [79] (Table 7.2).

7.2.2.4 Ectoine and Hydroxyectoine

Ectoine is a cyclic amino acid and a compatible solute, which serves as a protective substance against extreme osmotic stress by acting as an osmolyte. Ectoine is found in high concentrations in halophilic microorganisms such as *Marinococcus halophilus* or *Halomonas elongata* [150, 151]. Ectoine for use in cosmetic products have been sold for companies like Bitop AG (http://www.bitop.de) in Germany since 2001. The first commercial use of ectoine was as a skin care ingredient especially in sun protection and anti-aging products [152]. Moreover, it has been shown that ectoine is able to inhibit the UV-induced DNA single strand breaks [153]. The ectoine synthesis pathway, starting from the precursor L-aspartate-semialdehyde (ASA), comprises three enzymatic steps, which genes are normally organized in an operon (*ectABC*), which can also include the gene *ectD* for the synthesis of hydroxyectoine [154] (Fig. 7.3). During the first step, L-aspartate-β-semialdehyde is transaminated by the diaminobutyrate transaminase (EctB) to diaminobutyrate using glutamate as amino group donor. Diaminobutyrate is further acetylated by the diaminobutyrate acetyl transferase (EctA). Finally, acetyl diaminobutyrate is condensed to ectoine by the ectoine synthase (EctC) [155]. The expression of *ectABC* is typically induced in response to increased osmolarity and temperature [155]. The natural ectoine/hydroxyectoine-producing halophilic bacterium *H. elongata* is used to produce ectoine at industrial scale using a fermentation method called bacterial milking [156]. In this approach, *H. elongata* is grown in a hyperosmotic medium at 15% NaCl, and subsequently given a hypoosmotic downshock to 3% NaCl which triggers the bacteria to release compatible solutes into the medium [156]. Ectoine production in *E. coli* [82] and in *C. glutamicum* [157] under non-hyperosmotic conditions has been shown. The codon-optimized *ectABC* gene cluster from *Pseudomonas stutzeri* was expressed in the L-lysine *C. glutamicum* producer LYS-1 under the control of the constitutive *tuf* promoter. The L-lysine exporter gene was deleted, and the final ECT-2 could produce up to 4.5 g/L of ectoine in fed-batch fermentation [157]. In another study the *ectABC* genes from *Chromohalobacter salexigens* were heterologous expressed in the L-lysine *C. glutamicum* producer DM1729. The regulatory gene *sugR* was deleted to derepress glucose metabolism. In addition, the lactate dehydrogenase gene *ldhA* was also deleted to avoid L-lactate formation as by-product. The resulting strain Ecto5 was tested in fed-batch fermentation where it produced 22 g/L of ectoine after 69 h [80]. By fine optimization of the expression level of *ectABC* from *P. stutzeri*, the ectoine titer of 65 g/L was achieved in *C. glutamicum* [81] (Table 7.2). In *E. coli*, the *ectABC* operon from *Marinococcus halophilus* was heterologously expressed in *E. coli* DH5α. Co-expression of the gene of the feedback-resistant aspartate kinase from *C. glutamicum* improved ectoine formation [158]. In another study, the *ectABC* gene cluster from *H. elongata* was introduced into *E. coli* W3110. The *thrA* gene encoding the bifunctional aspartokinase/homoserine dehydrogenase was disrupted to weaken the competitive L-threonine branch. The promoter of the PEP carboxylase gene was replaced and the glyoxylate shunt transcriptional repressor gene *iclR* was deleted to reinforce the oxaloacetate supply. The final strain ECT05(pTrcECT, pSTVLysC-CG) produced 25.1 g/L of ectoine by fed-batch fermentation [82] (Table 7.2). *E. coli* has also been used as microbial platform for the production of hydroxyectoine. The whole *ectABCD-ask* gene cluster under the control of its salt-inducible native promoter from *P. stutzeri* DSM5190 was expressed in *E. coli* DH5α. For efficient hydroxyectoine production, the medium was supplemented with 0.34 M NaCl [159]. Recently, *E. coli* W3110 was engineered for high production of hydroxyectoine. First, a combinatorial optimization of the expression strength of the *ectABCD* genes from *H. elongata* was conducted to avoid formation of ectoine as by-product. Next, a quorum-sensing system was used to channel intracellular α-ketoglutarate for hydroxyectoine formation. Finally, the highest hydroxyectoine titer of 14.93 g/L as achieved by the strain HECT31 [83] (Table 7.2).

7.2.2.5 L-Carnitine

L-Carnitine is a bioactive compound derived from L-lysine and *S*-adenosyl-L-methionine (SAM). Carnitine can be used as dietary supplement, feed additive, and as L-carnitine or acyl-L-carnitine esters in pharmaceutical applications [84]. Biosynthesis of L-carnitine in the filamentous fungus *Neurospora crassa* starts with L-N^{ϵ}-trimethyllysine (TML). TML is converted to L-carnitine in four enzymatic steps (Fig. 7.3). First, TML is hydroxylated by the TML hydroxylase to yield (2*S*,3*S*)-3-hydroxy-TML (HTML). Second, HTML is cleaved into glycine and 4-trimethylaminobutyraldehyde (TMABA) by the HTML aldolase. Third, TMABA is oxidized to γ-butyroβine (γ-BB) by the NAD^+-dependent TMABA dehydrogenase. Finally, stereoselective hydroxylation of γ-BB by the enzyme γ-BB hydroxylase yields L-carnitine [84]. The genes of the L-carnitine biosynthetic pathway from *N. crassa* were characterized and functionally expressed in *E. coli*. Biotransformation of TML to L-carnitine was demonstrated by using a genetically encoded L-carnitine biosensor [160]. Additionally, *E. coli* was metabolically engineered to enable *de novo* production of L-carnitine. A concentration of 15.9μM of L-carnitine was detected by LC-MS in the supernatant after a cultivation supplemented with 1 mM of TML [84] (Table 7.2).

7.2.3 Value-Added N-Containing Chemicals Derived from L-Tryptophan

The aromatic amino acid L-tryptophane is of great importance in the food industry, but it is also a precursor of the biosynthesis of methylated, chlorinated and brominated compounds that are relevant in agriculture, food, and pharmaceutical industries. The antimicrobial violacein and the pigment Tyrian purple can also be synthesized from L-tryptophan (Fig. 7.4).

7.2.3.1 Violacein

Violacein is a purple-colored pigment that is biosynthesized by the condensation of two tryptophan molecules. Violacein was first isolated from *Chromobacterium violaceum* and have a broad range of biological activities, including anti-tumoral, bacteriostatic and antibiotic potential, antifungal, anti-protozoan, anti-cancer, and antiviral properties [161]. The biosynthetic pathway of violacein from L-tryptophan involves five enzymes named VioA, B, C, D, and E encoded by the *vioABCDE* operon (Fig. 7.4). First, L-tryptophan is converted to indole 3-pyruvic acid (IPA) imine by the enzyme flavin-dependent tryptophan-2 monooxygenase (VioA). IPA imine is dimerized by VioB. Then, the imine dimer is converted to protoviolaceinic acid (PDVA) by VioE. PDVA is converted into protoviolaceinic acid (PVA) by an NADP-dependent oxygenase VioD. Violaceinic acid (VA) is produced from PVA by another NADP-dependent oxygenase VioC. Finally, VA undergoes a spontaneous oxidative decarboxylation yielding violacein [161]. Heterologous expression of the *vioABCDE* gene cluster in model organisms and L-tryptophan over-production chassis strains enabled violacein production. These microbial cell factories include *E. coli*, *C. glutamicum* and *Citrobacter freundii*. The *E. coli* B8/TRPH1 strain is a L-tryptophan producer that was chosen as platform to express the *vioABCDE* gene cluster. The strain B8/pTRPH1-pVio-VioE accumulated 4.5 g/L of violacein after fed-batch fermentation [86]. The L-tryptophan producer *C. glutamicum* ATCC 21850 was used as platform for heterologous expression of a synthetic *vioABCDE* operon harboring optimal ribosome binding site (RBS) sequences. *C. glutamicum* 21,850 (pEC-C-vio1) was able to produce 5.4 g/L of violacein in bioreactor [85] (Table 7.2). Same strategy was applied in *Citrobacter freundii* in which heterologous expression of pCom10vio plasmid in *C. freundii* enabled violacein synthesis [87].

7.2.3.2 Hydroxytryptophan, Serotonin and Melatonin

5-Hydroxytryptophan (5-HTP), also known as oxitriptan, is used as an antidepressant, appetite suppressant, and sleep aid. Formation of 5-HTP takes place via one-step reaction by hydroxylation of L-tryptophan catalyzed by L-tryptophan hydroxylase (TPH) [88] (Fig. 7.4). A synthetic L-tryptophan hydroxylation pathway was implemented in *E. coli* by co-expressing human TPH and tetrahydrobiopterin biosynthesis and regeneration pathways. This allowed the formation of 5-HTP from L-tryptophan without exogenous supplementation of the pterin cofactor reaching the final 5-TPH titer of 1.3 g/L and 5.1 g/L in shake flask and fed-batch fermentation [88]. Next, the L- tryptophan biosynthetic pathway was optimized and integrated into the *E. coli* genome, increasing the 5-TPH production by 25% during flask cultivations [162] (Table 7.2). 5-HTP could also be produced when 5-hydroxyindole was fed to *C. glutamicum* cells that lack the α-subunit of tryptophan synthase and dynamically express the gene for the β-subunit of tryptophan synthase in response to the intracellular L-serine concentration [163].

Serotonin or 5-hydroxytryptamine is a neurotransmitter in animals that is involved in a variety of functions including mood, sleep cycles, appetite, and liver regeneration [164]. Serotonin is synthesized from 5-HTP via one decarboxylation step by a tryptophan decarboxylase (TDC) [165] (Fig. 7.4). In *E. coli*, recombinant TPH genes were coexpressed together with a TDC enabling the production of 24 mg/L of serotonin [165]. In another research in *E. coli*, the tryptophan pathway was extended by expressing the phenylalanine hydroxylase gene (ctAAAH) from *Cupriavidus taiwanensis* and an endogenous cofactor with an artificial regeneration system to produce 5-HTP. Next, 5-HTP was converted into serotonin with a tryptophan decarboxylase enzyme. Thought this approach, 154.3 ± 14.3 mg/L of serotonin were produced [89] (Table 7.2).

Melatonin is a hormone associated with control of the sleep–wake cycle and its main medical uses are related to sleep disorders. Further extension of the serotonin pathway led to melatonin biosynthesis [166] (Fig. 7.4). Two enzymatic steps are required for the biosynthesis of melatonin from serotonin. In the first reaction, the enzyme serotonin *N*-acetyltransferase (SNAT) catalyzes the conversion of serotonin to *N*-acetylserotonin, which is then followed by the action of *N*-acetylserotonin *O*-methyltransferase (ASMT) that synthetizes melatonin from *N*-acetylserotonin. In *E. coli*, a caffeic acid *O*-methyltransferase enzyme from rice with ASMT activity was coexpressed with sheep SNAT enabling production of 1.46 mg/L of melatonin [167]. In another study, the production of melatonin by *E. coli* was achieved by expressing of physostigmine biosynthetic genes from *Streptomyces albulus* together with the phenylalanine 4-hydroxylase gene from *Xanthomonas campestris* as well as the caffeic acid 3-*O*-methyltransferase from *Oryza sativa*. Further modifications yielded the strain EcMEL8 that produced 0.6 ± 0.1 g/L of melatonin with tryptophan supplementation [90] (Table 7.2).

7.2.3.3 Anthranilate and *N*-Methylanthranilate

Anthranilate is not directly derived from L-tryptophan, but it is a precursor within the L-tryptophan biosynthetic pathway (Fig. 7.4). Anthranilate production has been established in *E. coli* via expression feedback inhibition resistant version of the enzyme 3-deoxy-D-arabino-heptulosonate-7-phosphate synthase (*aroGfbr*), transketolase (*tktA*), glucokinase (*glk*) and galactose permease (*galP*), as well as inactive PTS. The best strain was evaluated in fed-batch fermentation using glucose as sole carbon source and producing 14 g/L of anthranilate in 34 h [91]. In *P. putida* the anthranilate phosphoribosyltransferase and an indole-3-glycerol phosphate synthase genes (*trpCD*) were deleted to avoid the conversion of anthranilate to L-tryptophan. Additionally, the chorismate mutase gene (*pheA*) was also deleted avoiding the step from chorismate to prephenate and L-phenylalanine. Finally, feedback resistant versions of the 3-deoxy-D-arabino-heptulosonate-7-phosphate synthase and an anthranilate synthase were heterologous overexpressed to increase anthranilate production. The best performing strain *P. putida ΔtrpDC* (pSEVA234_$aroG^{D146N}$_$trpE^{S40FG}$) produced 1.5 ± 0.3 g/L of anthranilate in a tryptophan-limited fed-batch process [168] (Table 7.2).

N-Methylanthranilate is an important precursor for bioactive compounds such as anticancer alkaloids, the flavor compounds, antinociceptive alkaloids, and as a building block for peptide-based drugs [169]. Expression of the methyltransferase 1 from *Zea mays* together with an increased anthranilate supply and enhanced the intracellular availability of the cofactor *S*-adenosyl methionine (SAM) enabled *N*-methylanthranilate synthesis in *E. coli* and *C. glutamicum*. The *N*-methylanthranilate titers achieved were 4.1 g/L and 4.5 g/L with the engineered *E. coli* and *C. glutamicum* strains, respectively, in bioreactor [92]. Also, in *C. glutamicum*, heterologous expression of the anthranilate *N*-methyltransferase gene from *Ruta graveolens* resulted in *N*-methylanthranilate production. Expression of the homologous adenosylhomocysteinase gene increased SAM regeneration and *N*-methylanthranilate production. The final titer of 0.5 g/L of *N*-methylanthranilate was achieved in bioreactor [169] (Table 7.2).

7.2.3.4 Chlorinated Tryptophan

Halogenated amino acids are sought after by the pharmaceutical, chemical and agrochemical industries since, e.g., 30% of agrochemicals are halogenated. They are found in antibiotics, the plant growth-regulators, antifungals and alkaloids [93]. The overexpression of the FADH-dependent halogenase gene *rebH* and the NADH-dependent flavin reductase gene *rebF* from *Lechevalieria aerocolonigenes* in the *C. glutamicum* L-tryptophan overproducing strain Tp679 (pCES208-*trpD*) [170] (Fig. 7.4) enabled halogenation of L-tryptophan to 7-chloro-L-tryptophan (7-Cl-Trp) [93] (Table 7.2).

7.2.3.5 Brominated Tryptophane and Tyrian Purple

Brominated compounds have applications in the agriculture, food, and pharmaceutical industries. In particular, 7-bromo-L-tryptophan (7-Br-Trp) can be used as precursor of the antimitotic agents, indole derivates, and protease inhibitors [94]. The *C. glutamicum* L-tryptophan overproducing strain Tp679(pCES208-

trpD) [170] was used as platform for 7-Br-Trp production. The FAD-dependent halogenase gene *rebH* and the NADH-dependent flavin reductase gene *rebF* from *L. aerocolonigenes* were heterologous overexpressed in Tp679(pCES208-*trpD*) (Fig. 7.4). The recombinant *C. glutamicum* strain was tested in bioreactors in batch and fed-batch mode. A titer of 1.2 g/L of 7-Br-Trp was produced in fed-batch fermentation in HSG complex medium [94]. In *E. coli*, production of 6-bromo-L-tryptophan (6-Br-Trp) was achieved by expressing the tryptophan 6-halogenase from *Streptomyces toxytricini* [95] (Table 7.2). Additional modifications expanded the pathway towards Tyrian purple production from 6-Br-Trp. Tyrian purple is a purple pigment traditionally extracted from sea snails. Expression of the endogenous tryptophanase from *E. coli* catalyzed the formation of 6-Br-indole from 6-Br-Trp. 6-Br-indole was further converted to Tyrian by the monooxygenase MaFMO from *Methylophaga aminisulfidivorans*. The recombinant *E. coli* strain could produce 315 mg/L of Tyrian purple [95] (Table 7.2).

7.2.3.6 Brominated Indoles and Tryptamines

In another work, production of brominated indoles and tryptamines was established via extension of the tryptophan biosynthesis pathway in *C. glutamicum* (Fig. 7.4). To this end, halogenation followed by decarboxylation to yield tryptamines or cleavage to yield indoles was needed. The tryptophanase genes from *E. coli* and *Proteus vulgaris* as well as the aromatic amino acid decarboxylase genes from *Bacillus atrophaeus, Clostridium sporogenes, and Ruminococcus gnavus* were heterologously expressed. In this work, 16 mg/L of 7-Cl-indole and 23 mg/L of 7-Br-indole were achieved. Regarding tryptamines, 0.15 g/L of 7-Br-tryptamine and 0.22 g/L of 7-Br-tryptamine were produced in flasks cultures. Scaling-up of the 7-Br-tryptamine producing strain yielded 0.36 g/L of 7-Br-tryptamine and a volumetric productivity of 8.3 mg/L/h in bioreactor cultivation [96] (Table 7.2).

7.2.4 *Value-Added N-Containing Chemicals Derived from L-Isoleucine*

L-Isoleucine, like other BCAA, is associated with insulin resistance. Additionally, L-isoleucine is the precursor of the non-proteinogenic amino acid 4-hydroxyisoleucine.

7.2.4.1 4-Hydroxyisoleucine

4-Hydroxyisoleucine (4-HIL) is a nonproteinogenic amino acid that exhibits insulinotropic biological activity [171]. The L-isoleucine dioxygenase (IDO) enzyme

Bacillus thuringiensis converts L-isoleucine to 4-HIL. An IDO variant with improved catalytic efficiency, thermal stability and catalytic rate was expressed in *E. coli ΔsucAΔaceA* the α-ketoglutarate dehydrogenase and *aceA*-encoding isocitrate lyase genes deleted. The final strain produced 22.4 g/L of 4-HIL [172] (Table 7.2). The L-isoleucine producer *C. glutamicum* YI was chosen as platform to produce 4-HIL. For the conversion of L-isoleucine to 4-HIL, the *ido* gene from *B. thuringiensis* TCCC 11826 was constitutively expressed. The pool of oxaloacetate was increased by modifications to the PEP-pyruvate-oxaloacetate node via *ppc* overexpression and *pyk2* deletion. By boosting the TCA cycle and interrupting the glyoxylate cycle via overexpression of *gltA* and *icd*, combined with deletion of *aceA* shifted the carbon flux from L-isoleucine to α-ketoglutarate. Finally, the activity of ODHC was dynamically modulated via regulation of transcription of *odhI* and *odhA*. The 4-HIL production of the final strain HIL18 reached 34.21 g/L at 64 h in fed-batch fermentation [97]. In another research with *C. glutamicum*, the use of dynamic control of 4-HIL biosynthesis by an L-isoleucine biosensor could regulate the three substrates involved in the process which are L-isoleucine, α-ketoglutarate, and O_2. This approach led to an improvement in 4-HIL formation of 92.3% reaching a final titer 19.9 g/L with the strain ST17 during flask cultivations [98] (Table 7.2).

7.3 Microbial Production of N-Containing Compounds from Renewable Substrates

Historical reliance on oil and other fossil fuels is dragging serious environmental problems, worsened by increased population growth. At large production scales, price, availability and competing uses of carbon sources are key factors for economic success of biotechnological processes [173] (Fig. 7.5). Hence, there is a need for renewable substrates and a portfolio of products which have varying market receptivity. Common microbial production hosts like *E. coli*, *C. glutamicum*, *P. putida* and *Bacillus subtilis* differ with respect to their native ability to consume carbon sources for growth and production. Thanks to the metabolic engineering achievements within strain development, the substrate scope of these organisms with respect to access to non-native carbon sources has been greatly broaden. Additionally, non-conventional biotechnology workhorses are gaining interest due to their use of specific carbon sources as, for instance, the methylotroph *Bacillus methanolicus* [175].

Renewable substrates provide different qualities with regard to the production of amine chemicals. As it is true for many fermentation processes, the absence of inhibitory or toxic compounds in the renewable substrates and their preparations, e.g., as hydrolysates, is important. Moreover, the kind of sugars that are available as carbon and energy sources is relevant since bacteria prefer certain sugars, mostly glucose, over other sugars since these sugars support growth and production with different yields and rates. Since the production of amine chemicals is considered

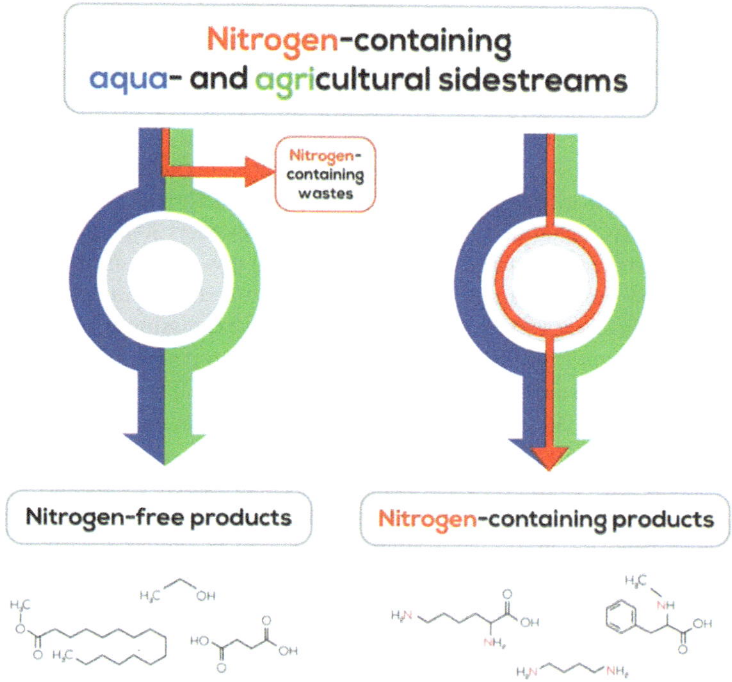

Fig. 7.5 Comparison of traditional carbon-centered biorefinery concepts to produce nitrogen-free biofuels and biochemicals with a nitrogen-centered biorefinery concept to produce amines (graphical abstract reprinted from Wendisch et al. [174])

here, the nitrogen content and quality of the renewable substrates has to be considered. Ideally, the proportion of carbon and nitrogen sources of the renewable substrates reflects the requirements to sustain growth (C:N ratio of biomass) as well as production (C:N ratio of the target product). Since this often is not the case, it will be important to add the limiting substrate (C or N source) to achieve production of amine chemicals from these renewables. Thus, as farming requires adjusted mixtures of fertilizers or aquaculture requires adjusted feed formulations, fermentation processes to amine chemicals from renewables require media preparations with adjusted ratios of carbon, nitrogen and energy sources.

7.3.1 Wood/Plant-Derived Substrates

Woody related sugars and biomass has been evaluated since decades as alternative microbial feedstock for bioprocesses. Within this group of substrates, the glucose-based polymers starch and cellulose as well as the sugar pentoses xylose and arabinose are included.

7.3.1.1 Starch

A wide variety of wood/plant-derived resources are available for conversion into added-value and N-containing products. These may include plant constituents like starch, processing by-products like hydrolysates, or wastes like spent sulfite liquor (SSL) from biorefineries. How to access recalcitrant polymers in order to enable consumption of their monomers by the microbial cell factories.

Starch is a polysaccharide comprising glucose monomers joined in α-1,4 glycosidic bonds. Starch can be extracted from agricultural raw materials, and it is present in food and non-food materials. A L-lysine producing *C. glutamicum* strain was engineered to get access to glucose from starch by expressing a α-amylase gene (*amyA*) from *Streptomyces griseus*. The enzyme was produced and secreted to the medium hydrolysing starch to glucose and dextrins [176]. The expression of *amyA* from *S. griseus* was later used to show L-PA, 5AVA and ectoine production in recombinant *C. glutamicum* strains [75, 78, 80] (Table 7.3). In a different approach, *C. glutamicum* was engineered for the utilization of starch by displaying in the surface a fusion protein containing the α-amylase from *Streptococcus bovis* and the protein PgsA from *B. subtilis*. This strategy was used to show L-lysine and cadaverine production in *C. glutamicum* [177, 178] (Table 7.3). Similarly, the α-amylase from *S. bovis* was expressed and surface displayed by fusing the enzyme with the endogenous NCgl1221 protein of *C. glutamicum* and showing L-glutamate from starch [198]. Finally, a binary *E. coli*-*C. glutamicum* synthetic consortia was established in which a lysine auxotroph *E. coli* expressing *amyA* from *S. griseus* enabled growth from starch for both bacteria while *C. glutamicum* supplemented L-lysine to *E. coli* as well produced the L-lysine derivatives cadaverine and L-PA [199] (Table 7.3).

7.3.1.2 Cellulose

Cellulose is the most abundant polysaccharide on Earth as well as is the major polymeric component of plant matter. Although cellulosic biomass is one of the most abundant waste materials, its utilization is challenging because of its complex structure. In order to efficiently degrade cellulose, the synergistic action of several enzymes it is needed. In *C. glutamicum* the β-glucosidase from the cellulose degrading *Saccharophagus degradans* was displayed on the cell surface allowing production of L-lysine from the disaccharide cellobiose [200]. β-Glucan is comprised of glucose residues which are linked by β-(1–4) and β-(1–3) glycosidic bonds. Synthesis and secretion of endoglucanase from *Clostridium thermocellum* with the signal peptide sequence of *torA* from *E. coli* was establish in *C. glutamicum*. In combination with the expression of the β-glucosidade from *Aspergillus oryzae*, *C. glutamicum* could growth and produce L-glutamate from β-glucan [201]. Besides, *C. glutamicum* strains co-expressing endoglucanase genes from *X. campestris* XCC3521 and *X. campestris* XCC2387 together with the β-glucosidase gene from

Table 7.3 Relevant microbial biofactories for the production of amino acids and amino acid derivatives from alternative carbon sources

Strain	Substrate	Product	Production values*	Reference
C. glutamicum CSS	Starch	L-lysine	T: 6.0, Y:18.9, P: 0.25	[177]
C. glutamicum 5AVA3 (pECXT99A-*amyA*)	Starch	5AVA	T: 0.8, Y: 0.07, P: 0.025**	[75]
C. glutamicum PIPE4 (pECXT99A-amyA)	Starch	L-PA	T: 1.52, Y: 0.12	[78]
C. glutamicum Ecto5 (pECXT99A-amyA)	Starch	Ectoine	T: 0.5, Y: 0.04, P: 0.04	[80]
C. glutamicum CSS-cadAF	Starch	Cadaverine	T: 2.3	[178]
C. glutamicum DM1729 (pVWEx1-TAT-XCC2387) (pEKEx3-TAT-Sde1394)	Cellulose	L-lysine	T: 0.86, Y: 0.04, P: 0.01	[179]
C. glutamicum DM1729 (pEKEx3-xylAB)	Xylose	L-lysine	T: 1.75, Y: 0.12, P: 0.03	[180]
C. glutamicum ATCC13032 (pEKEx3-xylAB)	Xylose	L-glutamate	T: 2.13, Y: 0.14, P: 0.03	[180]
C. glutamicum ORN1 (pEKEx3-xylAB)	Xylose	L-ornithine	T: 2.6, Y: 0.17, P: 0.04	[180]
C. glutamicum PUT21 (pEKEx3-xylAB)	Xylose	Putrescine	T: 1.3, Y: 0.09, P: 0.03	[180]
C. glutamicum 5AVA3 (pECXT99A-*xylAB*)	Xylose	5AVA	T: 1.1, Y: 0.11, P: 0.01**	[75]
C. glutamicum PIPE4 (pEKEx3-*xylAB*)	Xylose	L-PA	T: 0.52, Y: 0.05	[78]
C. glutamicum Ecto5 (pEKEx3-*xylAB*)	Xylose	Ectoine	T: 0.4, Y: 0.04, P: 0.01	[80]
C. glutamicum SAR3	Xylose	Sarcosine	T: 8.6, Y: 0.25, P: 0.12	[181]
C. glutamicum GABA13	Xylose	GABA	T: 1.2, Y: 0.08, P: 0.03	[63]
C. glutamicum DAP-Xyl2	Xylose	Cadaverine	T: 103, Y: 0.18, P: 1.37	[182]
C. glutamicum DM1729 (pVWEx1-*araBAD*)	Arabinose	L-lysine	T: 8.0, Y: 0.13	[183]
C. glutamicum ATCC13032 (pVWEx1-*araBAD*)	Arabinose	L-glutamate	T: 5.4, Y: 0.07	[183]
C. glutamicum ORN1 (pVWEx1-*araBAD*)	Arabinose	L-ornithine	T: 11.8, Y: 0.33	[183]
C. glutamicum ARG1 (pVWEx1-*araBAD*)	Arabinose	L-arginine	T: 4.4, Y: 0.37	[183]
C. glutamicum 5AVA3 (pECXT99A-*araBAD*)	Arabinose	5AVA	T: 0.8, Y: 0.08, P: 0.01**	[75]
C. glutamicum Ecto5 (pEKEx3-*araBAD*)	Arabinose	Ectoine	T: 0.4, Y: 0.04, P: 0.01	[80]

(continued)

Table 7.3 (continued)

Strain	Substrate	Product	Production values*	Reference
C. glutamicum DM1729Δ*pck* (pEKEx3-*glpFKD*)	Glycerol	L-lysine	T: 3.0, Y: 0.12	[184]
C. glutamicum ATCC13032 (pEKEx3-*glpFKD*)	Glycerol	L-glutamate	T: 1.1, Y: 0.05	[184]
C. glutamicum ARG1 (pEKEx3-*glpFKD*)	Glycerol	L-ornithine	T: 2.3, Y: 0.11	[184]
C. glutamicum ORN1 (pEKEx3-*glpFKD*)	Glycerol	L-arginine	T: 0.8, Y:0.04	[184]
C. glutamicum PUT21 (pEKEx3-*glpFKD*)	Glycerol	Putrescine	T: 0.4, Y: 0.02	[184]
C. glutamicum PIPE4 (pEKEx3-*glpKDF*)	Glycerol	L-PA	T: 1.42, Y: 0.14	[78]
C. glutamicum Ecto5 (pEKEx3-*glpFKD*)	Glycerol	Ectoine	T: 0.6, Y: 0.05, P: 0.03	[80]
C. glutamicum RiboSSL	Synthetic spent sulfite liquor	Riboflavin	T: 0.03, Y: 0.005, P: 0.003	[185]
C. glutamicum DM1729 (pEKEx3-*xylA*)(pVWEx1-*araBAD*)	Lignocellulosic hydrolysate	L-lysine	T: 6.5, Y: 0.16	[186]
C. glutamicum ATCC13032 (pEKEx3-*xylA*)(pVWEx1-*araBAD*)	Lignocellulosic hydrolysate	L-glutamate	T: 13.7, Y: 0.33	[186]
C. glutamicum AVA1-G	Lignocellulosic hydrolysate	5AVA	T: 4.0, Y: 0.15, V: 0.15	[187]
C. glutamicum PUT Xyl	Lignocellulosic hydrolysate	Putrescine	T: 0.09, Y: 0.002, V: 0.001	[188]
C. glutamicum CAD-XA	Lignocellulosic hydrolysate	Cadaverine	T: 7.6, Y: 0.16, V: 0.14	[187]
C. glutamicum LYS-XA	*N*-acetylglucosamine	L-lysine	T: 8.0, Y: 0.24 , V: 0.37	[187]
C. glutamicum GABA14	*N*-acetylglucosamine	GABA	T: 3.9, Y: 0.18, V: 0.15	[63]
C. glutamicum DM1946 (pVWEx1-*nagAB*)	Glucosamine	L-lysine	T: 3.4, Y: 0.10	[189]
C. glutamicum PUT21 (pEKEx3-*nagB*)	Glucosamine	Putrescine	T: 4.5, Y: 0.10	[189]
C. glutamicum GABA12	Glucosamine	GABA	T: 1.8, Y: 0.1, V: 0.07	[63]
C. glutamicum 5AVA3 (pECXT99A-*nanB*)	Glucosamine	5AVA	T: 1.25, Y: 0.13, V: 0.04**	[75]
C. glutamicum PIPE4 (pEKEx3-*nagB*)	Glucosamine	L-PA	T: 0.34, Y: 0.03	[78]

(continued)

Table 7.3 (continued)

Strain	Substrate	Product	Production values*	Reference
C. glutamicum Ecto5 (pEKEx3-*nagB*)	Glucosamine	Ectoine	T: 0.8, Y: 0.08, V: 0.045	[80]
B. methanolicus NOA2 #13A52-8A66 + MTL	Methanol	L-lysine	T: 65	[190]
B. methanolicus MGA3 (pHP13)	Methanol	L-glutamate	T: 33.9	[191]
B. methanolicus MGA3 (pTH1mp-*gad*St)	Methanol	GABA	T: 9.0, V: 0.13	[191]
B. methanolicus MGA3 (pTH1mp-*cadA*)	Methanol	Cadaverine	T: 6.5, V: 0.22	[192]
M. glycogenes A513(pTHD-30)	Methanol	L-threonine	T: 16.3, V: 0.223	[193]
Methylobacterium sp. MN43	Methanol	L-serine	T: 0.06, V:5.41E-04	[194]
B. methanolicus MGA3_PatA^{Ec}	Methanol	5AVA	T: 0.02	[195]
C. glutamicum SEA-7	Mannitol	L-lysine	T: 76, Y: 0.22, V: 2.1	[196]
C. glutamicum CgRibo4	Mannitol	Riboflavin	T: 0.47, Y: 0.05, V: 0.01	[197]
C. glutamicum SEA-7	Seaweed hydrolysate	L-lysine	T: 2.8, Y: 0.16, V: 0.21	[196]
C. glutamicum CgRibo4	Seaweed hydrolysate	Riboflavin	T: 1.1, Y: 0.08, V: 0.016	[197]
C. glutamicum CgRibo4	Seaweed extract	Riboflavin	T: 1.3, Y: 0.07, V: 0.017	[197]

**T* titer in g/L, *Y* Product yield per substrate in g/g, *V* Volumetric productivity in g/L/h **Approx. Values

S. degradans Sde1394 were constructed which could produce L-lysine from cellulose [179].

7.3.1.3 Xylose

The pentose xylose can be found in hemicelluloses. Xylose consumption by *C. glutamicum* was enabled by heterologously expressing xylose isomerase gene *xylA* from *E. coli*. Faster growth from xylose resulted from the overexpression of *xylA* from *X. campestris* in combination with the endogenous xylulokinase gene *xylB* [180]. This pathway is called isomerase pathway in which xylose is converted to xylulose, which is then phosphorylated to xylulose-5-phosphate, an intermediate of the pentose phosphate pathway. Several N-containing high-value compounds have been produced in *C. glutamicum* by introducing this pathway as, for instance,

L-lysine, L-glutamate, L-ornithine, putrescine, L-PA, 5AVA, ectoine, sarcosine (*N*-methyl-glycine), and dipicolinic acid (DPA) [75, 78, 80, 180, 181, 202] (Table 7.3). Furthermore, the expression of the *xylXABCD*-operon from *Caulobacter crescentus* in *C. glutamicum* enabled xylose utilization via the Weimberg pathway. It this pathway, xylose catabolism is oxidated to xylonolactone, catalyzed by a xylose dehydrogenase (encoded by *xylB*). Then, a xylonolactonase (encoded by *xylC*) opens the lactone ring yielding xylonate, which is dehydrated twice in two successive reactions by a xylonate dehydratase (encoded by *xylD*) and a 2-keto-3-desoxyxylonate dehydratase (encoded by *xylX*). Finally, α-ketoglutarate semialdehyde is oxidized to α-ketoglutarate by the α-ketoglutarate semialdehyde dehydrogenase (encoded by *xylA*) [203, 204].

7.3.1.4 Arabinose

Arabinose is also a pentose sugar present in hemicellulose. While *E. coli* and *B. subtilis* can utilize arabinose, *C. glutamicum* required modifications. In *C. glutamicum*, arabinose utilization became possible by heterologous expression of the *araBAD* operon of *E. coli* that encoded encode for L-ribulokinase, L-arabinose isomerase, and L-ribulose-5-phosphate 4-epimerase respectively [205]. Here, also production of several N-containing high-value compounds have been shown such us L-glutamate, L-lysine, L-arginine, ornithine, L-PA, 5AVA, and ectoine [75, 78, 80, 183] (Table 7.3). Co-consumption of glucose, xylose and arabinose was established in *C. glutamicum* for the fermentative production of the vitamin riboflavin in bioreactors [185].

7.3.2 Agricultural Residues

There are a number of agro-wastes or sidestreams from agricultural processes, e.g., straws, husks or shreds. The quality of these polymeric compounds with regard to fermentation depends on a number of parameters, e.g., the content of lignin as in woody plants or silicates as in grasses. Typically, agro-wastes are processed e.g. by hydrolysis as pretreatment procedure. The resulting lignocellulosic hydrolysates are attractive carbon and nitrogen sources for biotechnological processes. Regardless of hydrolysis methods, glucose and xylose are commonly the most abundant sugars in these hydrolysates. For instance, *E. coli* have been engineered to produce cadaverine from soybean hydrolysates which was used as sole nitrogen source [206] *C. glutamicum* strains have been developed to enable access to the sugars in the rice straw hydrolysate containing mainly glucose and xylose. Production of L-glutamate, L-lysine, 5AVA, putrescine and sarcosine were shown from this hydrolysate [181, 186, 188]. 5AVA production in *C. glutamicum* was also proved from *Miscanthus* hydrolysate [207] and wheat hydrolysate containing glucose, xylose and arabinose [187]. Besides, accumulation of L-glutamate from was shown from biotin-rich corn stover hydrolysate and recombinant *C. glutamicum* [208]. Lignocellulosic hydrolysates can contain hexuronic acids like D-galacturonic acid and D-glucuronic acid [209]. Heterologous expression of the *uxaCAB*, *uxuAB*, *kdgK*, *eda* and *exuT* from *E. coli* enabled growth of *C. glutamicum* from hexuronic acids and production of the N-containing compounds L-ornithine and L-lysine [209].

7.3.3 Side Streams from Industrial Processes

Commonly, industrial processes generate unwanted by-products or side streams that, in some cases, can be repurposed. For instance, those side streams that contain unutilized carbohydrates and/or nitrogen sources could be serve as microbial substrate for bioprocesses.

7.3.3.1 Glycerol

Glycerol is a by-product of the biodiesel synthesis process which is produced in large quantities [210]. Glycerol is a natural growth substrate for *E. coli*, *B. subtilis*, *P. putida*, but not for *C. glutamicum* [173]. Overexpression of endogenous glycerol kinase gene *glpK* with the glycerol-3-phosphate dehydrogenase gene *glpD* or of the respective *E. coli* genes enabled growth with glycerol in *C. glutamicum*. Accelerated growth was achieved with the additional expression of the glycerol facilitator gene *glpF* from *E. coli* [211]. Recombinant *C. glutamicum* strains were used for glycerol-based production of the N-containing compounds L-glutamate, L-lysine, L-ornithine, L-arginine, putrescine, L-PA, and ectoine [78, 80, 184, 211] (Table 7.3).

7.3.3.2 Spent Sulfite Liquor

Spent sulfite liquor (SSL) is a by-product obtained in the process of manufacturing dissolving pulp by the acid sulfite method. The main components present in this spent liquor are sugars from hemicelluloses and lignosulfonates. Hemicelluloses are composed of several sugars that can include the six-carbon (C6) sugars glucose, mannose and galactose, the C5 sugars xylose and arabinose, and the C6 deoxy sugar rhamnose [212, 213]. Synthetic SSL containing glucose, xylose and mannose was used as carbon source in bioreactor fermentations following a dynamic co-cultivation approach with engineered *C. glutamicum* strains. The process was coupled with the production of riboflavin [185]. Recent studies with *C. glutamicum* were focused on the development of a dynamic model describing biomass growth on spent sulfite liquor [214]. However, to the best of our knowledge, no more N-containing compounds have been obtained from fermentations with *C. glutamicum* utilizing SSL.

7.3.3.3 Amino Sugars

Chitin is one of the most abundant polymers on earth. Chitin from shellfish waste represents an important resource of substrate for potential applications in biorefinery. Degradation of chitin releases the amino sugar *N*-acetylglucosamine (GlcNAc) and its deacetylated derivate glucosamine (GlcN), which are monosaccharide derivatives of glucose [189]. *P. putida* can efficiently utilize glucosamine and *B. subtilis* and *E. coli* grows with GlcN and GlcNAc [215, 216]. *C. glutamicum* strains have been developed to enable access to GlcN and GlcNAc [189, 217]. In *C. glutamicum*, fast growth with glucosamine as sole source of carbon and nitrogen was achieved by overexpression or derepression of the endogenous *nagB* gene encoding glucosamine-6-phosphate deaminase while the uptake of glucosamine is carried out by the glucose specific permease PtsG [189]. To enable growth of *C. glutamicum* with GlcNAc, the GlcNAc-specific PTS gene *nagE* from the related *Corynebacterium glycinophilum* had to be expressed together with the endogenous deaminase *nagB* and native acetylase *nagA* [217]. The amino acid L-lysine was produced from GlcN and GlcNAc as carbon source [189, 217] (Table 7.3).

7.3.3.4 Residues from Food and Beverage Production

By-products from food and beverage industries can be revalorized via two well defined approaches. The whole-stream valorization makes use of the material and/or the energetic content of residues as, for example, utilization of bioresidues as animal feed, soil conditioner or fertilizer, mushroom cultivation, and bioadsorbents [218]. When focusing on the valorization of single fractions, targeting of high-quality fine chemicals for their extraction and purification as well as the creation of new added-value chains from those fractions via microbial fermentations are in the spotlight. Fruit peels and vegetable processing by-products are most promising with a view to recuperating high-value fractions, some of them with high content of fermentable substrates that could be used in bioprocesses [218]. For instance, D-galacturonic acid is abundant in pectin-rich waste streams such as peels. In this regard, *C. glutamicum* was engineered to produce the N-containing compounds L-ornithine and L-lysine from D-galacturonic [209]. Other residues that can also be attractive for exploitation in the biotechnological industry are the lees (spent yeasts) from fermentation and aging processes which contain abundant unused nutrients [219]. However, to the best of our knowledge, lees are only used in microbial fermentations to produce lactic acid, citric acid and xylitol [219].

7.3.4 Methanol as Representative C1 Substrate

Methanol is a reduced form of carbon dioxide (CO_2). This one-carbon (C1) compound can be used as non-food carbon source in biotechnological processes. The interest to produce high valuable products from methanol has risen in the past years. Methanol-dependent growth and partial synthetic methylotrophy has been achieved in *E. coli* [220–222]. In *C. glutamicum*, methanol-dependent growth was achieved via metabolic engineering and adaptative laboratory evolution (ALE) [223, 224]. Recombinant *C. glutamicum* was able to produce C13-labeled cadaverine from C13-labeled methanol as co-substrate [225]. Besides, L-glutamate production from methanol was also shown [223].

However, to date, native methylotrophs are the regular microbial platforms for the production of added-value compounds and, specially, N-containing compounds from methanol [226]. For instance, *B. methanolicus* had been engineered for the production of L-glutamate, L-lysine, GABA, cadaverine and 5AVA [191, 227–229] (Table 7.3). The amino acids L-threonine, L-glutamate and L-lysine were produced with *Methylobacillus glycogenes* [193, 230]. *Methylobacterium extorquens* was used to synthesize L-serine from methanol [194] (Table 7.3).

7.3.5 Marine Resources

Algal biomass contains various fermentable polysaccharides, for instance, galactose, mannitol, alginates or laminarin. However, seaweed has not been fully exploited as sustainable microbial feedstock. Mannitol is a native carbon source for some biotechnology workhorses like *E. coli*, *B. subtilis* and *B. methanolicus* [231–233].

C. glutamicum possesses an arabitol utilization operon. Deletion of the transcriptional repressor gene *atlR* or overexpression of the operon can enable mannitol consumption by *C. glutamicum* [234]. The operon is repressed by the regulatory protein AtlR and can be induced by arabitol, but not by mannitol [234]. This idea was implemented in *C. glutamicum* and L-lysine production was shown from mannitol. Further improvements combining systems metabolic engineering as well as the heterologous expression of a fructokinase led to a L-lysine titer of 76 g/L. Besides, the macroalgae *Durvillaea antarctica* or *Laminaria digitate* were tested as carbon sources [196] (Table 7.3). Alternatively, the mannitol-dependent phosphotransferase system from *B. subtilis* was implemented in *C. glutamicum* to enable mannitol utilization. Overproduction of riboflavin was coupled via constitutive overexpression of the endogenous biosynthetic riboflavin operon. The final strain CgRibo4 was tested with glucose, mannitol, glucose plus mannitol, as well as seaweed hydrolysate and seaweed extract from the brown macroalgae *Laminaria hyperborea*. Finally, seaweed-based fed-batch bioprocesses were established reaching 1.3 g/L of riboflavin from *L. hyperborea* extract as carbon source [197] (Table 7.3).

7.4 Perspectives for the Microbial Production of N-Containing Compounds

The flexibility of biological systems offers endless possibilities to cope with the problems associated with the increasing population's demands. Microbial-based bioprocesses can provide faster, and greener solutions as compared to other chemical-based approaches. New and innovative metabolic engineering tools are continuously in development, alternative microbial feedstocks are being explored every year, and new added-value chains are created by expanding the product portfolio from those microbial-based bioprocesses.

7.4.1 Trending Approaches in Metabolic Engineering

Strain construction is accelerated by an ever-increasing amount of synthetic biology tools. CRISPR-Cas systems are powerful mechanisms for generating cell factories with desirable properties. The genetic manipulation via CRISPR-Cas toolbox provides excellent opportunities for fine-tuning of metabolic pathways, improvement of metabolic robustness, or efficient utilization of new substrates [235]. For instance, a synthetic 5-methylpyrazine-2-carboxylic acid biosynthetic pathway was integrated in *E. coli* BL21 using CRISPR-Cas [236]. Another example is the creation of libraries of RBS, 5′ UTR regions, and promoters for several genes for *C. glutamicum* and *B. subtilis* [237]. Although the engineering possibilities of CRISPR are quite broad, the application of these tools in the field of N-compounds production is limited to a few research works. In *E. coli*, variants of the enzyme AroG with increased resistance to feedback inhibition were identified improving the L-tryptophan production [238]. In another study, CRISPR-assisted DNA polymerases were used to generate mutations in the ornithine aminotransferase gene in order to improve production of L-proline from L-arginine in *Corynebacterium crenatum* [239]. In *C. glutamicum,* CRISPRi was applied to reduced expression of the genes *pgi*, *pck* and *pyk* resulting in higher titers of L-lysine and L-glutamate [240]. Repurposing natural genetic control elements could be used for the development of biosensors, which can be applied in the construction of new producing strains. The LysR-type transcriptional regulator LysG activates transcription of *lysE,* which encodes the L-lysine exporter in *C. glutamicum* [46]. The L-lysine regulator LysG and the two amino acid-responsive transcriptional regulators Lrp from *C. glutamicum*, were used to develop genetically-encoded biosensors [241] Similarly, a L-lysine transcriptional factor-based biosensor was developed for *E. coli* fusing the promoter of the transcriptional regular gene *argP* to a reporter gene [241]. L-Valine, L-methionine and BCAA production activates the expression of the *brnFE* operon in *C. glutamicum*, which encodes an export system for these amino acids. The transcriptional fusion of the *brnF* promoter with the green fluorescent protein (*gfp*) gene allowed intracellular detection and quantification of these amino acids [242]. Transcriptional fusions of one of their target promoters with a fluorescence reporter gene allowed the monitoring of intracellular L-lysine, L-valine, L-methionine and BCAA concentrations of single cells by using fluorescence activated cell screening/sorting (FACS) [241–243]. A L-lysine biosensor was used to identify pyruvate carboxylase variants in *C. glutamicum* improving L-lysine production from glucose [244]. In *E. coli*, the native putrescine-responsive PuuR repressor protein was used as a sensing element to engineer biosensors specific for putrescine [245]. A riboswitch-based L-lysine biosensor was used to control gene expression according to the metabolic demand. The L-lysine riboswitches from *E. coli* and *B. subtilis* were used as L-lysine sensors to control the citrate synthase gene *gltA* in *C. glutamicum* L-lysine producing strains. Low intracellular L-lysine concentrations maintained the riboswitch conformation allowing access of the RBS of *gltA* and citrate synthase is synthesized [246]. Besides, positive on-demand control of L-lysine export protein was achieved by controlling the L-lysine export gene *lysE* with a L-lysine riboswitch [247]. Rational approaches for strain development can be hampered by the complexity of the biological systems, which is a bottleneck that can be skipped via forced evolution or evolutionary engineering. ALE requires microbial growth, normally repetitive batch cultivations or continuous cultivations in flasks or bioreactors. ALE strategies can be used to obtain better production strains, identify non-intuitive targets for strain engineering, and to gain understanding of pathway regulation [248]. As example of application, L-ornithine production was improved 20% in *C. glutamicum* after 70 days of repetitive batch cultivations in media containing glucose and L-ornithine [249]. ALE has been also used to increase the production of putrescine. After 11 days of repetitive batch cultivations in rich media containing putrescine and applying random mutagenesis, a *C. glutamicum* strain accumulated 78 single nucleotide polymorphisms (SNPs) [103]. Coupling of evolutionary engineering and biosensors allowed biosensor-driven adaptive laboratory evolution to improve L-valine with *C. glutamicum* [250]. While traditional synthetic biology focuses on a single bacterial population, synthetic microbial consortia focuses on several communities and tries to exploit their interactions. Engineering of microbial consortia is required to generate more complex dynamics and to increase the robustness of programmed behavior [251]. Intraspecific synthetic microbial consortia have been established for *E. coli* and *C. glutamicum*. For instance, a glucose positive *E. coli* overproducing L-lysine was co-cultivated with a glucose negative and glycerol positive *E. coli* strain converting L-lysine into cadaverine [252]. A glucose negative and L-phenylalanine auxotrophic *E. coli* strain that produced L-tyrosine and tyrosol was co-cultivated with a xylose-negative, L-tyrosine auxotrophic *E. coli* producing L-phenylalanine and salidroside [253]. Moreover, a co-cultivation of *C. glutamicum* strains was established with L-lysine overproducing and L-lysine auxotrophic strains in defined microenvironments [254]. On the other hand, interspecific synthetic microbial with different bacterial species have been also created. For instance, L-lysine, cadaverine and L-PA were produced from a starch-based consortia in which a L-lysine auxotrophic *E. coli* strain secreting α-amylase was co-cultivated with starch negative *C. glutamicum* overproducing strains [199].

7.4.2 Expanding the Substrate Spectra

The increasing global population and, hence the demand for food and energy production pushes the development of new sustainable strategies in biotechnology, including the sought-after new and environmentally friendly microbial feedstocks. C1 compounds, which include formate, methane, methanol, or CO_2, are gaining especial interest recently as microbial substrates. Archaea and bacteria with the capacity to utilize C1 molecules as carbon and energy sources are widespread across the planet. Various methanogenic archaea like *Methanobacterium formicum* can use formate electron donors and can grow with formate as sole carbon and energy source producing methane [255]. Synthetic formatotrophy and methylotrophy was recently demonstrated in *E. coli*. To enable *E. coli* growth on formate the reductive glycine pathway was targeted. In this pathway formate and CO_2 are directed assimilated into central metabolism. Further expression of methanol dehydrogenase gene enabled the conversion of methanol to formate, hence supporting growth on this methanol carbon source [222]. In another study with *E. coli*, the oxygen-sensitive enzyme pyruvate formate lyase was located in synthetic bacterial microcompartments converting formate and acetyl-phosphate into pyruvate [256]. Furthermore, synthetic autotrophy was also established in *E. coli*. The engineered *E. coli* strain used the Calvin-Benson Bassham cycle for carbon fixation and harvested energy and reducing power from the one-carbon molecule formate. The stepwise bioengineering process required the expression of Calvin cycle enzymes, rewiring of endogenous metabolic networks, and ALE approach. In the end, *E. coli* could generate biomass from CO_2 [257]. Methane is also a promising feedstock with high abundance and low cost for the sustainable production. Methane can be converted to methanol via the methane monooxygenase (MMO). Therefore, expression of a MMO could enabled methane utilization by methylotrophs or synthetic methylotrophs organisms [258]. Another promising microbial feedstock is seaweed. Rational utilization of seaweed (macroalgae) biomass in microbial fermentations is fast growing field in biotechnology, yet it remains quite unexploited. Seaweeds are promising feedstock for microbial bioprocesses due to their great biomass production yields and abundance of fermentable carbohydrates such as alginates, laminarin or mannitol in brown seaweeds [259]. Deletion of the arabitol repressor AtlR in the lysine producing strain *C. glutamicum* LYS-12, enabled mannitol utilization via arabitol uptake permease and arabitol dehydrogenase. Growth of engineered *C. glutamicum* was shown from acid pre-treated seaweed *D. antarctica* or *L. digitate* [196]. Furthermore, mannitol growth with *C. glutamicum* was shown via the mannitol-dependent phosphotransferase system from *B. subtilis* and bioprocesses in bioreactors were established using *L. hyperborea* substrates [197]. *B. methanolicus* was grown in brown seaweed extract from *Laminaria hyperborea* and production of cadaverine was demonstrated here from the mannitol and glucose present in the extract [260]. Brown seaweed are an important source of alginate. Different prokaryotic organisms like *Stenotrophomonas maltophilia* and *Vibrio splendidus* can use alginate as carbon and energy source [261, 262], however alginate is not a native substrate for regular biotechnology workhorses. Alginate consumption as non-native carbon source has been achieved in *E. coli* by overexpression of 8 different alginate lyases. Each of the alginate lyases were necessary to release monomers that could enter glycolysis [263]. Production of amines, amino acids and other N-containing compounds can be beneficed by the nitrogen present in certain agri- and aqua-culture sidestreams [174]. Inedible by-products like brewers' grains and yeast, malt sprouts or wheat mill feed [264] may contain unused proteins and amino acids that could be use as supply of nitrogen in microbial fermentations. Similarly, oil production by-products such as press cakes from processing oilseeds and oil meals are also an excellent source of nitrogen. For instance, peanut meal contains up to 51% of protein content and castor seed cake contains up to 36% [265]. Protein recovery from inedible tissues is an essential step in valorization of slaughterhouse waste. By-products like blood or bones are processed to recover partially hydrolyzed protein [266]. Algal proteins are of high quality and whole microalgae biomass can be used as a protein source in food and feed. Access to proteins from microalgae requires cell disruption and extraction [267]. Furthermore, dried and undried wastes from ruminants, poultry or swine, fishmeal, crab meal, and shrimp meal have been used as feed ingredients. The protein content ranges from 18% for cattle manure up to 48% for poultry manure, hence being source of nitrogen for microbial processes [268].

7.4.3 Expanding the Product Portfolio

Nowadays, the bio-based added-value product portfolio of microbial cell factories continues to diversify in order to bypass the dependence on fossil-based fuels and chemicals. For instance, most of the regular polyamides-based plastics are commonly derived from nonrenewable sources. Alternatively, the microbial production of polyamides precursors such diamines and lactams has been boosted in recent years to cope with this matter. Polyamides can be synthesized by condensation of dicarboxylates and diamines or from lactams by ring-opening polycondensation [269]. *E. coli* and *C. glutamicum* have proven to be outstanding platforms for the production of the diamines putrescine and cadaverine from native and alternative carbon sources [6, 173, 269]. However, it is expected that new insights will be generated with regard bio-based production of other relevant diamines such as 1,3-diaminopropane, 1,6-diaminohexane, and long-chain diamines. For instance, only one study reported production of 1,3-diaminopropane by engineered *E. coli*. In such study, the enzymes diaminobutyrate-2-oxoglutarate transaminase and the L-2,4-diaminobutyrate decarboxylase from *Acinetobacter baumannii* were used to enable 1,3-diaminopropane production in *E. coli* from L-aspartate semialdehyde [270]. Alternative 1,3-diaminopropane production hosts or alternative biosynthetic pathways could lead to better production performance of this three-carbon diamine. The diamine 1,6-diaminohexane is in high industrial demand since it is used in the production of polyamide 6,6 and 6,10. To date, not natural biosynthetic pathways for the synthesis of 1,6-diaminohexane have been discovered. Although 1,6-diaminohexane production in *E. coli* was established via a complex multi-step pathway [271]. Diamines with more than six-carbon backbones or long-chain diamines are also of great interest, however their fermentative production has not been reported yet. Instead, whole-cell biotransformation approaches were successfully applied for the synthesis of 1,8-octanediamine, 1,10-decanediamine, 1,12-dodecanediamine, and 1,14-tetradecanediamine [272, 273]. The cycle amides or lactams are also interesting building blocks for the production of polyamides. These lactams can be derived from omega amino acids such as 5AVA, GABA and caproic acid. Production of valerolactam, the lactam form of 5AVA, was enabled by expression of the β-alanine CoA transferase (encoded by *act* gene) from *Clostridium propionicum* in an *E. coli* 5AVA producing strain reaching 1.2 g/L of valerolactam after 69 h in a fed-batch cultivation. In addition, production of the butyrolactam from GABA has been also proven in *E. coli* by the overexpressing the *act* gene in a GABA producer. Similarly, caproic acid producing strain was chosen to express *act* and establish caprolactam production in *E. coli* [66]. Although important efforts have been made in the development of Ω-amino acids producing strains, only the report from Chae and colleagues in 2017 showed the fermentative production of these three lactams [66]. Therefore, this topic will be a hot spot for future research lines. In general, microbial production of non-proteinogenic amino acids such as ectoine or L-PA is also in the spotlight. Theses amino acids present interesting biological activities or can be used as precursors for pharmaceuticals and healthcare products [274, 275]. However, the production of this compounds via fermentative approach can be further improved and exploited. Particularly, production of hydrolysate forms of L-PA is quite unexplored. Recently, Luo and colleagues developed a six-steps pathway for the conversion of L-lysine to *N*-hydroxypipecolic acid [276]. Here, the use of preestablished L-lysine and L-PA producers will be quite advantageous.

Ideally, the portfolio of the N-containing compounds produced by state-of-the-art microbial cell factories will continue to expand in order to cope with the rising socioeconomic and environmental demands.

7.5 Conclusion and Future Outlook

Recent trends in the synthesis and production of N-containing compounds have shown increased interest in green bioprocesses. Particularly, there has been a movement toward implementation of fermentations approaches to cope with sustainability demands. In this regard, fermentative production of N-containing compounds using metabolically engineered microbial strains is a research field that is growing at an ever-increasing rate. The product portfolio increases every year and the processes for existing products are continuously improving which denotes the importance of this industry. It is expected to gather soon new insights regarding bio-based production of new and promising N-containing compounds such as short- and long-chain diamines or compounds with novel bioactivities. Furthermore, and aiming for a greener and sustainable society, the substrate spectra for microbial bioprocesses has been greatly expanded and it is nowadays moving towards reutilization of industrial related wastes and by-products as well as C1-compounds. Here, the trend is to increase bioprocesses activities using third generation feedstocks that rely on algae biomass, which also calls for the implementation of carbon neutral or carbon positive CO_2 reduction technologies. Although *E. coli* and *C. glutamicum* are predominantly dominating the microbial production of N-containing compounds, other non-conventional biotechnology workhorses like *B. methanolicus* are in the spotlight. Sidestreams are commonly harsh environments for bacteria due to the presence of growth inhibitors like furans. Here, bacteria with native mechanisms to detoxify or consume those inhibitors should be considered in the near future as alternative microbial hosts. This fast-moving evolution of the field is supported by the development of new technologies for the efficient full-scale genome editing such as CRISPR toolbox or ALE approaches and the new screening and selection methods such as riboswitch-based biosensors.

References

1. Gröger H. Biocatalytic concepts for synthesizing amine bulk chemicals: recent approaches towards linear and cyclic aliphatic primary amines and ω-substituted derivatives thereof. Appl Microbiol Biotechnol. 2019;103:83–95. https://doi.org/10.1007/s00253-018-9452-0.
2. Willke T. Methionine production—a critical review. Appl Microbiol Biotechnol. 2014;98:9893–914. https://doi.org/10.1007/s00253-014-6156-y.
3. Joo Y-C, Hyeon JE, Han SO. Metabolic design of *Corynebacterium glutamicum* for production of l-cysteine with consideration of Sulfur-supplemented animal feed. J Agric Food Chem. 2017;65:4698–707. https://doi.org/10.1021/acs.jafc.7b01061.
4. Wendisch VF. Metabolic engineering advances and prospects for amino acid production. Metab Eng. 2020;58:17–34. https://doi.org/10.1016/j.ymben.2019.03.008.
5. Ma Q, Zhang Q, Xu Q, Zhang C, Li Y, Fan X, Xie X, Chen N. Systems metabolic engineering strategies for the production of amino acids. Synth Syst Biotechnol. 2017;2:87–96. https://doi.org/10.1016/j.synbio.2017.07.003.
6. Becker J, Rohles CM, Wittmann C. Metabolically engineered *Corynebacterium glutamicum* for bio-based production of chemicals, fuels, materials, and healthcare products. Metab Eng. 2018;50:122–41. https://doi.org/10.1016/j.ymben.2018.07.008.
7. Asakura Y, Kimura E, Usuda Y, Kawahara Y, Matsui K, Osumi T, Nakamatsu T. Altered metabolic flux due to deletion of *odhA* causes l-glutamate overproduction in *Corynebacterium glutamicum*. Appl Environ Microbiol. 2007;73:1308–19. https://doi.org/10.1128/AEM.01867-06.
8. Wada M, Sawada K, Ogura K, Shimono Y, Hagiwara T, Sugimoto M, Onuki A, Yokota A. Effects of phosphoenolpyruvate carboxylase desensitization on glutamic acid production in *Corynebacterium glutamicum* ATCC 13032. J Biosci Bioeng. 2016;121:172–7. https://doi.org/10.1016/j.jbiosc.2015.06.008.
9. Zhang D, Guan D, Liang J, Guo C, Xie X, Zhang C, Xu Q, Chen N. Reducing lactate secretion by *ldhA* deletion in L-glutamate- producing strain *Corynebacterium glutamicum* GDK-9. Braz J Microbiol. 2014;45:1477–83. https://doi.org/10.1590/S1517-83822014000400044.
10. Blombach B, Hans S, Bathe B, Eikmanns BJ. Acetohydroxyacid synthase, a novel target for improvement of l-lysine production by *Corynebacterium glutamicum*. Appl Environ Microbiol. 2009;75:419–27. https://doi.org/10.1128/AEM.01844-08.

11. Becker J, Zelder O, Häfner S, Schröder H, Wittmann C. From zero to hero–design-based systems metabolic engineering of *Corynebacterium glutamicum* for L-lysine production. Metab Eng. 2011;13:159–68. https://doi.org/10.1016/j.ymben.2011.01.003.
12. Xu J-Z, Wu Z-H, Gao S-J, Zhang W. Rational modification of tricarboxylic acid cycle for improving l-lysine production in *Corynebacterium glutamicum*. Microb Cell Factories. 2018;17:105. https://doi.org/10.1186/s12934-018-0958-z.
13. Ying H, He X, Li Y, Chen K, Ouyang P. Optimization of culture conditions for enhanced lysine production using engineered *Escherichia coli*. Appl Biochem Biotechnol. 2014;172:3835–43. https://doi.org/10.1007/s12010-014-0820-7.
14. Diesveld R, Tietze N, Fürst O, Reth A, Bathe B, Sahm H, Eggeling L. Activity of exporters of *Escherichia coli* in *Corynebacterium glutamicum*, and their use to increase L-threonine production. Microb Physiol. 2009;16:198–207. https://doi.org/10.1159/000142530.
15. Livshits VA, Zakataeva NP, Aleshin VV, Vitushkina MV. Identification and characterization of the new gene *rhtA* involved in threonine and homoserine efflux in *Escherichia coli*. Res Microbiol. 2003;154:123–35. https://doi.org/10.1016/S0923-2508(03)00036-6.
16. Lee KH, Park JH, Kim TY, Kim HU, Lee SY. Systems metabolic engineering of *Escherichia coli* for L-threonine production. Mol Syst Biol. 2007;3:149. https://doi.org/10.1038/msb4100196.
17. Lee JH, Sung BH, Kim MS, Blattner FR, Yoon BH, Kim JH, Kim SC. Metabolic engineering of a reduced-genome strain of *Escherichia coli* for L-threonine production. Microb Cell Factories. 2009;8:2. https://doi.org/10.1186/1475-2859-8-2.
18. Yin L, Shi F, Hu X, Chen C, Wang X. Increasing l-isoleucine production in *Corynebacterium glutamicum* by overexpressing global regulator Lrp and two-component export system BrnFE. J Appl Microbiol. 2013;114:1369–77. https://doi.org/10.1111/jam.12141.
19. Vogt M, Krumbach K, Bang W-G, van Ooyen J, Noack S, Klein B, Bott M, Eggeling L. The contest for precursors: channelling l-isoleucine synthesis in *Corynebacterium glutamicum* without byproduct formation. Appl Microbiol Biotechnol. 2015;99:791–800. https://doi.org/10.1007/s00253-014-6109-5.
20. Ma W, Wang J, Li Y, Hu X, Shi F, Wang X. Enhancing pentose phosphate pathway in *Corynebacterium glutamicum* to improve l-isoleucine production. Biotechnol Appl Biochem. 2016;63:877–85. https://doi.org/10.1002/bab.1442.
21. Park JH, Oh JE, Lee KH, Kim JY, Lee SY. Rational design of *Escherichia coli* for L-isoleucine production. ACS Synth Biol. 2012;1:532–40. https://doi.org/10.1021/sb300071a.
22. Huang J-F, Liu Z-Q, Jin L-Q, Tang X-L, Shen Z-Y, Yin H-H, Zheng Y-G. Metabolic engineering of *Escherichia coli* for microbial production of L-methionine. Biotechnol Bioeng. 2017;114:843–51. https://doi.org/10.1002/bit.26198.
23. Qin T, Hu X, Hu J, Wang X. Metabolic engineering of *Corynebacterium glutamicum* strain ATCC13032 to produce l-methionine. Biotechnol Appl Biochem. 2015;62:563–73. https://doi.org/10.1002/bab.1290.
24. Li Y, Cong H, Liu B, Song J, Sun X, Zhang J, Yang Q. Metabolic engineering of *Corynebacterium glutamicum* for methionine production by removing feedback inhibition and increasing NADPH level. Antonie Van Leeuwenhoek. 2016;109:1185–97. https://doi.org/10.1007/s10482-016-0719-0.
25. Ikeda M, Katsumata R. Hyperproduction of tryptophan by *Corynebacterium glutamicum* with the modified pentose phosphate pathway. Appl Environ Microbiol. 1999;65:2497–502. https://doi.org/10.1128/AEM.65.6.2497-2502.1999.
26. Liu L, Duan X, Wu J. L-tryptophan production in *Escherichia coli* improved by weakening the Pta-AckA pathway. PLoS One. 2016;11:e0158200. https://doi.org/10.1371/journal.pone.0158200.
27. Chen L, Zeng A-P. Rational design and metabolic analysis of *Escherichia coli* for effective production of L-tryptophan at high concentration. Appl Microbiol Biotechnol. 2017;101:559–68. https://doi.org/10.1007/s00253-016-7772-5.

28. Chen Y, Liu Y, Ding D, Cong L, Zhang D. Rational design and analysis of an *Escherichia coli* strain for high-efficiency tryptophan production. J Ind Microbiol Biotechnol. 2018;45:357–67. https://doi.org/10.1007/s10295-018-2020-x.
29. Nakamura J, Hirano S, Ito H, Wachi M. Mutations of the *Corynebacterium glutamicum* NCgl1221 gene, encoding a mechanosensitive channel homolog, induce l-glutamic acid production. Appl Environ Microbiol. 2007;73:4491–8. https://doi.org/10.1128/AEM.02446-06.
30. Nottebrock D, Meyer U, Krämer R, Morbach S. Molecular and biochemical characterization of mechanosensitive channels in *Corynebacterium glutamicum*. FEMS Microbiol Lett. 2003;218:305–9. https://doi.org/10.1111/j.1574-6968.2003.tb11533.x.
31. Niebisch A, Kabus A, Schultz C, Weil B, Bott M. Corynebacterial protein kinase G controls 2-oxoglutarate dehydrogenase activity via the phosphorylation status of the OdhI protein. J Biol Chem. 2006;281:12300–7. https://doi.org/10.1074/jbc.M512515200.
32. Schultz C, Niebisch A, Gebel L, Bott M. Glutamate production by *Corynebacterium glutamicum*: dependence on the oxoglutarate dehydrogenase inhibitor protein OdhI and protein kinase PknG. Appl Microbiol Biotechnol. 2007;76:691–700. https://doi.org/10.1007/s00253-007-0933-9.
33. Peters-Wendisch PG, Schiel B, Wendisch VF, Katsoulidis E, Möckel B, Sahm H, Eikmanns BJ. Pyruvate carboxylase is a major bottleneck for glutamate and lysine production by *Corynebacterium glutamicum*. J Mol Microbiol Biotechnol. 2001;3:295–300.
34. Ikeda M, Takeno S. Recent advances in amino acid production. In: Inui M, Toyoda K, editors. *Corynebacterium glutamicum*: biology and biotechnology. Cham: Springer International Publishing; 2020. p. 175–226. https://doi.org/10.1016/j.copbio.2016.04.017.
35. Riedel C, Rittmann D, Dangel P, Möckel B, Petersen S, Sahm H, Eikmanns BJ. Characterization of the phosphoenolpyruvate carboxykinase gene from *Corynebacterium glutamicum* and significance of the enzyme for growth and amino acid production. J Mol Microbiol Biotechnol. 2001;3:573–83.
36. Ohnishi J, Mitsuhashi S, Hayashi M, Ando S, Yokoi H, Ochiai K, Ikeda M. A novel methodology employing *Corynebacterium glutamicum* genome information to generate a new L-lysine-producing mutant. Appl Microbiol Biotechnol. 2002;58:217–23. https://doi.org/10.1007/s00253-001-0883-6.
37. van Ooyen J, Noack S, Bott M, Reth A, Eggeling L. Improved L-lysine production with *Corynebacterium glutamicum* and systemic insight into citrate synthase flux and activity. Biotechnol Bioeng. 2012;109:2070–81. https://doi.org/10.1002/bit.24486.
38. Kind S, Becker J, Wittmann C. Increased lysine production by flux coupling of the tricarboxylic acid cycle and the lysine biosynthetic pathway–metabolic engineering of the availability of succinyl-CoA in *Corynebacterium glutamicum*. Metab Eng. 2013;15:184–95. https://doi.org/10.1016/j.ymben.2012.07.005.
39. Kalinowski J, Cremer J, Bachmann B, Eggeling L, Sahm H, Pühler A. Genetic and biochemical analysis of the aspartokinase from *Corynebacterium glutamicum*. Mol Microbiol. 1991;5:1197–204. https://doi.org/10.1111/j.1365-2958.1991.tb01893.x.
40. Georgi T, Rittmann D, Wendisch VF. Lysine and glutamate production by *Corynebacterium glutamicum* on glucose, fructose and sucrose: roles of malic enzyme and fructose-1,6-bisphosphatase. Metab Eng. 2005;7:291–301. https://doi.org/10.1016/j.ymben.2005.05.001.
41. Becker J, Klopprogge C, Herold A, Zelder O, Bolten CJ, Wittmann C. Metabolic flux engineering of L-lysine production in *Corynebacterium glutamicum*–over expression and modification of G6P dehydrogenase. J Biotechnol. 2007;132:99–109. https://doi.org/10.1016/j.jbiotec.2007.05.026.
42. Marx A, Hans S, Möckel B, Bathe B, de Graaf AA, McCormack AC, Stapleton C, Burke K, O'Donohue M, Dunican LK. Metabolic phenotype of phosphoglucose isomerase mutants of *Corynebacterium glutamicum*. J Biotechnol. 2003;104:185–97. https://doi.org/10.1016/s0168-1656(03)00153-6.

43. Becker J, Klopprogge C, Zelder O, Heinzle E, Wittmann C. Amplified expression of fructose 1,6-bisphosphatase in *Corynebacterium glutamicum* increases in vivo flux through the pentose phosphate pathway and lysine production on different carbon sources. Appl Environ Microbiol. 2005;71:8587–96. https://doi.org/10.1128/AEM.71.12.8587-8596.2005.
44. Kabus A, Georgi T, Wendisch VF, Bott M. Expression of the *Escherichia coli pntAB* genes encoding a membrane-bound transhydrogenase in *Corynebacterium glutamicum* improves l-lysine formation. Appl Microbiol Biotechnol. 2007;75:47–53. https://doi.org/10.1007/s00253-006-0804-9.
45. Eikmanns BJ, Metzger M, Reinscheid D, Kircher M, Sahm H. Amplification of three threonine biosynthesis genes in *Corynebacterium glutamicum* and its influence on carbon flux in different strains. Appl Microbiol Biotechnol. 1991;34:617–22. https://doi.org/10.1007/BF00167910.
46. Vrljic M, Sahm H, Eggeling L. A new type of transporter with a new type of cellular function: l-lysine export from *Corynebacterium glutamicum*. Mol Microbiol. 1996;22:815–26. https://doi.org/10.1046/j.1365-2958.1996.01527.x.
47. Li Y, Wei H, Wang T, Xu Q, Zhang C, Fan X, Ma Q, Chen N, Xie X. Current status on metabolic engineering for the production of l-aspartate family amino acids and derivatives. Bioresour Technol. 2017;245:1588–602. https://doi.org/10.1016/j.biortech.2017.05.145.
48. Kocabaş P, Çalık P, Özdamar TH. Fermentation characteristics of l-tryptophan production by thermoacidophilic *bacillus acidocaldarius* in a defined medium. Enzym Microb Technol. 2006;39:1077–88. https://doi.org/10.1016/j.enzmictec.2006.02.012.
49. Mindt M, Beyraghdar Kashkooli A, Suarez-Diez M, Ferrer L, Jilg T, Bosch D, Martins dos Santos V, Wendisch VF, Cankar K. Production of indole by *Corynebacterium glutamicum* microbial cell factories for flavor and fragrance applications. Microb Cell Factories. 2022;21:45. https://doi.org/10.1186/s12934-022-01771-y.
50. Eggeling L, Bott M, Bott M. Handbook of *Corynebacterium glutamicum*. Boca Raton: CRC Press; 2005. https://doi.org/10.1201/9781420039696.
51. Rodriguez A, Martnez JA, Flores N, Escalante A, Gosset G, Bolivar F. Engineering *Escherichia coli* to overproduce aromatic amino acids and derived compounds. Microb Cell Factories. 2014;13:126. https://doi.org/10.1186/s12934-014-0126-z.
52. Zhao Z-J, Zou C, Zhu Y-X, Dai J, Chen S, Wu D, Wu J, Chen J. Development of l-tryptophan production strains by defined genetic modification in *Escherichia coli*. J Ind Microbiol Biotechnol. 2011;38:1921–9. https://doi.org/10.1007/s10295-011-0978-8.
53. Xiong B, Zhu Y, Tian D, Jiang S, Fan X, Ma Q, Wu H, Xie X. Flux redistribution of central carbon metabolism for efficient production of l-tryptophan in *Escherichia coli*. Biotechnol Bioeng. 2021;118:1393–404. https://doi.org/10.1002/bit.27665.
54. Ger Y-M, Chen S-L, Chiang H-J, Shiuan D. A single Ser-180 mutation desensitizes feedback inhibition of the phenylalanine-sensitive 3-deoxy-D-arabino-Heptulosonate 7-phosphate (DAHP) synthetase in *Escherichia coli*1. J Biochem. 1994;116:986–90. https://doi.org/10.1093/oxfordjournals.jbchem.a124657.
55. Kikuchi Y, Tsujimoto K, Kurahashi O. Mutational analysis of the feedback sites of phenylalanine-sensitive 3-deoxy-D-arabino-heptulosonate-7-phosphate synthase of *Escherichia coli*. Appl Environ Microbiol. 1997;63:761–2. https://doi.org/10.1128/aem.63.2.761-762.1997.
56. Lee J-H, Wendisch VF. Production of amino acids - genetic and metabolic engineering approaches. Bioresour Technol. 2017;245:1575–87. https://doi.org/10.1016/j.biortech.2017.05.065.
57. Gu P, Yang F, Kang J, Wang Q, Qi Q. One-step of tryptophan attenuator inactivation and promoter swapping to improve the production of L-tryptophan in *Escherichia coli*. Microb Cell Factories. 2012;11:30. https://doi.org/10.1186/1475-2859-11-30.
58. Gu P, Yang F, Li F, Liang Q, Qi Q. Knocking out analysis of tryptophan permeases in *Escherichia coli* for improving L-tryptophan production. Appl Microbiol Biotechnol. 2013;97:6677–83. https://doi.org/10.1007/s00253-013-4988-5.

59. Pérez-García F, Wendisch VF. Transport and metabolic engineering of the cell factory *Corynebacterium glutamicum*. FEMS Microbiol Lett. 2018;365:fny166. https://doi.org/10.1093/femsle/fny166.
60. Nguyen AQD, Schneider J, Reddy GK, Wendisch VF. Fermentative production of the diamine putrescine: system metabolic engineering of *Corynebacterium glutamicum*. Meta. 2015;5:211–31. https://doi.org/10.3390/metabo5020211.
61. Schneider J, Eberhardt D, Wendisch VF. Improving putrescine production by *Corynebacterium glutamicum* by fine-tuning ornithine transcarbamoylase activity using a plasmid addiction system. Appl Microbiol Biotechnol. 2012;95:169–78. https://doi.org/10.1007/s00253-012-3956-9.
62. Qian Z-G, Xia X-X, Lee SY. Metabolic engineering of *Escherichia coli* for the production of putrescine: a four carbon diamine. Biotechnol Bioeng. 2009;104:651–62. https://doi.org/10.1002/bit.22502.
63. Jorge JMP, Nguyen AQD, Pérez-García F, Kind S, Wendisch VF. Improved fermentative production of gamma-aminobutyric acid via the putrescine route: systems metabolic engineering for production from glucose, amino sugars, and xylose. Biotechnol Bioeng. 2017;114:862–73. https://doi.org/10.1002/bit.26211.
64. Shi F, Luan M, Li Y. Ribosomal binding site sequences and promoters for expressing glutamate decarboxylase and producing γ-aminobutyrate in *Corynebacterium glutamicum*. AMB Express. 2018;8:61. https://doi.org/10.1186/s13568-018-0595-2.
65. Yu P, Chen K, Huang X, Wang X, Ren Q. Production of γ-aminobutyric acid in *Escherichia coli* by engineering MSG pathway. Prep Biochem Biotechnol. 2018;48:906–13. https://doi.org/10.1080/10826068.2018.1514519.
66. Chae TU, Ko Y-S, Hwang K-S, Lee SY. Metabolic engineering of *Escherichia coli* for the production of four-, five- and six-carbon lactams. Metab Eng. 2017;41:82–91. https://doi.org/10.1016/j.ymben.2017.04.001.
67. Chen J, Wang Y, Guo X, Rao D, Zhou W, Zheng P, Sun J, Ma Y. Efficient bioproduction of 5-Aminolevulinic acid, a promising biostimulant and nutrient, from renewable bioresources by engineered *Corynebacterium glutamicum*. Biotechnol Biofuels. 2020;13:41. https://doi.org/10.1186/s13068-020-01685-0.
68. Shih I-T, Yi Y-C, Ng I-S. Plasmid-free system and modular design for efficient 5-Aminolevulinic acid production by engineered *Escherichia coli*. Appl Biochem Biotechnol. 2021;193:2858–71. https://doi.org/10.1007/s12010-021-03571-3.
69. Hagihara R, Ohno S, Hayashi M, Tabata K, Endo H. Production of l-theanine by *Escherichia coli* in the absence of supplemental ethylamine. Appl Environ Microbiol. 2021;87:e00031–21. https://doi.org/10.1128/AEM.00031-21.
70. Benninghaus L, Walter T, Mindt M, Risse JM, Wendisch VF. Metabolic engineering of *pseudomonas putida* for fermentative production of l-theanine. J Agric Food Chem. 2021;69:9849–58. https://doi.org/10.1021/acs.jafc.1c03240.
71. Mindt M, Walter T, Risse JM, Wendisch VF. Fermentative production of *N*-Methylglutamate from glycerol by recombinant *pseudomonas putida*. Front Bioeng Biotechnol. 2018;6:159. https://doi.org/10.3389/fbioe.2018.00159.
72. Qian Z-G, Xia X-X, Lee SY. Metabolic engineering of *Escherichia coli* for the production of cadaverine: a five carbon diamine. Biotechnol Bioeng. 2011;108:93–103. https://doi.org/10.1002/bit.22918.
73. Kind S, Neubauer S, Becker J, Yamamoto M, Völkert M, Von AG, Zelder O, Wittmann C. From zero to hero - production of bio-based nylon from renewable resources using engineered *Corynebacterium glutamicum*. Metab Eng. 2014;25:113–23. https://doi.org/10.1016/j.ymben.2014.05.007.
74. Prell C, Vonderbank S-A, Meyer F, Pérez-García F, Wendisch VF. Metabolic engineering of *Corynebacterium glutamicum* for de novo production of 3-hydroxycadaverine. Curr Res Biotechnol. 2022;4:32–46. https://doi.org/10.1016/j.crbiot.2021.12.004.

75. Jorge JMP, Pérez-García F, Wendisch VF. A new metabolic route for the fermentative production of 5-Aminovalerate from glucose and alternative carbon sources. Bioresour Technol. 2017;245:1701–9. https://doi.org/10.1016/j.biortech.2017.04.108.
76. Rohles CM, Gießelmann G, Kohlstedt M, Wittmann C, Becker J. Systems metabolic engineering of *Corynebacterium glutamicum* for the production of the carbon-5 platform chemicals 5-aminovalerate and glutarate. Microb Cell Factories. 2016;15:154. https://doi.org/10.1186/s12934-016-0553-0.
77. Shin JH, Park SH, Oh YH, Choi JW, Lee MH, Cho JS, Jeong KJ, Joo JC, Yu J, Park SJ, Lee SY. Metabolic engineering of *Corynebacterium glutamicum* for enhanced production of 5-aminovaleric acid. Microb Cell Factories. 2016;15:174. https://doi.org/10.1186/s12934-016-0566-8.
78. Pérez-García F, Max Risse J, Friehs K, Wendisch VF. Fermentative production of L-pipecolic acid from glucose and alternative carbon sources. Biotechnol J. 2017;12:646. https://doi.org/10.1002/biot.201600646.
79. Xu X, Rao Z-M, Xu J-Z, Zhang W-G. Enhancement of l-pipecolic acid production by dynamic control of substrates and multiple copies of the *pipA* gene in the *Escherichia coli* genome. ACS Synth Biol. 2022;11(2):760–9. https://doi.org/10.1021/acssynbio.1c00467.
80. Pérez-García F, Ziert C, Risse JM, Wendisch VF. Improved fermentative production of the compatible solute ectoine by *Corynebacterium glutamicum* from glucose and alternative carbon sources. J Biotechnol. 2017;258:59–68. https://doi.org/10.1016/j.jbiotec.2017.04.039.
81. Gießelmann G, Dietrich D, Jungmann L, Kohlstedt M, Jeon EJ, Yim SS, Sommer F, Zimmer D, Mühlhaus T, Schroda M, Jeong KJ, Becker J, Wittmann C. Metabolic engineering of *Corynebacterium glutamicum* for high-level ectoine production: design, combinatorial assembly, and implementation of a transcriptionally balanced heterologous ectoine pathway. Biotechnol J. 2019;14:e1800417. https://doi.org/10.1002/biot.201800417.
82. Ning Y, Wu X, Zhang C, Xu Q, Chen N, Xie X. Pathway construction and metabolic engineering for fermentative production of ectoine in *Escherichia coli*. Metab Eng. 2016;36:10–8. https://doi.org/10.1016/j.ymben.2016.02.013.
83. Ma Q, Xia L, Wu H, Zhuo M, Yang M, Zhang Y, Tan M, Zhao K, Sun Q, Xu Q, Chen N, Xie X. Metabolic engineering of *Escherichia coli* for efficient osmotic stress-free production of compatible solute hydroxyectoine. Biotechnol Bioeng. 2022;119:89–101. https://doi.org/10.1002/bit.27952.
84. Kugler P, Trumm M, Frese M, Wendisch VF. L-carnitine production through biosensor-guided construction of the *neurospora crassa* biosynthesis pathway in *Escherichia coli*. Front Bioeng Biotechnol. 2021;9:671321. https://doi.org/10.3389/fbioe.2021.671321.
85. Sun H, Zhao D, Xiong B, Zhang C, Bi C. Engineering *Corynebacterium glutamicum* for violacein hyper production. Microb Cell Factories. 2016;15:148. https://doi.org/10.1186/s12934-016-0545-0.
86. Zhou Y, Fang M-Y, Li G, Zhang C, Xing X-H. Enhanced production of crude violacein from glucose in *Escherichia coli* by overexpression of rate-limiting key enzyme(S) involved in violacein biosynthesis. Appl Biochem Biotechnol. 2018;186:909–16. https://doi.org/10.1007/s12010-018-2787-2.
87. Yang C, Jiang P, Xiao S, Zhang C, Lou K, Xing X-H. Fed-batch fermentation of recombinant *Citrobacter freundii* with expression of a violacein-synthesizing gene cluster for efficient violacein production from glycerol. Biochem Eng J. 2011;57:55–62. https://doi.org/10.1016/j.bej.2011.08.008.
88. Wang H, Liu W, Shi F, Huang L, Lian J, Qu L, Cai J, Xu Z. Metabolic pathway engineering for high-level production of 5-hydroxytryptophan in *Escherichia coli*. Metab Eng. 2018;48:279–87. https://doi.org/10.1016/j.ymben.2018.06.007.
89. Mora-Villalobos J-A, Zeng A-P. Synthetic pathways and processes for effective production of 5-hydroxytryptophan and serotonin from glucose in *Escherichia coli*. J Biol Eng. 2018;12:3. https://doi.org/10.1186/s13036-018-0094-7.

90. Zhang Y, He Y, Zhang N, Gan J, Zhang S, Dong Z. Combining protein and metabolic engineering strategies for biosynthesis of melatonin in *Escherichia coli*. Microb Cell Factories. 2021;20:170. https://doi.org/10.1186/s12934-021-01662-8.
91. Balderas-Hernández VE, Sabido-Ramos A, Silva P, Cabrera-Valladares N, Hernández-Chávez G, Báez-Viveros JL, Martínez A, Bolívar F, Gosset G. Metabolic engineering for improving anthranilate synthesis from glucose in *Escherichia coli*. Microb Cell Factories. 2009;8:19. https://doi.org/10.1186/1475-2859-8-19.
92. Luo ZW, Cho JS, Lee SY. Microbial production of methyl anthranilate, a grape flavor compound. Proc Natl Acad Sci U S A. 2019;116:10749–56. https://doi.org/10.1073/pnas.1903875116.
93. Veldmann KH, Minges H, Sewald N, Lee J-H, Wendisch VF. Metabolic engineering of *Corynebacterium glutamicum* for the fermentative production of halogenated tryptophan. J Biotechnol. 2019;291:7–16. https://doi.org/10.1016/j.jbiotec.2018.12.008.
94. Veldmann KH, Dachwitz S, Risse JM, Lee J-H, Sewald N, Wendisch VF. Bromination of L-tryptophan in a fermentative process with *Corynebacterium glutamicum*. Front Bioeng Biotechnol. 2019;7:219. https://doi.org/10.3389/fbioe.2019.00219.
95. Lee J, Kim J, Song JE, Song W-S, Kim E-J, Kim Y-G, Jeong H-J, Kim HR, Choi K-Y, Kim B-G. Production of Tyrian purple indigoid dye from tryptophan in *Escherichia coli*. Nat Chem Biol. 2021;17:104–12. https://doi.org/10.1038/s41589-020-00684-4.
96. Kerbs A, Burgardt A, Veldmann KH, Schäffer T, Lee J-H, Wendisch VF. Fermentative Production of halogenated tryptophan derivatives with *Corynebacterium glutamicum* overexpressing tryptophanase or decarboxylase genes. Chem Bio Chem. 2022;23: e202200007. https://doi.org/10.1002/cbic.202200007.
97. Zhang C, Li Y, Ma J, Liu Y, He J, Li Y, Zhu F, Meng J, Zhan J, Li Z, Zhao L, Ma Q, Fan X, Xu Q, Xie X, Chen N. High production of 4-hydroxyisoleucine in *Corynebacterium glutamicum* by multistep metabolic engineering. Metab Eng. 2018;49:287–98. https://doi.org/10.1016/j.ymben.2018.09.008.
98. Tan S, Shi F, Liu H, Yu X, Wei S, Fan Z, Li Y. Dynamic control of 4-Hydroxyisoleucine biosynthesis by modified l-isoleucine biosensor in recombinant *Corynebacterium glutamicum*. ACS Synth Biol. 2020;9:2378–89. https://doi.org/10.1021/acssynbio.0c00127.
99. Lammens TM, Le Nôtre J, Franssen MCR, Scott EL, Sanders JPM. Synthesis of biobased succinonitrile from glutamic acid and glutamine. ChemSusChem. 2011;4:785–91. https://doi.org/10.1002/cssc.201100030.
100. Roerdink E, Warnier JMM. Preparation and properties of high molar mass nylon-4,6: a new development in nylon polymers. Polymer. 1985;26:1582–8. https://doi.org/10.1016/0032-3861(85)90098-9.
101. Schneider J, Wendisch VF. Putrescine production by engineered *Corynebacterium glutamicum*. Appl Microbiol Biotechnol. 2010;88:859–68. https://doi.org/10.1007/s00253-010-2778-x.
102. Nguyen AQD, Schneider J, Wendisch VF. Elimination of polyamine *N*-acetylation and regulatory engineering improved putrescine production by *Corynebacterium glutamicum*. J Biotechnol. 2015;201:75–85. https://doi.org/10.1016/j.jbiotec.2014.10.035.
103. Li Z, Shen Y-P, Jiang X-L, Feng L-S, Liu J-Z. Metabolic evolution and a comparative omics analysis of *Corynebacterium glutamicum* for putrescine production. J Ind Microbiol Biotechnol. 2018;45:123–39. https://doi.org/10.1007/s10295-018-2003-y.
104. Abdou AM, Higashiguchi S, Horie K, Kim M, Hatta H, Yokogoshi H. Relaxation and immunity enhancement effects of γ-aminobutyric acid (GABA) administration in humans. Biofactors. 2006;26:201–8. https://doi.org/10.1002/biof.5520260305.
105. Hayakawa K, Kimura M, Kasaha K, Matsumoto K, Sansawa H, Yamori Y. Effect of a γ-aminobutyric acid-enriched dairy product on the blood pressure of spontaneously hypertensive and normotensive Wistar–Kyoto rats. Br J Nutr. 2004;92:411–7. https://doi.org/10.1079/BJN20041221.

106. Tsuge Y, Matsuzawa H. Recent progress in production of amino acid-derived chemicals using *Corynebacterium glutamicum*. World J Microbiol Biotechnol. 2021;37:49. https://doi.org/10.1007/s11274-021-03007-4.
107. Jorge JMP, Leggewie C, Wendisch VF. A new metabolic route for the production of gamma-aminobutyric acid by *Corynebacterium glutamicum* from glucose. Amino Acids. 2016;48:2519–31. https://doi.org/10.1007/s00726-016-2272-6.
108. Zhang R, Yang T, Rao Z, Sun H, Xu M, Zhang X, Xu Z, Yang S. Efficient one-step preparation of γ-aminobutyric acid from glucose without an exogenous cofactor by the designed *Corynebacterium glutamicum*. Green Chem. 2014;16:4190–7. https://doi.org/10.1039/C4GC00607K.
109. Le Vo TD, Kim TW, Hong SH. Effects of glutamate decarboxylase and gamma-aminobutyric acid (GABA) transporter on the bioconversion of GABA in engineered *Escherichia coli*. Bioprocess Biosyst Eng. 2012;35:645–50. https://doi.org/10.1007/s00449-011-0634-8.
110. Pham VD, Lee SH, Park SJ, Hong SH. Production of gamma-aminobutyric acid from glucose by introduction of synthetic scaffolds between isocitrate dehydrogenase, glutamate synthase and glutamate decarboxylase in recombinant *Escherichia coli*. J Biotechnol. 2015;207:52–7. https://doi.org/10.1016/j.jbiotec.2015.04.028.
111. Huang J, Mei L, Wu H, Lin D. Biosynthesis of γ-aminobutyric acid (GABA) using immobilized whole cells of *lactobacillus brevis*. World J Microbiol Biotechnol. 2007;23:865–71. https://doi.org/10.1007/s11274-006-9311-5.
112. Li H, Qiu T, Huang G, Cao Y. Production of gamma-aminobutyric acid by *lactobacillus brevis* NCL912 using fed-batch fermentation. Microb Cell Factories. 2010;9:85. https://doi.org/10.1186/1475-2859-9-85.
113. Kang Z, Zhang J, Zhou J, Qi Q, Du G, Chen J. Recent advances in microbial production of δ-Aminolevulinic acid and vitamin B12. Biotechnol Adv. 2012;30:1533–42. https://doi.org/10.1016/j.biotechadv.2012.04.003.
114. Liu S, Zhang G, Li X, Zhang J. Microbial production and applications of 5-Aminolevulinic acid. Appl Microbiol Biotechnol. 2014;98:7349–57. https://doi.org/10.1007/s00253-014-5925-y.
115. Feng L, Zhang Y, Fu J, Mao Y, Chen T, Zhao X, Wang Z. Metabolic engineering of *Corynebacterium glutamicum* for efficient production of 5-Aminolevulinic acid. Biotechnol Bioeng. 2016;113:1284–93. https://doi.org/10.1002/bit.25886.
116. Ramzi AB, Hyeon JE, Kim SW, Park C, Han SO. 5-Aminolevulinic acid production in engineered *Corynebacterium glutamicum* via C5 biosynthesis pathway. Enzym Microb Technol. 2015;81:1–7. https://doi.org/10.1016/j.enzmictec.2015.07.004.
117. Su T, Guo Q, Zheng Y, Liang Q, Wang Q, Qi Q. Fine-tuning of hemB using CRISPRi for increasing 5-Aminolevulinic acid production in *Escherichia coli*. Front Microbiol. 2019;10:1731. https://doi.org/10.3389/fmicb.2019.01731.
118. Ge F, Wen D, Ren Y, Chen G, He B, Li X, Li W. Downregulating of *hemB* via synthetic antisense RNAs for improving 5-Aminolevulinic acid production in *Escherichia coli*. 3 Biotech. 2021;11:230. https://doi.org/10.1007/s13205-021-02733-8.
119. Liu Z-W, Wu Z-J, Li H, Wang Y-X, Zhuang J. L-theanine content and related gene expression: novel insights into theanine biosynthesis and hydrolysis among different tea plant (*Camellia sinensis L.*) tissues and cultivars. Front. Plant Sci. 2017;8:498. https://doi.org/10.3389/fpls.2017.00498.
120. Vuong QV, Bowyer MC, Roach PD. L-theanine: properties, synthesis and isolation from tea. J Sci Food Agric. 2011;91:1931–9. https://doi.org/10.1002/jsfa.4373.
121. Ma H, Fan X, Cai N, Zhang D, Zhao G, Wang T, Su R, Yuan M, Ma Q, Zhang C, Xu Q, Xie X, Chen N, Li Y. Efficient fermentative production of l-theanine by *Corynebacterium glutamicum*. Appl Microbiol Biotechnol. 2020;104:119–30. https://doi.org/10.1007/s00253-019-10255-w.

122. Gioia MLD, Leggio A, Malagrinò F, Romio E, Siciliano C, Liguori A. N-methylated α-amino acids and peptides: synthesis and biological activity. Mini-Rev Med Chem. 2016;16:683–90. https://doi.org/10.2174/1389557516666160322152457.
123. Damann FE, Carter DO. Human decomposition ecology and postmortem microbiology. In: Manual of forensic taphonomy. Boca Raton: CRC Press; 2013.
124. Mimitsuka T, Sawai H, Hatsu M, Yamada K. Metabolic engineering of *Corynebacterium glutamicum* for cadaverine fermentation. Biosci Biotechnol Biochem. 2007;71:2130–5. https://doi.org/10.1271/bbb.60699.
125. Miller-Fleming L, Olin-Sandoval V, Campbell K, Ralser M. Remaining mysteries of molecular biology: the role of polyamines in the cell. J Mol Biol. 2015;427:3389–406. https://doi.org/10.1016/j.jmb.2015.06.020.
126. Kusano T, Berberich T, Tateda C, Takahashi Y. Polyamines: essential factors for growth and survival. Planta. 2008;228:367–81. https://doi.org/10.1007/s00425-008-0772-7.
127. Samartzidou H, Mehrazin M, Xu Z, Benedik MJ, Delcour AH. Cadaverine inhibition of porin plays a role in cell survival at acidic pH. J Bacteriol. 2003;185:13–9. https://doi.org/10.1128/JB.185.1.13-19.2003.
128. Kind S, Wittmann C. Bio-based production of the platform chemical 1,5-diaminopentane. Appl Microbiol Biotechnol. 2011;91(5):1287–96. https://doi.org/10.1007/s00253-011-3457-2.
129. Huang Y, Ji X, Ma Z, Łężyk M, Xue Y, Zhao H. Green chemical and biological synthesis of cadaverine: recent development and challenges. RSC Adv. 2021;11:23922–42. https://doi.org/10.1039/D1RA02764F.
130. Kind S, Jeong WK, Schröder H, Wittmann C. Systems-wide metabolic pathway engineering in *Corynebacterium glutamicum* for bio-based production of diaminopentane. Metab Eng. 2010;12:341–51. https://doi.org/10.1016/j.ymben.2010.03.005.
131. Lubitz D, Jorge JMP, Pérez-García F, Taniguchi H, Wendisch VF. Roles of export genes *cgmA* and *lysE* for the production of L-arginine and L-citrulline by *Corynebacterium glutamicum*. Appl Microbiol Biotechnol. 2016;100:8465–74. https://doi.org/10.1007/s00253-016-7695-1.
132. Li M, Li D, Huang Y, Liu M, Wang H, Tang Q, Lu F. Improving the secretion of cadaverine in *Corynebacterium glutamicum* by cadaverine–lysine antiporter. J Ind Microbiol Biotechnol. 2014;41:701–9. https://doi.org/10.1007/s10295-014-1409-4.
133. Kang S-B, Choi J-I. Production of cadaverine in recombinant *Corynebacterium glutamicum* overexpressing lysine decarboxylase (*ldcC*) and response regulator *dr1558*. Appl Biochem Biotechnol. 2022;194:1013–24. https://doi.org/10.1007/s12010-021-03685-8.
134. Adkins J, Jordan J, Nielsen DR. Engineering *Escherichia coli* for renewable production of the 5-carbon polyamide building-blocks 5-aminovalerate and glutarate. Biotechnol Bioeng. 2013;110:1726–34. https://doi.org/10.1002/bit.24828.
135. Revelles O, Espinosa-Urgel M, Fuhrer T, Sauer U, Ramos JL. Multiple and interconnected pathways for l-lysine catabolism in *pseudomonas putida* KT2440. J Bacteriol. 2005;187:7500–10. https://doi.org/10.1128/JB.187.21.7500-7510.2005.
136. Park SJ, Kim EY, Noh W, Park HM, Oh YH, Lee SH, Song BK, Jegal J, Lee SY. Metabolic engineering of *Escherichia coli* for the production of 5-aminovalerate and glutarate as C5 platform chemicals. Metab Eng. 2013;16:42–7. https://doi.org/10.1016/j.ymben.2012.11.011.
137. Li Z, Xu J, Jiang T, Ge Y, Liu P, Zhang M, Su Z, Gao C, Ma C, Xu P. Overexpression of transport proteins improves the production of 5-aminovalerate from L-lysine in *Escherichia coli*. Sci Rep. 2016;6:30884. https://doi.org/10.1038/srep30884.
138. Pérez-García F, Peters-Wendisch P, Wendisch VF. Engineering *Corynebacterium glutamicum* for fast production of L-lysine and L-pipecolic acid. Appl Microbiol Biotechnol. 2016;100:8075–90. https://doi.org/10.1007/s00253-016-7682-6.
139. Fujii T, Aritoku Y, Agematu H, Tsunekawa H. Increase in the rate of L-pipecolic acid production using *lat*-expressing *Escherichia coli* by *lysP* and *yeiE* amplification. Biosci Biotechnol Biochem. 2002;66:1981–4. https://doi.org/10.1271/bbb.66.1981.

140. Bernsdorff F, Döring A-C, Gruner K, Schuck S, Bräutigam A, Zeier J. Pipecolic acid orchestrates plant systemic acquired resistance and Defense priming via salicylic acid-dependent and -independent pathways. Plant Cell. 2016;28:102–29. https://doi.org/10.1105/tpc.15.00496.
141. Návarová H, Bernsdorff F, Döring A-C, Zeier J. Pipecolic acid, an endogenous mediator of Defense amplification and priming, is a critical regulator of inducible plant immunity[W]. Plant Cell. 2012;24:5123–41. https://doi.org/10.1105/tpc.112.103564.
142. Misono H, Hashimoto H, Uehigashi H, Nagata S, Nagasaki S. Properties of L-lysine ε-dehydrogenase from *agrobacterium tumefaciens*. J Biochem. 1989;105:1002–8. https://doi.org/10.1093/oxfordjournals.jbchem.a122757.
143. Gouesbet G, Jebbar M, Talibart R, Bernard T, Blanco C. Pipecolic acid is an osmoprotectant for *Escherichia coli* taken up by the general osmoporters ProU and ProP. Microbiol Read Engl. 1994;140(Pt 9):2415–22. https://doi.org/10.1099/13500872-140-9-2415.
144. Pérez-García F, Brito LF, Wendisch VF. Function of L-pipecolic acid as compatible solute in *Corynebacterium glutamicum* as basis for its production under hyperosmolar conditions. Front Microbiol. 2019;10:340. https://doi.org/10.3389/fmicb.2019.00340.
145. Kadouri-Puchot C, Comesse S. Recent advances in asymmetric synthesis of pipecolic acid and derivatives. Amino Acids. 2005;29:101–30. https://doi.org/10.1007/s00726-005-0193-x.
146. Eichhorn E, Roduit J-P, Shaw N, Heinzmann K, Kiener A. Preparation of (S)-piperazine-2-carboxylic acid, (R)-piperazine-2-carboxylic acid, and (S)-piperidine-2-carboxylic acid by kinetic resolution of the corresponding racemic carboxamides with stereoselective amidases in whole bacterial cells. Tetrahedron Asymmetry. 1997;8:2533–6. https://doi.org/10.1016/S0957-4166(97)00256-5.
147. Gupta RN, Spenser ID. Biosynthesis of the piperidine nucleus: the mode of incorporation of lysine into pipecolic acid and into piperidine alkaloids. J Biol Chem. 1969;244:88–94. https://doi.org/10.1016/S0021-9258(19)78195-2.
148. Miller DL, Rodwell VW. Metabolism of basic amino acids in *pseudomonas putida*: catabolism of lysine by cyclic and acyclic intermediates. J Biol Chem. 1971;246:2758–64. https://doi.org/10.1016/S0021-9258(18)62249-5.
149. Ying H, Tao S, Wang J, Ma W, Chen K, Wang X, Ouyang P. Expanding metabolic pathway for de novo biosynthesis of the chiral pharmaceutical intermediate l-pipecolic acid in *Escherichia coli*. Microb Cell Factories. 2017;16:52. https://doi.org/10.1186/s12934-017-0666-0.
150. Cánovas D, Vargas C, Calderón MI, Ventosa A, Nieto JJ. Characterization of the genes for the biosynthesis of the compatible solute ectoine in the moderately halophilic bacterium *Halomonas elongata* DSM 3043. Syst Appl Microbiol. 1998;21:487–97. https://doi.org/10.1016/S0723-2020(98)80060-X.
151. Louis P, Galinski EA. Characterization of genes for the biosynthesis of the compatible solute ectoine from *Marinococcus halophilus* and osmoregulated expression in *Escherichia coli*. Microbiol Read Engl. 1997;143(Pt 4):1141–9. https://doi.org/10.1099/00221287-143-4-1141.
152. Graf R, Anzali S, Buenger J, Pfluecker F, Driller H. The multifunctional role of ectoine as a natural cell protectant. Clin Dermatol. 2008;26:326–33. https://doi.org/10.1016/j.clindermatol.2008.01.002.
153. Botta C, Di Giorgio C, Sabatier A-S, De Méo M. Genotoxicity of visible light (400-800 nm) and photoprotection assessment of ectoin, L-ergothioneine and mannitol and four sunscreens. J Photochem Photobiol B. 2008;91:24–34. https://doi.org/10.1016/j.jphotobiol.2008.01.008.
154. Lo C-C, Bonner CA, Xie G, D'Souza M, Jensen RA. Cohesion group approach for evolutionary analysis of aspartokinase, an enzyme that feeds a branched network of many biochemical pathways. Microbiol Mol Biol Rev. 2009;73:594–651. https://doi.org/10.1128/MMBR.00024-09.
155. García-Estepa R, Argandoña M, Reina-Bueno M, Capote N, Iglesias-Guerra F, Nieto JJ, Vargas C. The *ectD* gene, which is involved in the synthesis of the compatible solute hydroxyectoine, is essential for Thermoprotection of the halophilic bacterium

Chromohalobacter salexigens. J Bacteriol. 2006;188:3774–84. https://doi.org/10.1128/JB.00136-06.
156. Sauer T, Galinski EA. Bacterial milking: a novel bioprocess for production of compatible solutes. Biotechnol Bioeng. 1998;57:306–13.
157. Becker J, Schäfer R, Kohlstedt M, Harder BJ, Borchert NS, Stöveken N, Bremer E, Wittmann C. Systems metabolic engineering of *Corynebacterium glutamicum* for production of the chemical chaperone ectoine. Microb Cell Factories. 2013;12:110. https://doi.org/10.1186/1475-2859-12-110.
158. Bestvater T, Louis P, Galinski EA. Heterologous ectoine production in *Escherichia coli*: by-passing the metabolic bottle-neck. Saline Syst. 2008;4:12. https://doi.org/10.1186/1746-1448-4-12.
159. Seip B, Galinski EA, Kurz M. Natural and engineered hydroxyectoine production based on the *pseudomonas stutzeri ectABCD-ask* gene cluster. Appl Environ Microbiol. 2011;77:1368–74. https://doi.org/10.1128/AEM.02124-10.
160. Kugler P, Fröhlich D, Wendisch VF. Development of a biosensor for Crotonobetaine-CoA ligase screening based on the elucidation of *Escherichia coli* carnitine metabolism. ACS Synth Biol. 2020;9:2460–71. https://doi.org/10.1021/acssynbio.0c00234.
161. Ahmed A, Ahmad A, Li R, Al-Ansi W, Fatima M, Mushtaq BS, Basharat S, Li Y, Bai Z. Recent advances in synthetic, industrial and biological applications of violacein and its heterologous production. J Microbiol Biotechnol. 2021;31:1465–80. https://doi.org/10.4014/jmb.2107.07045.
162. Xu D, Fang M, Wang H, Huang L, Xu Q, Xu Z. Enhanced production of 5-hydroxytryptophan through the regulation of L-tryptophan biosynthetic pathway. Appl Microbiol Biotechnol. 2020;104:2481–8. https://doi.org/10.1007/s00253-020-10371-y.
163. Ferrer L, Elsaraf M, Mindt M, Wendisch VF. L-serine biosensor-controlled fermentative production of l-tryptophan derivatives by Corynebacterium glutamicum. Biology. 2022;11:744. https://doi.org/10.3390/biology11050744.
164. Mohammad-Zadeh LF, Moses L, Gwaltney-Brant SM. Serotonin: a review. J Vet Pharmacol Ther. 2008;31:187–99. https://doi.org/10.1111/j.1365-2885.2008.00944.x.
165. Park S, Kang K, Lee SW, Ahn M-J, Bae J-M, Back K. Production of serotonin by dual expression of tryptophan decarboxylase and tryptamine 5-hydroxylase in *Escherichia coli*. Appl Microbiol Biotechnol. 2011;89:1387–94. https://doi.org/10.1007/s00253-010-2994-4.
166. Zhao D, Yu Y, Shen Y, Liu Q, Zhao Z, Sharma R, Reiter RJ. Melatonin synthesis and function: evolutionary history in animals and plants. Front Endocrinol. 2019;10:249. https://doi.org/10.3389/fendo.2019.00249.
167. Byeon Y, Back K. Melatonin production in *Escherichia coli* by dual expression of serotonin *N*-acetyltransferase and caffeic acid *O*-methyltransferase. Appl Microbiol Biotechnol. 2016;100:6683–91. https://doi.org/10.1007/s00253-016-7458-z.
168. Kuepper J, Dickler J, Biggel M, Behnken S, Jäger G, Wierckx N, Blank LM. Metabolic engineering of *pseudomonas putida* KT2440 to produce anthranilate from glucose. Front Microbiol. 2015;6:1310. https://doi.org/10.3389/fmicb.2015.01310.
169. Walter T, Al Medani N, Burgardt A, Cankar K, Ferrer L, Kerbs A, Lee J-H, Mindt M, Risse JM, Wendisch VF. Fermentative N-Methylanthranilate production by engineered *Corynebacterium glutamicum*. Microorganisms. 2020;8:866. https://doi.org/10.3390/microorganisms8060866.
170. Syukur Purwanto H, Kang M-S, Ferrer L, Han S-S, Lee J-Y, Kim H-S, Lee J-H. Rational engineering of the shikimate and related pathways in *Corynebacterium glutamicum* for 4-hydroxybenzoate production. J Biotechnol. 2018;282:92–100. https://doi.org/10.1016/j.jbiotec.2018.07.016.
171. Smirnov SV, Kodera T, Samsonova NN, Kotlyarova VA, Rushkevich NY, Kivero AD, Sokolov PM, Hibi M, Ogawa J, Shimizu S. Metabolic engineering of *Escherichia coli* to produce (2S, 3R, 4S)-4-hydroxyisoleucine. Appl Microbiol Biotechnol. 2010;88:719–26. https://doi.org/10.1007/s00253-010-2772-3.

172. Zhang C, Ma J, Li Z, Liang Y, Xu Q, Xie X, Chen N. A strategy for L-isoleucine dioxygenase screening and 4-hydroxyisoleucine production by resting cells. Bioengineered. 2017;9:72–9. https://doi.org/10.1080/21655979.2017.1304872.
173. Wendisch VF, Brito LF, Gil Lopez M, Hennig G, Pfeifenschneider J, Sgobba E, Veldmann KH. The flexible feedstock concept in industrial biotechnology: metabolic engineering of *Escherichia coli, Corynebacterium glutamicum, pseudomonas, bacillus* and yeast strains for access to alternative carbon sources. J Biotechnol. 2016;234:139–57. https://doi.org/10.1016/j.jbiotec.2016.07.022.
174. Wendisch VF, Nampoothiri KM, Lee J-H. Metabolic engineering for valorization of agri- and aqua-culture Sidestreams for production of nitrogenous compounds by *Corynebacterium glutamicum*. Front Microbiol. 2022;13:835131. https://doi.org/10.3389/fmicb.2022.835131.
175. Wendisch VF, Kosec G, Heux S, Brautaset T. Aerobic utilization of methanol for microbial growth and production. Berlin: Springer; 2021. p. 1–44. https://doi.org/10.1007/10_2021_177.
176. Seibold G, Auchter M, Berens S, Kalinowski J, Eikmanns BJ. Utilization of soluble starch by a recombinant *Corynebacterium glutamicum* strain: growth and lysine production. J Biotechnol. 2006;124:381–91. https://doi.org/10.1016/j.jbiotec.2005.12.027.
177. Tateno T, Fukuda H, Kondo A. Production of L-lysine from starch by *Corynebacterium glutamicum* displaying alpha-amylase on its cell surface. Appl Microbiol Biotechnol. 2007;74:1213–20. https://doi.org/10.1007/s00253-006-0766-y.
178. Tateno T, Hatada K, Tanaka T, Fukuda H, Kondo A. Development of novel cell surface display in *Corynebacterium glutamicum* using porin. Appl Microbiol Biotechnol. 2009;84:733–9. https://doi.org/10.1007/s00253-009-2021-9.
179. Anusree M, Wendisch VF, Nampoothiri KM. Co-expression of endoglucanase and β-glucosidase in *Corynebacterium glutamicum* DM1729 towards direct lysine fermentation from cellulose. Bioresour Technol. 2016;213:239–44. https://doi.org/10.1016/j.biortech.2016.03.019.
180. Meiswinkel TM, Gopinath V, Lindner SN, Nampoothiri KM, Wendisch VF. Accelerated pentose utilization by *Corynebacterium glutamicum* for accelerated production of lysine, glutamate, ornithine and putrescine. Microb Biotechnol. 2013;6:131–40. https://doi.org/10.1111/1751-7915.12001.
181. Mindt M, Heuser M, Wendisch VF. Xylose as preferred substrate for sarcosine production by recombinant *Corynebacterium glutamicum*. Bioresour Technol. 2019;281:135–42. https://doi.org/10.1016/j.biortech.2019.02.084.
182. Buschke N, Schäfer R, Becker J, Wittmann C. Metabolic engineering of industrial platform microorganisms for biorefinery applications–optimization of substrate spectrum and process robustness by rational and evolutive strategies. Bioresour Technol. 2013;135:544–54. https://doi.org/10.1016/j.biortech.2012.11.047.
183. Schneider J, Niermann K, Wendisch VF. Production of the amino acids l-glutamate, l-lysine, l-ornithine and l-arginine from arabinose by recombinant *Corynebacterium glutamicum*. J Biotechnol. 2011;154:191–8. https://doi.org/10.1016/j.jbiotec.2010.07.009.
184. Meiswinkel TM, Rittmann D, Lindner SN, Wendisch VF. Crude glycerol-based production of amino acids and putrescine by *Corynebacterium glutamicum*. Bioresour Technol. 2013;145:254–8. https://doi.org/10.1016/j.biortech.2013.02.053.
185. Pérez-García F, Burgardt A, Kallman DR, Wendisch VF, Bar N. Dynamic co-cultivation process of *Corynebacterium glutamicum* strains for the fermentative production of riboflavin. Fermentation. 2021;7:11. https://doi.org/10.3390/fermentation7010011.
186. Gopinath V, Meiswinkel TM, Wendisch VF, Nampoothiri KM. Amino acid production from rice straw and wheat bran hydrolysates by recombinant pentose-utilizing *Corynebacterium glutamicum*. Appl Microbiol Biotechnol. 2011;92:985–96. https://doi.org/10.1007/s00253-011-3478-x.

187. Burgardt A, Prell C, Wendisch VF. Utilization of a wheat sidestream for 5-Aminovalerate production in *Corynebacterium glutamicum*. Front Bioeng Biotechnol. 2021;9:3389. https://doi.org/10.3389/fbioe.2021.732271.
188. Sasikumar K, Hannibal S, Wendisch VF, Nampoothiri KM. Production of biopolyamide precursors 5-amino valeric acid and putrescine from rice straw hydrolysate by engineered *Corynebacterium glutamicum*. Front Bioeng Biotechnol. 2021;9:635509. https://doi.org/10.3389/fbioe.2021.635509.
189. Uhde A, Youn J-W, Maeda T, Clermont L, Matano C, Krämer R, Wendisch VF, Seibold GM, Marin K. Glucosamine as carbon source for amino acid-producing *Corynebacterium glutamicum*. Appl Microbiol Biotechnol. 2013;97:1679–87. https://doi.org/10.1007/s00253-012-4313-8.
190. Brautaset T, Jakobsen ØM, Degnes KF, Netzer R, Naerdal I, Krog A, Dillingham R, Flickinger MC, Ellingsen TE. *Bacillus methanolicus* pyruvate carboxylase and homoserine dehydrogenase I and II and their roles for L-lysine production from methanol at 50 degrees C. Appl Microbiol Biotechnol. 2010;87:951–64. https://doi.org/10.1007/s00253-010-2559-6.
191. Irla M, Nærdal I, Brautaset T, Wendisch VF. Methanol-based γ-aminobutyric acid (GABA) production by genetically engineered *bacillus methanolicus* strains. Ind Crop Prod. 2017;106:12–20. https://doi.org/10.1016/j.indcrop.2016.11.050.
192. Nærdal I, Pfeifenschneider J, Brautaset T, Wendisch VF. Methanol-based cadaverine production by genetically engineered *bacillus methanolicus* strains. Microb Biotechnol. 2015;8:342–50. https://doi.org/10.1111/1751-7915.12257.
193. Motoyama H, Yano H, Terasaki Y, Anazawa H. Overproduction of l-lysine from methanol by *Methylobacillus glycogenes* derivatives carrying a plasmid with a mutated *dapA* gene. Appl Environ Microbiol. 2001;67:3064–70. https://doi.org/10.1128/AEM.67.7.3064-3070.2001.
194. Hagishita T, Yoshida T, Izumi Y, Mitsunaga T. Efficient l-serine production from methanol and glycine by resting cells of *Methylobacterium* sp. strain MN43. Biosci Biotechnol Biochem. 1996;60:1604–7. https://doi.org/10.1271/bbb.60.1604.
195. Brito LF, Irla M, Nærdal I, Le SB, Delépine B, Heux S, Brautaset T. Evaluation of heterologous biosynthetic pathways for methanol-based 5-Aminovalerate production by thermophilic *bacillus methanolicus*. Front Bioeng Biotechnol. 2021;9:686319. https://doi.org/10.3389/fbioe.2021.686319.
196. Hoffmann SL, Jungmann L, Schiefelbein S, Peyriga L, Cahoreau E, Portais J-C, Becker J, Wittmann C. Lysine production from the sugar alcohol mannitol: design of the cell factory *Corynebacterium glutamicum* SEA-3 through integrated analysis and engineering of metabolic pathway fluxes. Metab Eng. 2018;47:475–87. https://doi.org/10.1016/j.ymben.2018.04.019.
197. Pérez-García F, Klein VJ, Brito LF, Brautaset T. From Brown seaweed to a sustainable microbial feedstock for the production of riboflavin. Front Bioeng Biotechnol. 2022;10:863690. https://doi.org/10.3389/fbioe.2022.863690.
198. Yao W, Chu C, Deng X, Zhang Y, Liu M, Zheng P, Sun Z. Display of alpha-amylase on the surface of *Corynebacterium glutamicum* cells by using NCgl1221 as the anchoring protein, and production of glutamate from starch. Arch Microbiol. 2009;191:751–9. https://doi.org/10.1007/s00203-009-0506-7.
199. Sgobba E, Stumpf AK, Vortmann M, Jagmann N, Krehenbrink M, Dirks-Hofmeister ME, Moerschbacher B, Philipp B, Wendisch VF. Synthetic *Escherichia coli-Corynebacterium glutamicum* consortia for l-lysine production from starch and sucrose. Bioresour Technol. 2018;260:302–10. https://doi.org/10.1016/j.biortech.2018.03.113.
200. Adachi N, Takahashi C, Ono-Murota N, Yamaguchi R, Tanaka T, Kondo A. Direct l-lysine production from cellobiose by *Corynebacterium glutamicum* displaying beta-glucosidase on its cell surface. Appl Microbiol Biotechnol. 2013;97:7165–72. https://doi.org/10.1007/s00253-013-5009-4.

201. Tsuchidate T, Tateno T, Okai N, Tanaka T, Ogino C, Kondo A. Glutamate production from β-glucan using endoglucanase-secreting *Corynebacterium glutamicum*. Appl Microbiol Biotechnol. 2011;90:895–901. https://doi.org/10.1007/s00253-011-3116-7.
202. Schwardmann LS, Dransfeld AK, Schäffer T, Wendisch VF. Metabolic engineering of *Corynebacterium glutamicum* for sustainable production of the aromatic dicarboxylic acid dipicolinic acid. Microorganisms. 2022;10:730. https://doi.org/10.3390/microorganisms10040730.
203. Radek A, Krumbach K, Gätgens J, Wendisch VF, Wiechert W, Bott M, Noack S, Marienhagen J. Engineering of Corynebacterium glutamicum for minimized carbon loss during utilization of D-xylose containing substrates. J Biotechnol. 2014;192:156–60. https://doi.org/10.1016/j.jbiotec.2014.09.026.
204. Radek A, Tenhaef N, Müller MF, Brüsseler C, Wiechert W, Marienhagen J, Polen T, Noack S. Miniaturized and automated adaptive laboratory evolution: evolving Corynebacterium glutamicum towards an improved d-xylose utilization. Bioresour Technol. 2017;245:1377–85. https://doi.org/10.1016/j.biortech.2017.05.055.
205. Kawaguchi H, Sasaki M, Vertès AA, Inui M, Yukawa H. Engineering of an L-arabinose metabolic pathway in *Corynebacterium glutamicum*. Appl Microbiol Biotechnol. 2008;77:1053–62. https://doi.org/10.1007/s00253-007-1244-x.
206. Guo X, Li M, Li H, Xu S, He X, Ouyang P, Chen K. Enhanced cadaverine production by engineered *Escherichia coli* using soybean residue hydrolysate (SRH) as a sole nitrogen source. Appl Biochem Biotechnol. 2021;193:533–43. https://doi.org/10.1007/s12010-020-03444-1.
207. Joo JC, Oh YH, Yu JH, Hyun SM, Khang TU, Kang KH, Song BK, Park K, Oh M-K, Lee SY, Park SJ. Production of 5-aminovaleric acid in recombinant *Corynebacterium glutamicum* strains from a *miscanthus hydrolysate* solution prepared by a newly developed *miscanthus* hydrolysis process. Bioresour Technol. 2017;245:1692–700. https://doi.org/10.1016/j.biortech.2017.05.131.
208. Wen J, Bao J. Engineering *Corynebacterium glutamicum* triggers glutamic acid accumulation in biotin-rich corn stover hydrolysate. Biotechnol Biofuels. 2019;12:86. https://doi.org/10.1186/s13068-019-1428-5.
209. Hadiati A, Krahn I, Lindner SN, Wendisch VF. Engineering of *Corynebacterium glutamicum* for growth and production of L-ornithine, L-lysine, and lycopene from hexuronic acids. Bioresour Bioprocess. 2014;1:25. https://doi.org/10.1186/s40643-014-0025-5.
210. Khanna S, Goyal A, Moholkar VS. Microbial conversion of glycerol: present status and future prospects. Crit Rev Biotechnol. 2012;32(3):235–62. https://doi.org/10.3109/07388551.2011.604839.
211. Rittmann D, Lindner SN, Wendisch VF. Engineering of a glycerol utilization pathway for amino acid production by *Corynebacterium glutamicum*. Appl Environ Microbiol. 2008;74:6216–22. https://doi.org/10.1128/AEM.00963-08.
212. Ebringerová A, Hromádková Z, Heinze T. Hemicellulose. In: Heinze T, editor. Polysaccharides I: structure, characterization and use. Berlin: Springer; 2005. p. 1–67. https://doi.org/10.1007/b136816.
213. Rueda C, Calvo PA, Moncalián G, Ruiz G, Coz A. Biorefinery options to valorize the spent liquor from sulfite pulping. J Chem Technol Biotechnol. 2015;90:2218–26. https://doi.org/10.1002/jctb.4536.
214. Sinner P, Stiegler M, Herwig C, Kager J. Noninvasive online monitoring of *Corynebacterium glutamicum* fed-batch bioprocesses subject to spent sulfite liquor raw material uncertainty. Bioresour Technol. 2021;321:124395. https://doi.org/10.1016/j.biortech.2020.124395.
215. Álvarez-Añorve LI, Calcagno ML, Plumbridge J. Why does *Escherichia coli* grow more slowly on glucosamine than on N-acetylglucosamine? effects of enzyme levels and allosteric activation of GlcN6P deaminase (NagB) on growth rates. J Bacteriol. 2005;187:2974–82. https://doi.org/10.1128/JB.187.9.2974-2982.2005.

216. Gaugué I, Oberto J, Putzer H, Plumbridge J. The use of amino sugars by *Bacillus subtilis*: presence of a unique operon for the catabolism of glucosamine. PLoS One. 2013;8:e63025. https://doi.org/10.1371/journal.pone.0063025.
217. Matano C, Uhde A, Youn J-W, Maeda T, Clermont L, Marin K, Krämer R, Wendisch VF, Seibold GM. Engineering of *Corynebacterium glutamicum* for growth and L-lysine and lycopene production from *N*-acetyl-glucosamine. Appl Microbiol Biotechnol. 2014;98:5633–43. https://doi.org/10.1007/s00253-014-5676-9.
218. Kusch-Brandt S, Mumme J, Nashalian O, Girotto F, Lavagnolo MC, Udenigwe C. 13-valorization of residues from beverage production. In: Grumezescu AM, Holban AM, editors. Processing and sustainability of beverages. Woodhead Publishing; 2019. p. 451–94. https://doi.org/10.1016/B978-0-12-815259-1.00013-6.
219. Pérez-Bibbins B, Torrado-Agrasar A, Salgado JM, de Souza Oliveira RP, Domínguez JM. Potential of lees from wine, beer and cider manufacturing as a source of economic nutrients: an overview. Waste Manag. 2015;40:72–81. https://doi.org/10.1016/j.wasman.2015.03.009.
220. Meyer F, Keller P, Hartl J, Gröninger OG, Kiefer P, Vorholt JA. Methanol-essential growth of *Escherichia coli*. Nat Commun. 2018;9:1508. https://doi.org/10.1038/s41467-018-03937-y.
221. Chen FY-H, Jung H-W, Tsuei C-Y, Liao JC. Converting *Escherichia coli* to a synthetic methylotroph growing solely on methanol. Cell. 2020;182:933–946.e14. https://doi.org/10.1016/j.cell.2020.07.010.
222. Kim S, Lindner SN, Aslan S, Yishai O, Wenk S, Schann K, Bar-Even A. Growth of *E. coli* on formate and methanol via the reductive glycine pathway. Nat Chem Biol. 2020;16:538–45. https://doi.org/10.1038/s41589-020-0473-5.
223. Tuyishime P, Wang Y, Fan L, Zhang Q, Li Q, Zheng P, Sun J, Ma Y. Engineering *Corynebacterium glutamicum* for methanol-dependent growth and glutamate production. Metab Eng. 2018;49:220–31. https://doi.org/10.1016/j.ymben.2018.07.011.
224. Hennig G, Haupka C, Brito LF, Rückert C, Cahoreau E, Heux S, Wendisch VF. Methanol-essential growth of *Corynebacterium glutamicum*: adaptive laboratory evolution overcomes limitation due to methanethiol assimilation pathway. Int J Mol Sci. 2020;21:3617. https://doi.org/10.3390/ijms21103617.
225. Leßmeier L, Pfeifenschneider J, Carnicer M, Heux S, Portais J-C, Wendisch VF. Production of carbon-13-labeled cadaverine by engineered *Corynebacterium glutamicum* using carbon-13-labeled methanol as co-substrate. Appl Microbiol Biotechnol. 2015;99:10163–76. https://doi.org/10.1007/s00253-015-6906-5.
226. Irla M, Wendisch VF. Efficient cell factories for the production of N-methylated amino acids and for methanol-based amino acid production. Microb Biotechnol. 2022;15(8):2145–59. https://doi.org/10.1111/1751-7915.14067.
227. Brautaset T, Williams MD, Dillingham RD, Kaufmann C, Bennaars A, Crabbe E, Flickinger MC. Role of the *bacillus methanolicus* citrate synthase II gene, city, in regulating the secretion of glutamate in l-lysine-secreting mutants. Appl Environ Microbiol. 2003;69:3986–95. https://doi.org/10.1128/AEM.69.7.3986-3995.2003.
228. Gunji Y, Yasueda H. Enhancement of l-lysine production in methylotroph *Methylophilus methylotrophus* by introducing a mutant LysE exporter. J Biotechnol. 2006;127:1–13. https://doi.org/10.1016/j.jbiotec.2006.06.003.
229. Irla M, Heggeset TMB, Nærdal I, Paul L, Haugen T, Le SB, Brautaset T, Wendisch VF. Genome-based genetic tool development for *bacillus methanolicus*: theta- and rolling circle-replicating plasmids for inducible gene expression and application to methanol-based cadaverine production. Front Microbiol. 2016;7:1481. https://doi.org/10.3389/fmicb.2016.01481.
230. Motoyama H, Anazawa H, Katsumata R, Araki K, Teshiba S. Amino acid production from methanol by *Methylobacillus glycogenes* mutants: isolation of L-glutamic acid hyper-producing mutants from *M. glycogenes* strains, and derivation of L-threonine and L-lysine-producing mutants from them. Biosci Biotechnol Biochem. 1993;57:82–7. https://doi.org/10.1271/bbb.57.82.

231. Figge RM, Ramseier TM, Saier MH. The mannitol repressor (MtlR) of *Escherichia coli*. J Bacteriol. 1994;176:840–7. https://doi.org/10.1128/jb.176.3.840-847.1994.
232. Heravi KM, Altenbuchner J. Regulation of the *Bacillus subtilis* mannitol utilization genes: promoter structure and transcriptional activation by the wild-type regulator (MtlR) and its mutants. Microbiol Read Engl. 2014;160:91–101. https://doi.org/10.1099/mic.0.071233-0.
233. López MG, Irla M, Brito LF, Wendisch VF. Characterization of D-Arabitol as newly discovered carbon source of *bacillus methanolicus*. Front Microbiol. 2019;10:1725. https://doi.org/10.3389/fmicb.2019.01725.
234. Laslo T, von Zaluskowski P, Gabris C, Lodd E, Rückert C, Dangel P, Kalinowski J, Auchter M, Seibold G, Eikmanns BJ. Arabitol metabolism of *Corynebacterium glutamicum* and its regulation by AtlR. J Bacteriol. 2012;194:941–55. https://doi.org/10.1128/JB.06064-11.
235. Shi S, Qi N, Nielsen J. Microbial production of chemicals driven by CRISPR-Cas systems. Curr Opin Biotechnol. 2022;73:34–42. https://doi.org/10.1016/j.copbio.2021.07.002.
236. Gu L, Yuan H, Lv X, Li G, Cong R, Li J, Du G, Liu L. High-yield and plasmid-free biocatalytic production of 5-methylpyrazine-2-carboxylic acid by combinatorial genetic elements engineering and genome engineering of *Escherichia coli*. Enzym Microb Technol. 2020;134:109488. https://doi.org/10.1016/j.enzmictec.2019.109488.
237. Wang Y, Cheng H, Liu Y, Liu Y, Wen X, Zhang K, Ni X, Gao N, Fan L, Zhang Z, Liu J, Chen J, Wang L, Guo Y, Zheng P, Wang M, Sun J, Ma Y. In-situ generation of large numbers of genetic combinations for metabolic reprogramming via CRISPR-guided base editing. Nat Commun. 2021;12:678. https://doi.org/10.1038/s41467-021-21003-y.
238. Chen M, Chen L, Zeng A-P. CRISPR/Cas9-facilitated engineering with growth-coupled and sensor-guided in vivo screening of enzyme variants for a more efficient chorismate pathway in *E. coli*. Metab Eng Commun. 2019;9:e00094. https://doi.org/10.1016/j.mec.2019.e00094.
239. Long M, Xu M, Qiao Z, Ma Z, Osire T, Yang T, Zhang X, Shao M, Rao Z. Directed evolution of ornithine Cyclodeaminase using an EvolvR-based growth-coupling strategy for efficient biosynthesis of l-proline. ACS Synth Biol. 2020;9:1855–63. https://doi.org/10.1021/acssynbio.0c00198.
240. Cleto S, Jensen JV, Wendisch VF, Lu TK. *Corynebacterium glutamicum* metabolic engineering with CRISPR interference (CRISPRi). ACS Synth Biol. 2016;5:375–85. https://doi.org/10.1021/acssynbio.5b00216.
241. Binder S, Schendzielorz G, Stäbler N, Krumbach K, Hoffmann K, Bott M, Eggeling L. A high-throughput approach to identify genomic variants of bacterial metabolite producers at the single-cell level. Genome Biol. 2012;13:R40. https://doi.org/10.1186/gb-2012-13-5-r40.
242. Mustafi N, Grünberger A, Kohlheyer D, Bott M, Frunzke J. The development and application of a single-cell biosensor for the detection of l-methionine and branched-chain amino acids. Metab Eng. 2012;14:449–57. https://doi.org/10.1016/j.ymben.2012.02.002.
243. Mustafi N, Grünberger A, Mahr R, Helfrich S, Nöh K, Blombach B, Kohlheyer D, Frunzke J. Application of a genetically encoded biosensor for live cell imaging of L-valine production in pyruvate dehydrogenase complex-deficient *Corynebacterium glutamicum* strains. PLoS One. 2014;9:e85731. https://doi.org/10.1371/journal.pone.0085731.
244. Kortmann M, Mack C, Baumgart M, Bott M. Pyruvate carboxylase variants enabling improved lysine production from glucose identified by biosensor-based high-throughput fluorescence-activated cell sorting screening. ACS Synth Biol. 2019;8:274–81. https://doi.org/10.1021/acssynbio.8b00510.
245. Chen X-F, Xia X-X, Lee SY, Qian Z-G. Engineering tunable biosensors for monitoring putrescine in *Escherichia coli*. Biotechnol Bioeng. 2018;115:1014–27. https://doi.org/10.1002/bit.26521.
246. Zhou L-B, Zeng A-P. Exploring lysine riboswitch for metabolic flux control and improvement of l-lysine synthesis in *Corynebacterium glutamicum*. ACS Synth Biol. 2015;4:729–34. https://doi.org/10.1021/sb500332c.

247. Zhou L-B, Zeng A-P. Engineering a lysine-ON riboswitch for metabolic control of lysine production in *Corynebacterium glutamicum*. ACS Synth Biol. 2015;4:1335–40. https://doi.org/10.1021/acssynbio.5b00075.
248. Stella RG, Wiechert J, Noack S, Frunzke J. Evolutionary engineering of *Corynebacterium glutamicum*. Biotechnol J. 2019;14:1800444. https://doi.org/10.1002/biot.201800444.
249. Jiang L-Y, Chen S-G, Zhang Y-Y, Liu J-Z. Metabolic evolution of *Corynebacterium glutamicum* for increased production of L-ornithine. BMC Biotechnol. 2013;13:47. https://doi.org/10.1186/1472-6750-13-47.
250. Mahr R, Gätgens C, Gätgens J, Polen T, Kalinowski J, Frunzke J. Biosensor-driven adaptive laboratory evolution of l-valine production in *Corynebacterium glutamicum*. Metab Eng. 2015;32:184–94. https://doi.org/10.1016/j.ymben.2015.09.017.
251. Smith RP, Tanouchi Y, You L. Chapter 13-synthetic microbial consortia and their applications. In: Zhao H, editor. Synthetic biology. Boston: Academic; 2013. p. 243–58. https://doi.org/10.1016/B978-0-12-394430-6.00013-3.
252. Wang J, Lu X, Ying H, Ma W, Xu S, Wang X, Chen K, Ouyang P. A novel process for cadaverine bio-production using a consortium of two engineered *Escherichia coli*. Front Microbiol. 2018;9:1312. https://doi.org/10.3389/fmicb.2018.01312.
253. Liu X, Li X-B, Jiang J, Liu Z-N, Qiao B, Li F-F, Cheng J-S, Sun X, Yuan Y-J, Qiao J, Zhao G-R. Convergent engineering of syntrophic *Escherichia coli* coculture for efficient production of glycosides. Metab Eng. 2018;47:243–53. https://doi.org/10.1016/j.ymben.2018.03.016.
254. Burmeister A, Hilgers F, Langner A, Westerwalbesloh C, Kerkhoff Y, Tenhaef N, Drepper T, Kohlheyer D, von Lieres E, Noack S, Grünberger A. A microfluidic co-cultivation platform to investigate microbial interactions at defined microenvironments. Lab Chip. 2018;19:98–110. https://doi.org/10.1039/c8lc00977e.
255. Chua HB, Robinson JP. Formate-limited growth of *Methanobacterium formicium* in steady-state cultures. Arch Microbiol. 1983;135:158–60. https://doi.org/10.1007/BF00408027.
256. Kirst H, Ferlez BH, Lindner SN, Cotton CAR, Bar-Even A, Kerfeld CA. Toward a glycyl radical enzyme containing synthetic bacterial microcompartment to produce pyruvate from formate and acetate. Proc Natl Acad Sci. 2022;119:e2116871119. https://doi.org/10.1073/pnas.2116871119.
257. Gleizer S, Ben-Nissan R, Bar-On YM, Antonovsky N, Noor E, Zohar Y, Jona G, Krieger E, Shamshoum M, Bar-Even A, Milo R. Conversion of *Escherichia coli* to generate all biomass carbon from CO2. Cell. 2019;179:1255–1263.e12. https://doi.org/10.1016/j.cell.2019.11.009.
258. Gregory GJ, Bennett RK, Papoutsakis ET. Recent advances toward the bioconversion of methane and methanol in synthetic methylotrophs. Metab Eng. 2021;71:99–116. https://doi.org/10.1016/j.ymben.2021.09.005.
259. Barbot YN, Al-Ghaili H, Benz R. A review on the valorization of macroalgal wastes for biomethane production. Mar Drugs. 2016;14:120. https://doi.org/10.3390/md14060120.
260. Hakvåg S, Nærdal I, Heggeset TMB, Kristiansen KA, Aasen IM, Brautaset T. Production of value-added chemicals by *bacillus methanolicus* strains cultivated on mannitol and extracts of seaweed *saccharina latissima* at 50°C. Front Microbiol. 2020;11:680. https://doi.org/10.3389/fmicb.2020.00680.
261. Shin JW, Lee OK, Park HH, Kim HS, Lee EY. Molecular characterization of a novel oligoalginate lyase consisting of AlgL- and heparinase II/III-like domains from *Stenotrophomonas maltophilia* KJ-2 and its application to alginate saccharification. Korean J Chem Eng. 2015;32:917–24. https://doi.org/10.1007/s11814-014-0282-1.
262. Badur AH, Jagtap SS, Yalamanchili G, Lee J-K, Zhao H, Rao CV. Alginate lyases from alginate-degrading *Vibrio splendidus* 12B01 are endolytic. Appl Environ Microbiol. 2015;81:1865–73. https://doi.org/10.1128/AEM.03460-14.
263. Wargacki AJ, Leonard E, Win MN, Regitsky DD, Santos CNS, Kim PB, Cooper SR, Raisner RM, Herman A, Sivitz AB, Lakshmanaswamy A, Kashiyama Y, Baker D, Yoshikuni Y. An engineered microbial platform for direct biofuel production from brown macroalgae. Science. 2012;335:308–13. https://doi.org/10.1126/science.1214547.

264. Naik SN, Goud VV, Rout PK, Dalai AK. Production of first and second generation biofuels: a comprehensive review. Renew Sust Energ Rev. 2010;14:578–97. https://doi.org/10.1016/j.rser.2009.10.003.
265. Hicks TM, Verbeek CJR. Chapter 1-protein-rich by-products: production statistics, legislative restrictions, and management options. In: Singh Dhillon G, editor. Protein byproducts. Academic; 2016. p. 1–18. https://doi.org/10.1016/B978-0-12-802391-4.00001-X.
266. Adhikari BB, Chae M, Bressler DC. Utilization of slaughterhouse waste in value-added applications: recent advances in the development of wood adhesives. Polymers. 2018;10:176. https://doi.org/10.3390/polym10020176.
267. Amorim ML, Soares J, Coimbra JS, Dos R, de Oliveira Leite M, Albino LFT, Martins MA. Microalgae proteins: production, separation, isolation, quantification, and application in food and feed. Crit Rev Food Sci Nutr. 2021;61:1976–2002. https://doi.org/10.1080/10408398.2020.1768046.
268. Chen S, Liao W, Liu C, Wen Z, Kincaid RL, Harrison JH, Elliott DC, Brown MD, Solana AE, Stevens DJ. Value-added chemicals from animal manure. Richland., Environmental Molecular Sciences Laboratory (US): Pacific Northwest National Lab; 2003. https://doi.org/10.2172/15009485.
269. Wendisch VF, Mindt M, Pérez-García F. Biotechnological production of mono- and diamines using bacteria: recent progress, applications, and perspectives. Appl Microbiol Biotechnol. 2018;102:3583–94. https://doi.org/10.1007/s00253-018-8890-z.
270. Chae TU, Kim WJ, Choi S, Park SJ, Lee SY. Metabolic engineering of *Escherichia coli* for the production of 1,3-diaminopropane, a three carbon diamine. Sci Rep. 2015;5:13040. https://doi.org/10.1038/srep13040.
271. Wang L, Li G, Deng Y. Diamine biosynthesis: research progress and application prospects. Appl Environ Microbiol. 2020;86:e01972–20. https://doi.org/10.1128/AEM.01972-20.
272. Klatte S, Wendisch VF. Redox self-sufficient whole cell biotransformation for amination of alcohols. Bioorg Med Chem. 2014;22:5578–85. https://doi.org/10.1016/j.bmc.2014.05.012.
273. Sung S, Jeon H, Sarak S, Murshidul Ahsan M, Patil MD, Kroutil W, Kim B-G, Yun H. Parallel anti-sense two-step cascade for alcohol amination leading to ω-amino fatty acids and α,ω-diamines. Green Chem. 2018;20:4591–5. https://doi.org/10.1039/C8GC02122H.
274. Maddess ML, Tackett MN, Ley SV. Total synthesis studies on macrocyclic pipecolic acid natural products: FK506, the antascomicins and rapamycin. In: Petersen F, Amstutz R, editors. Natural compounds as drugs, vol. 2. Basel: Birkhäuser; 2008. p. 13–186. https://doi.org/10.1007/978-3-7643-8595-8_2.
275. Pastor JM, Salvador M, Argandoña M, Bernal V, Reina-Bueno M, Csonka LN, Iborra JL, Vargas C, Nieto JJ, Cánovas M. Ectoines in cell stress protection: uses and biotechnological production. Biotechnol Adv. 2010;28:782–801. https://doi.org/10.1016/j.biotechadv.2010.06.005.
276. Luo Z, Wang Z, Wang B, Lu Y, Yan L, Zhao Z, Bai T, Zhang J, Li H, Wang W, Cheng J. An artificial pathway for N-hydroxy-pipecolic acid production from L-lysine in *Escherichia coli*. Front Microbiol. 2022;13:842804. https://doi.org/10.3389/fmicb.2022.842804.

Part III
Application of N-containing Biomass to Manufacture of Chemicals and Materials

Chapter 8
Engineering Biochar-Based Materials for Carbon Dioxide Adsorption and Separation

Shuai Gao, Jack Shee, Wei Chen, Lujiang Xu, Chengyu Dong, and Bing Song

Abstract Increasing emissions of carbon dioxide, the primary greenhouse gas, are the main contributor to climate change. Developing an effective carbon capture, storage and utilization approach is paramount to overcoming global warming. Emerging research in engineered biochars provides a promising means of utilizing highly abundant lignocellulosic biomass as a precursor for carbon capture. Given appropriate production and modification of its physiochemical properties, biochar can be used as a cost-effective and selective adsorbent for CO_2 capture. In this chapter, the engineering of biochar for CO_2 capture is reviewed. The effects of different modification processes on the material properties of biochars (i.e. specific surface area, pore volume, pore size, hierarchical pore structure and surface chemistry) and their impacts to CO_2 uptake are discussed. Feedstock type, thermochemical conditions of pyrolysis and surface chemical modification via functional groups all play significant roles in determining the texture, porosity, aromaticity and hydrophobicity of biochar, which are key factors to increase CO_2 adsorption capacity.

Keywords Climate change · Carbon capture · Engineered biochar · Porous structure · Functional groups

---

S. Gao · W. Chen · L. Xu · C. Dong
Biomass Group, College of Engineering, Nanjing Agricultural University, Nanjing, China

J. Shee
School of Biomedical Engineering, University of Melbourne, Parkville, VIC, Australia

B. Song (✉)
Scion, Titokorangi Drive, Rotorua, New Zealand
e-mail: bing.song@scionresearch.com

Z. Fang et al. (eds.), *Production of N-containing Chemicals and Materials from Biomass*, Biofuels and Biorefineries 12,
https://doi.org/10.1007/978-981-99-4580-1_8

8.1 Introduction

The increasing emission of greenhouse gases, especially CO_2, is responsible for various unwanted environmental changes such as the rise of sea level, drought, desertification, global warming and ocean acidification. Based on NOAA's Global Monitoring Lab's observation, atmospheric carbon dioxide concentration sharply increased from around 300 ppm in the 1960s to more than 400 ppm in the 2020s. The International Panel on Climate Change (IPCC) further predicts that atmospheric CO_2 concentration will reach 570 ppm by 2100, resulting in a mean temperature increase of 1.9 °C [1]. This will lead to a significant impact on the global environment and further exacerbate adverse conditions. Thus, developing sustainable technologies for mitigating CO_2 emissions to overcome global warming has become a priority as stated in the fifth assessment report of the IPCC. Carbon capture has been highlighted as a straightforward means that can directly decrease the distribution of CO_2 in the atmosphere if appropriate absorbent and scalable techniques are employed [2].

In 2005, IPCC identified three possible intervention phases during the process of carbon emissions by significant stationary sources: precombustion carbon capture, oxy-fuel process, and post-combustion carbon capture [3]. Among them, post-combustion capture is the only technology that can be applied to existing large point sources (e.g., power plants and cement manufacturing plants) and therefore is a promising approach in the short and medium term with low technological risk. More specifically, CO_2 separation technologies involve most commonly four processes, (1) absorption; (2) adsorption; (3) gas separation membranes; and (4) cryogenic distillation [3]. Adsorption, due to the low energy consumption, easy regeneration of adsorbent, adaptability at a wide range of temperatures and pressures, as well as lack of unfavourable by-product formation, has been considered a feasible process for CO_2 capture at an industrial scale. Materials used for adsorption sit at the heart of any adsorption process. Zeolites, activated carbons, porous silicas, engineered carbon nanomaterials and metal-organic frameworks have been studied for carbon capture in the last few years [4]. However, none of the above-mentioned materials can be practically applied for CO_2 capture without improving their selectivity of CO_2 adsorption in cost-effective ways via appropriate engineering processes.

Biochar is defined as a carbon-rich, porous solid produced by the thermal decomposition of biomass in a process with the absence or limited access to oxygen at moderate temperatures (normally below 700 °C) [5]. As biochars can be prepared from abundant biomass such as crop residues, wood wastes, animal manure, food wastes, municipal solid wastes, and sewage sludge, it is sustainable and widely accessible to end users at low cost. Given the variety of raw materials and tuneable preparation conditions, biochars can be purposely prepared with abundant surface functional groups (C–O, C=O, –COOH, and –OH, etc.), which provide modifiable sites for the synthesis of various functionalized carbon materials, making biochar a versatile precursor amenable for many uses. Hence, engineering biochars is more promising than developing other carbon-based materials. For instance,

Table 8.1 Main distinctions between biochar and activated carbon (Adapted from [6–8])

	Biochar	Activated carbon
Main source	Biomass	Coal, asphalt, biomass
Carbon content	40–90%	80–95%
Structure	Amorphous carbon with abundant surface functional groups, nanostructures, or porosity	Amorphous carbon with abundant porosity
Production method	Pyrolysis of biomass at medium temperature (400–600 °C) and then modification with physical or chemical methods	Carbonization of main source at high temperature (700–1000 °C) under physical or chemical activation
Average energy demand	6.1 MJ/kg	97 MJ/kg
Average unit price (2017)	US $5/kg	US $5.60/kg
Estimated break-even price (2009)	US $246/ton	US $1500/ton

compared with activated carbon, biochar exhibits significant advantages in energy demand and production cost (Table 8.1),

In the past few years, the great potential of using biochars and related engineered materials for environmental applications has garnered much attention. Carbon capture by biochars as well as enhancing the adsorption capacity by structural modification and surface functionalization have been of great interest to the research community. In this chapter, the advantages of using biochar and related engineered materials for CO_2 adsorption and separation have been evaluated. The factors related to the CO_2 uptake of biochar as well as the engineering strategies targeting CO_2 separation are also discussed. Upon overview and discussion of the recent advances, the chapter concludes with key challenges and perspectives to advise development towards a sustainable approach to reducing carbon emissions.

8.2 Recent Advances in Using Biochar as an Adsorbent for Carbon Capture

Biochar is a low-cost and sustainable adsorption medium generated from natural biomass or agricultural waste. In industrial applications, biochars can be produced as a major product or by-product, accompanied by various liquid and gas flows (e.g., syngas and bio-oil). At present, the estimated amount of biochar produced by fast pyrolysis of biomass is 220 million tons/year (with an estimated 10% of biochar). With the continuously growing demand for biorefinery products, this amount is expected to increase sharply in the next 30 years [9]. Table 8.2 presents an overview

Table 8.2 Production capacity of selected types of biomass/biochar (Adapted from [11])

Type of Sample	Country/Region	Year	Production Capacity (tons/year)
Lignin	Global	2019	100,000,000
Oil tea	China	2018	56,250,000
Wood pellets	Global	2020	52,700,000
Animal waste (manure)	Vietnam	2017	26,361,000
Coconut	Indonesia	2019	17,130,000
Whitewood	Finland	2019	11,800,000 m^3/year
Olive	Spain	2020	6,560,000
Oats	Russian Federation	2020	4,761,000
Almond	U. S. A.	2020	2,003,000
Coconut shell	Malaysia	2005	577,000
Pine sawdust	Mexico	2019	280,000
Pine cones	Andalusia	2012	55,000

of the production capacity of some significant types of biomass or biochars throughout the world. Due to the wide availability of biomass, biochars are about 10 times cheaper than other CO_2 adsorbents [10]. However, raw biochars have limited CO_2 adsorption capacity and various modification techniques are required to improve the performance of biochars for carbon capture.

Many studies have demonstrated that having a microporous structure allows for more active sites on the biochar; enhancing the CO_2 adsorption capacity. For instance, a novel activated carbon prepared from bamboo by KOH activation with enhanced specific surface area (SSA) showed a 15% CO_2 adsorption capacity up to 1.50 mmol g^{-1} at 25 °C [12]. In addition, biochars prepared from food waste, wood waste gasification and KOH + CO_2 activation showed that the development of a microporous structure contributed to a high CO_2 adsorption capacity (q_{max} = 3.23 mmol g^{-1} at 25 °C) [13]. The CO_2 adsorption performances of wood pellet-derived biochar with steam and KOH activation were compared [14]. The KOH-activated biochar exhibited a higher CO_2 uptake (50.8 mg g^{-1}) than that of steam-activated biochar (38.3 mg g^{-1}) which supports the statement that a higher percentage of micropores contributes to higher CO_2 uptake. It was also found that carbon capture from microporous biochar produced via KOH activation may be lower in high humidity conditions since water molecules occupy available micropore sites, acting as a competitive inhibitor. Thus, it is important to remove water from the gas stream before the CO_2 adsorption process, while it is even more preferred to enhance the stability and adsorption efficiency of biochar in conditions with high relative humidity.

It has been widely accepted that surface functionalization of biochars will improve the capture of acidic CO_2. Specifically, biochars with N-functional groups provide active sites for the uptake of CO_2 through both chemical and physical adsorption. For instance, pyridinic-N, pyrrolic-N, and primary, secondary and tertiary amines can all react with CO_2. The presence of these N-functional groups (Fig. 8.1) increases the alkalinity of biochar, promoting Lewis acid-base interactions,

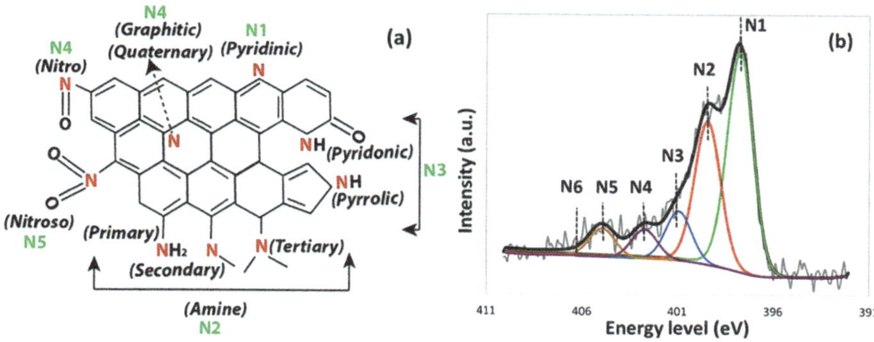

Fig. 8.1 (**a**) Schematic of nitrogen functionality on carbon surface and (**b**) XPS peak deconvolution results of nitrogen (N-1 s). Reprinted with permission from [17]. Copyright © 2022, Elsevier

thus enhancing the CO_2 adsorption [15]. Coffee grounds were used as the raw material for the preparation of biochars for CO_2 adsorption through ammoxidation and KOH activation. The biochars with nitrogen-doping by melamine and KOH activation exhibited high microporosity and developed more pyrrolic nitrogen as active adsorption sites contributing to high CO_2 uptake (2.67 mmol g^{-1} at 35 °C). In addition, the prepared biochar showed high stability (total decrease of ca. 6% in 10 cycles) and high selectivity of CO_2 over N_2 ($CO_2/N_2 = 74.2$) [16]. This excellent performance was attributed to a high proportion of micropore volume ($V_{micro}/V_{total} = 82\%$), increased nitrogen doping (5.1 wt%) and specific pyrrolic nitrogen (active adsorption sites).

Lignin was used as a precursor for the preparation of hierarchical porous carbon by KOH activation and NH_3 was selected as the source for N-doping [17]. The obtained biochar showed a large SSA (1631–2922 m^2 g^{-1}) and a highly developed porous structure with both micropores and mesopores. NH_3 activation introduced basic functional groups including mainly pyridinic as well as amino and pyrrolic/pyridonic type of functionality with a total nitrogen content of 5.6–7.1 at.%. All these properties are beneficial for CO_2 adsorption (5.48 mmol g^{-1} at 25 °C) and separation (CO_2/N_2 = 37–433), as well as a consistent working capacity (up to 10 cycles).

Recently, a simple method to prepare N-doped biochar for carbon capture was reported where the importance of amine and nitrile content was highlighted [18]. Bagasse and hickory chips were used as the raw materials and processed with pyrolysis at 450 or 650 °C under an N_2 atmosphere followed by ball milling with $NH_3 \cdot H_2O$. The dehydration of oxygen-containing functional groups including carboxyl and hydroxyl transferred to nitrogen species such as $-NH_2$ and $C \equiv N$ on the surface of the prepared biochar. Meanwhile, a lower pyrolysis temperature (450 °C) was found to contribute to doping more nitrogen on the biochar surface compared to a higher pyrolysis temperature (600 °C). One possible explanation is that low pyrolysis temperature may result in higher concentration of oxygen-containing functional groups, thus more nitrogen-containing functional groups can be retained. Nevertheless, biochars (pyrolysis at 450 °C) with nitrogen-containing functional groups enhanced the CO_2 uptake by 31.6–55.2% compared with those of green and ball-milled biochars.

Recent efforts have also evaluated biochars as a potential adsorbent for cyclic adsorption processes. Karimi et al. [11] comprehensively reviewed studies on cyclic adsorption processes (e.g., pressure/temperature/vacuum swing adsorption) which highlighted the application of biochars as a raw material used for engineered adsorbents in large-scale applications of gas adsorption and separation. Biomass waste (olive and cheery stone) was used as the raw material for preparing adsorbent for the separation of CO_2 and CH_4 [19]. The precursor was calcinated with CO_2 in the atmosphere at a high temperature followed by loading in a bench-scale pressure swing adsorption (PSA) setup equipped with a fixed-bed adsorption column. The result showed that the olive stone-derived adsorbent was comparable with the commercial carbon adsorbent and presented a high sorbent selection value for efficient separation of CO_2/CH_4 gas mixture at given conditions. Karimi et al. [20] evaluated municipal solid wastes-derived biochars for carbon capture. The samples were activated using H_2SO_4 and thermal treatment at 400 °C and 800 °C, respectively. It has been found that the combination of H_2SO_4 treatment and activation at 800 °C contributed to the best CO_2 adsorption capacity with an uptake (2.6 mmol g^{-1} at 40 °C and 2.5 bar) comparable with commercial carbon materials. The following scale-up process carried out in a conceptual PSA unit confirmed that municipal solid wastes derived biochars exhibited excellent performance as a source of material for carbon capture under conditions of post-combustion operation.

8.3 Key Engineering Strategies Targeting Biochar for Carbon Dioxide Adsorption and Separation

The CO_2 adsorption capacity of biochar is the amount of CO_2 uptake per unit weight of biochar, which mainly depends on its physio-chemical properties including composition, SSA, pore volume, pore size, surface functional groups, presence of alkali and alkali earth metals and hydrophobicity. These properties of biochar are significantly influenced by the type of feedstocks and the thermal treatment conditions as well as the use of additives and activation methods, which are further discussed in the following subsections.

8.3.1 Strategies for Modifying the Physical Properties of Biochar

The van der Waals force and pore filling mechanism play important roles between CO_2 molecules and the solid phase of biochar [21, 22], thus increasing the SSA and pore volume as well as optimising the pore size and distribution (hierarchical pore structure) are the main strategies for modifying the physical properties of biochar.

8.3.1.1 Increasing the Specific Surface Area

Increasing the SSA means providing biochars with a wider surface for active sites responsible for carbon capture, thus contributing to a larger adsorption capacity. A positive relationship can be seen between the SSA ($R^2 = 0.6475$), micropore area ($R^2 = 0.9032$), and the CO_2 adsorption capacity of biochars (Fig. 8.2). Many reports have concluded that a high SSA of adsorbents presents a superior CO_2 adsorption capacity.

Creamer et al. [23] investigated the effect of SSA of biochar on CO_2 adsorption performance. Sugarcane bagasse and hickory wood were used as the precursor for the preparation of biochars at higher pyrolysis temperatures with a large SSA and excellent CO_2 capture performance (73.50 mg g^{-1} and 61.0 mg g^{-1} at 25 °C). Physical adsorption played an important role by bonding CO_2 to the surface of biochar, thus SSA was an important contributor to CO_2 adsorption. Huang et al. [24] prepared biochars by microwave pyrolysis of rice straw for CO_2 capture and reported the high correlation (correlation coefficient: 0.84) of the SSA with the CO_2 uptake of biochar. The prepared biochar exhibited a CO_2 adsorption capacity of up to 80 mg g^{-1} at 20 °C. Sun et al. [25] prepared engineered biochar from wood waste by pyrolysis in molten salts. It was found that high temperature (800 °C) and ternary carbonate contributed to the largest SSA (476 m^2 g^{-1}) and highest CO_2 capacity

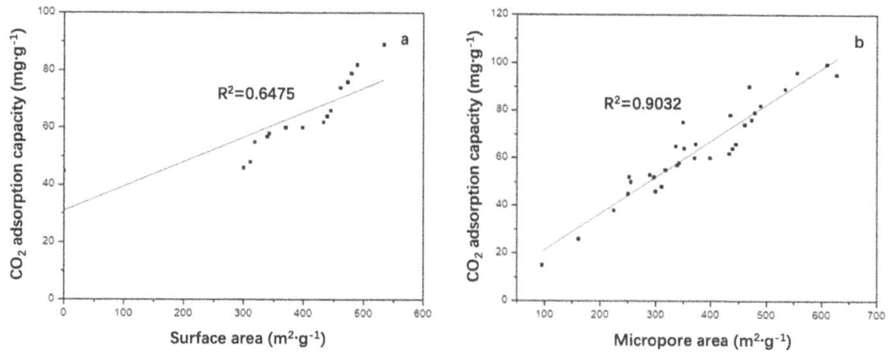

Fig. 8.2 Relationship between the SSA (**a**), micropore area (**b**), and CO_2 adsorption capacity of biochar. Reprinted with permission from [26]. Copyright © 2022, Elsevier

(4.5 mmol g^{-1}, 0 °C, 100 kPa) of the resultant biochar. The high selection parameter (S) in both PSA and vacuum swing adsorption (VSA) processes and high recovery rate (89%) of the molten salt showed great potential for repurposing biowaste as green and efficient adsorbents for carbon capture.

8.3.1.2 Increasing Pore Volume and Optimising Pore Size

During biomass pyrolysis, the release of volatile organic matter causes the formation of porous structures in the resulting biochar. This behaviour is related to the above-mentioned SSA. In addition, the specific pore structure (e.g., volume and size) can also alter the carbon capture capacity. The CO_2 uptake was found to closely follow a linear correlation with the size of pores ≤1.5 nm at the ambient pressure [27]. At 1 bar, pores <0.8 nm contributed the most to the CO_2 uptake and at 0.1 bar pores ≤0.5 nm were preferred. The coefficients of determination between the CO_2 uptake at a certain pressure and the cumulative pore volume of pores less than or equal to a certain pore diameter are shown in Fig. 8.3a. It agrees well with the CO_2 adsorption performance of bamboo-derived activated carbon [28], where the volume of micropores in the range of 0.33–0.82 nm demonstrates a positive correlation with CO_2 adsorption in Fig. 8.3b.

The type of biomass or feedstock is one of the main factors that may influence the pore volume and pore size of biochar. Specifically, biochars produced from biomass materials that are high in cellulose are known to develop more pores that are less than 2 nm during the pyrolysis, while pyrolysis of biomass materials that are high in lignin results in pores larger than 50 nm [29]. In addition, higher pyrolysis and activation temperatures result in the widening of the pore size [30]. It is worth noting that over-pyrolysis and activation of biochar may lead to excess widening of the pore

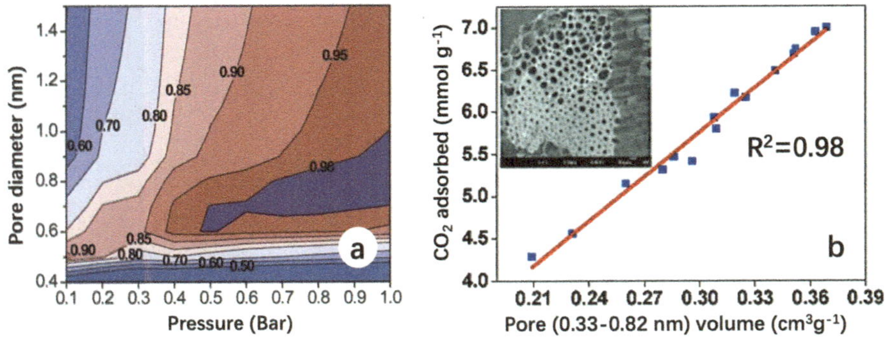

Fig. 8.3 (**a**) Coefficient of determination between CO_2 uptake between 0.1 bar and 1.0 bar (x-axis) and the volume of pores smaller to threshold size (y-axis) and (**b**) volume of micropores in the range of 0.33–0.82 nm has a good linear relationship with CO_2 adsorption on bamboo-derived activated carbon. Reprinted with permission from [27]. Copyright © 2022, RSC. Reprinted with permission from [28]. Copyright © 2023, John Wiley and Sons

which may be detrimental to the CO_2 adsorption capacity as micropores are more desirable for gas adsorption.

Igalavithana et al. [31] prepared different types of biochars from pine sawdust and paper mill sludge and investigated the effect of porous structure on carbon capture performance. It was found that the SSA and microporosity, especially the micropores of 0.5 nm diameter played an important role in CO_2 adsorption (positively correlated to carbon capture with $r = 0.960$ and $p = 0.039$). The study demonstrated that the feedstock, pyrolysis temperature and activation method (steam activation) had a great effect on the textural, surface, and structural properties of biochar as well as the resulting carbon capture performance.

8.3.1.3 Developing a Hierarchical Pore Structure

The International Union of Pure and Applied Chemistry defines pores with a diameter > 50 nm as macropores, those with a diameter between 2 nm and 50 nm as mesopores, and those with a diameter of <2 nm as micropores [32]. Micropores with a pore size of <1 nm predominantly affect CO_2 adsorption. However, hierarchical structures with balanced and cross-linked macro-, meso- and micro-porosity also play a significant role in the adsorption process [33]. Feng et al. prepared biochars from corn straw with a hierarchical-functionalized structure and explored its adsorption properties of CO_2 with both experimental data analysis and molecular dynamics simulation [34]. The results showed strong evidence of a link between one-step activation with H_2O and the construction of functional groups on the biochar surface, however, the two-step activation with H_2O and NH_3 was beneficial to the enhancement of pore formation and aromatization of biochar. The adsorption sites of CO_2 were mainly generated from micropores with a diameter of less than 0.7 nm, while the mesopores provided a significant contribution to the total adsorption capacity due to the creation of a concentration gradient for diffusion based mass transfer. In another study, hierarchically porous carbon was prepared from an ocean pollutant (*Enteromorpha prolifera*) via hydrothermal carbonization, KOH activation and HF washing [35]. The intrinsic sponge structure of the precursor, hydrolysis of hemicellulose, carbon gasification, KOH activation and inorganic compounds together contributed to the hierarchical carbon with micropores, mesopores and macropores (Fig. 8.4). As a result, the diffusion of CO_2 into the bulk carbon skeleton was accelerated by the presence of meso- and macropores in the adsorbent, thus improving the kinetics of adsorption.

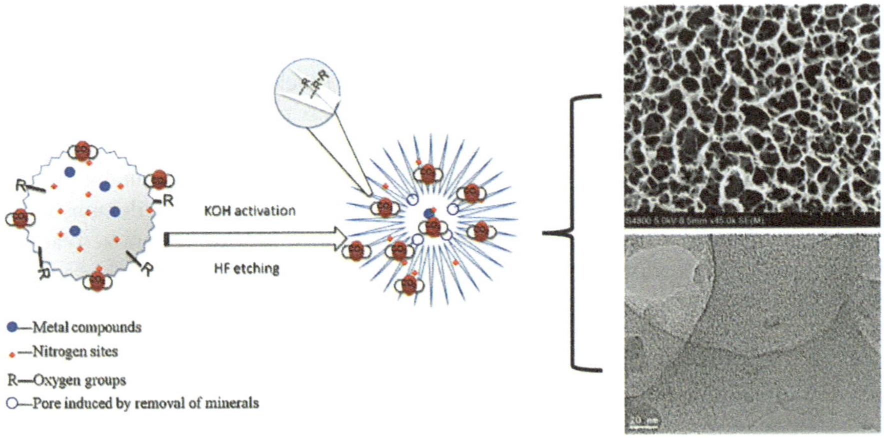

Fig. 8.4 Diagram of porosity formation and enhanced CO_2 capacity of KOH activated and hydrothermally treated ocean waste-derived biochar. Reprinted with permission from [35]. Copyright © 2022, Elsevier

8.3.2 Strategies for Functionalizing Biochar for High-Performance CO_2 Adsorption and Separation

The uptake of CO_2 by biochar surface is also influenced by the functionality (chemical properties) of the biochar including alkalinity, presence of surface functional groups, mineral composition, hydrophobicity, etc. Thus, the CO_2 adsorption capacity of biochar can be optimised by introducing functional groups and elements with a strong affinity for acidic gas.

8.3.2.1 Introducing Basic Functional Groups

The reaction between the acidic CO_2 molecules and basic nitrogen-containing functional groups is well accepted as the main contribution to CO_2 capture. Pyridone, pyrrole, amine, pyridine-N-oxide, quaternary-N, cyanide, and pyridine groups with different basic strengths are common N species on the carbon surface. Among them, pyridone, pyridine and pyrrole are well reported to enhance CO_2 adsorption due to various interactions including hydrogen bonding and Lewis acid-base interactions [36, 37]. Figure 8.5 summarises different methods (agents) for introducing various N-containing functional groups.

The direct pyrolysis of nitrogen-rich precursors is a sustainable method for doping nitrogen atoms into the carbonaceous matrix of biochar, which is conducive to the application of the carbon capture [39, 40]. Nitrogen-doped microporous carbon was prepared from chitosan and K_2CO_3 as a nitrogen-containing precursor and activation agent, respectively [41]. The high CO_2 uptake by the porous carbon was attributed to the large pore volume, well-defined microporosity and high

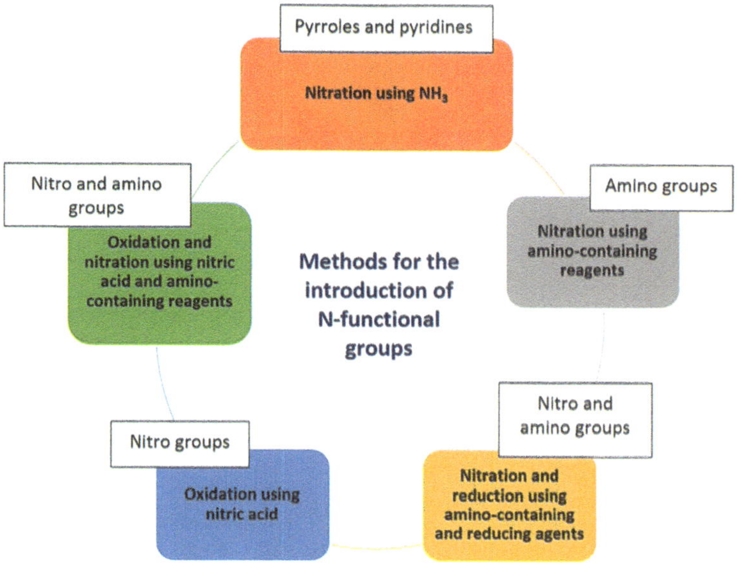

Fig. 8.5 Different methods for introducing various N-containing functional groups. Reprinted with permission from [38]. Copyright © 2022, Elsevier

nitrogen content. In the XPS analysis of N-containing functional groups, the dominant peak belonged to pyridone, and pyridine contributed less. As for pyridone nitrogen, the *p* orbital in –OH produces a *p*–*π* conjugation effect with the *π* bond and as consequence, the Lewis basicity of the nitrogen located in the *ortho*-position of –OH with a stronger negative charge is enhanced. Thus, pyridonic nitrogen was attributed as the main active site for CO_2 capture. Density functional theories calculation also confirmed that the pyridone groups exhibited the maximum energy of adsorption contributed by the H bond between the H atom of the –OH and the O atom of the CO_2 with high electronegativity [37].

Sawdust was impregnated with a urea phosphate solution and kept for 24 h, then carbonized at a high temperature (600–900 °C) under N_2 followed by hot water washing and drying [42]. The obtained N-doped biochar exhibited a large SSA (1189 $m^2\ g^{-1}$), high micropore volume (0.433 $cm^3\ g^{-1}$), high nitrogen content (4.24 at.%) with the main basic functional groups including pyridinic and pyrrolic N, and decent CO_2 adsorption capacity (1.34 mmol g^{-1} at 30 °C and 1 bar). These results confirmed that the adsorption of CO_2 was related to the synergistic effect of SSA, micropore volume and N-containing functional groups [43]. Specifically, due to the strong polarity and improvement for acid-base reaction, pyridinic N was more conducive to CO_2 adsorption than other types of N species. On the other hand, pyrrolic N can promote the hydrogen-bond interaction between the biochar surface and CO_2 (Fig. 8.6). In addition, a high pyrolysis temperature (900 °C) was found to transfer some nitrogen groups into aromatic nitrogen-containing groups and weaken the alkalinity of the biochar.

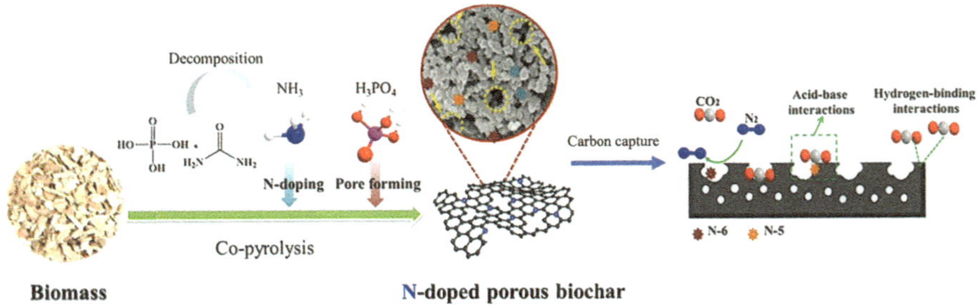

Fig. 8.6 One-step method for preparation of microporous N-doped biochar from plant biomass using urea phosphate (N-5: pyrrolic N; N-6: pyridinic N). Adapted from [42]. Copyright © 2022, Elsevier

CO_2 adsorbent was prepared from biomass waste by three ammoxidation methods and KOH activation (400 °C for 1 h and 600 °C for 1 h under an N_2 atmosphere) [16]. In particular, the process of using melamine contributed to the smallest SSA (990 $m^2\ g^{-1}$) but the highest nitrogen content (5.1%) and the largest CO_2 capacity (117.5 mg g^{-1} at 35 °C) of the resultant biochar. The study suggests that N-functional groups (mainly pyrrolic nitrogen) rather than the SSA are considered to be the dominant factor for CO_2 adsorption. The high value (49.9 kJ mol^{-1}) of adsorption enthalpy of biochar doped with pyrrolic nitrogen indicates that the interaction (e.g., hydrogen bonding or intermolecular force) between adsorbent and CO_2 was enhanced, resulting in the improved carbon capture performance.

Xie et al. [44] investigated the bean dregs-based porous carbon for CO_2 capture. A significant increase of C and N contents in the hydrothermally treated biochar was found and the pyridinic-N and pyrrolic-N accounted for more than 60% of the total N-containing groups. As a result, the isosteric heat of adsorption at the low CO_2 uptake stage was much higher than the value of interaction between CO_2 and pure carbon-based materials, indicating that the adsorption of CO_2 on N-rich biochar is not completely physical and may occur via intermolecular forces and covalent bonds. Nuclear magnetic resonance spectroscopy data verifies that interactions between N-containing species and CO_2 result in the formation of new chemical bonds (Fig. 8.7).

Apart from pyridone, pyridine and pyrrole, other N-containing groups including Amine-N (– NH_2 and C–N) and N=N [45, 46] were also investigated for their performance of carbon capture. It was found that all these functional groups presented the ability to interact with CO_2, and most of them can be regenerated except for the C-N.

8.3.2.2 Introducing Oxygenated Functional Groups

It has been reported that hydrogen bonding plays a critical role in the interaction between adsorbate and adsorbent when oxygenated functional groups are presented

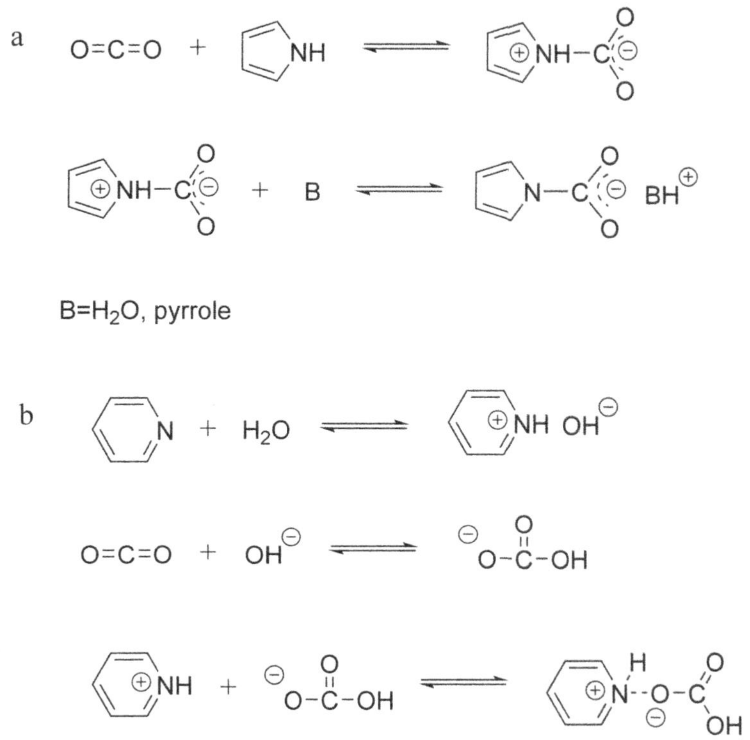

Fig. 8.7 Adsorption mechanism of CO_2 on (**a**) pyrrolic-N and (**b**) pyridine-N. Reprinted with permission from [44]. Copyright © 2023, John Wiley and Sons

on the surface of the adsorbent [38, 47]. Although most oxygen-containing functional groups (e.g., –COOH, –OH, C=O, phenolic, lactone, and peroxide groups) are acidic which would inhibit the capturing of acidic CO_2, following studies found that the adsorption of CO_2 may be enhanced by the functional groups including –OH and –COOH, when they were presented on the carbon skeleton. Xing et al. [48] reported the influence of introducing O-containing functional groups to the adsorbent matrix (by chlorination and followed by HNO_3 oxidation) on the CO_2 adsorption capacity of carbide-derived carbon (CDC). It was found that the CO_2 adsorption capacity per unit SSA was positively related to the oxygen content of the carbons, suggesting that the CO_2 uptake was significantly facilitated by the doping of O-containing groups. Quantum chemical calculation further supports that the introduction of oxygen atoms into the carbon surface promotes hydrogen bonding between the CO_2 molecules and the carbon surface (Fig. 8.8). In addition, the binding energy difference between oxygenated and non-oxygenated complexes roughly agreed with the difference in CO_2 adsorption heat between the pristine CDC and oxygenated CDC. The findings indicate that acidic O-containing groups of carbon adsorbents can also facilitate CO_2 adsorption.

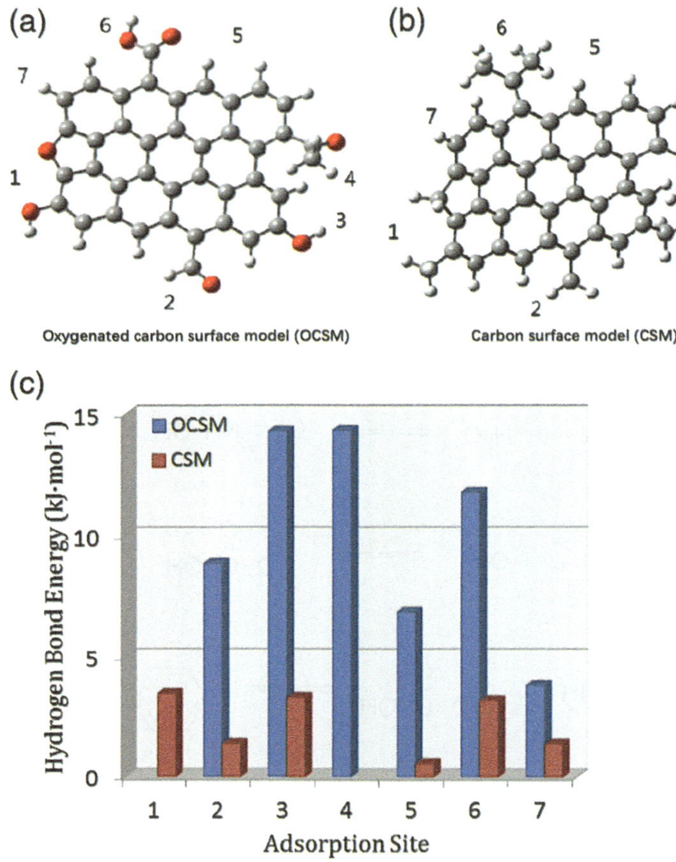

Fig. 8.8 Theoretical models for (**a**) oxygen-containing carbon surface and (**b**) pure carbon surface (red ball: oxygen atom; grey ball: carbon atom; small white ball: hydrogen atom). (**c**) Hydrogen bond energies at different adsorption sites. Reprinted with open access from [48] under the terms of the creative commons attribution 4.0 International License (https://creativecommons.org/licenses/by/4.0)

8.3.2.3 Loading Alkaline and Alkaline Earth Metals

It has been demonstrated that the carbon frameworks with open metal sites tend to have a high affinity for CO_2 [49]. The basic nature of metal oxide and carbon framework may improve the selectivity of adsorbent on CO_2 and suppress the adsorption of other gases such as N_2 and CH_4. This is because metal oxyhydroxides tend to bond with acidic CO_2 when polar surfaces are in contact. Creamer et al. [50] prepared biochar/metal (Al, Mg and Fe) oxyhydroxide composites using cottonwood as the precursor and evaluated their capacity to uptake CO_2. The treatment with a sequence of metal ions impregnation and pyrolysis at 600 °C prepared biochar loaded with metal oxyhydroxide nanoparticles. The CO_2 uptake by biochar/metal oxyhydroxide composites was attributed to the surface adsorption mechanisms contributed by both carbon matrix and metal oxyhydroxide particles. The high CO_2 uptake by AlOOH–biochar (71 mg g^{-1} at 25 °C) is roughly equal to commercial adsorbents, and together with the low-cost regeneration at 120 °C (90% and 99% desorption) suggests that metal oxyhydroxide–biochar composites can be used for carbon capture with high efficiency at a low cost. Lahijani et al. [51] further investigated the impact of pyrolysis temperature (500 °C, 700 °C and 900 °C) and type of metal oxide (Mg, AL, Fe, Ni, Na and Ca) on the carbon capture ability of the modified biochar prepared from walnut shell. It was found that increasing the pyrolysis temperature was effective to enhance the SSA and porosity of the biochar. In particular, the largest SSA (397 m^2g^{-1}) and highest pore volume (0.198 cm^3 g^{-1}) were obtained by preparing the sample at 900 °C. Meanwhile, the loading of basic metal sites onto the metal/biochar composite surface enhanced the adsorption of acidic CO_2 gas in the order of Mg > Al > Fe > Ni > Ca > green-biochar > Na. In addition, Mg-loaded biochar exhibited high stability and reusability upon 10 cycles of CO_2 adsorption and desorption with almost no compromise in its carbon capture performance.

8.3.3 Summary

In addition to the above-mentioned strategies to engineer biochars for improved CO_2 adsorption and separation, studies have indicated that biochars with hydrophobic and non-polar characteristics may enhance CO_2 uptake by restraining the interference of H_2O (polar molecule) [8, 52]. The carbonization process can reduce the concentration of O, H, N and S in the biochar matrix, thus increasing the degree of hydrophobicity by increasing the degree of aromaticity, resulting in a high potential to adsorb CO_2 when competing with water molecules in humid conditions [29]. In summary, CO_2 adsorption on biochars is a complex interaction that may be affected by multiple factors including van der Waals forces, hydrogen bonds, micropore filling phenomenon as well as Lewis acid-base reaction when functional groups (N and O) presented on the surface of adsorbents. Therefore, tuning the properties of biochars and developing their surface functionality to enhance the above-mentioned interactions is critical to engineering biochars with improved CO_2 adsorption capacity and selectivity.

8.4 Challenges and Perspectives

The use of biochars as a widely accessible and low-cost adsorbent for CO_2 capture promises chances for valuable application of biomass resources and an alternative approach for reducing carbon emissions, thus contributing to the achievement of our sustainable development goals. As an emerging research area of interest, the practical use of biochars for carbon capture is yet to be realised with efforts addressing challenges in all aspects from the initial biochar preparation and engineering to the application of CO_2-saturated biochar in real practices. Based on the above overview, some key challenges and corresponding perspectives are given as follows.

Biochar is a complex material with variants in raw materials and preparation methodologies. This provides possibilities for preparing versatile products for multiple purposes including CO_2 adsorption. However, there is a key challenge to developing standardised protocols for biochar preparation and engineering towards targeted purposes. Addressing this challenge will require a fundamental understanding of the correlations among feedstock of biomass, engineering strategies (e.g., activation and modification), the physiochemical properties of biochar and the CO_2 uptake capability. The emerging application of artificial intelligence (AI) techniques such as machine learning in biomass valorization is promising to assist in identifying these correlations based on a considerable amount of experimental data.

Overall, achieving ideal adsorption performance will require engineering biochar by both physical and chemical means. This also means extra costs and the possible use and/or emission of toxic chemicals. From the perspective of life-cycle analysis, the process of biochar preparation and engineering is carbon emissive, despite the facts (1) producing and storing biochar is considered a negative emission process [53] and (2) the adsorption and storage of CO_2 in biochar further contribute to emission reduction. Life-cycle analysis is thus highly recommended to detail the emission portfolios.

Enabling and promoting the practical application of biochar as a CO_2 adsorbent also expects a shift from Carbon Capture & Storage to Carbon Capture & Utilization, whilst re-utilizing engineered biochar for carbon capture in a circular manner. This calls for efforts in (1) improving the reusability of biochar, and (2) developing proof-of-concept of integrated biochar carbon capture and CO_2 utilization or valorization. Recent drastically progressing technologies such as electrochemical CO_2 reduction [54] and the use of CO_2 for algae cultivation [55] have provided wide opportunities for the post-adsorption valorization of CO_2.

8.5 Conclusions and Future Outlook

Engineered biochars with their enhanced inherent properties are promising sustainable materials for CO_2 capture. Specifically, the specific surface area, micropore area and volume, hierarchical pore structure, presence of functional groups (e.g., N-containing functional groups) and hetero elements are great contributing factors to CO_2 adsorption capacity. Thus, strategies of biochar modification through physical and chemical processes will significantly improve the aforementioned properties, thereby enhancing the CO_2 uptake of biochar.

On the other hand, the practical use of biochar is yet to be realised in all aspects from the initial biochar preparation and engineering to the application of CO_2

capture in large-scale production. Hence, future studies should focus on developing standardised biochar preparation and engineering protocols towards specific carbon capture scenarios with a complete life-cycle analysis. The high-value utilization of biochar as byproducts from biorefinery or lignin residue after valerization of cellulose and hemicellulose is suggested to increase the overall competitiveness of bioenergy. Novel technologies and biochar-based materials such as nano-biochar composites and biomass-derived carbon foam should be developed to enhance the CO_2 adsorption capacity of biochar. Study on the adsorption and reaction mechanism of CO_2 with specific functional groups of biochar is preferred as it not only helping understand the adsorption process but also paving the way for CO_2 conversion on carbon capture and utilization. Moreover, industrial-scale application of biochar for CO_2 capture should also be highlighted as well as developing proof-of-concept of integrated biochar carbon capture and CO_2 utilization.

References

1. Field CB, Barros VR. Climate change 2014–impacts, adaptation and vulnerability: regional aspects. Cambridge: Cambridge University Press; 2014. https://doi.org/10.1017/CBO9781107415386.
2. Wu F, Argyle MD, Dellenback PA, Fan M. Progress in O_2 separation for oxy-fuel combustion–a promising way for cost-effective CO_2 capture: a review. Prog Energy Combust Sci. 2018;67: 188–205. https://doi.org/10.1016/j.pecs.2018.01.004.
3. Pires JCM, Martins FG, Alvim-Ferraz MCM, Simões M. Recent developments on carbon capture and storage: an overview. Chem Eng Res Des. 2011;89:1446–60. https://doi.org/10.1016/j.cherd.2011.01.028.
4. Farmahini AH, Krishnamurthy S, Friedrich D, Brandani S, Sarkisov L. Performance-based screening of porous materials for carbon capture. Chem Rev. 2021;121:10666–741. https://doi.org/10.1021/acs.chemrev.0c01266.
5. Joseph S, Peacocke C, Lehmann J, Munroe P. Developing a biochar classification and test methods. In: Lehmann J, Joseph S, editors. Biochar for environmental management: science and technology. 1st ed. London: Routledge; 2009. p. 107–26. https://doi.org/10.4324/9781849770552.
6. Alhashimi HA, Aktas CB. Life cycle environmental and economic performance of biochar compared with activated carbon: a meta-analysis. Resour Conserv Recycl. 2017;118:13–26. https://doi.org/10.1016/j.resconrec.2016.11.016.
7. Liu W-J, Jiang H, Yu H-Q. Development of biochar-based functional materials: toward a sustainable platform carbon material. Chem Rev. 2015;115:12251–85. https://doi.org/10.1021/acs.chemrev.5b00195.
8. Ahmad M, Lee SS, Dou X, Mohan D, Sung J-K, Yang JE, Ok YS. Effects of pyrolysis temperature on soybean stover- and peanut shell-derived biochar properties and TCE adsorption in water. Bioresour Technol. 2012;118:536–44. https://doi.org/10.1016/j.biortech.2012.05.042.
9. Song B, Cao X, Gao W, Aziz S, Gao S, Lam C-H, Lin R. Preparation of nano-biochar from conventional biorefineries for high-value applications. Renew Sust Energ Rev. 2022;157: 112057. https://doi.org/10.1016/j.rser.2021.112057.
10. Chatterjee R, Sajjadi B, Mattern DL, Chen W-Y, Zubatiuk T, Leszczynska D, Leszczynski J, Egiebor NO, Hammer N. Ultrasound cavitation intensified amine functionalization: a feasible strategy for enhancing CO_2 capture capacity of biochar. Fuel. 2018;225:287–98. https://doi.org/10.1016/j.fuel.2018.03.145.

11. Karimi M, Shirzad M, Silva JAC, Rodrigues AE. Biomass/biochar carbon materials for CO_2 capture and sequestration by cyclic adsorption processes: a review and prospects for future directions. J CO2 Util. 2022;57:101890. https://doi.org/10.1016/j.jcou.2022.101890.
12. Ji Y, Zhang C, Zhang XJ, Xie PF, Wu C, Jiang L. A high adsorption capacity bamboo biochar for CO_2 capture for low temperature heat utilization. Sep Purif Technol. 2022;293:121131. https://doi.org/10.1016/j.seppur.2022.121131.
13. Igalavithana AD, Choi SW, Dissanayake PD, Shang J, Wang C-H, Yang X, Kim S, Tsang DCW, Lee KB, Ok YS. Gasification biochar from biowaste (food waste and wood waste) for effective CO_2 adsorption. J Hazard Mater. 2020;391:121147. https://doi.org/10.1016/j.jhazmat.2019.121147.
14. Zhang C, Sun S, Xu S, Wu C. CO_2 capture over steam and KOH activated biochar: effect of relative humidity. Biomass Bioenergy. 2022;166:106608. https://doi.org/10.1016/j.biombioe.2022.106608.
15. Leng L, Xu S, Liu R, Yu T, Zhuo X, Leng S, Xiong Q, Huang H. Nitrogen containing functional groups of biochar: an overview. Bioresour Technol. 2020;298:122286. https://doi.org/10.1016/j.biortech.2019.122286.
16. Liu S-H, Huang Y-Y. Valorization of coffee grounds to biochar-derived adsorbents for CO_2 adsorption. J Clean Prod. 2018;175:354–60. https://doi.org/10.1016/j.jclepro.2017.12.076.
17. Saha D, Van Bramer SE, Orkoulas G, Ho H-C, Chen J, Henley DK. CO_2 capture in lignin-derived and nitrogen-doped hierarchical porous carbons. Carbon. 2017;121:257–66. https://doi.org/10.1016/j.carbon.2017.05.088.
18. Xu X, Zheng Y, Gao B, Cao X. N-doped biochar synthesized by a facile ball-milling method for enhanced sorption of CO_2 and reactive red. Chem Eng J. 2019;368:564–72. https://doi.org/10.1016/j.cej.2019.02.165.
19. Álvarez-Gutiérrez N, García S, Gil MV, Rubiera F, Pevida C. Towards bio-upgrading of biogas: biomass waste-based adsorbents. Energy Procedia. 2014;63:6527–33. https://doi.org/10.1016/j.egypro.2014.11.688.
20. Karimi M, Zafanelli LFAS, Almeida JPP, Ströher GR, Rodrigues AE, Silva JAC. Novel insights into activated carbon derived from municipal solid waste for CO_2 uptake: synthesis, adsorption isotherms and scale-up. J Environ Chem Eng. 2020;8:104069. https://doi.org/10.1016/j.jece.2020.104069.
21. Sun L, Yang L, Zhang Y-D, Shi Q, Lu R-F, Deng W-Q. Accurate van der Waals force field for gas adsorption in porous materials. J Comput Chem. 2017;38:1991–9. https://doi.org/10.1002/jcc.24832.
22. Wedler C, Span R. Micropore analysis of biomass chars by CO_2 adsorption: comparison of different analysis methods. Energy Fuel. 2021;35:8799–806. https://doi.org/10.1021/acs.energyfuels.1c00280.
23. Creamer AE, Gao B, Zhang M. Carbon dioxide capture using biochar produced from sugarcane bagasse and hickory wood. Chem Eng J. 2014;249:174–9. https://doi.org/10.1016/j.cej.2014.03.105.
24. Huang Y-F, Chiueh P-T, Shih C-H, Lo S-L, Sun L, Zhong Y, Qiu C. Microwave pyrolysis of rice straw to produce biochar as an adsorbent for CO_2 capture. Energy. 2015;84:75–82. https://doi.org/10.1016/j.energy.2015.02.026.
25. Sun M, Zhu X, Wu C, Masek O, Wang C-H, Shang J, Ok YS, Tsang DCW. Customizing high-performance molten salt biochar from wood waste for CO_2/N_2 separation. Fuel Process Technol. 2022;234:107319. https://doi.org/10.1016/j.fuproc.2022.107319.
26. Dissanayake PD, You S, Igalavithana AD, Xia Y, Bhatnagar A, Gupta S, Kua HW, Kim S, Kwon J-H, Tsang DCW, Ok YS. Biochar-based adsorbents for carbon dioxide capture: a critical review. Renew Sust Energ Rev. 2020;119:109582. https://doi.org/10.1016/j.rser.2019.109582.
27. Presser V, McDonough J, Yeon S-H, Gogotsi Y. Effect of pore size on carbon dioxide sorption by carbide derived carbon. Energy Environ Sci. 2011;4:3059–66. https://doi.org/10.1039/C1EE01176F.

28. Wei H, Deng S, Hu B, Chen Z, Wang B, Huang J, Yu G. Granular bamboo-derived activated carbon for high CO_2 adsorption: the dominant role of narrow micropores. ChemSusChem. 2012;5:2354–60. https://doi.org/10.1002/cssc.201200570.
29. Sajjadi B, Chen W-Y, Egiebor NO. A comprehensive review on physical activation of biochar for energy and environmental applications. Rev Chem Eng. 2019;35:735–76. https://doi.org/10.1515/revce-2017-0113.
30. Feng D, Zhao Y, Zhang Y, Zhang Z, Che H, Sun S. Experimental comparison of biochar species on in-situ biomass tar H_2O reforming over biochar. Int J Hydrog Energy. 2017;42:24035–46. https://doi.org/10.1016/j.ijhydene.2017.08.013.
31. Igalavithana AD, Choi SW, Shang J, Hanif A, Dissanayake PD, Tsang DCW, Kwon J-H, Lee KB, Ok YS. Carbon dioxide capture in biochar produced from pine sawdust and paper mill sludge: effect of porous structure and surface chemistry. Sci Total Environ. 2020;739:139845. https://doi.org/10.1016/j.scitotenv.2020.139845.
32. Thommes M, Kaneko K, Neimark AV, Olivier JP, Rodriguez-Reinoso F, Rouquerol J, Sing KSW. Physisorption of gases, with special reference to the evaluation of surface area and pore size distribution. Pure Appl Chem. 2015;87:1051–69. https://doi.org/10.1515/pac-2014-1117.
33. Shi Q, Zhang X, Shen B, Ren K, Wang Y, Luo J. Enhanced elemental mercury removal via chlorine-based hierarchically porous biochar with $CaCO_3$ as template. Chem Eng J. 2021;406: 126828. https://doi.org/10.1016/j.cej.2020.126828.
34. Feng D, Guo D, Zhang Y, Sun S, Zhao Y, Chang G, Guo Q, Qin Y. Adsorption-enrichment characterization of CO_2 and dynamic retention of free NH_3 in functionalized biochar with H_2O/$NH_3{\cdot}H_2O$ activation for promotion of new ammonia-based carbon capture. Chem Eng J. 2021;409:128193. https://doi.org/10.1016/j.cej.2020.128193.
35. Zhang Z, Wang K, Atkinson JD, Yan X, Li X, Rood MJ, Yan Z. Sustainable and hierarchical porous Enteromorpha prolifera based carbon for CO_2 capture. J Hazard Mater. 2012;229-230: 183–91. https://doi.org/10.1016/j.jhazmat.2012.05.094.
36. Chiang Y-C, Yeh C-Y, Weng C-H. Carbon dioxide adsorption on porous and functionalized activated carbon fibers. Appl Sci. 2019;9:1977. https://doi.org/10.3390/app9101977.
37. Lim G, Lee KB, Ham HC. Effect of N-containing functional groups on CO_2 adsorption of carbonaceous materials: a density functional theory approach. J Phys Chem C. 2016;120:8087–95. https://doi.org/10.1021/acs.jpcc.5b12090.
38. Shafawi AN, Mohamed AR, Lahijani P, Mohammadi M. Recent advances in developing engineered biochar for CO_2 capture: an insight into the biochar modification approaches. J Environ Chem Eng. 2021;9:106869. https://doi.org/10.1016/j.jece.2021.106869.
39. Nguyen T-D, Shopsowitz KE, MacLachlan MJ. Mesoporous nitrogen-doped carbon from nanocrystalline chitin assemblies. J Mater Chem A. 2014;2:5915–21. https://doi.org/10.1039/C3TA15255C.
40. Mishra RK, Misra M, Mohanty AK. Value-added biocarbon production through slow pyrolysis of mixed bio-oil wastes: studies on their physicochemical characteristics and structure–property–processing co-relation. Biomass Convers Biorefin. 2022;6:2. https://doi.org/10.1007/s13399-022-02906-2.
41. Fan X, Zhang L, Zhang G, Shu Z, Shi J. Chitosan derived nitrogen-doped microporous carbons for high performance CO_2 capture. Carbon. 2013;61:423–30. https://doi.org/10.1016/j.carbon.2013.05.026.
42. Ma Q, Chen W, Jin Z, Chen L, Zhou Q, Jiang X. One-step synthesis of microporous nitrogen-doped biochar for efficient removal of CO_2 and H_2S. Fuel. 2021;289:119932. https://doi.org/10.1016/j.fuel.2020.119932.
43. Zhang S, Zhou Q, Jiang X, Yao L, Jiang W, Xie R. Preparation and evaluation of nitrogen-tailored hierarchical meso−/micro-porous activated carbon for CO_2 adsorption. Environ Technol. 2020;41:3544–53. https://doi.org/10.1080/09593330.2019.1615131.
44. Xie W-H, Yao X, Li H, Li H-R, He L-N. Biomass-based N-rich porous carbon materials for CO_2 capture and in-situ conversion. ChemSusChem. 2022;15:e202201004. https://doi.org/10.1002/cssc.202201004.

45. Sutar PN, Jha A, Vaidya PD, Kenig EY. Secondary amines for CO_2 capture: a kinetic investigation using N-ethylmonoethanolamine. Chem Eng J. 2012;207-208:718–24. https://doi.org/10.1016/j.cej.2012.07.042.
46. Zhang X, Zhang S, Yang H, Shao J, Chen Y, Liao X, Wang X, Chen H. Generalized two-dimensional correlation infrared spectroscopy to reveal mechanisms of CO_2 capture in nitrogen enriched biochar. Proc Combust Inst. 2017;36:3933–40. https://doi.org/10.1016/j.proci.2016.06.062.
47. Liu Y, Wilcox J. Effects of surface heterogeneity on the adsorption of CO_2 in microporous carbons. Environ Sci Technol. 2012;46:1940–7. https://doi.org/10.1021/es204071g.
48. Xing W, Liu C, Zhou Z, Zhou J, Wang G, Zhuo S, Xue Q, Song L, Yan Z. Oxygen-containing functional group-facilitated CO_2 capture by carbide-derived carbons. Nanoscale Res Lett. 2014;9:189. https://doi.org/10.1186/1556-276X-9-189.
49. D'Alessandro DM, Smit B, Long JR. Carbon dioxide capture: prospects for new materials. Angew Chem Int Ed. 2010;49:6058–82. https://doi.org/10.1002/anie.201000431.
50. Creamer AE, Gao B, Wang S. Carbon dioxide capture using various metal oxyhydroxide–biochar composites. Chem Eng J. 2016;283:826–32. https://doi.org/10.1016/j.cej.2015.08.037.
51. Lahijani P, Mohammadi M, Mohamed AR. Metal incorporated biochar as a potential adsorbent for high capacity CO_2 capture at ambient condition. J CO2 Util. 2018;26:281–93. https://doi.org/10.1016/j.jcou.2018.05.018.
52. Shen Y, Linville JL, Ignacio-de Leon PAA, Schoene RP, Urgun-Demirtas M. Towards a sustainable paradigm of waste-to-energy process: enhanced anaerobic digestion of sludge with woody biochar. J Clean Prod. 2016;135:1054–64. https://doi.org/10.1016/j.jclepro.2016.06.144.
53. Glaser B, Parr M, Braun C, Kopolo G. Biochar is carbon negative. Nat Geosci. 2009;2:2. https://doi.org/10.1038/ngeo395.
54. Liu J, Cai Y, Song R, Ding S, Lyu Z, Chang Y-C, Tian H, Zhang X, Du D, Zhu W, Zhou Y, Lin Y. Recent progress on single-atom catalysts for CO_2 electroreduction. Mater Today. 2021;48:95–114. https://doi.org/10.1016/j.mattod.2021.02.005.
55. Alami AH, Alasad S, Ali M, Alshamsi M. Investigating algae for CO_2 capture and accumulation and simultaneous production of biomass for biodiesel production. Sci Total Environ. 2021;759:143529. https://doi.org/10.1016/j.scitotenv.2020.143529.

Chapter 9
Producing N-Containing Chemicals from Biomass for High Performance Thermosets

Jiahui Li, Qi Cao, and Zhihuan Weng

Abstract Chemicals containing nitrogen atoms or N-containing chemicals (NCCs) are pivotal building blocks for polymers and composites that have high interest in technological fields, especially when they can be endowed with special properties and new functionalities. To alleviate the dependence on petroleum resources, efforts have focused on sustainable production of NCCs from biomass and its derivatives. In this chapter, production of NCCs from biomass for thermosets are introduced. Thermosets can form highly cross-linked network after curing reactions, and they have been widely applied in the fields of coatings, adhesives, advanced composites and electronic packaging due to their remarkable integrated properties.

This chapter provides a systematic overview regarding recent advances in sustainable high-performance thermosets derived from NCCs. Firstly, the origins and access of bio-based feedstocks for NCCs applied in thermoset are discussed. Then, the synthesis and structure-property relationship of epoxy resin, benzoxazine, polyurethane and other typical bio-based thermosets with N-containing chemicals are reviewed. Finally, some thoughts about the future of the synthesis of bio-based thermosets bearing NCCs and their various applications are presented. The objective of this chapter is to help us to have a deeper understanding of the impact of the introduction of NCCs on the properties of polymer materials, thus promoting the rapid development of this field.

Keywords Bio-based · N-containing chemicals · High performance · Thermoset

---

J. Li · Q. Cao · Z. Weng (✉)
State Key Laboratory of Fine Chemicals, Frontiers Science Center for Smart Materials Oriented Chemical Engineering, Dalian University of Technology, Dalian, China
e-mail: zweng@dlut.edu.cn

Z. Fang et al. (eds.), *Production of N-containing Chemicals and Materials from Biomass*, Biofuels and Biorefineries 12,
https://doi.org/10.1007/978-981-99-4580-1_9

9.1 Introduction

Nitrogen-containing compounds (NCCs) are essential building blocks that are widely used in advanced functional materials, pharmaceuticals, pesticides, nutritions, textiles, polymers and surfactants [1]. In particular, 170 of the top 200 drugs contained nitrogen and/or amino groups, and most of the top pesticides contain nitrogen skeleton structures [2]. The global market for NCCs is worth more than $50 billion per year, but unfortunately, most of the industrially relevant organic nitrogen compounds are currently manufactured from petrochemical resources [3]. Considering rising fossil fuel prices and falling inventories, as well as growing environmental concerns, there is a pressing need to explore alternative routes to access organic nitrogen chemicals from sustainable biomass.

In the past decade, a few dozen valuable NCCs including amines, amino acids, nitriles, and N-heterocyclic compounds have been synthesized from various biomass components [4]. This includes several strategies such as conversion of nitrogenous carbohydrates, catalytic rapid pyrolysis of carbohydrates in ammonia, and reductive amination of biomass-derived platform molecules. Obviously, using biomass as a carbon feedstock to prepare functional materials can not only reduce the carbon footprint, but also simplify the synthetic route by utilizing the inherent functions of biomass resources [5]. Up to date, NCCs have been extensively applied in the synthesis of thermosetting resins such as epoxy resin, benzoxazine, phthalonitrile, polyurethane and isocyanate. However, although renewable bio-based platform chemicals can alleviate the over-reliance on petroleum resources, some of them still require structural design and chemical modification to obtain bio-based high-performance resins that reaching comparable using level to their petroleum-based counterparts.

9.2 Overview of Nitrogen-Containing Compounds Derived from Renewable Platform Chemicals

Recently, production of value-added N-containing chemicals from renewable biomass resources has gained significant interest. Two strategies are proposed to obtain organic nitrogen-containing bio-based chemicals. For bio-based platform compounds derived from lignin and carbohydrates, nitrogen-containing functional groups such as amino, cyano, nitro, isocyanate group can be introduced into the molecular structure through various reactions. The second strategy involves the use of inherently nitrogen-containing bio-based compounds such as chitin, amino acids and their derivatives as the starting materials.

9.2.1 Nitrogenous Compounds Derived from Nitrogen-Free Biobased Platform Compounds

Lignin, which accounts for about 15–30% of the most abundant biomass resource on earth [6], is a phenolic biopolymer formed by the free radical polymerization of monolignol and is considered the largest renewable resource of aromatics [7]. Over the past few decades, great progress has been made in the field of converting lignin into chemicals with high added value, such as vanillin, guaiacol, eugenol, ferulic acid etc. [8]. Benefiting from a tailored synthesis strategy, various nitrogen-containing products have been obtained from a range of biomass feedstock either directly or via intermediate platform compounds.

9.2.1.1 From Vanillin

Although lignin depolymerization can yield a wide variety of aromatic monomers, only vanillin is currently produced on an industrial scale, making it a particularly attractive component in the formulation of valuable organic nitrogen chemicals [9].

Using vanillin derivatives as raw materials, Savonnet et al. identified two synthetic pathways for the synthesis of bio-based aromatic amines (MDVA and DMAN) through the reduction of oxime and the acyl azide rearrangement moieties, leading to bis-benzylamine and bis-aniline moieties, respectively [10], as shown in Scheme 9.1a1, a2. Whereas these reactions suffered for some drawbacks, such as lengthy fabrication procedures, and relatively harsh reaction conditions.

Mora et al. promoted diglycidyl ethers derived from vanillin to prepare a new bio-based amine monomer (DHAMHY) by direct amination with aqueous ammonia, which was inexpensive and a non-toxic reactant [11]. The reaction method is shown in Scheme 9.1a3. However, syn- and anti- attacks can occur, and reverse amine side products can be formed despite the configuration, which to some extent will affect the further application of the compound.

Not only amino groups, but also nitrile-containing compounds using vanillin as a feedstock have also been successfully prepared. Qi et al. transformed the aldehyde group of vanillin into the nitrile group by employing a mild one-pot method using hydroxylamine hydrochloride and iron III chloride, followed by a cyclotrimerization reaction to obtain an aromatic triazine product (THMT) [12], as shown in Scheme 9.1a4.

9.2.1.2 From Guaiacol

Aromatic N-heterocycles are an important chemical substance currently mainly produced from petroleum. Unfortunately, their synthesis from biomass-derived starting materials remains challenging. A series of different N-heterocycles were prepared by Bhusal's team from dimethyl itaconic acid and pyrrole, which are

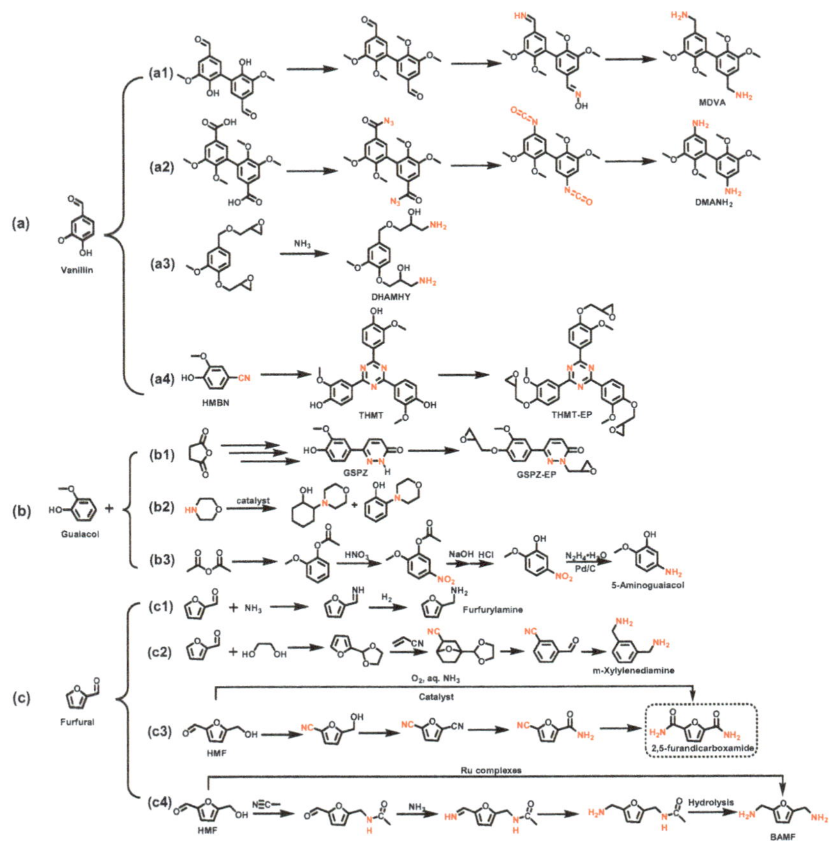

Scheme 9.1 Preparations of nitrogen-containing compounds derived from (**a**) vanillin, (**b**) guaiacol and (**c**) furfural

available from biomass [13]. However, the lengthy steps and harsh conditions of this reaction make it difficult to scale up. Kallmeier et al. produced a facile and sustainable synthesis of pyrroles from alcohols and aminoalcohols via manganese catalysis, however, since only alcohols could be obtained from digestible and abundantly available lignocellulose, the resulted pyrroles were partially bio-based [14]. It seems that a simple synthetic route and a high bio-based content are difficult to balance. Recently, Qi et al. provided a general protocol that relied on Friedel-Crafts acylation of guaiacol and succinic anhydride, followed by cyclization and dehydrogenation with hydrazine hydrate and sodium 3-nitrobenzenesulfonate, respectively, to obtain a fully renewable pyridazine-based aromatic N-heterocycle (GSPZ) [15], the synthetic route is shown in Scheme 9.1b1.

Currently, there is also a lack of sustainable strategies for the production of alkoxy-functionalized cyclohexylamines. Synthesis of alkoxy-functionalized cyclohexylamines from renewable biomass resources would be an interesting alternative to fossil-based pathways. As shown in Scheme 9.1b2, Zheng et al. synthesized methoxy-functionalized cyclohexylamines in excellent yields by reductively coupling guaiacol with amines and H_2 catalyzed by Pd/C with high selectivity of 88.1% [16].

In addition, Li et al. synthesized an aromatic amine (5-aminoguaiacol) with 100% bio-based carbon resources using guaiacol as a raw material through the reduction of nitroaromatic hydrocarbons [17]. The synthetic route is shown in Scheme 9.1b3. In order to weaken the positioning effect of the phenolic hydroxyl group and increase the orthosteric hindrance of -OCH_3, they adopted the protection/deprotection step of the phenolic hydroxyl group to obtain 5-aminoguaiacol by the hydrogenation reduction reaction of the nitrated guaiacol.

9.2.1.3 Furan-Derived Nitrogen-Containing Compounds

Furan derivatives, such as 5-hydroxymethylfurfural (HMF) and furfural, are promising sustainable alternatives to petroleum-based chemicals. Furfural is a precursor for furfuryl alcohol, furoic acid, methylfuran, tetrahydrofuran and γ-valerolacton [18]. Furfural can be obtained from acid-catalyzed tandem hydrolytic dehydration of C-5 carbohydrates from hemicellulose or acid-promoted transformation of cellulose-derived hexoses [19]. The reductive amination of furfural with ammonia or primary amines is a versatile method for the synthesis of valuable nitrogen-containing chemicals including furfurylamine and its derivatives. The reductive amination of furfural with amines is a cascade reaction involving the formation of imines and successive hydrogenations to produce furfurylamine or its derivatives [20]. Ammonia is an inexpensive and readily available nitrogen source that has attracted much attention in the field of reductive amination. By using homogeneous $RuCl_2(PPh_3)_3$ as catalyst, Jagadeesh et al. performed the reductive amination of furfural with ammonia, giving 85% yield of furfurylamine [21], the reaction route is shown in Scheme 9.1c1. In addition to the above strategies, as shown in Scheme 9.1c2, Scodeller and co-workers selectively synthesized m-xylylenediamine via 100% carbon-economical Diels-Alder/aromatization starting from furfural and acrylonitrile, which has the potential to produce functionalized biomass-derived aromatics (benzaldehyde, benzylamine, etc.) [5].

The presence of hydroxyl and aldehyde groups makes HMF a versatile candidate for the synthesis of various nitrogen-containing compounds. Similar to furfural, reductive amination of HMF is an important fundamental pathway for the sustainable production of nitrogen-containing compounds from renewable biomass [22]. Recently, catalytic ammoxidation of HMF was explored to be an efficient strategy to obtain bio-based diimine and diamidine under mild conditions. Li et al. used Al-cation-doped cryptomelanes as catalyst for the aerobic ammoxidationhydration tandem reaction of HMF to 2,5-furandicarboxamide [23], and the synthetic route is shown in Scheme 9.1c3.

2,5-Bis(aminomethyl)furan, a versatile feedstock for the production of polymer materials, was successfully synthesized via a new reaction pathway in which the hydroxyl group of 5-hydroxymethylfurfural was aminated via a Ritter reaction, followed by direct reductive amination of the aldehyde group [24]. The yield of intermediate N-acyl-5-aminomethylfurfural can reach up to 89.1 wt%. The optimized yield of 2,5-bis(aminomethyl)furan (BAMF) can reach 45.7 wt%. In addition, 5-HMF can also be directly aminated by a one-step method under the catalysis of ruthenium complexes [25], as shown in Scheme 9.1c4.

9.2.2 *Nitrogenous Biomass Found in Nature*

9.2.2.1 Chitin

Chitin, the second most abundant biomass in the world, is widely found in the shells of insects, crustaceans (such as shrimps, crabs, etc.), the bones of mollusks, and the cell walls of some algae and fungi in nature [26]. It possesses a similar chemical structure to cellulose, but has biologically nitrogen fixed in an amide functional form. As shown in Scheme 9.2a, chitin is an ideal resource for the preparation of organic nitrogen-containing chemicals, which, in general, start from highly energy-consuming ammonia synthesis [4, 27]. As the monomer of chitin, N-acetylglucosamine (GlcNAc) is a key building block for the production of organonitrogen chemicals from nitrogen-containing biomass, and can be used as a raw material for the synthesis of nitrogen-containing chemicals [28]. In addition, chitosan obtained via partial deacetylation of chitin is more suitable for useful bioapplications [29].

On the one hand, the complete deoxygenation of chitin or its monomeric GlcNAc is of great significance for the synthesis of commercial chemicals such as amines. There are also many related reports, which are summarized in Scheme 9.2a. However, reports of selective deoxygenation of GlcNAc are limited. The key challenge is that the amide group is easily detached from the main chain during deoxygenation. To address this issue, Liu et al. demonstrated a new catalytic method to produce bio-based amines from GlcNAc by selectively removing oxygen-containing groups via supported noble metal catalysts. The amino groups were retained by phosphoric acid-catalyzed protonation [27].

9.2.2.2 Amino Acid

Besides chitin, amino acids obtained from biomass fermentation or isolated from protein waste hydrolysates, are also an important class of starting materials for the synthesis of nitrogen-containing compounds [30]. The literature on the utilization and transformation of amino acids is extremely diverse. However, most of these studies have focused on the application of amino acids for nutritional, medical or physiological functions. A common feature of amino acids is the presence of carboxyl and amino functions, in fact, most (bio)chemical information focuses on

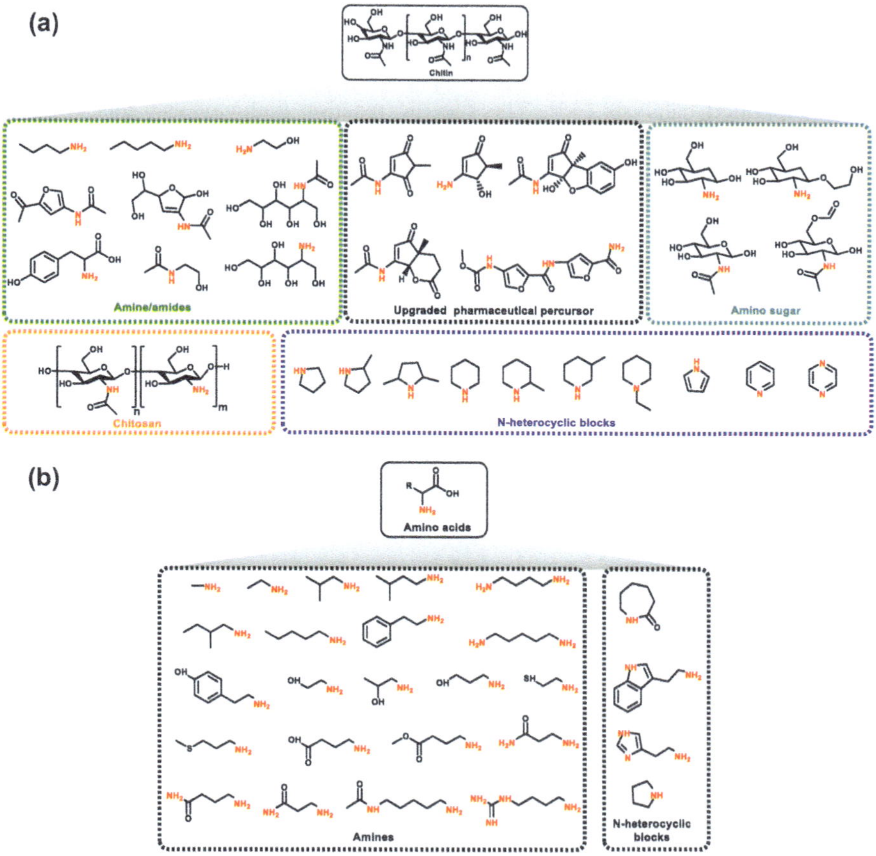

Scheme 9.2 Preparations of nitrogen-containing compounds derived from (**a**) chitin and (**b**) amino acids

the reactivity and transformation of these groups. Selective modification or elimination of carboxylic acid moieties on the α-carbon of amino acids provides direct access to a range of value-added nitrogen-containing chemicals [31]. Several representative examples of the production of nitrogen-containing compounds from biological amino acids are reviewed in Scheme 9.2b.

Claes and co-authors reported on a chemocatalytic, metal-free approach for decarboxylation of amino acids, thereby providing a direct access to primary amines [32]. They reported that various amino acids, including phenylalanine, glycine, alanine, valine, leucine, isoleucine and norleucine were converted to the corresponding primary amines in excellent yields (>98%). In addition, the authors found that in N,N-dimethylformamide, the acidity of the amino acid itself can catalyze the N-formylation of the amino group.

Among various amino acids, lysine is the most abundant amino acid industrially produced by fermentation of glucose. Due to its relatively low price and ideal $-NH_2$

and –COOH functional groups in the molecule, it is an ideal raw material for the preparation of valuable organonitrogen chemicals. Sebastian et al. converted L-lysine to caprolactam by one-pot hydrogen decomposition over bifunctional metal-supported catalysts [33]. The specific reaction pathway was that L-lysine was first cyclized and n-methylated to form intermediate α-dimethyl amino caprolactam, then $CN(CH_3)_2$ bond hydrogenation to obtain caprolactam. Furthermore, L-Lysine contains two amino groups and can be refined into diamines, which can be further used in the synthesis of bio-based nylons or as epoxy curing agents. Since the mechanism of lysine deoxygenation remains unclear, selective deoxygenation of lysine without changing the amino group is challenging. Lin's group used the combination of Ru/C and phosphoric acid to selectively deoxygenate lysine [34]. The conversion of L-lysine within 2 h was 100%, the overall yield of amine was 94%, and the selectivity to diamine was 51%.

9.3 Bio-based Nitrogen-Containing Epoxy Resin

9.3.1 Heat Resistant Bio-based Epoxy Resin

Heat resistance is a priority for epoxy resins in many applications, and the glass transition temperature of cross-linked polymers is a key indicator for determining uppermost operating temperature, which is mainly determined by the backbone stiffness and cross-link density of the network. There are two main strategies to improve the rigidity of the cross-linked network: The first is to introduce rigid groups such as aromatic rings or polyaromatic structures and aromatic heterocyclic structures into the epoxy resin molecular skeleton. As the stiffness of the molecular chain increases, the movement of the segments in the cured product system is slowed down, which increases the bulk density of the cured product, it is beneficial to improve the thermal performance of the cured product.

Qi et al. reported an a pyridazine-based epoxy precursor from guaiacol and succinic anhydride, and the twisted non-coplanar molecular structure of all aromatic rings endowed the polymer with excellent comprehensive properties such as high temperature resistance [15]. The synthetic route is shown in Scheme 9.1b1. Compared with diglycidyl ether of bisphenol A (DGEBA), the cured resin exhibits higher T_g (187 °C vs. 173 °C) and storage modulus, indicating that the aromatic structure makes the system exceptionally rigid. It is noteworthy that the presence of nitrogen can promote the formation of hydrogen bonds and enhance the intermolecular forces, thereby further improving the storage modulus of the system. Furthermore, the presence of tertiary amines in the network endows the material with excellent flame-retardant properties.

Some researchers have also introduced a biphenyl structure into the epoxy resin backbone to achieve the purpose of improving thermal properties. Savonnet et al. synthesized biphenyl-containing epoxy precursors and aromatic diamines using vanillin as a substrate, respectively [10]. The cured epoxy resin exhibited a glass transition temperature as high as 212 °C, which could satisfy most of the application requirements of epoxy resins. In addition, Liu et al. reported a bio-based epoxy resin derived from a rosin acid [35]. Due to the large hydrogenated phenanthrene ring structure, the cured resin possessed high glass transition temperature (T_g = 153.8 ° C), high storage modulus at room temperature (G' = 2.4 GPa) and good thermal stability.

In addition to the above methods, improving the crosslinking density by developing and synthesizing multifunctional epoxy resins containing three or more epoxy groups is another important way to improve the thermal properties of epoxy resins. Garrison's group synthesized trans-resveratrol trisepoxy (*I*) and trifunctional amine (*II*) from resveratrol and the cured system had T_g as high as 246 °C, approximately 70 °C superior than network based on DGEBA due to the high cross-link density promoted by the trifunctionality of monomers [36], the synthetic route is shown in Scheme 9.3a.

9.3.2 Intrinsically Flame-Retardant Bio-based Epoxy Resin

Nowadays, with the continuous development of production economy, epoxy resin can be used in almost all aspects of industrial production and human life, including electrical and electronic, household furniture, aerospace and other fields [37]. However, due to the high hydrocarbon content of the molecular skeleton structure, conventional epoxy resins are extremely flammable and release a lot of heat and smoke during combustion, which poses a high fire risk [38]. Therefore, the development of flame-retardant high-performance epoxy resins has become a research hotspot. Traditional halogenated flame retardants are gradually abandoned due to their toxicity and environmental pollution, and the research on flame retardants tends to be non-toxic, efficient and multifunctional. As shown in Scheme 9.3b, Qian's team prepared a furfural-derived flame retardant (*III*) containing nitrogen and phosphorus elements via the addition reaction between 9,10-dihydro-9-oxa-10-phosphaphenanthrene-10-oxide (DOPO) and furfural-derived Schiff bases [39]. Due to the synergistic flame retardancy of nitrogen and phosphorus elements, the resin with only 5 wt% compound *III* added has a limiting oxygen index (LOI) value of 35.7% and passed UL-94 V0 rating. In addition, He et al. prepared a lignin flame retardant (*IV*) by chemically grafting polyethyleneimine and polyphosphoric acid through a mild two-step method [40], the reaction process is shown in Scheme 9.3c. When the added amount of lignin-based flame retardant in the resin reaches 8 wt%, the LOI value of the resin is as high as 31.4% and the resin passes the V0 level of the UL-94 test. It suggested that the lignin-based flame retardants with nitrogen and phosphorus modification can effectively improve the flame retardancy of the resin and reduce smoke emission.

Usually, the addition of functional additives to the resin matrix is accompanied by the weakening of other properties, so the design of epoxy resins with intrinsic flame retardant functions from the molecular structure has gradually become a research

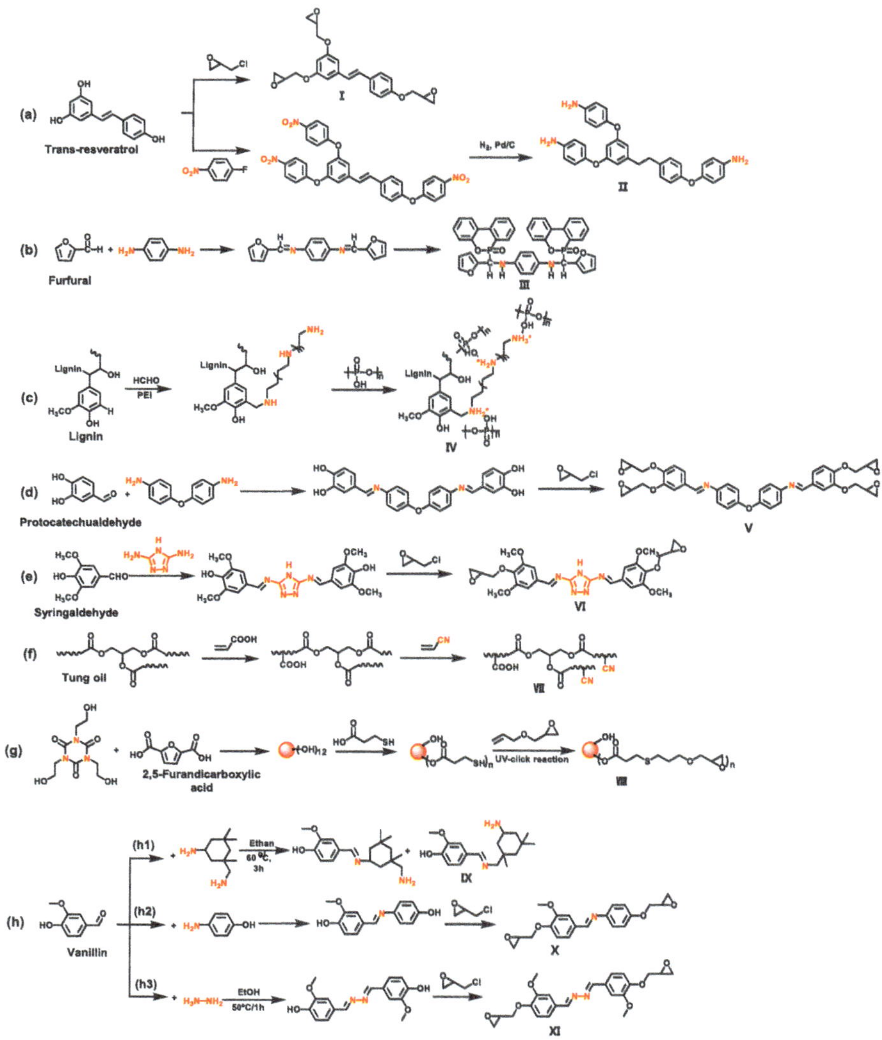

Scheme 9.3 Synthesis route of multifunctional bio-based nitrogen-containing epoxy resin from (**a**) trans-reveratrol, (**b**) furfural, (**c**) lignin, (**d**) protocatechualdehyde, (**e**) syringaldehyde, (**f**) toung oil, (**g**) 2,5-furandicarboxylic acid and (**h**) vanillin

hotspot. In recent years, studies have found that Schiff bases have an interesting antagonistic response to an elevated temperature/fire scenario [38, 41, 42]. Thus, the introduction of Schiff base structures into polymer networks is an effective strategy to prepare intrinsically flame-retardant resins. As shown in Scheme 9.3d, Zhao and co-authors reported a UL-94 V0, with an LOI of 32% for protocatechualdehyde-derived diimine-functionalized epoxy resin (*V*) cured with 4,4′-diaminodiphenyl methane (DDM) combined with high T_g (205 °C) [43]. The excellent comprehensive performance is attributed to the fact that the aromatic Schiff base structure can not only help to reduce molecular chain motion but also be self-crosslinked at high temperature to form a stable network structure during the combustion process [44].

However, most of the bio-based epoxy thermosets researches to date use either bio-based epoxy monomers or bio-based curing agents. In order to increase the bio-based content of the system, Hu's group prepared an epoxy monomer (*VI*) containing a Schiff base and a triazole structure through the reaction of syringaldehyde and 3,5-diamino-1,2,4-triazole [45], the synthetic route is shown in Scheme 9.3e. The resin which cured with a furfural-derived diamine combined high bio-based content, higher T_g and better fire-safety performance than DGEBA/DDM due to the excellent carbonization ability induced by the Schiff base and triazole structures.

9.3.3 Toughening of Bio-based Epoxy Resins

Existing high-performance epoxy resins limit their industrial applications due to their inherent brittleness and poor resistance to crack propagation [46]. To improve the toughness of conventional epoxy resins, numerous attempts were made for improving the brittleness and fracture energy of conventional epoxy resins by adding bio-based modifiers. However, it should be noted that the key of these methods is to achieve the purpose of toughening without reducing the tensile modulus, mechanical strength and thermal stability. The sacrificial bond is a kind of interaction that breaks before the covalent bond network under the action of external force and can effectively dissipate mechanical energy. Sacrificial bonds can not only dissipate energy, but also help to eliminate stress concentration and promote the orientation of polymer molecular chains, thereby significantly improving the strength and toughness of polymers [47]. As shown in Scheme 9.3f, Nie and co-workers prepared a renewable toughener (*VII*) by incorporating cyano groups into tung oil, which could form sacrificial bonds by the dipole-dipole interaction and hydrogen bonds [48]. The results suggested great improvements in terms of impact strength (76.7 kJ m^{-2}), elongation at break (25.0%), and toughness (9.6 MJ m^{-3}) compared to the pure DGEBA resin. These weak interactions are able to rupture before the covalent bond, which dissipates a large amount of energy, while the length of the hidden flexible chain is released, leading to higher elongation before breaking.

In addition to the above-mentioned introduction of sacrificial bonds, it is also possible to use modifiers to form a phase-separated structure with the matrix for toughening. The critical factor influencing the degree of improvement of strength and toughness is the compatibility of the modified substance with the epoxy resin and the regulation of the phase structure. Zhang's team reported a serious of hyperbranched epoxy resins (*VII*) by an esterification and a thiol-ene click reaction based on tris(2-hydroxyethyl)isocyanurate and bio-based 2,5-furandicarboxylic acid [49], as shown in Scheme 9.3g. The ellipsoidal topological structure of *VII* can disentangle the molecular chain of DGEBA, penetrate each other and reduce chain aggregation thus achieving a homogeneous structure and good compatibility.

Compared with the pure resin, the impact strength of composite with 12 wt% addition was nearly doubled, and the mechanical strength was also improved, realizing the in situ reinforcing and toughening of the epoxy resin.

9.3.4 Biodegradable and Recycled Bio-based Epoxy Resin

Because of irreversible permanently cross-linked networks, traditional epoxy resins cannot be degraded, reshaped or reprocessed after reaching the end of serve life. The current treatment methods are mainly through typical landfills or incineration, which undoubtedly results serious environmental pollution and waste of resources [50]. Therefore, in order to realize a circular economy, there is an urgent requirement to develop an efficient method to recycle thermosets with low-energy consumption. To date, dynamic covalent adaptable networks (CANs) have attracted extensive attention by incorporating labile bonds during the cross-linking process including esters [51], disulfide bonds [52], Diels-Alder (D-A) /retro-D-A reaction [53], siloxane [54], Schiff base [55, 56], carbamate [57], vinylogous urethanes/ureas [58], hindered urea bond [59], 1,2,3-triazolium [60], diketoenamine [61] and hemiaminals [62]. Under external stimuli such as heat, light, or pH, the reversible covalent linkages in epoxy networks can be rearranged by exchange reactions, corresponding to the decomposition of the epoxy networks. For example, Liu et al. prepared a bio-based imine-containing epoxy hardener (*IX*) from vanillin and isophorone diamine (IPDA), as shown in Scheme 9.3h1, the cured bisphenol A-type epoxy resin (epoxy/*IX*) exhibited high performance and closed loop recyclability in the solution of IPDA and ethanol at 60 °C [63]. Abu-Omar' team synthesized a vanillin-derived recyclable and malleable epoxy resin (*X*) bearing aromatic imine bonds [64], as shown in Scheme 9.3h2. The reported epoxy resin could be easily degraded and recycled by acid-aided hydrolysis of the imine linkages. Besides, Zeng et al. used epoxidized soybean oil (ESO) to react with a Schiff base-containing amine hardener for making fiber-reinforced polymer composites [52]. Although the resin can be degraded in acid solution to achieve non-destructive recovery of carbon fiber, since the network contains flexible molecular chains, the above resins exhibit modest thermodynamic properties, which limits the application in the fields where high-performance materials are required. Ma and co-authors prepared a Schiff base-embedded epoxy resin cross-linking with DDM [65]. On account of the stabilization from the conjugated benzene rings, the Schiff base epoxy networks exhibited combining superior temperature and acidity-controlled degradability, and excellent thermodynamic properties.

Besides Schiff base, ester bond is also one of the most commonly used dynamic covalent bonds. However, catalysts with CAN matrices limit their application because of their toxicity, incompatibility, and corrosiveness [66]. Chen et al. reported a turpentine-derived epoxidized menthane diamine (EMDA) with tertiary amines [67]. After cured by adipic acid (AA), the resin EMDA-AA shows excellent reprocessing, shape memory and self-adhesive properties under the autocatalysis of inherent tertiary amines. Besides, Ma et al. developed an amide bond-containing amino acid PDA-MAH from bio-based 1,5-pentanediamine (PDA) and maleic anhydride (MAH) as a new type of latent curing agent for epoxy resin [68]. The ester bonds in the network endow the system with favorable degradability under alkaline conditions especially in 50 °C 1 M NaOH acetone/methyl ethyl ketone (5/5, v/v) solution. Not only chemical degradation, but also biodegradation has been widely concerned by researchers. Takada et al. employed a biobased curing reagent poly(L-lysine) (PL) to crosslink with polyglycerol polyglycidyl ether (PGPE) preparing a bio-based epoxy system with aerobic biodegradability [69].

9.3.5 Bio-based Epoxy Resins with Other Functions

With the deepening of research on nitrogen-containing chemicals, bio-based epoxy resins with special properties have been gradually developed, such as antibacterial and aging resistance properties.

Ma et al. synthesized a vanillin-based epoxy resin (*XI*) containing Schiff base bonds, and the Schiff base in the resin possessed excellent antibacterial effect on Gram-negative bacteria (E. coli) with a killing rate of 91% [55]. Further, as shown in Scheme 9.3h3, they designed another epoxy system based on the dihydrazone bond from vanillin, and the antibacterial rate was further improved to 95.8% due to the increase of the Schiff base structure [65]. For other bacteria commonly found in life, S. aureus, Hu's group reported a Schiff base-embedded syringaldehyde-based epoxy resin exhibiting a clear and large inhibition zone with a diameter of 16 mm against S. aureus suggesting excellent antibacterial activities [41].

Anda et al. prepared a cyclic diamine-limonene (DA-LIM) and an aromatic diamine-allyl eugenol (DA-AE), from D-limonene and allyl-eugenol via thiol-ene addition [70]. Recently, employing the above diamines and hexamethylene diamine (HMDA), which also can be derived from bio-based sources [71], as curing agent, they prepared three aging resistance matrices (i.e., RE/HMDA, RE/DA-LIM and RE/DA-AE) which possessed remarkable fatigue crack resistance properties [72]. The result suggested that three fully bio-based systems showed a good compromise between toughness and creep resistance a better resistance to crack propagation under dynamic loading compared to DGEBA/HMDA. Besides, Illy et al. employed a bio-based chitosan as a crosslinking agent with waterborne DGEBA which facilitated the processing [73].

In addition, extensive cases with interesting functions have been reported based on the D-A reaction. Lu et al. reported a modified epoxidized soybean oil (ESOM) bearing epoxy and maleimide group and a bio-based curing agent (CFMA) derived from furfural, cardanol and 1,8-p-menthane-diamine [74]. The dual-cure network form by epoxy-amine reaction and D-A reaction between furan ring and maleimide could significantly enhance the toughness and thermal property of epoxy thermosets, besides, the stable and outstanding photothermal conversion property. Based on epoxy-amine ring opening and D-A reactions as the supporting frame, Liu et al.

prepared a novel bio-based recyclable phase change materials using bio-based magnolol and epichlorohydrin, bio-based 2-furanmethanamine (FA) and C36-alkylenediamines bismaleimide (MA). The D-A and retro-D-A reactions provides recyclability of the crosslinked network. The 1-octadecanethiol and beeswax provide phase change latent heat up to 119.1 J g^{-1} [75].

9.4 Bio-based Nitrogen-Containing Benzoxazine Resin

As a new type of thermosetting resin, benzoxazine is usually prepared by Mannich polymerization with phenols, primary amines and aldehydes in a molar ratio of 1: 2:1, and then under certain conditions, benzoxazine can undergo ring-opening polymerization to form polybenzoxazine [76]. Therefore, benzoxazine resins can be designed by selecting raw materials containing specific functional groups to satisfy practicality requirements. What's more, due to the excellent comprehensive properties, such as good dimensional stability, chemical resistance and electrical resistance, low water absorption, high char yield and no need of additional catalysts, benzoxazine resins have been rapidly recognized by not only from academia but also from industry and exhibit great potential in the fields of communications, automobiles, and aerospace [77].

9.4.1 Bio-based Benzoxazines with High Thermal Property

Thermal properties of bio-based benzoxazine resins can be improved by introducing precursors with aromatic to increase structural rigidity [78] or containing functional groups that can provide additional crosslinking sites [79]. Lochab's team reported the utilization of a rigid carbohydrate-based isomannide diamine (ima) and cardanol/guaiacol as raw materials for the synthesis of fully naturally sourced bifunctional benzoxazines (*I–III*) [80], the synthetic route is shown in Scheme 9.4a. Both the synthetic benzoxazines revealed a much higher char yield and reasonably close 10% weight loss temperature ($T_{d10\%}$) value to the petro-based benzoxazines. As shown in Scheme 9.4b, Dubois and co-workers synthesized a bio-based benzoxazine monomer(*IV*) with allyl functional group using chavicol as phenol source [81]. The additional allylic crosslinks increase the crosslink density of the network, interconnecting all molecular moieties and enhancing their thermal stability. In addition, as shown in Scheme 9.4c, Ishida et al. prepared a benzoxazine (*V*) utilizing furfurylamine and coumarin as raw materials [82]. The transesterification of the coumarin moiety at high temperature and the substitution of the furan ring during curing synergistically increase the crosslink density and contribute to high thermal stability which have 5% weight loss temperature ($T_{d5\%}$) and char yield at 800 °C (C_{y800}) value of 420 °C and 60%, respectively.

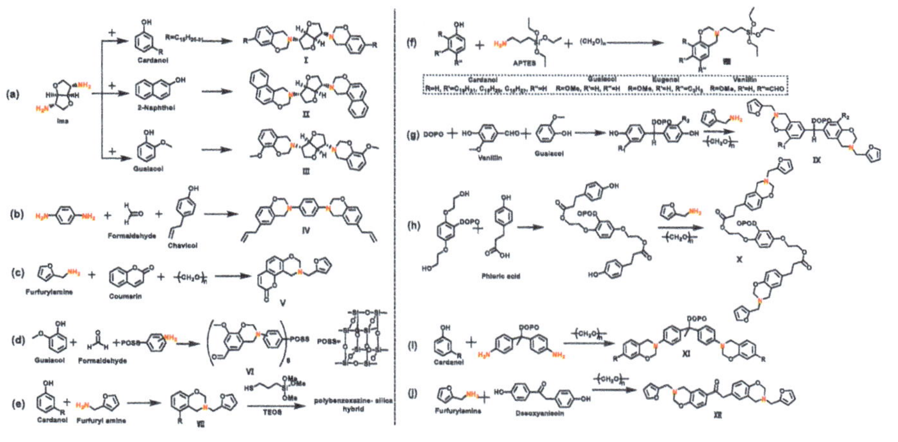

Scheme 9.4 Synthesis route of multifunctional bio-based benzoxazine resin from (**a**) carbohydrate-based isomannide diamine (ima), (**b**) chavicol, (**c**) coumarin, (**d**) guaiacol, (**e**) cardanol, (**f**) cardanol/guaiacol/eugenol/vanillin, (**g, h, i**) DOPO-containing benzoxazine resin and (**j**) desoxyanisoin

Alternatively, introducing polyhedral oligomeric silsesquioxanes (POSS) into polymer system is an effective method to improve thermal properties of polymers. For example, as shown in Scheme 9.4d, Thirukumaran et al. prepared polybenzoxazine-tethered POSS nanocomposites (*VI*) by the reaction of bio-phenols: eugenol, guaiacol, vanillin with POSS-octaamine [83]. The cured resin showed good thermal stability and excellent dielectric properties. Later, Devaraju et al. incorporate 3-mercaptopropyltrimethoxysilane into cardanol benzoxazine (*VII*) by thermally initiated thiol-ene click reaction [84], the synthetic route is shown in Scheme 9.4e. The results showed that the thermal decomposition temperature and C_{y800} of polybenzoxazine-silica hybrid increased with the increase in the weight content of silica. Additionally, 3-aminopropyltriethoxysilane (APTES) has been explored as a co-reactant that undergoes Mannich-type condensation with phenols to generate silane-functionalized benzoxazine monomers. Recently, as shown in Scheme 9.4f, APTES was employed as amine source to prepare a polybenzoxazine-silica network (*VII*) with cardanol, guaiacol, eugenol, and vanillin, respectively [85]. Silane functional groups are expected to copolymerize with benzoxazine rings to form thermally stable siloxane bonds, thereby improving thermal stability.

9.4.2 Bio-based Benzoxazines with Flame Retardancy

In order to improve the flame retardant properties, the introduction of flame retardants into polymers through physical blending or covalent bonds is currently a widely adopted strategy [86]. However, as the flame retardant added by physical blending cannot react with the epoxy resin, it is likely to cause the flame-retardant performance to decline because of precipitation during long-term service life. In addition, since most of the flame retardants are rigid macromolecules, they are easily unevenly dispersed or difficult to dissolve, which in turn leads to insignificant flame-retardant effects. Therefore, in recent years, many researchers have turned their attention to reactive flame retardants. 9,10-dihydro-9-oxa-10-phosphaphenanthrene-10-oxide (DOPO) is a very common organophosphorus flame retardant [87]. It has excellent heat resistance and flame-retardant properties due to the special biphenyl and phosphaphenanthrene structure [88]. As shown in Scheme 9.4g, Juang et al. synthesized a benzoxazine (*IX*) from vanillin-derived DOPO-containing bisphenols and furfurylamine [89]. The cured resin exhibits excellent flame-retardant properties combined with top thermal properties, the initial decomposition temperature and C_{y800} are 371 °C and 63%, respectively.

Koschek et al. reacted the chain-extended DOPO with phloric acid to obtain a bisphenol compound, which was used as a phenol source to react with furfurylamine and paraformaldehyde to obtain benzoxazine monomers (*X*) [90], the synthetic route is shown in Scheme 9.4h. The cured resin passes the V0 level of the UL-94 test.

As shown in Scheme 9.4i, Hu's team synthesized a bio-based phosphorus-containing benzoxazine monomer (*XI*) through the reaction between cardanol and DOPO-based diamine [91]. After copolymerization with epoxy, boron-doped graphene (BG) nanosheets were added to further improve the flame retardancy. The results showed that when combining 8 wt% *XI* and 2 wt% BG, the sample successfully passes UL-94 V0 rating tests under the synergistic effect of *XI* and BG. Meanwhile, the flexible aliphatic chain in cardanol improved the impact strength of the composite.

However, phosphorus-containing polymers are prone absorb moisture and hydrolyze, which has negative effects on the application of electronic field. Therefore, the synthesis of phosphorus-free flame-retardant benzoxazine thermosets is an interesting yet challenging task. Hu et al. synthesized an anti-flammable benzoxazine monomer (*XII*) derived from furfurylamine and desoxyanisoin [86], the synthetic route is shown in Scheme 9.8j. Benefitting from the superior charring ability of the desoxyanisoin structure and furan ring, the cured resin exhibited a high LOI value (40.0%), a UL-94 V0 rating, and an extremely low heat release capacity (35.5 J/(g K)).

As shown in Scheme 9.5a, Yan et al. prepared three full bio-based benzoxazines (*I–III*) using phloretic acid, p-coumaric acid, and ferulic acid as a phenol source and furfurylamine as an amine source [92]. The C_{y800} of the cured resin is up to 56%, and micro-scale combustion calorimetry (MCC) results showed that the resin exhibited excellent flame retardancy.

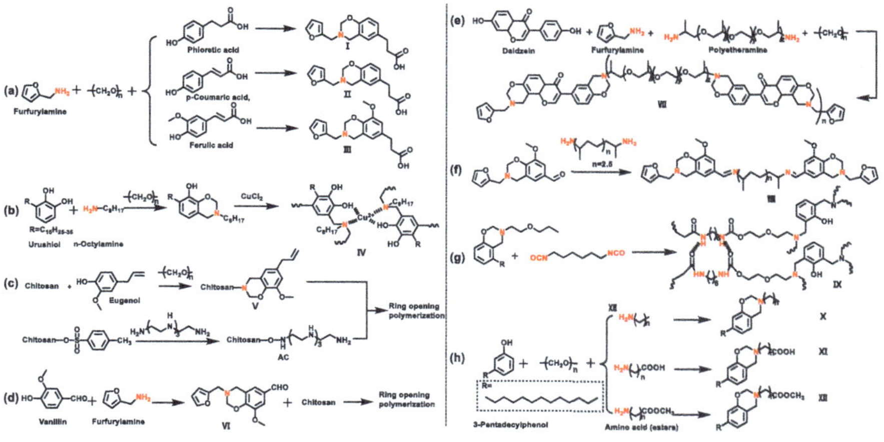

Scheme 9.5 Synthesis route of multifunctional bio-based benzoxazine resin from (**a**) phloretic acid/p-coumaric acid/ ferulic acid, (**b**) urushiol, (**c**) chitosan, (**d**)(**f**) vanillin, (**e**) daidzein, (**g**) cardanol and (**h**) 3-pentadecylphenol

9.4.3 Bio-based Benzoxazines with Antibacterial and Algaecidal Properties

Marine biofouling is a worldwide problem in marine systems which is mainly caused by the adhesion of barnacles, macroalgae and microbial slime. In the case of ships, the detrimental effects of this biological settlement are well known. According to the mainstream view of the biofouling mechanism, preventing bacterial surface colonization is the key to solving biofouling inhabitation [93]. As shown in Scheme 9.5b, Xu et al. reported a bio-based benzoxazine copper polymer (*IV*) using urushiol, n-octylamine and paraformaldehyde as raw materials [94]. Metal ions (i.e., Cu^{2+}) have antibacterial and algae-inhibiting properties, and the controllable release of metal ions is beneficial to improve the electrostatic antifouling effect. The antibacterial properties of *IV* against typical Gram-negative bacteria (*E. coli*) and Gram-positive bacteria (*S. aureus*) are almost 100%, and its ability to inhibit the activity of algae is more than 99%.

Besides, Kim et al. prepared bio-films from chitosan based polybenzoxazine (*V*) and amino cellulose (AC) [95], the synthetic route is shown in Scheme 9.5c. These biofilms had a better inhibitory effect on S. aureus and Candida albicans (C. albicans) confirming that the films can effectively prevent biofilm-related infections and avoid the formation of fungal infections.

Yadav et al. grafted a fully bio-based benzoxazine monomer (*VI*) onto chitosan (CS) obtaining films combined with good mechanical properties and improved antibacterial performance [96], as shown in Scheme 9.5d. Expansion of CS galleries and leaching out of phenolic species from bio-based polymer films contributed to an improved antibacterial activity against S. aureus.

As shown in Scheme 9.5e, Lu and co-workers synthesized a variety of mainchain-type benzoxazine polymers (*VII*) from the renewable daidzein, furfurylamine, polyetheramine, and paraformaldehyde [97]. These cured MCBP films exhibited satisfied antibacterial and algicidal properties due to the presence of daidzein and furan units. The highest elimination ratio of these polybenzoxazine films on E. coli and B. cereus. Was 99.9% and 70.6%, respectively. In addition, these polybenzoxazine films also exhibited a toxic effect on both algae like Phaeodactylum tricornutum and Navicula sp. and had good algae resistance.

9.4.4 Bio-based Benzoxazine Resins with Other Functions

With the unremitting exploration of benzoxazine resins, more and more functions have been endowed, such as self-healing, shape memory, and phase-change energy storage. Bifunctional benzoxazines (*VII*) were prepared by introducing Schiff bases into benzoxazine monome synthesized from vanillin, furfurylamine and paraformaldehyde [98], the synthetic route is shown in Scheme 9.5f. The cured resin exhibits excellent shape memory properties, with a shape recovery rate of 98% within 40 s. Recently, as shown in cardanol route (Scheme 9.5g), Devaraju et al. reported a sustainable poly(benzoxazine-co-urethane) (*IX*) that can self-heal under a mild external force due to supramolecular hydrogen-bonding interactions [99]. After the repair process repeating for 3–4 trials, the material still showed good self-healing behavior.

Organic phase change materials have received extensive attention as an effective heat storage material, but their sustainability is restricted due to the non-renewable, poor stability, and insecurity of the materials. Liu's team developed a series of benzoxazine bio-based phase change materials (*X–XII*) from renewable 3-pentadecylphenol (a cardanol derivative), amino acids and amino acid esters [100], the synthetic route is shown in Scheme 9.5h. The prepared benzoxazines exhibited tunable latent heats (48.78–147.02 J g^{-1}) by selecting raw materials and displayed satisfactory recycle performance after 100 times repeated use. The introduction of oxazine rings solves the problems of low durability, flammability and poor thermal stability of traditional phase change materials.

In addition, polybenzoxazine is also an inherent hydrophobicity substance, and have been studied as water blocking materials for oil-water separation in academia. Xin and co-authors synthesized a superhydrophobic/superoleophilic polybenzoxazine/SiO_2 films (PC-a/SiO_2) [101]. PC-a/SiO_2 fabric exhibited excellent separation performance for several types of heavy oil/water under gravity drive, the oil flux was around 6500–9500 L m^{-2} h^{-1}, and the separation efficiency was above 90%, with good recyclability.

9.5 Other Bio-based Nitrogen-Containing Thermosetting Resins

9.5.1 Bio-based Phthalonitrile Resin

Phthalonitrile resin is a general term for a class of high-performance thermosetting resins end-capped with a phthalonitrile structure. A cross-linked network structure containing aromatic heterocycles is formed by addition polymerization of the cyano group in the phthalonitrile monomer. The cured resin can form a cross-linked network structure [102], and the main structures include: triazine ring, phthalocyanine ring, isoindole ring, dehydrophthalocyanine ring. Compared with epoxy resins and polybenzoxazines, phthalonitrile resins exhibits better properties [103], including: extremely high T_g, excellent thermal and oxidative stability, good mechanical performance and low water absorption. These properties make them useful as carbon precursors, composite matrices, and binders. With the development of bio-based materials research in recent years, the research on bio-based phthalonitrile resin has also made a lot of progress.

As the resveratrol routes shown in Scheme 9.6a, Laskoski et al. used renewable bio-based compounds resveratrol and dihydroresveratrol as raw materials to synthesize two bio-based phthalonitrile monomers (*I* and *II*) [104]. The prepared monomer exhibited good rheological viscosity and is suitable for resin transfer molding and resin infusion molding. The cured resin possessed excellent thermal stability and a high the initial decomposition temperature (above 500 °C). Due to the high cross-linking density, the material has a very low water absorption rate (1 ± 0.2%), and no glass transition temperature was found in the test temperature range. After the

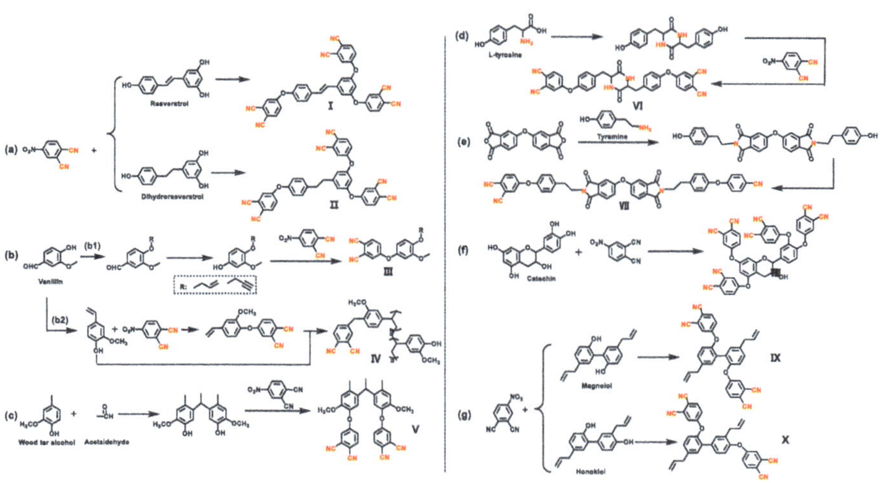

Scheme 9.6 Synthesis route of bio-based phthalonitrile resin from (**a**) resveratrols, (**b**) vanillin, (**c**) wood tar alcohol, (**d**) L-tyrosine, (**e**) tyramine, (**f**) catechin and (**g**) magnolol/honokiol

dielectric property test, the dielectric constant of the resin is close to 3.0. These results indicate that resveratrol-derived phthalonitrile resins show great application prospects in marine and aerospace fields.

Vanillin has been a renewable monomer of great interest in bio-based compounds. Zhao's team synthesized a vanillin-based phthalonitrile precursor (*III*) with good processing properties using a three-step method [105], the structure is shown in Scheme 9.6b1. First, vanillin was functionalized by allyl or propargyl, then the aldehyde group was oxidized to phenol, and finally the vanillin-based phthalonitrile precursor was synthesized by nucleophilic substitution with 4-nitrophthalonitrile. Both monomers possessed excellent processing characteristics such as low viscosity, low melting temperature and a wide processing window.

Besides, Fang et al. designed and synthesized phthalocyanine monomer (*IV*) from vanillin [106]. The synthetic route is shown in Scheme 9.6b2. Firstly, the aldehyde group of vanillin was converted to a vinyl group via the witting reaction, and then treating the alkylated vanillin with nitrophthalonitrile. Secondly, the abstained products were thermally polymerized with different molar ratios to obtain a series of vanillin-based phthalocyanine resins. The resulted resin possessed a char yield of up to 76% and a T_g over of 400 °C.

In addition, in the wood tar alcohol route (Scheme 9.6c), Yang's team employed the lignin derivative wood tar alcohol and acetaldehyde to generate an intermediate product in hydrochloric acid solution, and then the intermediate product and 4-nitrophthalonitrile were subjected to nucleophilic substitution reaction to generate bio-based phthalonitrile resin (*V*), with a biomass content of 54% [107]. The results show that BPN has a high glass transition temperature (T_g = 400 °C) and good thermal stability, with a mass retention rate of 95% in nitrogen and air at around 420 °C.

In addition to the common use of lignin-derived bio-based compounds in bio-based materials, amino acids were also selected to prepare bio-based phthalonitrile resins. L-Tyrosine and monomers derived from protein building blocks. In the L-tyrosine route (Scheme 9.6d), Yang's group reported a bio-based phthalonitrile monomer (*VI*) derived from L-tyrosine [108]. The results showed that the resin has higher glass transition temperature (T_g = 455 °C) and better thermal stability ($T_{d5\%}$ = 450 °C) than bisphenol A phthalonitrile resin.

Recently, as shown in the tyramine route (Scheme 9.6e), Yang et al. synthesized a new type of phthalonitrile-based resin (*VII*) with bioamine tyramine as raw material, and its melting point was much lower than that of bio-based phthalonitrile resin in hydrogen bond system [109]. Compared with dimethyl nitrile, the resin has higher glass transition temperature (T_g = 480 °C) and comparable thermal properties ($T_{d5\%}$ = 505 °C, C_{y800} = 72%). These results demonstrate the potential of tyramide as a building block for constructing high-performance polymers.

The author's research group has also done some exploratory research on the synthesis of bio-based phthalonitrile. For example, the reaction of 4-nitrophthalonitrile with catechin [110], and the synthetic route is shown in Scheme 9.6f. Since the structure contains an adjustable number of phenolic hydroxyl groups, the phthalonitrile precursor (*VII*) can realize self-curing reaction without adding a curing agent, avoiding the defects caused by the volatilization or decomposition of the small-molecule curing agent at high temperature. Under the same curing condition, the catechin-containing phthalonitrile resin possessed better thermal properties than the petroleum-based bisphenol A phthalonitrile resin. Recently, the author's group also reported two phthalonitrile resins (*IX* and *X*) preparing from magnolol and honokiol [102], and the synthetic route as shown in Scheme 9.6g. Self-curing reactions can be carried out with allyl groups without sacrificing the thermal stabilities of the thermoset resins. The resulted resin possessed an initial decomposition temperature of up to 527 °C and a T_g over of 500 °C, which were superior to those of petroleum-based counterpart.

9.5.2 Bio-based Polyurethane Resin

Polyurethanes (PUs) are one of the most versatile polymers and have been used in a wide variety of applications such as leather, coatings, foams and inks. PUs is a general term for a class of polymer compounds containing urethane groups (-NH-COO-) in the main chain. Usually, it is formed by the polyaddition reaction of isocyanate (R-N=C=O) and polyol (R'-OH), and it is a typical block copolymer [111]. The simplest formula for a PU is linear. However, traditional PU products usually contain large amounts of organic solvents, which are harmful to human health and the environment [112]. Growing concerns about emissions of volatile organic compounds and hazardous air pollutants have led to strong interest in the development of waterborne polyurethane product.

Karak et al. modified tannic acid-based waterborne hyperbranched polyurethane with different mass fractions of glycerol-based epoxy resins [113], and the reaction mechanism is shown in the Scheme 9.7. The results showed that the when 30 wt% epoxy resin was added to the thermosetting resin, the modified polyurethane possessed significantly improved mechanical properties such as tensile strength (~3.4 times), scratch hardness (~2 times), impact resistance (~1.3 times) and toughness (~1.7 times). Furthermore, the modification did not adversely affect the inherent biodegradability of the original polymer system.

Waterborne polyurethane is environmentally friendly and exhibits excellent properties such as tunable mechanical properties and adhesion to various substrates [114]. Typically, polyurethanes formed by the step-growth polymerization of diisocyanates and polyols are hydrophobic. Therefore, in order to disperse the polymer in water, it is necessary to use an emulsifier. Zhang et al. synthesized a fully bio-based internal emulsifier from epoxidized soybean oil and glutaric acid by a solvent-free and catalyst-free method [115]. The hydroxyl group of the obtained emulsifier is employed as the cross-linking agent, and the carboxylic acid is used as the ion block. After that, using the emulsifier and typical bio-based and petroleum-based polyols as raw materials, a series of anionic bio-based aqueous polyurethane

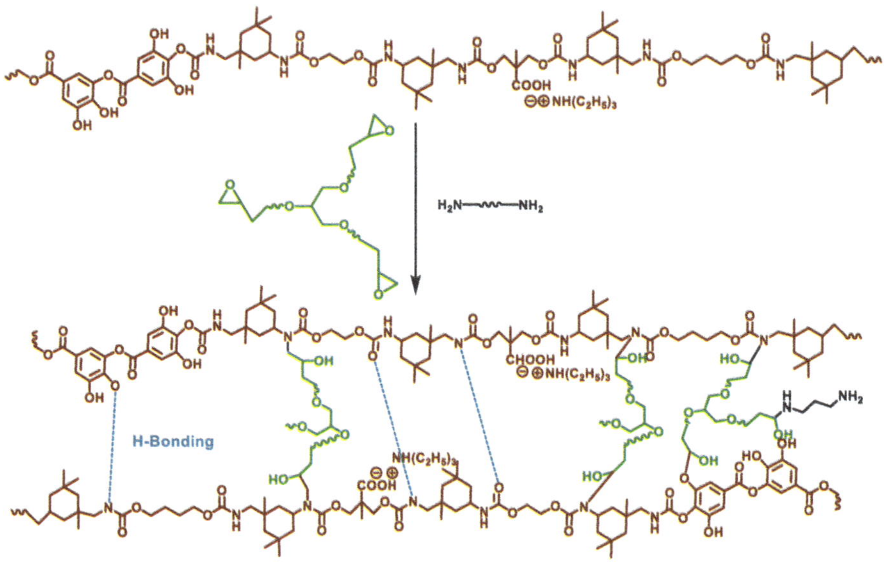

Scheme 9.7 Synthesis route of bio-based hyperbranched polyurethane from tannic acid

dispersions were prepared, the castor oil route is shown in Scheme 9.8a. The bio-based content of the obtained polyurethane film reached 74%, and it possessed superior tensile strength, toughness and thermal stability than those of the solvent-based polyurethane film without the emulsifier.

Synthesis of isocyanate, which is one of the raw materials of polyurethane, requires the use of toxic phosgene [116]. Thus, the reaction between cyclic carbonates and amines to form non-isocyanate polyhydroxyurethanes is receiving attention as well as becoming an important alternative to current industrial synthetic routes. Cramail et al. succeeded in the catalytic oxidation of a series of bioderived alcohols to nitriles, which were then reduced to primary amines [117]. The obtained primary amines were polymerized with fatty acid-based bicyclic carbonates to design fully bio-based poly(hydroxyurethanes), the bioderived alcohol route is shown in Scheme 9.8b. In addition, using sebacic acid and 1,2-carbonated glycerol as raw materials, Jacobsen and co-workers synthesized sebacic acid bicyclic carbonate catalyzed by lipase in a solvent-free reaction system [118]. These cyclic carbonates can be employed subsequently as NIPU monomers.

9.5.3 *Bio-based Cyanate Ester Resin*

Cyanate ester resins are widely used in many fields such as microelectronics, aerospace and coatings due to their excellent processing properties, dielectric properties, mechanical properties and thermal properties, even surpassing epoxy resins,

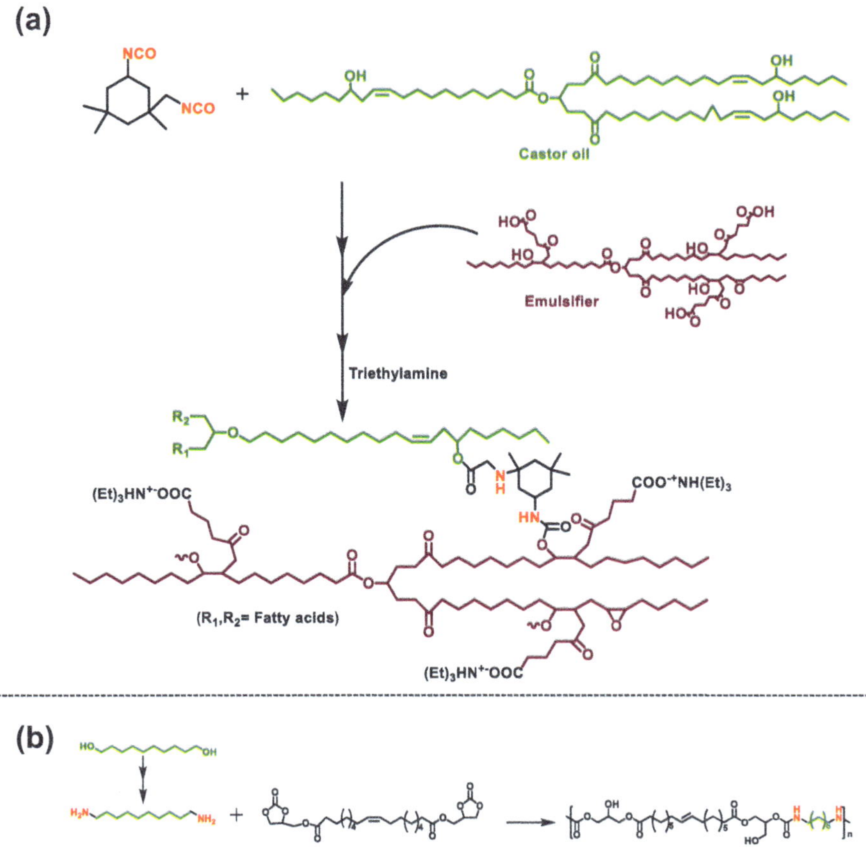

Scheme 9.8 Synthesis route of bio-based polyurethane resins from (**a**) epoxidized soybean oil, (**b**) bioderived alcohol

phenolic resins and polyester resins [119]. Bisphenol A cyanate, the most commercially produced polycyanurate thermoset, is obtained through the condensation reaction of bisphenol A (BPA) and cyanogen chloride in the presence of alkali [120]. In the presence of metal salt catalysts, cyanate monomers are trimerized to form a cyanurate ring structure at relatively high temperatures. Due to the environmental and health issues of BPA, bio-based cyanate resins have been developing continuously in the past decade. It has been reported that the presence of an ortho-methoxyl functional group relative to the phenolic hydroxyl group in BPA strongly attenuates or even eliminates the health threat [121]. Lignin-derived ethoxylated alkylphenols such as vanillin, guaiacol, and eugenol naturally possess this favorable design making them potential candidates for BPA. Among the candidate phenols, vanillin and 2-methoxy-4-methylphenol (cresol) exhibit great promise.

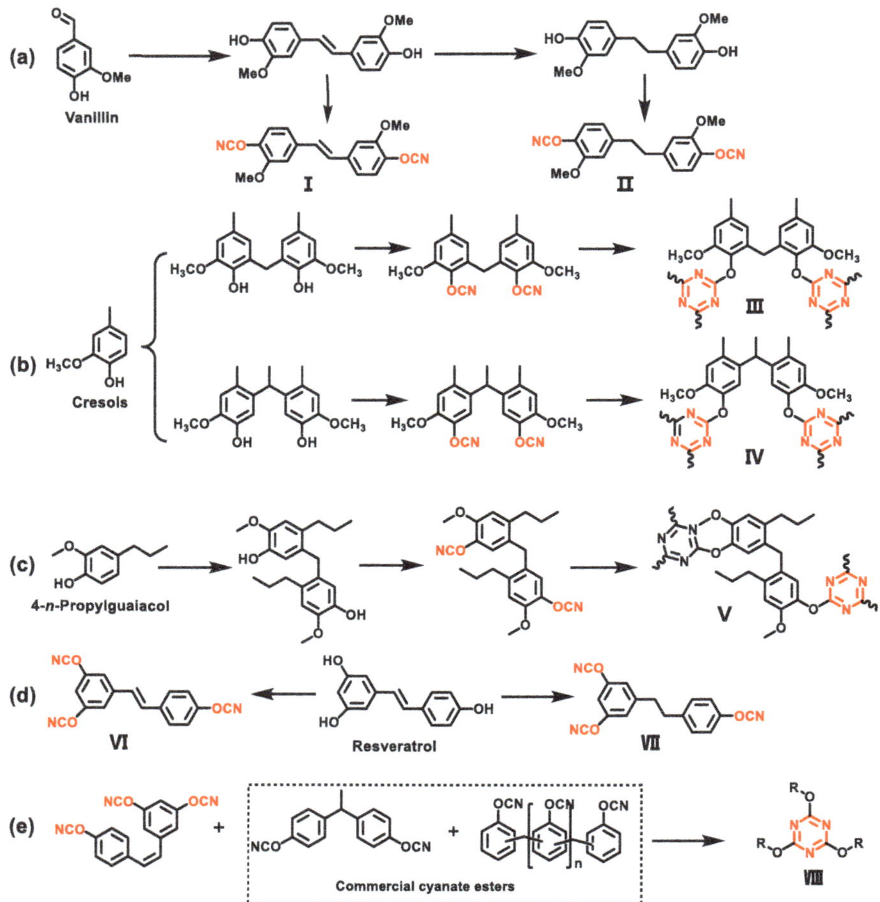

Scheme 9.9 Synthesis route of bio-based cyanate ester resin from (**a**) vanillin, (**b**) cresols, (**c**) guaiacol derives and (**d**) (**e**) resveratrols

Harvey et al. synthesized two cyanate resins (*I* and *II*) from vanillin [122]. In the vanillin route (Scheme 9.9a), firstly, the bisphenol precursor was prepared by McMurry coupling reaction with vanillin and $TiCl_4$/Mg. Secondly, under the catalysis of PtO_2, the stilbene was reduced to obtain saturated bisphenol. The cyanate esters were then prepared by conventional methods. The completely cured resin possessed a T_g of 202 °C, initial decomposition temperature of 335 °C and decomposed to form isocyanates and phenols.

Cresols can be produced from lignin, or easily obtained from vanillin by catalytic hydrogenation. In the cresol route (Scheme 9.9b), Harvey's team also selectively synthesized bisphenols linked through the carbon meta to the hydroxyl group by the condensation of cresols with formaldehyde, acetaldehyde, and propionaldehyde, respectively [123]. In addition, ortho-coupled bisphenol was also prepared. Then bisphenol reacts with cyanogen bromide to prepare a series of bis(cyanate) esters (*III* and *IV*). The T_g of the cured resin was 181–257 °C, the $T_{d5\%}$ in a nitrogen atmosphere was 317–360 °C, and the char yield at 600 °C is 27–35%.

In addition to the above bio-based chemicals, as shown in the guaiacol route (Scheme 9.9c), Sels et al. prepared a carbon atom-linked bisphenol precursor by acid-catalyzed condensation with formaldehyde using 4-n-propylguaiacol as a raw material [121]. The corresponding bis(cyanate ester) was synthesized via the reaction of the bisphenol with cyanogen bromide. The cured resin (*V*) exhibited a T_g of 193 °C, $T_{d5\%}$ of 389 °C and a water absorption of only 1.18% after soaking in 85 °C water for 4 days.

Likewise, resveratrol is also a good choice for the preparation of cyanate esters. Davis and co-workers synthesized two tricyanate esters (*VI* and *VII*) from the natural product resveratrol (Scheme 9.9d) [124]. The thermal decomposition temperature, glass transition temperature and char yield at 600 °C of the cured resins can reach up to 412 °C, 340 °C and 71%, respectively. The excellent comprehensive performances contributed to the stilbene backbone of the resin as well as the high crosslink. However, the above precursor possesses a high melting point of 165 °C which limits the processing window. Trans-resveratrol can be isomerized to cis-resveratrol (*VI*) under ultra-violet light, and then cyanated to form a new compound tricyanate with a melting point of only 77 °C [125]. To further improve processability, Harvey et al. blended *VI* with commercial cyanate esters [126], as shown in Scheme 9.9e. The cured resins (*VII*) exhibited higher T_g, improved thermal stability and lower moisture uptake than one-component resin.

9.6 Conclusions and Perspectives

The utilization of biomass for creating sustainable nitrogen-containing chemicals and high-performance thermosets has been an attractive and promising topic in recent years. Dozens of valuable nitrogen-containing products including amines, amino acids, nitriles, and N-heterocyclic compounds have been synthesized from lignin, chitin, protein and their derivatives. Taking advantage of the active nitrogen-containing groups, the bio-based epoxy resins, polybenzoxazine, phthalonitrile resin, polyurethane, cyanate ester and other resins possessed comparable performance to or even higher than their petroleum counterparts, and meanwhile various functionalities. Despite the huge potential, the development of bio-based nitrogenous thermosets is limited by the following issues.

First of all, the conversion of bio-based platform molecules into N-containing chemicals with high added value is still in its infancy in many aspects due to limited technology transfer from academic research to commercial market applications. This leads to the high cost of bio-based high-performance thermosetting resins, which is also a major bottleneck restricting the development of bio-based materials to commercialization. Secondly, the synthesis and usage of some bio-based NCCs still involve the disadvantages of petrochemical, low durability and toxicity which lead to the reduction of the bio-based content of materials. Additionally, despite the great number of bio-based materials proposed to date, there is very limited information about their environmental impact. The variety of NCCs is relatively scarce compared to petroleum-based products especially aromatic compounds, the key building blocks of high-performance resins. Finally, for non-degradable bio-based thermosets, it still exists the problem of environmental pollution and resource waste after reaching their service life, so it is urgent to expand the way of secondary utilization, such as being used as carbon materials in adsorption materials or supercapacitors. Therefore, the next generation bio-based thermosetting resins should implement the green chemistry principle during the whole synthesis process and ensure high performance or versatile functionality, to achieve the sustainability of resins at the source, process and end product.

In all of the above issues, sufficient attention is being given and great progress is being made. We have hopes for the near future in which sustainable materials will make up the majority of our industrial production and daily life.

Acknowledgements This work was supported by the National Natural Science Foundation of China (nos. 51873027, 52073038), the Fundamental Research Funds for the Central Universities (DUT20TD114, DUT22LAB605).

References

1. Maschmeyer T, Luque R, Selva M. Upgrading of marine (fish and crustaceans) biowaste for high added-value molecules and bio(nano)-materials. Chem Soc Rev. 2020;49:4527–63. https://doi.org/10.1039/c9cs00653b.
2. Jagadeesh Rajenahally V, Murugesan K, Alshammari Ahmad S, Neumann H, Pohl M-M, Radnik J, Beller M. MOF-derived cobalt nanoparticles catalyze a general synthesis of amines. Science. 2017;358(6361):326–32. https://doi.org/10.1126/science.aan6245.
3. Wu H, Li H, Fang Z. Hydrothermal amination of biomass to nitrogenous chemicals. Green Chem. 2021;23(18):6675–97. https://doi.org/10.1039/d1gc02505h.
4. Chen X, Song S, Li H, Gozaydin G, Yan N. Expanding the boundary of biorefinery: organonitrogen chemicals from biomass. Acc Chem Res. 2021;54(7):1711–22. https://doi.org/10.1021/acs.accounts.0c00842.
5. Scodeller I, Mansouri S, Morvan D, Muller E, de Oliveira VK, Wischert R, Jerome F. Synthesis of renewable meta-xylylenediamine from biomass-derived furfural. Angew Chem Int Ed Engl. 2018;57(33):10510–4. https://doi.org/10.1002/anie.201803828.
6. Ragauskas Arthur J, Williams Charlotte K, Davison Brian H, Britovsek G, Cairney J, Eckert Charles A, Frederick William J, Hallett Jason P, Leak David J, Liotta Charles L, Mielenz Jonathan R, Murphy R, Templer R, Tschaplinski T. The path forward for biofuels and biomaterials. Science. 2006;311(5760):484–9. https://doi.org/10.1126/science.1114736.
7. Pelckmans M, Renders T, Van de Vyver S, Sels BF. Bio-based amines through sustainable heterogeneous catalysis. Green Chem. 2017;19(22):5303–31. https://doi.org/10.1039/c7gc02299a.
8. Llevot A, Grau E, Carlotti S, Grelier S, Cramail H. From lignin-derived aromatic compounds to novel biobased polymers. Macromol Rapid Commun. 2016;37(1):9–28. https://doi.org/10.1002/marc.201500474.

9. Fache M, Boutevin B, Caillol S. Vanillin production from lignin and its use as a renewable chemical. ACS Sustain Chem Eng. 2015;4(1):35–46. https://doi.org/10.1021/acssuschemeng.5b01344.
10. Savonnet E, Le Coz C, Grau E, Grelier S, Defoort B, Cramail H. Divanillin-based aromatic amines: synthesis and use as curing agents for fully vanillin-based epoxy thermosets. Front Chem. 2019;7:606. https://doi.org/10.3389/fchem.2019.00606.
11. Mora AS, Tayouo R, Boutevin B, David G, Caillol S. Vanillin-derived amines for bio-based thermosets. Green Chem. 2018;20(17):4075–84. https://doi.org/10.1039/c8gc02006j.
12. Qi Y, Weng Z, Kou Y, Song L, Li J, Wang J, Zhang S, Liu C, Jian X. Synthesize and introduce bio-based aromatic s-triazine in epoxy resin: enabling extremely high thermal stability, mechanical properties, and flame retardancy to achieve high-performance sustainable polymers. Chem Eng J. 2021;406:126881. https://doi.org/10.1016/j.cej.2020.126881.
13. Bhusal RP, Sperry J. Flexible synthesis of diverse N-heterocycles from substrates attainable from biomass. Green Chem. 2016;18(8):2453–9. https://doi.org/10.1039/c5gc02837j.
14. Kallmeier F, Dudziec B, Irrgang T, Kempe R. Manganese-catalyzed sustainable synthesis of pyrroles from alcohols and amino alcohols. Angew Chem Int Ed Engl. 2017;56(25):7261–5. https://doi.org/10.1002/anie.201702543.
15. Qi Y, Wang J, Kou Y, Pang H, Zhang S, Li N, Liu C, Weng Z, Jian X. Synthesis of an aromatic N-heterocycle derived from biomass and its use as a polymer feedstock. Nat Commun. 2019;10(1):2107. https://doi.org/10.1038/s41467-019-10178-0.
16. Zheng B, Wu H, Song J, Wu W, Mei X, Zhang K, Xu C, Xu J, He M, Han B. Production of alkoxyl-functionalized cyclohexylamines from lignin-derived guaiacols. Green Chem. 2021;23(21):8441–7. https://doi.org/10.1039/d1gc02790e.
17. Li J, Weng Z, Cao Q, Qi Y, Lu B, Zhang S, Wang J, Jian X. Synthesis of an aromatic amine derived from biomass and its use as a feedstock for versatile epoxy thermoset. Chem Eng J. 2022;433:134512. https://doi.org/10.1016/j.cej.2022.134512.
18. Zhang T, Li W, Xu Z, Liu Q, Ma Q, Jameel H, Chang HM, Ma L. Catalytic conversion of xylose and corn stalk into furfural over carbon solid acid catalyst in gamma-valerolactone. Bioresour Technol. 2016;209:108–14. https://doi.org/10.1016/j.biortech.2016.02.108.
19. Wang Y, Yang X, Zheng H, Li X, Zhu Y, Li Y. Mechanistic insights on catalytic conversion fructose to furfural on beta zeolite via selective carbon-carbon bond cleavage. Mol Catal. 2019;463:130–9. https://doi.org/10.1016/j.mcat.2018.11.022.
20. Mariscal R, Maireles-Torres P, Ojeda M, Sádaba I, López GM. Furfural: a renewable and versatile platform molecule for the synthesis of chemicals and fuels. Energy Environ Sci. 2016;9(4):1144–89. https://doi.org/10.1039/c5ee02666k.
21. Senthamarai T, Murugesan K, Schneidewind J, Kalevar NV, Baumann W, Neumann H, Kamer PCJ, Beller M, Jagadeesh RV. Simple ruthenium-catalyzed reductive amination enables the synthesis of a broad range of primary amines. Nat Commun. 2018;9(1):4123. https://doi.org/10.1038/s41467-018-06416-6.
22. Petri A, Masia G, Piccolo O. Biocatalytic conversion of 5-hydroxymethylfurfural: synthesis of 2,5-bis (hydroxymethyl)furan and 5-(hydroxymethyl)furfurylamine. Catal Commun. 2018;114:15–8. https://doi.org/10.1016/j.catcom.2018.05.011.
23. Li X, Ma J, Jia X, Xia F, Huang Y, Xu Y, Xu J. Al-doping promoted aerobic Amidation of 5-hydroxymethylfurfural to 2,5-furandicarboxamide over cryptomelane. ACS Sustain Chem Eng. 2018;6(6):8048–54. https://doi.org/10.1021/acssuschemeng.8b01617.
24. Wang X, Chen W, Li Z, Zeng X, Tang X, Sun Y, Lei T, Lin L. Synthesis of bis(amino)furans from biomass based 5-hydroxymethyl furfural. J Energy Chem. 2018;27(1):209–14. https://doi.org/10.1016/j.jechem.2017.06.015.
25. Pingen D, Schwaderer JB, Walter J, Wen J, Murray G, Vogt D, Mecking S. Diamines for polymer materials via direct amination of lipid- and lignocellulose-based alcohols with NH_3. ChemCatChem. 2018;10(14):3027–33. https://doi.org/10.1002/cctc.201800365.
26. Drover MW, Omari KW, Murphy JN, Kerton FM. Formation of a renewable amide, 3-acetamido-5-acetylfuran, via direct conversion of N-acetyl-d-glucosamine. RSC Adv. 2012;2(11):4642–4. https://doi.org/10.1039/c2ra20578e.

27. Xie S, Jia C, Go Ong SS, Wang Z, Zhu MJ, Wang Q, Yang Y, Lin H. A shortcut route to close nitrogen cycle: bio-based amines production via selective deoxygenation of chitin monomers over Ru/C in acidic solutions. iScience. 2020;23(5):101096. https://doi.org/10.1016/j.isci.2020.101096.
28. Cao S, Liu Y, Shi L, Zhu W, Wang H. N-Acetylglucosamine as a platform chemical produced from renewable resources: opportunity, challenge, and future prospects. Green Chem. 2022;24(2):493–509. https://doi.org/10.1039/d1gc03725k.
29. Dash M, Chiellini F, Ottenbrite RM, Chiellini E. Chitosan-a versatile semi-synthetic polymer in biomedical applications. Prog Polym Sci. 2011;36(8):981–1014. https://doi.org/10.1016/j.progpolymsci.2011.02.001.
30. Claes L, Verduyckt J, Stassen I, Lagrain B, De Vos DE. Ruthenium-catalyzed aerobic oxidative decarboxylation of amino acids: a green, zero-waste route to biobased nitriles. Chem Commun (Camb). 2015;51(30):6528–31. https://doi.org/10.1039/c5cc00181a.
31. Scott E, Peter F, Sanders J. Biomass in the manufacture of industrial products--the use of proteins and amino acids. Appl Microbiol Biotechnol. 2007;75(4):751–62. https://doi.org/10.1007/s00253-007-0932-x.
32. Claes L, Janssen M, De Vos DE. Organocatalytic decarboxylation of amino acids as a route to bio-based amines and amides. ChemCatChem. 2019;11(17):4297–306. https://doi.org/10.1002/cctc.201900800.
33. Sebastian J, Zheng M, Jiang Y, Zhao Y, Wang H, Song Z, Li X, Pang J, Zhang T. One-pot conversion of lysine to caprolactam over Ir/H-Beta catalysts. Green Chem. 2019;21(9):2462–8. https://doi.org/10.1039/c9gc00415g.
34. Xie S, Jia C, Wang Z, Ong SSG, Zhu M, Lin H. Mechanistic insight into selective deoxygenation of l-lysine to produce biobased amines. ACS Sustain Chem Eng. 2020;8(31):11805–17. https://doi.org/10.1021/acssuschemeng.0c04052.
35. Liu X, Zhang J. High-performance biobased epoxy derived from rosin. Polym Int. 2010;59(5):607–9. https://doi.org/10.1002/pi.2781.
36. Garrison MD, Savolainen MA, Chafin AP, Baca JE, Bons AM, Harvey BG. Synthesis and characterization of high-performance, bio-based epoxy-amine networks derived from resveratrol. ACS Sustain Chem Eng. 2020;8(37):14137–49. https://doi.org/10.1021/acssuschemeng.0c04816.
37. Froidevaux V, Negrell C, Caillol S, Pascault JP, Boutevin B. Biobased amines: from synthesis to polymers; present and future. Chem Rev. 2016;116(22):14181–224. https://doi.org/10.1021/acs.chemrev.6b00486.
38. Xie W, Huang S, Liu S, Zhao J. Imine-functionalized biomass-derived dynamic covalent thermosets enabled by heat-induced self-crosslinking and reversible structures. Chem Eng J. 2021;404:126598. https://doi.org/10.1016/j.cej.2020.126598.
39. Yao Z, Qian L, Qiu Y, Chen Y, Xu B, Li J. Flame retardant and toughening behaviors of bio-based DOPO-containing curing agent in epoxy thermoset. Polym Adv Technol. 2019;31(3):461–71. https://doi.org/10.1002/pat.4782.
40. Zhou S, Tao R, Dai P, Luo Z, He M. Two-step fabrication of lignin-based flame retardant for enhancing the thermal and fire retardancy properties of epoxy resin composites. Polym Compos. 2020;41(5):2025–35. https://doi.org/10.1002/pc.25517.
41. Nabipour H, Wang X, Song L, Hu Y. A high performance fully bio-based epoxy thermoset from a syringaldehyde-derived epoxy monomer cured by furan-derived amine. Green Chem. 2021;23(1):501–10. https://doi.org/10.1039/d0gc03451g.
42. Niu H, Nabipour H, Wang X, Song L, Hu Y. Phosphorus-free vanillin-derived intrinsically flame-retardant epoxy thermoset with extremely low heat release rate and smoke emission. ACS Sustain Chem Eng. 2021;9(15):5268–77. https://doi.org/10.1021/acssuschemeng.0c08302.
43. Xie W, Huang S, Tang D, Liu S, Zhao J. Biomass-derived Schiff base compound enabled fire-safe epoxy thermoset with excellent mechanical properties and high glass transition temperature. Chem Eng J. 2020;394:123667. https://doi.org/10.1016/j.cej.2019.123667.

44. Wu JN, Chen L, Fu T, Zhao HB, Guo DM, Wang XL, Wang YZ. New application for aromatic Schiff base: high efficient flame-retardant and anti-dripping action for polyesters. Chem Eng J. 2018;336:622–32. https://doi.org/10.1016/j.cej.2017.12.047.
45. Nabipour H, Niu H, Wang X, Batool S, Hu Y. Fully bio-based epoxy resin derived from vanillin with flame retardancy and degradability. React Funct Polym. 2021;168:105034. https://doi.org/10.1016/j.reactfunctpolym.2021.105034.
46. Tripathy R, Ojha U, Faust R. Polyisobutylene modified bisphenol a diglycidyl ether based epoxy resins possessing improved mechanical properties. Macromolecules. 2011;44(17): 6800–9. https://doi.org/10.1021/ma201081y.
47. Tang Z, Huang J, Guo B, Zhang L, Liu F. Bioinspired engineering of sacrificial metal-ligand bonds into elastomers with supramechanical performance and adaptive recovery. Macromolecules. 2016;49(5):1781–9. https://doi.org/10.1021/acs.macromol.5b02756.
48. Xiao L, Huang J, Wang Y, Chen J, Liu Z, Nie X. Tung oil-based modifier toughening epoxy resin by sacrificial bonds. ACS Sustain Chem Eng. 2019;7(20):17344–53. https://doi.org/10.1021/acssuschemeng.9b04284.
49. Chen X, Chen S, Xu Z, Zhang J, Miao M, Zhang D. Degradable and recyclable bio-based thermoset epoxy resins. Green Chem. 2020;22(13):4187–98. https://doi.org/10.1039/d0gc01250e.
50. Fang Z, Nikafshar S, Hegg EL, Nejad M. Biobased divanillin as a precursor for formulating biobased epoxy resin. ACS Sustain Chem Eng. 2020;8(24):9095–103. https://doi.org/10.1021/acssuschemeng.0c02351.
51. Wu J, Yu X, Zhang H, Guo J, Hu J, Li M-H. Fully biobased vitrimers from glycyrrhizic acid and soybean oil for self-healing, shape memory, weldable, and recyclable materials. ACS Sustain Chem Eng. 2020;8(16):6479–87. https://doi.org/10.1021/acssuschemeng.0c01047.
52. Liu YY, He J, Li YD, Zhao XL, Zeng JB. Biobased, reprocessable and weldable epoxy vitrimers from epoxidized soybean oil. Ind Crop Prod. 2020;153:112576. https://doi.org/10.1016/j.indcrop.2020.112576.
53. Li QT, Jiang MJ, Wu G, Chen L, Chen SC, Cao YX, Wang YZ. Photothermal conversion triggered precisely targeted healing of epoxy resin based on thermoreversible Diels-Alder network and amino-functionalized carbon nanotubes. ACS Appl Mater Interfaces. 2017;9(24): 20797–807. https://doi.org/10.1021/acsami.7b01954.
54. Wu X, Yang X, Yu R, Zhao XJ, Zhang Y, Huang W. A facile access to stiff epoxy vitrimers with excellent mechanical properties via siloxane equilibration. J Mater Chem A. 2018;6(22): 10184–8. https://doi.org/10.1039/c8ta02102c.
55. Xu X, Ma S, Wang S, Wu J, Li Q, Lu N, Liu Y, Yang J, Feng J, Zhu J. Dihydrazone-based dynamic covalent epoxy networks with high creep resistance, controlled degradability, and intrinsic antibacterial properties from bioresources. J Mater Chem A. 2020;8(22):11261–74. https://doi.org/10.1039/d0ta01419b.
56. Wang S, Ma S, Li Q, Xu X, Wang B, Huang K, liu Y, Zhu J. Facile preparation of polyimine vitrimers with enhanced creep resistance and thermal and mechanical properties via metal coordination. Macromolecules. 2020;53(8):2919–31. https://doi.org/10.1021/acs.macromol.0c00036.
57. Liu WX, Zhang C, Zhang H, Zhao N, Yu ZX, Xu J. Oxime-based and catalyst-free dynamic covalent polyurethanes. J Am Chem Soc. 2017;139(25):8678–84. https://doi.org/10.1021/jacs.7b03967.
58. Denissen W, Droesbeke M, Nicolaÿ R, Leibler L, Winne JM, Du Prez FE. Chemical control of the viscoelastic properties of vinylogous urethane vitrimers. Nat Commun. 2017;8(1):14857. https://doi.org/10.1038/ncomms14857.
59. Ying H, Zhang Y, Cheng J. Dynamic urea bond for the design of reversible and self-healing polymers. Nat Commun. 2014;5:3218. https://doi.org/10.1038/ncomms4218.
60. Tang J, Wan L, Zhou Y, Pan H, Huang F. Strong and efficient self-healing adhesives based on dynamic quaternization cross-links. J Mater Chem A. 2017;5(40):21169–77. https://doi.org/10.1039/c7ta06650c.

61. Christensen PR, Scheuermann AM, Loeffler KEH, BA. Closed-loop recycling of plastics enabled by dynamic covalent diketoenamine bonds. Nat Chem. 2019;11(5):442–8. https://doi.org/10.1038/s41557-019-0249-2.
62. Lei H, Wang S, Liaw DJ, Cheng Y, Yang X, Tan J, Chen X, Gu J, Zhang Y. Tunable and processable shape-memory materials based on solvent-free, catalyst-free polycondensation between formaldehyde and diamine at room temperature. ACS Macro Lett. 2019;8(5):582–7. https://doi.org/10.1021/acsmacrolett.9b00199.
63. Memon H, Liu H, Rashid MA, Chen L, Jiang Q, Zhang L, Wei Y, Liu W, Qiu Y. Vanillin-based epoxy vitrimer with high performance and closed-loop recyclability. Macromolecules. 2020;53(2):621–30. https://doi.org/10.1021/acs.macromol.9b02006.
64. Zhao S, Abu-Omar MM. Recyclable and malleable epoxy thermoset bearing aromatic imine bonds. Macromolecules. 2018;51(23):9816–24. https://doi.org/10.1021/acs.macromol.8b01976.
65. Xu X, Ma S, Wu J, Yang J, Wang B, Wang S, Li Q, Feng J, You S, Zhu J. High-performance, command-degradable, antibacterial Schiff base epoxy thermosets: synthesis and properties. J Mater Chem A. 2019;7(25):15420–31. https://doi.org/10.1039/c9ta05293c.
66. Liu T, Zhang S, Hao C, Verdi C, Liu W, Liu H, Zhang J. Glycerol induced catalyst-free curing of epoxy and vitrimer preparation. Macromol Rapid Commun. 2019;40(7):e1800889. https://doi.org/10.1002/marc.201800889.
67. Xu Y, Dai S, Bi L, Jiang J, Zhang H, Chen Y. Catalyst-free self-healing bio-based vitrimer for a recyclable, reprocessable, and self-adhered carbon fiber reinforced composite. Chem Eng J. 2022;429:132518. https://doi.org/10.1016/j.cej.2021.132518.
68. Yu Z, Ma S, Tang Z, Liu Y, Xu X, Li Q, Zhang K, Wang B, Wang S, Zhu J. Amino acids as latent curing agents and their application in fully bio-based epoxy resins. Green Chem. 2021;23(17):6566–75. https://doi.org/10.1039/d1gc02126e.
69. Takada Y, Shinbo K, Someya Y, Shibata M. Preparation and properties of bio-based epoxy montomorillonite nanocomposites derived from polyglycerol polyglycidyl ether and ε-polylysine. J Appl Polym Sci. 2009;113(1):479–84. https://doi.org/10.1002/app.30015.
70. Mattar N, de Anda AR, Vahabi H, Renard E, Langlois V. Resorcinol-based epoxy resins hardened with limonene and eugenol derivatives: from the synthesis of renewable diamines to the mechanical properties of biobased thermosets. ACS Sustain Chem Eng. 2020;8(34): 13064–75. https://doi.org/10.1021/acssuschemeng.0c04780.
71. Harmsen PFH, Hackmann MM, Bos HL. Green building blocks for bio-based plastics. Biofuels Bioprod Biorefin. 2014;8(3):306–24. https://doi.org/10.1002/bbb.1468.
72. Mattar N, Hübner F, Demleitner M, Brückner A, Langlois V, Renard E, Ruckdäschel H, Rios de Anda A. Multiscale characterization of creep and fatigue crack propagation resistance of fully bio-based epoxy-aamine resins. ACS Appl Polym Mater. 2021;3(10):5134–44. https://doi.org/10.1021/acsapm.1c00894.
73. Illy N, Benyahya S, Durand N, Auvergne R, Caillol S, David G, Boutevin B. The influence of formulation and processing parameters on the thermal properties of a chitosan-epoxy prepolymer system. Polym Int. 2014;63(3):420–6. https://doi.org/10.1002/pi.4516.
74. Lu C, Liu Y, Wang C, Yong Q, Wang J, Chu F. An integrated strategy to fabricate bio-based dual-cure and toughened epoxy thermosets with photothermal conversion property. Chem Eng J. 2022;433:134582. https://doi.org/10.1016/j.cej.2022.134582.
75. Liu Z, Zhu X, Tian Y, Zhou K, Cheng J, Zhang J. Bio-based recyclable form-stable phase change material based on thermally reversible Diels-Alder reaction for sustainable thermal energy storage. Chem Eng J. 2022;448:137749. https://doi.org/10.1016/j.cej.2022.137749.
76. Wang YX, Ishida H. Cationic ring-opening polymerization of benzoxazines. Polymer. 1999;40(16):4563–70. https://doi.org/10.1016/S0032-3861(99)00074-9.
77. Zhang K, Liu Y, Han M, Froimowicz P. Smart and sustainable design of latent catalyst-containing benzoxazine-bio-resins and application studies. Green Chem. 2020;22(4):1209–19. https://doi.org/10.1039/c9gc03504d.

78. Wang C, Sun J, Liu X, Sudo A, Endo T. Synthesis and copolymerization of fully bio-based benzoxazines from guaiacol, furfurylamine and stearylamine. Green Chem. 2012;14(10): 2799–806. https://doi.org/10.1039/c2gc35796h.
79. Monisha M, Yadav N, Lochab B. Sustainable framework of chitosan-benzoxazine with mutual benefits: low curing temperature and improved thermal and mechanical properties. ACS Sustain Chem Eng. 2019;7(4):4473–85. https://doi.org/10.1021/acssuschemeng.8b06515.
80. Amarnath N, Shukla S, Lochab B. Isomannide-derived chiral rigid fully biobased polybenzoxazines. ACS Sustain Chem Eng. 2019;7(22):18700–10. https://doi.org/10.1021/acssuschemeng.9b05305.
81. Dumas L, Bonnaud L, Olivier M, Poorteman M, Dubois P. Chavicol benzoxazine: ultrahigh T_g biobased thermoset with tunable extended network. Eur Polym J. 2016;81:337–46. https://doi.org/10.1016/j.eurpolymj.2016.06.018.
82. Froimowicz P, C RA, Han L, Ishida H. Smart, sustainable, and ecofriendly chemical design of fully bio-based thermally stable thermosets based on benzoxazine chemistry. Chem Sus Chem. 2016;9(15):1921–8. https://doi.org/10.1002/cssc.201600577.
83. Periyasamy T, Asrafali SP, Muthusamy S. New benzoxazines containing polyhedral oligomeric silsesquioxane from eugenol, guaiacol and vanillin. New J Chem. 2015;39(3): 1691–702. https://doi.org/10.1039/c4nj02047b.
84. Devaraju S, Krishnadevi K, Sriharshitha S, Alagar M. Design and development of environmentally friendly polybenzoxazine-silica hybrid from renewable bio-resource. J Polym Environ. 2018;27(1):141–7. https://doi.org/10.1007/s10924-018-1327-z.
85. Sharma P, Dutta P, Nebhani L. Sustainable approach towards enhancing thermal stability of bio-based polybenzoxazines. Polymer. 2019;184:121905. https://doi.org/10.1016/j.polymer.2019.121905.
86. Wang X, Niu H, Huang J, Song L, Hu Y. A desoxyanisoin-and furfurylamine-derived high-performance benzoxazine thermoset with high glass transition temperature and excellent anti-flammability. Polym Degrad Stab. 2021;189:109604. https://doi.org/10.1016/j.polymdegradstab.2021.109604.
87. Lin CH, Lin HT, Sie JWH, Tu KY, AP. Facile, one-pot synthesis of aromatic diamine-based phosphinated benzoxazines and their flame-retardant thermosets. J Polym Sci A Polym Chem. 2010;48(20):4555–66. https://doi.org/10.1002/pola.24247.
88. Gaan S, Liang S, Mispreuve H, Perler H, Naescher R, Neisius M. Flame retardant flexible polyurethane foams from novel DOPO-phosphonamidate additives. Polym Degrad Stab. 2015;113:180–8. https://doi.org/10.1016/j.polymdegradstab.2015.01.007.
89. Lin CM, Chen CH, Lin CH, Juang TY. High-performance bio-based benzoxazines derived from phosphinated biphenols and furfurylamine. Eur Polym J. 2018;108:48–56. https://doi.org/10.1016/j.eurpolymj.2018.08.024.
90. Haubold TS, Puchot L, Adjaoud A, Verge P, Koschek K. Bio-based bisbenzoxazines with flame retardant linker. Polymers. 2021;13:4330.
91. Guo W, Wang X, Gangireddy CSR, Wang J, Pan Y, Xing W, Song L, Hu Y. Cardanol derived benzoxazine in combination with boron-doped graphene toward simultaneously improved toughening and flame retardant epoxy composites. Compos Part A Appl Sci Manuf. 2019;116: 13–23. https://doi.org/10.1016/j.compositesa.2018.10.010.
92. Yan H, Zhan Z, Wang H, Cheng J, Fang Z. Synthesis, curing, and thermal stability of low-temperature-cured benzoxazine resins based on natural renewable resources. ACS Appl Polym Mater. 2021;3(7):3392–401. https://doi.org/10.1021/acsapm.1c00361.
93. Yebra DM, Kiil S, Dam-Johansen K. Antifouling technology-past, present and future steps towards efficient and environmentally friendly antifouling coatings. Prog Org Coat. 2004;50 (2):75–104. https://doi.org/10.1016/j.porgcoat.2003.06.001.
94. Chen J, Jian R, Yang K, Bai W, Huang C, Lin Y, Zheng B, Wei F, Lin Q, Xu Y. Urushiol-based benzoxazine copper polymer with low surface energy, strong substrate adhesion and antibacterial for marine antifouling application. J Clean Prod. 2021;318:128527. https://doi.org/10.1016/j.jclepro.2021.128527.

95. Periyasamy T, Asrafali S, Shanmugam M, Kim SC. Development of sustainable and antimicrobial film based on polybenzoxazine and cellulose. Int J Biol Macromol. 2021;170:664–73. https://doi.org/10.1016/j.ijbiomac.2020.12.087.
96. Yadav N, Monisha M, Niranjan R, Dubey A, Patil S, Priyadarshini R, Lochab B. Antibacterial performance of fully biobased chitosan-grafted-polybenzoxazine films: elaboration and properties of released material. Carbohydr Polym. 2021;254:117296. https://doi.org/10.1016/j.carbpol.2020.117296.
97. Lu G, Dai J, Liu J, Tian S, Xu Y, Teng N, Liu X. A new sight into bio-based polybenzoxazine: from tunable thermal and mechanical properties to excellent marine antifouling performance. ACS Omega. 2020;5(7):3763–73. https://doi.org/10.1021/acsomega.0c00025.
98. Sha X-L, Yuan L, Liang G, Gu A. Preparation of high performance bio-based benzoxazine resin through a green solvent-free strategy for shape memory application. Polymer. 2020;202:122673. https://doi.org/10.1016/j.polymer.2020.122673.
99. Sriharshitha S, Krishnadevi K, Devaraju S, Srinivasadesikan V, Lee SL. Eco-friendly sustainable poly(benzoxazine-co-urethane) with room-temperature-assisted self-healing based on supramolecular interactions. ACS Omega. 2020;5(51):33178–85. https://doi.org/10.1021/acsomega.0c04840.
100. Wuliu Y, Liu J, Zhang L, Wang S, Liu Y, Feng J, Liu X. Design of bio-based organic phase change materials containing a "safety valve". Green Chem. 2021;23(21):8643–56. https://doi.org/10.1039/d1gc02618f.
101. Yao H, Lu X, Xin Z, Zhang H, Li X. A durable bio-based polybenzoxazine/SiO2 modified fabric with superhydrophobicity and superoleophilicity for oil/water separation. Sep Purif Technol. 2019;229:115792. https://doi.org/10.1016/j.seppur.2019.115792.
102. Weng Z, Song L, Qi Y, Li J, Cao Q, Liu C, Zhang S, Wang J, Jian X. Natural magnolol derivatives as platform chemicals for bio-based phthalonitrile thermoset: achieving high performances without an external curing agent. Polymer. 2021;226:123814. https://doi.org/10.1016/j.polymer.2021.123814.
103. Keller TM, Dominguez DD. High temperature resorcinol-based phthalonitrile polymer. Polymer. 2005;46(13):4614–8. https://doi.org/10.1016/j.polymer.2005.03.068.
104. Laskoski M, Clarke JS, Neal A, Harvey BG, Ricks-Laskoski HL, Hervey WJ, Daftary MN, Shepherd AR, Keller TM. Sustainable high-temperature phthalonitrile resins derived from resveratrol and dihydroresveratrol. ChemistrySelect. 2016;1(13):3423–7. https://doi.org/10.1002/slct.201600304.
105. Han Y, Tang DH, Wang GX, Sun YN, Guo Y, Zhou H, Qiu WF, Zhao T. Phthalonitrile resins derived from vanillin: synthesis, curing behavior, and thermal properties. Chinese J Polym Sci. 2019;38(1):72–83. https://doi.org/10.1007/s10118-019-2311-3.
106. Wang C, Shi M, Fang L, Dai M, Huang G, Sun J, Fang Q. The bio-based phthalocyanine resins with high Tg and high char yield derived from vanillin. Polymer. 2021;224:123723. https://doi.org/10.1016/j.polymer.2021.123723.
107. Liu Y, Liu Z-z, Peng W-f, Lu Z, Hu J-h, Zeng K, Yang G. Inspiration from a new lignin-derived phthalonitrile resin: unique curing behavior, and thermal properties. Eur Polym J. 2019;121:109351. https://doi.org/10.1016/j.eurpolymj.2019.109351.
108. Peng W, Yao F, Hu J, Liu Y, Lu Z, Liu Y, Liu Z, Zeng K, Yang G. Renewable protein-based monomer for thermosets: a case study on phthalonitrile resin. Green Chem. 2018;20(22):5158–68. https://doi.org/10.1039/c8gc01824c.
109. Chen M, He X, Guo Y, Hu J, Liang B, Zeng K, Yang G. A new molecular design platform for high-performance polymers from versatile bio-based tyramine: a case study of tyramine-derived phthalonitrile resin. Polym Chem. 2021;12(3):408–22. https://doi.org/10.1039/d0py01322f.
110. Qi Y, Weng Z, Wang J, Zhang S, Zong L, Liu C, Jian X. A novel bio-based phthalonitrile resin derived from catechin: synthesis and comparison of curing behavior with petroleum-based counterpart. Polym Int. 2018;67(3):322–9. https://doi.org/10.1002/pi.5507.

111. Zhang C, Garrison TF, Madbouly SA, Kessler MR. Recent advances in vegetable oil-based polymers and their composites. Prog Polym Sci. 2017;71:91–143. https://doi.org/10.1016/j.progpolymsci.2016.12.009.
112. Xie F, Zhang T, Bryant P, Kurusingal V, Colwell JM, Laycock B. Degradation and stabilization of polyurethane elastomers. Prog Polym Sci. 2019;90:211–68. https://doi.org/10.1016/j.progpolymsci.2018.12.003.
113. Gogoi S, Karak N. Bio-based high-performance waterborne hyperbranched polyurethane thermoset. Polym Adv Technol. 2015;26(6):589–96. https://doi.org/10.1002/pat.3490.
114. Kang SY, Ji Z, Tseng LF, Turner SA, Villanueva DA, Johnson R, Albano A, Langer R. Design and synthesis of waterborne polyurethanes. Adv Mater. 2018;30(18):e1706237. https://doi.org/10.1002/adma.201706237.
115. Liu L, Lu J, Zhang Y, Liang H, Liang D, Jiang J, Lu Q, Quirino RL, Zhang C. Thermosetting polyurethanes prepared with the aid of a fully bio-based emulsifier with high bio-content, high solid content, and superior mechanical properties. Green Chem. 2019;21(3):526–37. https://doi.org/10.1039/c8gc03560a.
116. Karami Z, Kabiri K, Zohuriaan-Mehr MJ. Non-isocyanate polyurethane thermoset based on a bio-resourced star-shaped epoxy macromonomer in comparison with a cyclocarbonate fossil-based epoxy resin: a preliminary study on thermo-mechanical and antibacterial properties. J CO2 Util. 2019;34:558–67. https://doi.org/10.1016/j.jcou.2019.08.009.
117. Hibert G, Lamarzelle O, Maisonneuve L, Grau E, Cramail H. Bio-based aliphatic primary amines from alcohols through the 'nitrile route' towards non-isocyanate polyurethanes. Eur Polym J. 2016;82:114–21. https://doi.org/10.1016/j.eurpolymj.2016.07.007.
118. Wunschik DS, Ingenbosch KN, Zähres M, Horst J, Mayer C, Jäger M, Strehmel V, Dornbusch M, Hoffmann-Jacobsen K. Biocatalytic and solvent-free synthesis of a bio-based biscyclocarbonate. Green Chem. 2018;20(20):4738–45. https://doi.org/10.1039/c8gc02267d.
119. Ramdani N, Zaimeche H, Derradji M. Biobased thermally-stable aromatic cyanate ester thermosets: a review. React Funct Polym. 2021;168:105037. https://doi.org/10.1016/j.reactfunctpolym.2021.105037.
120. Barde M, Edmunds CW, Labbé N, Auad ML. Fast pyrolysis bio-oil from lignocellulosic biomass for the development of bio-based cyanate esters and cross-linked networks. High Perform Polym. 2019;31(9–10):1140–52. https://doi.org/10.1177/0954008319829517.
121. Koelewijn SF, Van den Bosch S, Renders T, Schutyser W, Lagrain B, Smet M, Thomas J, Dehaen W, Van Puyvelde P, Witters H, Sels BF. Sustainable bisphenols from renewable softwood lignin feedstock for polycarbonates and cyanate ester resins. Green Chem. 2017;19(11):2561–70. https://doi.org/10.1039/c7gc00776k.
122. Harvey BG, Guenthner AJ, Meylemans HA, Haines SRL, Lamison KR, Groshens TJ, Cambrea LR, Davis MC, Lai WW. Renewable thermosetting resins and thermoplastics from vanillin. Green Chem. 2015;17(2):1249–58. https://doi.org/10.1039/c4gc01825g.
123. Meylemans HA, Harvey BG, Reams JT, Guenthner AJ, Cambrea LR, Groshens TJ, Baldwin LC, Garrison MD, Mabry JM. Synthesis, characterization, and cure chemistry of renewable bis(cyanate) esters derived from 2-methoxy-4-methylphenol. Biomacromolecules. 2013;14(3):771–80. https://doi.org/10.1021/bm3018438.
124. Cash JJ, Davis MC, Ford MD, Groshens TJ, Guenthner AJ, Harvey BG, Lamison KR, Mabry JM, Meylemans HA, Reams JT, Sahagun CM. High T_g thermosetting resins from resveratrol. Polym Chem. 2013;4(13):3859–65. https://doi.org/10.1039/c3py00438d.
125. Cambrea LR, Davis MC, Garrison MD, Groshens TJ, Lyon RE, Safronava N. Processable cyanate ester resin from Cisresveratrol. J Polym Sci A Polym Chem. 2017;55(6):971–80. https://doi.org/10.1002/pola.28457.
126. Garrison MD, Harvey BG. Structure-property relationships of cis-resveratrol cyanate ester blends. Polymer. 2021;213:123194. https://doi.org/10.1016/j.polymer.2020.123194.

Chapter 10
Preparation of N-Doped Carbon Materials from Lignocellulosic Biomass Residues and Their Application to Energy Storage and Conversion Devices

Jessica Chaparro-Garnica, David Salinas-Torres, Miriam Navlani-García, Emilia Morallón, and Diego Cazorla-Amorós

Abstract The use of biomass in the energy scenario has generated great interest due to the variety of high value-added products that can be obtained from it. Among these products, carbon materials stand out since they have very interesting properties, highlighting the versatility that they offer by virtue of the modification of their surface chemistry through the incorporation of heteroatoms. The present chapter aims at covering the main aspects related to the synthesis of nitrogen-doped carbon materials from lignocellulosic biomass residues together with some representative examples of their use in energy-related applications. The focus has been paid to the most important synthetic routes for the preparation of biomass-derived nitrogen-doped carbon materials, by using a general classification to sort them into in situ and post-synthesis doping strategies. The applicability of such carbon materials is emphasized in: (1) electrocatalytic and catalytic applications, and (2) electrodes in supercapacitors.

Keywords Lignocellulosic biomass residues · N-doped carbon materials · Oxygen reduction reaction · Hydrogen production · Supercapacitor

---

Jessica Chaparro-Garnica and David Salinas-Torres contributed equally to this work.

J. Chaparro-Garnica · M. Navlani-García · D. Cazorla-Amorós (✉)
Department of Inorganic Chemistry and Materials Institute, University of Alicante, Alicante, Spain
e-mail: cazorla@ua.es

D. Salinas-Torres
Department of Chemical and Environmental Engineering, Polytechnical University of Cartagena, Cartagena, Spain

E. Morallón
Department of Physical Chemistry and Materials Institute, University of Alicante, Alicante, Spain

Z. Fang et al. (eds.), *Production of N-containing Chemicals and Materials from Biomass*, Biofuels and Biorefineries 12,
https://doi.org/10.1007/978-981-99-4580-1_10

10.1 Introduction

The implementation of renewable energies is necessary to achieve sustainable development. Biomass has received great attention because it is possible to produce energy from it in a renewable and sustainable way. Biomass is non-fossilized organic matter that can be generated from a spontaneous or provoked biological process. Biomass has very diverse resources and multiple uses for humanity, since it is a raw material in various chemical industries, including energy production. In addition, biomass production systems have a rapid regeneration; for this reason, biomass can be considered as a renewable raw material available in abundance that can be applied to obtain various products, materials, and carbon-based fuels [1, 2]. In the energy context, the transformation of biomass can be considered a renewable process because its energy content comes from the solar energy fixed during the photosynthesis process, which is accumulated in the formed organic molecules [3].

Biomass has been used by human beings as a source of energy in their daily tasks. However, the increased use of fossil fuels after the Industrial Revolution relegated biomass to a less important material and it did not have a significant contribution to the global energy system [1]. Currently, the dependence on fossil fuels to meet global energy demand entails major economic and environmental problems. The extraction of fossil fuels has become more complicated and the depletion of these resources has caused a large increase in their price [4]. Regarding environmental problems, fossil fuels contribute to air pollution due to the high release of greenhouse gases [5], causing an increase in the average temperature of the earth and causing major natural disasters (desertification, thawing of the poles, etc.). Considering the aforementioned issues, it is necessary to search for energy generation technologies and the development of processes, that lead to high-value chemical compounds, using renewable resources. In this way, a transition will be achieved towards a sustainable energy system based on inexhaustible energy sources [6, 7]. At this point, it is important to remember that a sustainable system must have an integrated and balanced approach between economic, environmental, and social aspects. The lack of any of these aspects will affect the performance of the others [8]. That is why biomass has acquired an important role in the global energy scenario and has positioned itself as one of the most reasonable alternatives to make the economy profitable and reduce the environmental problems caused by traditional energy sources, being the only source on the earth that contains carbon and hydrogen, from which it is possible to generate energy and obtain various chemical products with high added value [1, 5, 9].

Biomass is a renewable natural material that has diverse chemical compositions. Lignocellulosic biomass is mainly composed of three biopolymers mixed on a microscopic scale: cellulose, hemicellulose, and lignin. The composition and/or distribution of these three biopolymers changes according to the type of plant. Cellulose microfibers are formed from ordered polymeric chains containing crystalline regions that are found in a hemicellulose matrix, while lignin is located between the junctions formed by cellulose and hemicellulose [10–12]. Its potential for energy production varies in the process used, which can involve elementary or highly sophisticated technologies [13]. Biomass can be transformed into energy, such as electricity, heat, and biofuels [14, 15].

The use of lignocellulosic biomass to obtain compounds and/or chemical products began in the 1930s by William Jay Hale, who used carbohydrates from an agro-industrial origin as a platform to obtain commercial products (plastics, paints, gasoline, etc.). This revealed that agriculture could be considered as a potential source to obtain a wide variety of chemical compounds, which not only include those intended for the food industry [16]. These organic chemicals produced from biomass correspond to approximately 10% of world production [17]. Therefore, the development of new biomass valorization processes and the synthesis of chemical products with different applications at an industrial level have raised great interest [3]. These processes to convert biomass into high-added-value products are included in the biorefinery concept.

Biorefineries are facilities where biomass is transformed into a series of chemical products employing sustainable processes. Currently, most of the materials, chemical products, and fuels that are obtained from fossil fuels can be sustainably obtained from biomass [18]. For this reason, they can be considered analogous to petrochemical refineries and produce fuels, energy, heat, and other value-added chemical products using biomass as raw material [19, 20]. In addition, biorefineries have advantages as they have less environmental impact, unlimited reserves, and greater efficiency in the use of resources, thus having an important role in global energy development, as well as in the chemical industry [18, 21].

In summary, the main goal of processing biomass is to replace fossil fuels to achieve a sustainable energy system, considering the rapid depletion of traditional fuels and the renewable nature of biomass. At this point, lignocellulosic biomass can be considered a viable and sustainable alternative to the use of fossil fuels, since it can lead to biofuels, energy, and a large number of chemical products with high added value through different processes [22, 23]. For this reason, the design of efficient processes for the valorization of biomass is currently a research topic to which the scientific community is dedicating strong efforts. Biomass processing is usually aimed to obtain gaseous and liquid products, while solid products have generally been discarded without obtaining additional benefits from them. However, some biomass conversion processes generate solid by-products from which carbon materials with many potential applications can be obtained [24–26], achieving a greater valorization of the biomass.

In this sense, the preparation of carbon materials from biomass residues for their use in energy and environmental protection applications is receiving important attention [27–30]. Gasification and pyrolysis are the most used processes for the preparation of carbon materials using biomass as precursors. However, in the last two decades, many investigations have been published to obtain carbon materials from biomass precursors by hydrothermal carbonization (HTC) [12, 31–33].

HTC is a well-known and scientifically established treatment. However, it is currently not used on an industrial scale and mainly remains on a laboratory and pilot-plant scales [33, 34]. HTC is an effective environmentally friendly thermal treatment for the transformation of biomass into products with high added value at relatively low temperatures (180–250 °C). The treatment is carried out in a sealed container with self-generated pressure using aqueous solutions, which act as a reaction medium [35]. Regarding the products obtained during this process, three fractions are differentiated: a solid fraction (material rich in carbon), a liquid fraction, and a small part of gases. It is important to know that the properties and amounts of each fraction are directly related to the raw material and the process conditions [36]. Carbon-rich solid material (denoted as hydrochar) is the main product of HTC. In addition, given its homogeneity and hydrophobicity, it can be easily separated from the suspension [37]. The chemical composition and physical appearance of the resulting solid product depend on the raw material used [38]. The experimental variables of the HTC process (i.e. temperature, precursor, pressure, residence time, heating rate, water/biomass ratio, and catalyst used) are largely responsible for the physicochemical properties of the final products, which determine their potential applications. However, the resulting hydrochars have low porosity, low electrical conductivity, etc. Therefore, it is necessary to carry out a subsequent activation process to produce activated carbons with promising properties and applications.

Activated carbons are characterized by their high surface area (apparent surface area from 500 $m^2\ g^{-1}$ to 3000 $m^2\ g^{-1}$). These materials have a heterogeneous surface chemistry, high thermal resistance and chemical stability in both acidic and alkaline aqueous solutions, among other properties [39–41]. Likewise, their surface chemistry can be easily modified by introducing different functional groups, making them very interesting materials for different applications [42].

Biomass, specifically biomass residues, as porous carbon precursors generate great interest since they offer profitable alternatives by involving diverse renewable resources. At the same time, they have a high carbon content, and low ash content and the activated carbon preparation procedures are simple and can be cost-effective [43–45]. The characteristics of the obtained carbon materials depend largely on the precursor, being necessary a proper selection of it [46]. Many studies can be found about the preparation of activated carbons from different precursors. For example, we can find the use of biomass such as hemp [47], olive stones [48], cotton stems [49], and fruit peels [50], among others [51, 52], to obtain activated carbons that meet the necessary characteristics for their use in a given application.

The preparation of activated carbons consists of the reaction between a precursor and an activating agent, which leads to the formation of porosity. The so-called physical activation is carried out at elevated temperatures (usually above 800 °C) and through gasification reactions using gases such as CO_2 or steam [41, 53]. Chemical activation is carried out by impregnating the precursor with a chemical agent. Phosphoric acid (H_3PO_4) is the most used activating agent at the industrial level, but alkaline hydroxides such as KOH and NaOH are also used, among others [54, 55]. The use of harmless chemical agents would be the best alternative to reduce the environmental concerns that conventional activating agents present. For instance, Sevilla et al. reviewed the chemical activation strategies and they emphasized the use of alkali metal carbonates and bicarbonates (K_2CO_3, $KHCO_3$, Na_2CO_3), alkali metal organic salts (potassium acetate, potassium citrate, potassium oxalate), and alkali metal sulphates and thiosulphates (K_2SO_4, $Na_2S_2O_3$, etc.) [55]. These chemical agents promote the development of porosity by reactions with the precursor [56]. Chemical activation has certain advantages over physical activation such as lower activation temperatures, higher activated carbon yields, and better porosity development [54, 56, 57]. However, it presents disadvantages such as the need to perform washing steps to eliminate products of the activation reactions, as well as the greater corrosive nature of the process compared to physical activation [56].

H_3PO_4 and $ZnCl_2$ are the most used activating agents for the activation of lignocellulosic biomass. However, comparing them, H_3PO_4 is the most used, since it does not have such a polluting nature and it is easily reusable when recovered in the activated carbon washing step [58, 59]. Activation with H_3PO_4 produces wider pore size distribution, resulting in activated carbons that contain micropores and mesopores [60, 61]. Through activation with H_3PO_4, phosphorus-containing groups can be introduced in the carbon material, and these groups can provide acidity to the surface of the activated carbon [62]. In addition, several studies have shown that the presence of these groups influences the electrochemical stability of carbon materials [63, 64].

In particular, activated carbons derived from biomass residues have generated great interest in different energy storage and production applications due to their interesting characteristics such as high specific surface area, high electrical conductivity, electrochemical stability, low cost and the possibility to fine-tuning their surface chemistry [42, 65, 66]. The resulting porous carbon materials after the activation process usually contain an important amount of heteroatoms. Oxygen functional groups are predominant, but other heteroatoms-containing functional groups can be also formed either during the activation process or applying further post-synthetic treatments, or by their presence in the precursor used [67–70]. Among them, nitrogen-doped carbon materials have received great interest, since they exhibit promising behavior in a wide range of electrocatalytic and catalytic applications as well as in energy storage applications [71–74]. Among the multiple potential applications of lignocellulosic biomass-derived nitrogen-doped activated carbons, the present chapter focuses on covering representative breakthroughs achieved in catalytic and energy storage applications.

The most common nitrogen functionalities in nitrogen-containing carbon materials are pyridinic, pyrrolic, and quaternary groups [71, 75]. However, oxidized nitrogen and amine groups might be produced at the edges of the graphene layers. These nitrogen functionalities have different environments or configurations, which in turn implies different electronic states that can affect the final properties of the resulting carbon material.

The speciation of different nitrogen functionalities can be performed by several techniques. X-ray Photoelectron Spectroscopy (XPS) is commonly used, and a proper deconvolution of the N 1 s spectrum can give valuable information [76–78]. Nuclear Magnetic Resonance (NMR) spectroscopy is also a powerful technique [79, 80]. X-ray absorption spectroscopy (XAS) has been less used to ascertain the chemical state of nitrogen, although it can be also a valuable technique [81, 82]. The importance of identifying the nitrogen groups is vital because a specific nitrogen group may provide the resulting carbon material with particular properties for a given application.

10.2 Synthetic Routes for the Preparation of N-Doped Carbon Materials

As previously reported [71], establishing a clear classification of the synthetic routes for preparation of N-doped carbon materials might be a difficult task. However, dividing such routes into post-synthesis methods and in situ doping methods is generally accepted. It should be noted that even though N sources are used in both post-synthesis and in situ synthesis strategies, the experimental procedure is quite different. In the post-synthesis strategy, the carbon material is prepared in the first step and then it is modified by incorporation of the N-containing molecules, such for example NH_3. However, in the in situ approach, both the precursor of carbon material and N source are involved simultaneously (i.e. the carbonization process and the doping process occur simultaneously) [71, 74, 83]. The self-doped approach can be included within the in situ doping methods. In that case, nitrogen is naturally present in the biomass precursor (see Fig. 10.1).

10.2.1 Post-synthesis Strategies

These strategies involve the direct reaction between carbon materials previously synthesized and nitrogen-containing sources (ammonia, urea, cyanamide, aniline,

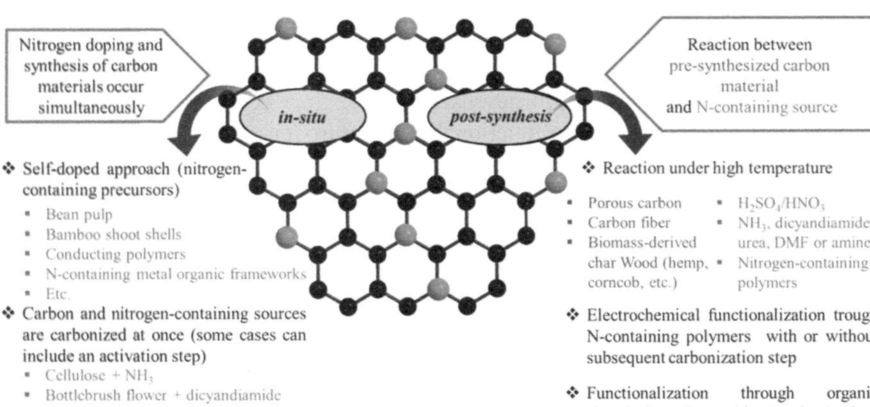

Fig. 10.1 Outline of strategies to synthesize nitrogen-containing carbon materials. Adapted with permission from [71]. Copyright © 2019, Elsevier

etc.) [84–87]. The incorporation of nitrogen on pre-synthesised carbon materials using ammonia (NH_3) has been widely used but this is a time and energy-consuming strategy and, in some cases, the carbon surface needs to be pre-treated to favor the incorporation of nitrogen atoms into the carbon network. As an example, Liang et al. prepared N-doped activated carbon using NH_3 as the nitrogen precursor, and treating the carbon material at 900 °C. The as-prepared nitrogen micro/mesoporous carbon materials displayed a N-doping level that ranged from 2.69 at.% to 9.50 at.% [88]. Urea is another precursor commonly used. Ma et al. reported the preparation of hierarchical porous carbons derived from tobacco stem by chemical activation at 800 °C, resulting in activated carbons with high porosity development ($S_{BET} > 2500$ m^2 g^{-1}) and nitrogen contents between 0.77 at.% and 1.73 at.% [89]. Other compounds such as melamine or dicyandiamide have been extensively used [90, 91].

A reliable prediction of the nitrogen groups formed, as well as their relative content, when using these methods is complex, since many experimental conditions (temperature and time of heat-treatment, initial carbon precursor, etc.) affect them. Previous studies revealed that temperatures around 550 °C might result in high nitrogen contents when NH_3 acted as the nitrogen precursor [92]. Yang et al. prepared walnut shell-derived N-doped activated carbons using urea as a nitrogen source [93]. They observed that pyridinic and pyrrolic groups are the most abundant, and quaternary groups appeared as the temperature increased. Another nitrogen-containing source which can incorporate nitrogen atoms in carbon materials is dicyandiamide [94, 95]. Plavniece et al. used dicyandiamide to incorporate nitrogen into activated carbons derived from alder wood char [95]. The char was activated with sodium hydroxide, and then the as-synthesized activated carbons were suspended in a dicyandiamide solution. Once the solvent was evaporated, the doping step was performed at 800 °C. The main nitrogen groups were pyrrolic and pyridinic, although quaternary and oxidized N groups could be also detected (the sum of quaternary and oxidized N species >20 at.%). In light of the above, nitrogen functionalities can differ even though the synthesis approach is similar, indicating that the initial precursor plays a key role in the final nitrogen functionalities.

Post-synthesis nitrogen doping approaches also encompass the chemical and electrochemical functionalization of carbon materials using nitrogen precursors like aniline or pyrrole, which produce conducting polymers such as polyaniline and polypyrrole and other monomers such as 4-aminophenylphosphonic acid, or by organic chemistry routes [77, 96–99]. As an example for biomass-derived carbon materials, Chaparro-Garnica et al. modified biomass waste-derived activated carbons prepared by HTC and further H_3PO_4 activation, through organic chemistry reactions under mild conditions (70 °C for 65 h). Briefly, the strategy consisted of mixing hemp residue-derived activated carbons with NH_4NO_3 in DMF, and then pyridine was added dropwise. The N-doped material showed a nitrogen content of 1.7 at.%. Furthermore, pristine activated carbon was heat-treated in an inert atmosphere and subsequently modified following the same organic chemistry route, leading to N-doped activated carbons with a nitrogen content of 1.8 at.%. Despite the heat treatment of the starting carbon material, both nitrogen-containing activated carbons presented the same amount of nitrogen groups, which were mainly pyridinic, amide/amine, pyrrolic/pyridine groups. Regarding the use of nitrogen-containing polymers, polyaniline and polypyrrole are commonly employed to functionalize carbon materials [77, 100, 101]. Such a nitrogen-containing conducting polymer can be deposited onto the carbon surface by chemical or electrochemical polymerization methods. The N-doped carbon materials are obtained after heat treatment in an inert atmosphere. As an example, the polymerization of aniline was performed in an activated carbon derived from celery stems using both chemical and electrochemical methods [102]. Then, the materials were carbonized at 850 °C under a nitrogen atmosphere, obtaining their carbonized counterparts. The obtained nitrogen content ranged from 2.66 at.% to 7.90 at.%, showing the presence of different nitrogen contributions, mainly pyridinic and N-oxidised groups. Pyrrole is another nitrogen-containing monomer also used to prepare N-doped carbon materials. Yu et al. through a simple and inexpensive two-step synthesis approach prepared N-doped carbon materials. First, biomass-derived carbon aerogels were prepared by pyrolysis of raw cattail fibers. Second, the carbon aerogel was modified by polymerization of pyrrole [103]. The main nitrogen group detected was pyrrolic, although positively charged nitrogen was also observed by XPS. The electrochemical characterization revealed good reversibility in charge-discharge processes and great areal capacitance. The resulting carbon material consisted of an interconnected carbon fiber aerogel, which makes it an interesting electrode in supercapacitors because it is a binder-free carbon material.

10.2.2 In Situ *Strategies*

Regarding *in situ* strategies to obtain nitrogen-containing carbon materials from lignocellulosic biomass [104–107], the easiest strategy is the carbonization of nitrogen-containing sources. However, a unique carbonization step might result in carbon materials with a low porosity development or inappropriate properties. For this reason, an additional activation step is required to improve such properties. Some biomass residues naturally contain nitrogen in their composition, which could be present in the resulting final carbon material [108, 109]. Among them, bean pulp has a high content of proteins (40–48 wt%), and also amino acids (~4 wt%). These nitrogen-containing molecules give rise to a lignocellulosic biomass waste with high nitrogen content (up to 8 wt%). It could lead to N-doped activated carbons, avoiding the use of harmful nitrogen-containing chemical compounds. Ding et al. utilized bean pulp (BP) as a N-containing biomass precursor, which was dried and carbonized at 500 °C under an Ar atmosphere. Then, the temperature was increased up to 800 °C (BPC), and the atmosphere was changed to 25 vol% of CO_2 in Ar to activate the carbon material [104]. Additionally, two additional samples were synthesized following the same procedure with different vol% of CO_2. It was observed that CO_2 concentration affected both the porous texture and the nitrogen content. The final nitrogen content ranged from 5 at.% to 10 at.%. The deconvolution of XPS N 1 s spectra revealed that pyridinic and pyrrolic groups are the most abundant for all BPC samples. These groups are electroactive and enhance the pseudocapacitance of the carbon material by redox reactions. Furthermore, the amounts of oxidized N and quaternary groups are also significant. The latter type of group was not affected by % vol CO_2, being beneficial to boost the electrical conductivity. Cao et al. also used bean sprouts as a biomass precursor and silica as the template [105]. The mixture was pyrolyzed at different temperatures for 2 h in a nitrogen atmosphere. The nitrogen content of the best-performing sample, that prepared at 800 °C, was around 4.5 at.%. The authors stated that the nitrogen groups in this material could provide pseudocapacitance.

Another strategy to prepare N-doped carbon materials from a carbon precursor that does not contain nitrogen is the heat treatment of such precursor in presence of a N-containing source. Luo et al. heat-treated a commercial cellulose precursor under an ammonia atmosphere [110]. The effect of both temperature and time was assessed and they observed that the nitrogen content was kept close to 10 at.% for temperatures between 500 °C and 700 °C regardless of the time of heat-treatment, while it was lower than 5 at.% at temperatures above 700 °C. Pyrrolic and pyridinic groups were the most abundant in the samples, and the content of pyridinic groups decayed as the temperature increased. However, another study revealed that the optimal temperature to reach the maximum nitrogen content is above 700 °C. In that study, biomass corncob-derived N-doped carbons were prepared using NH_3 as both activating agent and a nitrogen source [111]. The resulting N-doped activated carbons (NACs) contained a nitrogen amount over 10 wt% at temperatures higher than 700 °C. From the XPS N 1 s spectrum, NACs presented different N-containing groups, which included pyridinic, amine, oxidized N species, pyrrolic, and quaternary groups. Almost the same type of nitrogen groups was observed by Ma et al. when they used urea and tobacco stem as nitrogen source and biomass precursor, respectively [89]. Tian et al. used dicyandiamide to get N-doped activated carbon derived from bottlebrush flowers [112]. The biomass precursor was soaked with an aqueous solution of sodium bicarbonate/dicyandiamide. The resulting dried mixture was pyrolyzed at 700 °C for 3 h. The weight ratios of bicarbonate/dicyandiamide used were 6, 3, 1.5, and 1, and the N-doped activated carbons were named NPC-1, NPC-2, NPC-3, and NPC-4 respectively. As expected, the nitrogen contents increased as dicyandiamide concentration increased (the amount of $NaHCO_3$ was constant). NPC-3 and NPC-4 showed the highest nitrogen contents ($>$ 20 at.%). On the contrary, their apparent surface areas were lower than 120 $m^2\ g^{-1}$ which could be unsuitable for a given application. However, NPC-1 and NPC-2 (low amount of dicyandiamide) presented nitrogen contents of 3.41 at.% and 9.57 at.%, respectively. Interestingly, NPC-1 displayed an apparent surface area much higher than that obtained from the other N-doped activated carbon. The main nitrogen groups were pyrrolic and pyridinic, although quaternary and oxidized N groups could be also detected (the sum of quaternary and oxidized N species $<$10 at.%). Xu et al. used melamine to synthesize N-doped char through lycium chinensis as the biomass precursor [113]. Several lycium chinensis/melamine mass ratios were used to prepare carbonized samples at 600 °C. Then, N-doped chars were immersed in a KOH aqueous solution and these were heat-treated at 600 °C, 700 °C, and 800 °C. The XPS spectra revealed that nitrogen doping process was effective, which improved the electrical conductivity and contributed to the pseudocapacitance. Four types of nitrogen groups were found, being quaternary, pyridinic, and pyrrolic the most abundant.

10.3 Nitrogen-Doped Carbon Materials Derived from Lignocellulosic Biomass Residues in Energy-Related Applications

Due to the relevance of the topic, this section is dedicated to reviewing representative examples of nitrogen-doped carbon materials derived from lignocellulosic biomass residues used in electrocatalytic and catalytic applications, as well as electrodes in supercapacitors.

10.3.1 Electrocatalytic and Catalytic Applications

Electrochemical devices are known to play a key role in the energy scenario. However, the search for efficient electrocatalysts is the cornerstone of the fruitful implementation of electrochemical systems such as fuel cells and metal-air batteries [114, 115]. Biomass-derived nitrogen-doped carbon materials have received great attention in that field of research. In particular, they are of great interest in the development of electrocatalysts for the oxygen reduction reaction (ORR). Many kinds of lignocellulosic biomass have been used as the starting material for the preparation of both metal-free and metal-containing ORR electrocatalysts.

As for the metal-free electrocatalysts, Wang et al. explored the performance of amaranthus-derived nitrogen-doped carbon by using different kinds of residues (leaf, stalk, and extract remains). The developed materials were denoted as NDC-X-Y, X standing for the kinds of residue (L is the leaf, S is the stalk, and E is the remaining extract of natural amaranthus red), and Y stands for the annealing temperature used in the synthesis procedure (600, 700, 800, and 900 °C) [116]. The resulting carbon materials were washed with HCl and hot deionized water to remove any possible metal in the samples. Figure 10.2 contains information of the characterization of the prepared samples. XPS characterization was used to know the elemental composition and it was seen that only C, N and O were present in the samples. The N contents ranged from 1.42 at.% to 2.47 at.%. The detected N functional groups were pyridinic, quaternary, and graphitic, and, as expected, the contribution of graphitic nitrogen increased with the increasing annealing temperature. Raman spectra and XRD analysis evidenced the presence of graphitic phases for all the catalysts, with subsequent good conductivity (conductivity of 0.492 S cm^{-1}). The activity of the

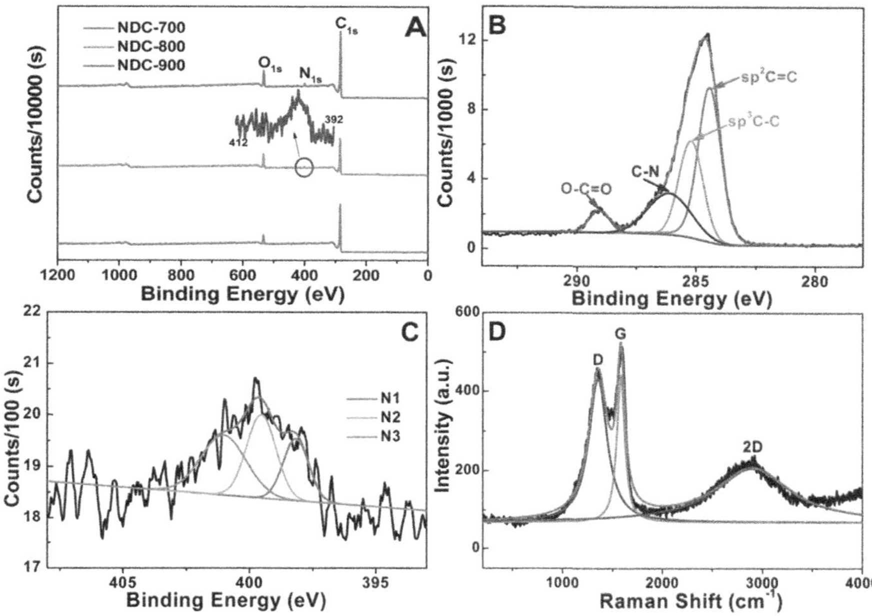

Fig. 10.2 (**a**) Full-scan XPS of NDC-L-700, NDC-L-800, and NDC-L-900. (**b** and **c**) C 1 s and high-resolution N 1 s XPS scan of NDC-L-800. (**d**) Raman spectrum of NDC-L-800. Reprinted with permission from [116]. Copyright © 2015, Royal Society of Chemistry

catalysts in the ORR was evaluated using cyclic voltammetry (CV) in a 0.1 M KOH solution saturated with N_2 or O_2 at the scan rate of 10 mV s^{-1} and rotating disk electrode (RDE). Linear sweep voltammetry (LSV) measurements were carried out to determine the onset potential. It was measured to be (0.24, 0.25, 0.27, 0.23) V vs. Hg/HgO for NDC-L-600, NDC-L-700, NDC-L-800, and NDC-L-900, respectively. Those values were close to the value obtained for the commercial Pt/C (0.30 V vs. Hg/HgO). It was also confirmed that NDC-L-800 had a 100% selectivity towards the $4e^-$ pathway. That electrocatalyst also displayed good stability compared to the reference material (Pt/C), which was claimed to result from the strength of the covalent C-N bond and the absence of activity decay originated in Pt/C from the sintering and leaching of Pt nanoparticles.

Chen et al. also prepared metal-free biomass-derived electrocatalysts for the ORR [117]. In that case, the authors used water hyacinth, a biomass feedstock with a high content of proteins that are mainly in the form of amino acids, such as leucine, valine, lysine, and tyrosine. From such feedstock, pyrolysis at various temperatures (i.e., 600–800 °C), and with $ZnCl_2$ as the activation agent, was carried out to attain biomass-derived nitrogen self-doped porous carbon. After that, the samples were rinsed with HNO_3 and HCl to remove any possible metal. XPS analysis confirmed the presence of only C, O, and N in the samples. The as-synthesized materials had a spongy and porous structure and, as expected, the structural order increased as the pyrolysis temperature increased. The nitrogen content was determined to be 5.02 at

%, 4.76 at%, and 2.80 at% for the samples prepared at 600 °C, 700 °C, and 800 °C, respectively, which was in the form of pyridinic, pyrrolic, and graphitic N and the last was the most abundant species in all the electrocatalysts. The samples were electrochemically characterized in a 0.1 M KOH solution saturated with N_2 or O_2 at a scan rate of 10 mV s^{-1}. It was seen that, based on the peak potential, the sample prepared at 700 °C was the best among the investigated, and the onset potential of that catalyst (+0.98 V vs. RHE) was comparable to that determined for the commercial Pt/C. Such performance was attributed to the nitrogen functionalization, high surface area and mesoporous structure of that electrocatalyst, which also showed excellent resistance against methanol crossover and long-term stability in continuous operation (30,000 s).

Yu et al. synthesized N-doped nanoporous carbon nanosheets derived from Typha orientalis by using a simple approach that consisted of the hydrothermal carbonization of the biomass precursor followed by an annealing in NH_3 atmosphere at 750 °C, 800 °C and 850 °C [118]. A reference sample was prepared under a N_2 atmosphere. The resulting material had abundant micropores and a high nitrogen content (9.1 at.% for the sample with the highest content). XPS analysis evidenced that the samples did not have metals. The developed electrocatalysts were assessed towards the ORR in both alkaline (0.1 KOH) and acidic (0.5 M H_2SO_4) medium. It was observed that, in both electrolytes, the sample prepared at 800 °C displayed the most promising activity. In an alkaline medium, that sample showed similar catalytic activity to the commercial 20% Pt/C catalysts and superior tolerance to methanol. In an acidic medium, it had an excellent catalytic ability, stability and tolerance to methanol, and the onset potential was only 85 mV lower than that of the reference Pt/C catalyst. However, this result is somewhat surprising since the electrocatalytic activity of N-doped carbons is very much dependent on the pH and it decreases when decreasing the pH. This has been explained in detail elsewhere [119]. That electrocatalyst also presented great tolerance to methanol poisoning and better durability than the commercial Pt/C benchmark sample. The good performance of the optimum catalyst was attributed to three aspects: (1) The presence of a large surface area with abundant micropores; (2) the high content of pyridine and pyrrolic functional groups, and (3) the 3D carbon structure that accelerated the reactant and electron transport.

Many examples of metal-containing biomass-derived electrocatalysts used in the ORR have been reported in the literature. For instance, Wark et al. used rye straw and coconut shells for the preparation of Fe-N-C catalysts [120]. In that study, phosphoric acid was used as the activation agent since the presence of P has been claimed to stabilize the resulting carbon material against electrochemical corrosion. Such a process can facilitate the loss of active Fe-N_x sites. After the activation of the biomass precursors, Fe-N-C catalysts were prepared by using iron(II) acetate and cyanamide as Fe and N precursors, respectively. A reference electrocatalyst based on commercial carbon black was also prepared. The resulting electrocatalysts were denoted as Fe-N-ox-BP, Fe-N-aRS, and Fe-N-aCoco, for the carbon black, rye straw, and coconut shell-derived materials, respectively. It was observed that biomass-based catalysts displayed better stability and much lower activity loss during potential cycling, which was attributed to the presence of phosphorus species and a higher content of pyrrolic N species. Li et al. recently reported on the preparation of efficient Fe-N-C electrocatalysts for ORR using a trunk of koelreuteria paniculate as the carbon precursor [121]. In that case, the nitrogen-containing carbon material (denoted as N-C_{wood}) was prepared with CH_3COOH, H_2O_2 and H_2O in the hydrothermal reaction and the resulting material was heat-treated in presence of Na_2CO_3 and urea under an inert atmosphere at 800 °C. The Fe-containing counterpart was synthesized from an iron phthalocyanine, which was mixed with the N-C_{wood} and the mixture was pyrolyzed at 900 °C (Fe-N-C_{wood}). A reference sample was prepared using a commercial carbon black (XC-72) instead of N-C_{wood}. SEM micrographs confirmed that the morphology of N-C_{wood} did not change upon the incorporation of Fe. It was observed that the iron nanoparticles were homogeneously dispersed and that the lattice distance was 0.204 nm, which corresponds to the Fe_3C (220) planes or Fe (110) planes. The morphology of the reference material was also checked for comparison purposes. XPS analysis confirmed the presence of zero-valent Fe, Fe(II) and Fe(III) species, while pyridinic, pyrrolic, graphitic, and oxidized nitrogen were detected in the XPS N 1 s spectrum. The electrocatalytic tests evidenced the excellent activity of Fe-N-$C_{wood,}$ which had a half-wave potential of 0.90 V and an onset potential of 0.98 V. For comparison purposes, the half-wave potential and onset potential were also determined for Pt/C, Fe-N-C_{XC-72}, XC-72 and N-C_{wood} (half-wave potential of 0.83 V, 0.80 V, 0.67 V and 0.78 V vs RHE, respectively; onset potential of 0.96 V, 0.90 V, 0.78 V and 0.88 V vs RHE, respectively). The superior electrocatalytic activity of Fe-N-C_{wood} was assigned to the efficient Fe active sites and the hierarchical porous structure of the developed biomass-derived materials, which had micropores ($S_{micro} = 378\ m^2\ g^{-1}$), mesopores, ($S_{extern} = 462\ m^2\ g^{-1}$, centered at about 4 nm) and macropores (centered at about 1.3 μm).

As indicated above, many kinds of lignocellulosic biomass have been used for the preparation of electrocatalysts for the ORR. Table 10.1 lists representative examples of those electrocatalysts as well as the parameters that determine the electrocatalytic performance.

Aside from the use of biomass-derived carbon materials in ORR, other interesting applications have been reported in the literature. For instance, Li et al. reported on the preparation of nitrogen-containing bamboo shoots-derived carbon materials, which were used as electrocatalysts for the electro-synthesis of NH_3 [146]. Such an application is of great interest since NH_3 is an auspicious candidate for hydrogen storage. In that case, the nitrogen that is naturally present in the raw biomass used, which is in the form of proteins and amino acids, was decomposed to N-functionalities into the carbon matrix by following a HTC approach at 180 °C and subsequent pyrolysis at different temperatures (i.e. 600 °C, 800 °C, and 1000 °C, materials denoted as NC-T, where T represents the pyrolysis temperature). The sample treated at 800 °C (See Fig. 10.3) indicated that the sample has an interconnected and loose structure formed by graphitic carbon. The homogeneous distribution of N atoms was also confirmed by High-Angle Annular Dark-Field Scanning TEM (HAADF-STEM). Raman analysis confirmed the existence of

Table 10.1 Representative examples of ORR performance of lignocellulosic biomass-derived N-doped carbon materials

Biomass precursor	N-source	Metal	E_{onset}	Limiting current	N° of electrons	Stability	Reference
Juncus	[a]	–	0.96 V vs RHE	4.51 mA cm^{-2} at 0.4 V	~4.00 (K-L)	100.0% CV after 1000 cycles	[122]
Soybean straw	NH_3	Co	0.87 V vs RHE	–	4.00 (K-L) 3.75 (RRDE)	–	[123]
Typha orientalis	NH_3	–	~0.85 V vs RHE	–	4.00 (K-L) 3.85 (RRDE)	84.0% CA after 10,000 s	[118]
Malachium aquaticum	[a]	–	0.83 V vs RHE	–	4.00 (K-L)	100.0% CV after 3000 cycles	[124]
Stem taro	Melamine	–	~0.87 V vs RHE	–	~3.97 (K-L)	96.5% CA after 20,000 s	[125]
Corn stalks	Melamine	Fe	0.99 V vs RHE	–	~3.72 (K-L)	~100.0% CV After 5000 cycles	[126]
Wood	NH_3	–	~0.95 V vs RHE	4.80 mA cm^{-2} at 0.8 V	~4.00 (K-L)	86.2% CA after 30,000 s	[127]
Sawdust	NH_3	Co	0.98 V vs RHE	4.87 mA cm^{-2} at 0.2 V	~3.95 (RRDE)	98.1% CA after 10,000 s	[128]
Bamboo powder	Melamine	Fe	0.97 V vs RHE	–	~4.00 (K-L)	92.0% CA after 40,000 s	[129]
Alder wood	Dicyandiamide	–	0.09 V vs SCE	5.50 mA cm^{-2} at 1.2 V	4.00 (K-L)	–	[130]
Paper towel	NH_3	–	0.88 V vs RHE	4.77 mA cm^{-2} at 0.4 V	~4.00 (K-L)	90.0% CA after 2000 s	[131]
Peanut biomass	Vitamin B12	Co	0.87 V vs RHE	6.16 mA cm^{-2} at 0.4 V	3.81 (K-L)	94.4% CA after 15,000 s	[132]
Coconut shells	Urea	–	0.98 V vs RHE	–	~3.70 (K-L)	75.0% CA after 50,000 s	[133]
Spermacoce latifolia	[a]	–	~0.90 V vs RHE	4.80 mA cm^{-2} at 0.0 V	3.92 (K-L)	–	[134]
Alder wood	Dicyandiamide	–	0.085 V vs SCE	–	4.00 (K-L)	93.0% CA after 72,000 s	[135]

Sunflower seed husks	Dicyandiamide	Cu	~0.95 V vs RHE	–	3.92 (K-L)	93.0% CA after 30,000 s	[136]
Sugarcane	NH_3	FeCo	0.91 V vs RHE	2.60 mA cm^{-2} at 0.3 V	~3.82 (RRDE)	~75.0% CA after 10,000 s	[137]
Palm kernel shell	NH_3	FeCo	0.90 V vs RHE	3.10 mA cm^{-2} at 0.3 V	~3.83 (RRDE)	~80.0% CA after 10,000 s	[137]
Eichhornia crassipes	Hydrazine	–	0.82 V vs RHE	7.50 mA cm^{-2} at 0.2 V	3.60 (K-L)	–	[138]
Sanbai melon peel	Thiourea	–	0.92 V vs RHE	–	~4.00 (K-L)	77.3% CA after 15,000 s	[139]
Soybean	[a]	–	~0.98 V vs RHE	5.00 mA cm^{-2} at 0.8 V	~4.00 (K-L)	–	[140]
Soybean shells	NH_3	–	~0.97 V vs RHE	5.24 mA cm^{-2} at 0.8 V	3.88 (K-L)	92.0% CA after 10,000 s	[141]
Pulsatilla chinensis regel	[a]	–	~0.98 V vs RHE	4.36 mA cm^{-2} at 0.8 V	~3.50 (K-L)	87.0% CA after 20,000 s	[142]
Tobacco stem	Dicyandiamide	CoNi	0.92 V vs RHE	5.60 mA cm^{-2} at 0.0 V	3.70 (K-L)	~100.0% CV After 5000 cycles	[143]
Okara/catkins	[a]	–	0.82 V vs RHE	3.79 mA cm^{-2} at 0.3 V	~3.80 (K-L)	~95.0% CA after 10,000 s	[144]
Aubergines peel	[a]	–	~0.85 V vs RHE	5.60 mA cm^{-2} at 0.5 V	3.80 (RRDE)	~100.0% CV After 5000 cycles	[145]

K-L koutecky-levich. RRDE rotating ring disk electrode
[a]Nitrogen naturally present in biomass precursor

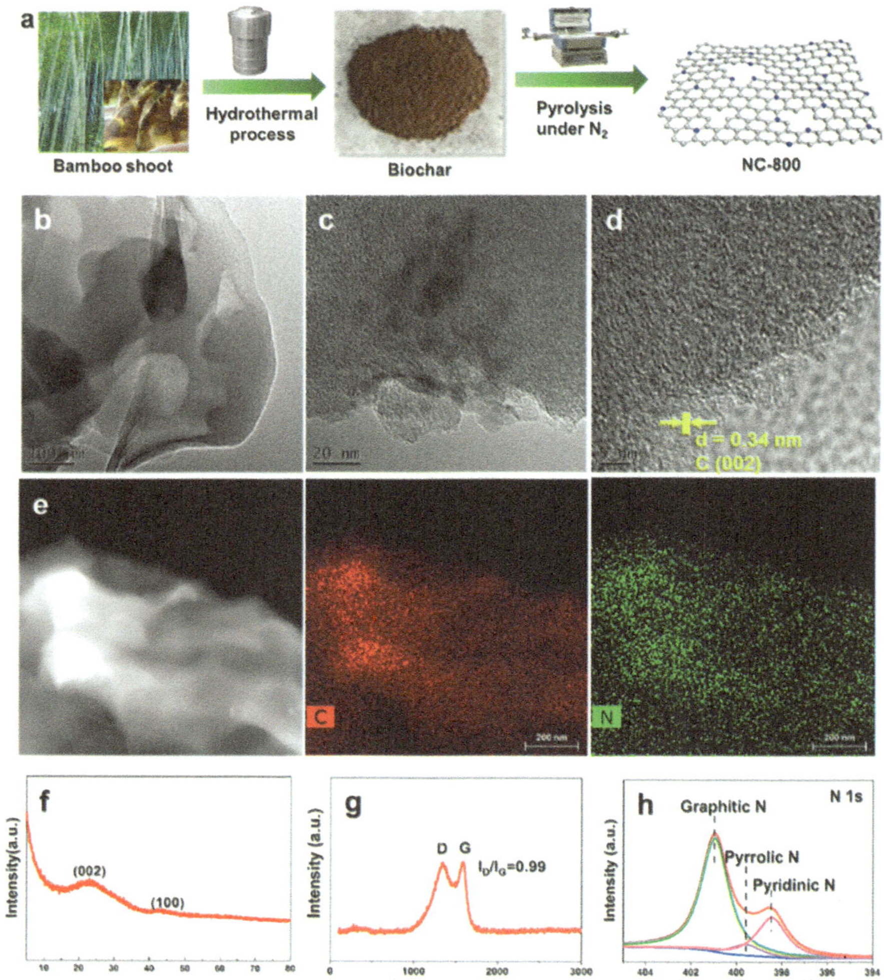

Fig. 10.3 Bamboo shoot-derived NC-800: (**a**) schematic illustration of fabrication process; (**b**), (**c**), and (**d**) HRTEM images of the catalyst NC-800; (**e**) HAADF-HRTEM image and corresponding elemental mapping; (**f**) XRD pattern; (**g**) Raman spectra; (h) N 1 s XPS spectra of NC-800 catalyst. Reprinted from [146]

graphitic order with a certain degree of defects, and XPS results indicated that the materials contained pyridinic, pyrrolic and graphitic nitrogen atoms. The electrochemical measurements indicated that the developed materials achieved a high NH_3 yield (16.3 $\mu gh^{-1}\ mg^{-1}$) and Faradaic efficiency (27.5%), in 0.1 M HCl solution at ambient conditions and at −0.35 V vs. RHE. The excellent electrochemical performance shown by the sample prepared at 800 °C was attributed to two aspects: (1) the large specific surface area and abundance of micropores, which make active sites more accessible and accelerate the electron and mass transfer process; (2) the presence of pyridinic nitrogen, which facilitated the activation of N_2 molecules.

Biomass-derived nitrogen-doped carbon materials have been used in other catalytic energy applications of great interest. For instance, it has been found that such materials can be used as supports of metal-based catalysts and that the presence of nitrogen prevents the sintering of the metal nanoparticles by serving as anchoring sites. Excellent catalysts have been developed for important applications such as the production of hydrogen from hydrogen carrier molecules. Our research group has significantly contributed to this field of research. We have recently developed biomass-derived activated carbons prepared from various biomass residues, such as hemp residues and almond shells, using a H_3PO_4-assisted HTC followed by carbonization and subsequent nitrogen functionalization. In this case, the nitrogen functionalization was done through an organic chemical reaction under mild conditions. For that, the activated carbon was mixed with a solution of NH_4NO_3 in DMF and pyridine was added dropwise to the resulting mixture. After that, the temperature was kept at 70 °C for 65 h. The resulting N-containing activated carbons were used as support for the preparation of catalysts for the dehydrogenation of formic acid in the liquid phase. This is a very hot topic since formic acid has been postulated as one of the most promising hydrogen carrier molecules. The production of hydrogen from formic acid takes place by the following reaction: $HCOOH \leftrightarrow H_2 + CO_2$.

In our first study, a hemp residue was used as the starting material for the preparation of the carbon supports. Pd was the metal phase used, and it was incorporated by using two synthetic approaches: (1) Impregnation of the support with the metal precursor followed by reduction with sodium borohydride, and (2) Impregnation of the support with the metal precursor and in situ reduction of the metal species under reaction conditions [97]. To assess the effect of nitrogen in the final performance, N-free and N-containing catalysts were prepared, characterized, and evaluated in the catalytic reaction. The results of the characterization indicated that the developed activated carbon was a microporous solid, with an apparent surface area above 1500 $m^2\ g^{-1}$, and that most of the pores were between 0.5 nm and 3 nm. It was observed that the porous texture of the activated carbon was neither modified after the incorporation of nitrogen functional groups nor after the incorporation of Pd nanoparticles. From XPS results, it was determined that nitrogen was in the form of pyridone/pyrrole, amine/amide, and pyridine groups, with a relative proportion of 26%, 45%, and 29%, respectively. The synthesized catalysts were evaluated in six consecutive reaction cycles by monitoring the gas evolution ($H_2 + CO_2$) during 30 min of reaction. The studied materials displayed excellent stability under reaction conditions, which is a very remarkable result since most of the catalysts used in this application show poor stability and frequently display a significant activity decay after one or two consecutive catalytic cycles. In the samples with non-pre reduced nanoparticles, an induction time was needed in the first reaction cycle for the gas to start evolving, the time in which the reduction of Pd precursor to attaining Pd nanoparticles took place.

Fresh and spent catalysts were characterized using Transmission Electron Microscopy (TEM), which suggested that, even though an increase in the average nanoparticle size was observed in all samples, the presence of nitrogen served to prevent the nanoparticles from sintering. That observation, together with the results seen from Pd XPS analysis, confirmed the important Pd-N interaction in the N-containing samples. The beneficial effect of the presence of N-functional groups was also evidenced by the catalytic results. Among investigated, the most promising catalyst was the sample with nitrogen in which the Pd nanoparticles were reduced under reaction conditions (denoted as Pd/NHTC (n.r.)). The stability of that catalyst was further assessed by performing six additional reaction cycles so that the gas evolution profiles were registered for a total of 12 consecutive cycles. It was seen that the initial activity of that catalyst was preserved during the cycles, which, together with the high value achieved for the Turnover Frequency number (TOF = 8365 h^{-1} calculated for the second reaction cycle and based on the surface Pd atoms), made the studied catalytic system a promising catalyst for the dehydrogenation of formic acid.

The interesting results achieved in that study motivated us to deepen into the catalytic behavior of catalysts based on Pd nanoparticles supported on biomass residue-derived activated carbon. In a second study, we synthesized similar catalysts and we further explored the effect of the properties of the carbon support by modification of their surface chemistry [147]. Such modification of the raw activated carbon (denoted as BC) was achieved by following three different strategies: (1) incorporation of nitrogen functional groups (support denoted as N-BC); (2) modification of the surface chemistry by performing a thermal treatment at 900 °C (support denoted as BC_TT) and (3) combination on both thermal treatment and nitrogen functionalization (support denoted as N-BC_TT). As indicated by the temperature-programmed desorption (TPD) measurements an important decrease in the amount of CO-type groups compared to the pristine activated carbon was detected for the heat-treated support, confirming the decrease in oxygen functional groups in the treated samples. The pH at the point of zero charge (pH_{PZC}) was measured to check the acid-base character of the different supports and the results of the measurement indicated that the pristine carbon material had an acid character, with pH_{PZC} of 4.8, while the pH_{PZC} increased upon thermal treatment (pH_{PZC} of 6.6) and nitrogen incorporation (pH_{PZC} of 5.6). That value increased further in the sample that was modified by both thermal treatment and nitrogen functionalization (pH_{PZC} of 7.1). Those activated carbons were loaded with Pd nanoparticles and the resulting catalysts (denoted as Pd/BC, Pd/N-BC, Pd/BC_TT, and Pd/N-BC_TT, according to the nomenclature used for the supports) were evaluated in the dehydrogenation of formic acid in the liquid phase. Figure 10.4 shows a summary of the catalytic activity of the studied samples, in which the gas evolution profiles achieved in the first, third, and sixth cycles are plotted.

As can be seen, the gas evolution profiles attained demonstrated the great stability of the catalysts, which preserved good activity after six consecutive reaction cycles. Among the studied samples, the catalyst based on N-containing heat-treated support (Pd/N-BC_TT) was the most promising sample, which suggested the potential of tuning the surface chemistry of the carbon support in attaining enhanced catalysts for the dehydrogenation of formic acid. This approach was also used in bimetallic PdAg

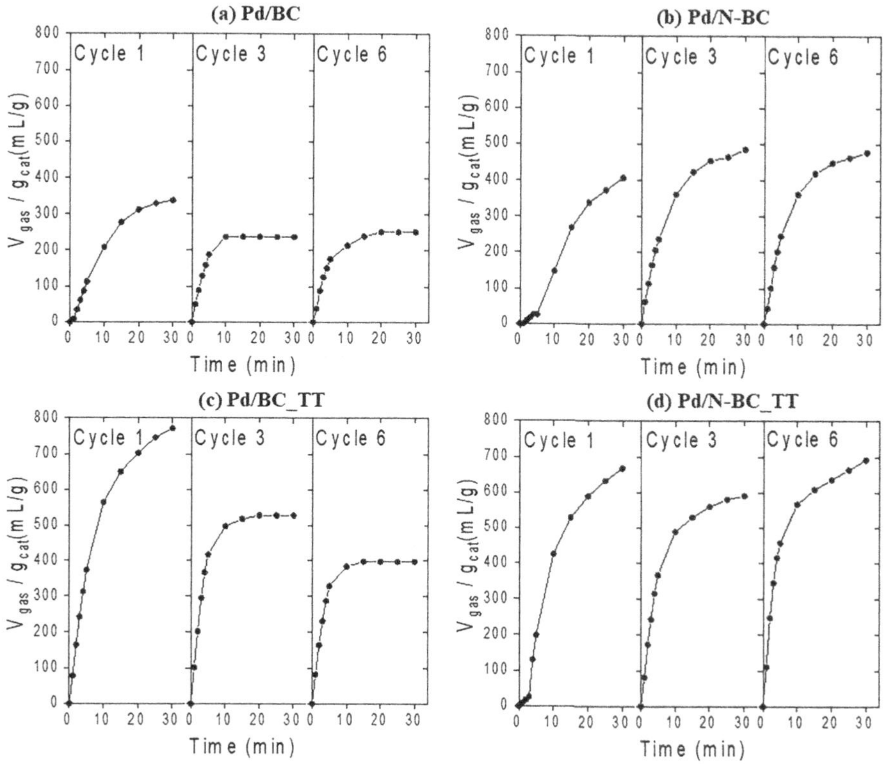

Fig. 10.4 Gas evolution profiles achieved for the first, third, and sixth cycle for: (**a**) Pd/BC, (**b**) Pd/N-BC, (**c**) Pd/BC_TT, and (d) Pd/N-BC_TT. Reprinted from [147]

systems supported on nitrogen-containing activated carbons, which also exhibited great activity and stability. In these catalysts, the nitrogen functional groups endowed the carbon supports with a basic character that is beneficial for the dehydrogenation of formic acid, since it increases the local concentration of formic acid molecules close to the metal active sites. Such nitrogen functionalities can also modify the electronic properties of the metal species and stabilize the metal nanoparticles preventing them from sintering. As observed in our studies [97, 147, 148], the incorporation of nitrogen in the support favors the formation of electron-deficient Pd species, which are reported to enhance the catalytic performance since such Pd^{2+} species interact with the formate ion.

Xiang et al. also explored the performance of catalysts based on N-doped biomass-derived carbon materials [149]. In that study, poplar catkins were used as the starting material for the preparation of 2D carbon material, prepared by carbonization at 800 °C, followed by ultrasonic crushing and subsequent modification processes using melamine. The resulting N-doped carbon microsheets (NPC) were loaded with Pd and Co_2P nanoparticles. The developed catalysts displayed well-dispersed Pd-Co_2P nanoparticles and had a TOF of 2980 h^{-1} at 50 °C. The most remarkable aspect was the good stability that those samples presented during seven consecutive recyclability cycles, maintaining 97.6% of the initial activity in terms of hydrogen production, which was attributed to the effect of Co_2P that improves the anti-poisoning resistance of the catalyst.

Some other studies addressing the use of biomass-derived carbon materials, both N-doped [150, 151] and N-free as catalytic supports for the dehydrogenation of formic acid [152, 153] can be found in the recent literature. However, it is still a relatively unexplored field that well deserves further efforts to fully exploit their potential as promising candidates for the preparation of catalytic supports.

10.3.2 Electrodes in Supercapacitors

Another field of application in which biomass wastes have been widely used as precursors of carbon materials is in energy storage devices, particularly, electrochemical capacitors (also named supercapacitors), whose electrodes are mainly made of porous carbon materials. The search for alternative precursors to reduce the cost of the electrodes has increased the interest in the use of biomass residues, and the wide variety of lignocellulosic biomass residues makes them an important resource. Many studies can be found in the literature, and some representative examples are listed in Table 10.2.

Liu et al. reported the preparation of nitrogen-rich activated carbons using $CaCl_2$ and urea in a one-step process using sugarcane bagasse as the biomass precursor [154]. The mixture was heated at the optimum temperature of 800 °C for 2 h (S_{122}). For comparison purposes, carbon materials were synthesized following the same procedure as for S_{122} but without $CaCl_2$ (S_{102}) or without urea (S_{120}). Elemental analysis (EA) and XPS demonstrated that the high nitrogen doping level for S_{122} (8.92 wt% and 8.34 at.%). In addition, the authors of that study reduced by half the amount of urea (S_{121}), and the resulting N-doped carbon material still showed 6.70 wt% and 5.59 at.% N content determined from EA and XPS analysis, respectively. The results suggest that the combination of $CaCl_2$ with urea is a very interesting strategy to prepare N-rich porous carbon materials. The generated nitrogen groups were mainly pyridinic (25–30 at.%) and pyrrolic (48–69 at.%), while quaternary and oxidized nitrogen groups were quite less abundant. The authors indicated that the high amount of pyridinic, pyrrolic and oxidized nitrogen groups compared to the quaternary nitrogen groups for the as-prepared samples could play an interesting role in the supercapacitor performance since those nitrogen species on the surface can contribute to the pseudocapacitance. The electrochemical performance in an alkaline electrolyte (6 M KOH) using a three-electrode system revealed that S_{122} and S_{121} displayed remarkable electrochemical behavior for their use as electrodes in supercapacitor. The authors suggested that the high and suitable nitrogen doping endowed the carbon materials with hydrophilicity and polarity, which in turn, resulted in the generation of pseudocapacitance. The two-electrode

Table 10.2 Representative examples of supercapacitors based on lignocellulosic biomass-derived N-doped carbon materials

Biomass precursor	N-source	Configuration/electrolyte	E_{max} (Wh kg^{-1})[b]	P (kW kg^{-1})[c]	$C_{retention}$ (%)[d]	Reference
Bean dregs and rice husk	[a]	Symmetric/6 M KOH	15.3(1 A g^{-1}, 1 V)	0.17	92.6(10,000, 4 A g^{-1})	[161]
Elm flower	[a]	Symmetric/6 M KOH	16.8(4 A g^{-1}, 1.2 V)	0.6	87.2(5000, 4 A g^{-1})	[162]
White clover	[a]	Symmetric/0.5 M Na_2SO_4	30.0(1 A g^{-1}, 2 V)	0.50	–	[163]
Bean pulp	[a]	Symmetric/6 M KOH	3.6(–, 1 V)	0.06	93.0(20,000, 10 A g^{-1})	[104]
Celery	[a]	Symmetric/6 M KOH	7.49(1 A g^{-1}, 1 V)	0.15	80.6(5000, –)	[164]
Celery	[a]	Symmetric/3 M $LiNO_3$	10.1(0.5 A g^{-1}, 1.5 V)	0.23	97.6(5000, 2 A g^{-1})	[164]
Cherry calyces	[a]	Symmetric/6 M KOH	12.2(0.1 A g^{-1}, 1 V)	0.03	94.8(10,000, 10 A g^{-1})	[165]
Cherry calyces	[a]	Symmetric/1 M Li_2SO4	22.8(0.5 A g^{-1}, 1.6 V)	0.20	94.9(10,000, 10 A g^{-1})	[165]
Cherry calyces	[a]	Symmetric/ionic liquid EMIM-TFSI	81.4(0.5 A g^{-1}, 3.6 V)	0.45	92.7(6000, 10 A g^{-1})	[165]
Sugarcane bagasse	Urea	Symmetric/6 M KOH	10.4(0.25 A g^{-1}, 1 V)	0.25	~100(10,000, 5 A g^{-1})	[154]
Cornstalk	Urea	Symmetric/1 M Na_2SO_4	10.0(0.5 A g^{-1}, 1 V)	0.25	99.8(10,000, –)	[155]
Persimmon branches	Urea	Li-ion capacitor/1 M $LiPF_6$ in ethylene and dimethyl carbonate	105.5(0.1 A g^{-1}, 1.7 V)	13.3	92.0(8000, 0.1 A g^{-1})	[166]
Paddy	Dicyandiamide	Symmetric/6 M KOH	14.4(20 A g^{-1}, 1.2 V)	–	93.3(10,000, 4 A g^{-1})	[156]
Hemp stem	Ammonia	Symmetric/6 M KOH	7.3(0.2 A g^{-1}, 1 V)	–	–	[157]
Hemp stem	Ammonia	Symmetric/ionic liquid EMIM-TFSI	99.5(0.5 A g^{-1}, 3.8 V)	0.47	94(2500, 10 A g^{-1})	[157]
Ginkgo leaf	Ammonia	Asymmetric/1 M KOH	42.2(1 A g^{-1}, 1.6 V)	0.70	70(10,000, 4 A g^{-1})	[167]

(continued)

Table 10.2 (continued)

Biomass precursor	N-source	Configuration/electrolyte	E_{max} (Wh kg^{-1})[b]	P (kW kg^{-1})[c]	$C_{retention}$ (%)[d]	Reference
Ricinus communis seed	Ammonia	Hybrid/3 M KOH	101.6(1 A g^{-1}, 1.8 V)	1.6	78.5(12,000, –)	[168]
Dragon fruit peels	Melamine	Symmetric/1 M $TEABF_4$/acetonitrile	112.0(–, 4 V)	3.21	109 (3.5 V)(5000, 50 mA cm^{-2})	[169]
Pine nut shells	Melamine	Symmetric/1 MNa_2SO_4	11.9(1 A g^{-1}, 1.8 V)	0.46	–	[170]
Wastes orange peel	Melamine	Symmetric/6 M KOH	23.3(0.7 A g^{-1}, 1 V)	2.33	–	[171]
Rice straw	Melamine	Symmetric/ionic liquid EMIM-TFSI	48.9(1 A g^{-1}, 3 V)	0.75	51(10,000, 5 A g^{-1})	[90]
Palm flowers	Melamine	Symmetric/ionic liquid [EMIM][ESO_4]	8.6(0.1 A g^{-1}, 3 V)	0.04	65(50,000, 5 A g^{-1})	[172]
Pueraria	Melamine	Symmetric/6 M KOH	8.5(0.5 A g^{-1}, 1 V)	0.12	74(10,000, 1 A g^{-1})	[91]
Houttuynia	Melamine	Symmetric/6 M KOH	15.9(–, 1 V)	0.50	95.74(10,000, 10 A g^{-1})	[173]
Hemp residue	NH_4NO_3/DMF	Symmetric/1 M TEMA-BF_4 in propyl-ene carbonate	17.9(0.25 A g^{-1}, 2.5 V)	0.28	86(5000, 1 A g^{-1})	[96]
Wood	NH_4NO_3/DMF	Symmetric/1 M TEMA-BF_4 in propyl-ene carbonate	18.6(0.25 A g^{-1}, 2.5 V)	0.30	90(5000, 1 A g^{-1})	[96]
Poplar sawdust	PANI	Symmetric/6 M KOH	15.5(0.5 A g^{-1}, 1.4 V)	3.50	–	[174]
Sugarcane bagasse	PANI	Symmetric/1 M H_2SO_4	27.7(1 A g^{-1}, 1.5 V)	0.77	95.5(5000, 5 A g^{-1})	[159]
Kapok fibers	PPy	Symmetric/6 M KOH	8.51(1 A g^{-1}, 1 V)	0.50	–	[160]
Cocozelle	PPy	Symmetric/2 M NaCl	70.5(0.8 mA cm^{-2}, 0.8 V)	0.40	76.9(1000, 0.8 mA cm^{-2})	[175]

[a]Nitrogen naturally present in biomass precursor
[b](current density, voltage)
[c]Power density supplied at $Voltage_{max}$
[d](cycles, current density)

Fig. 10.5 Strategy to synthesize paddy-derived N-doped hierarchical activated carbons by one-pot process using $KHCO_3$/dicyandiamide. Reprinted from [156]

configuration based on S_{122} was also assessed in 6 M KOH solution, and the symmetric capacitor supplied an energy output of 10.41 Wh kg^{-1} at 250 W kg^{-1}, and an outstanding stability was observed after 10,000 cycles of galvanostatic charge-discharge (~100%).

Urea was also used by Yu et al. for the synthesis of cornstalk-based N-doped porous carbon [155]. To do that, cornstalk powder was mixed with urea and the activating agent (KOH) in water. After centrifugation, the resulting slurry was dried and ground, and finally, the dried powder was thermally treated at 800 °C, leading to a N-doped activated carbon (NCSAC). The same procedure was used in absence of urea, and the activated carbon was denoted as CSAC, while the carbonized precursor was denoted as CSC. As expected, the electrochemical characterization of the different carbon materials in sodium sulphate revealed that the activated samples exhibited better rate performance than CSC due to their higher porosity development. Although the activated carbon samples displayed similar rate performance, NCSAC showed a higher gravimetric capacitance compared to non-doped activated carbon. Such an enhancement was assigned to the presence of nitrogen groups. XPS indicated that pyridinic and pyrrolic nitrogen were the main groups. To gain further insight into the potential applications of NCSAC as supercapacitor electrodes, a symmetric capacitor (NCSAC as positive and negative electrodes) was tested in the same electrolyte, giving an energy density of 10.0 Wh kg^{-1} at a power density of 249.9 W kg^{-1}. Moreover, this capacitor displayed outstanding capacitance retention (>99%) after 5000 cycles.

Dicyandiamide was used in a one-pot strategy to prepare N-doped hierarchical activated carbons from paddy [156]. Yuan et al. mixed puffed rice, $KHCO_3$ and dicyandiamide in water. After a drying step, such a mixture was heated at 800 °C for 2 h (PRK3N1) (see scheme in Fig. 10.5). Additionally, carbon materials were prepared under the same conditions using only $KHCO_3$ (PRK3) or dicyandiamide (PRN1).

A larger intensity ratio of D-band to G-band (I_D/I_G) was observed in the Raman spectra for the N-containing sample, which is in agreement with the defects introduced in the carbon materials upon nitrogen incorporation. The assessment of the chemical composition showed that the nitrogen content was 13.38% and 4.26% for PRN1 and PRK3N1, respectively. The deconvolution of XPS of PRK3N1 gave rise to three contributions: pyrrolic and pyridinic groups (which can provide pseudocapacitance and are electron-donating groups), and quaternary nitrogen (which might enhance the electrical conductivity and wettability). Such nitrogen composition together with a high porosity development of PRK3N1, makes it a suitable candidate as electrode material. A PRK3N1-based symmetric capacitor was assessed in 6 M KOH electrolyte using a potential window of 1.2 V, which is a high value in an alkaline medium. Regarding the capacitor performance, the energy density of this capacitor was 15.7 Wh kg^{-1} at 1 A g^{-1}, keeping 91.7% of this value at 20 A g^{-1}. This symmetric capacitor exhibited outstanding stability (93.3%) after 10,000 cycles.

Guo et al. reported the preparation of N-doped hierarchical porous carbon from hemp stem using simultaneous activation (with KOH) and doping processes (with NH_3) [157]. A simple impregnation of the residue with a KOH aqueous solution and a subsequent heat treatment at 750 °C for 2 h was performed under a NH_3 containing atmosphere (AKPC-750), obtaining a N-doped activated carbon with a high surface area (1950 m^2 g^{-1}). Furthermore, N-doped carbon materials without chemical activation (APC-750) and non-doped activated carbon (KPC-750) were prepared under the same experimental conditions. Elemental analysis indicated that the nitrogen content was 4.4 wt%, 2.8 wt%, and 1.1 wt%, for AKPC-750, APC-750, and KPC-750, respectively. Interestingly, AKPC-750 not only exhibited a high apparent surface area but also a high nitrogen content, which is quite difficult to obtain simultaneously. AKPC-750 displayed the best capacitance performance in 6 M KOH which was related to its porous texture and nitrogen content (mainly pyridinic and pyrrolic groups). The symmetric capacitor based on this carbon material supplied an energy density of 7.3 Wh kg^{-1}. To increase this energy density, the authors tested the symmetric capacitor in an ionic liquid electrolyte (1-Ethyl-3-methylimidazolium bis(trifluoromethylsulfonyl)imide, denoted as EMIM-TFSI), which could operate in a wider voltage window, obtaining an increase up to 99.5 Wh kg^{-1} in energy density and a power density of 473 W kg^{-1}, and at higher power density (21,600 W kg^{-1}), the energy density was 27.7 Wh kg^{-1}. However, the drawback of this strategy was its low yield (~10%) due to the high etching produced using NH_3 and KOH simultaneously.

Our group recently developed a synthesis approach that overcome such an issue related to the low yield [96]. In that case, a hemp residue was used to synthesize N-doped activated carbon following an organic chemistry route under mild conditions. In that case, a H_3PO_4-assisted hydrothermal carbonization using a low concentration of H_3PO_4 (25 wt%), and subsequent heat treatment at 450 °C was performed for the preparation of the activated carbon. The resulting carbon material (HTC_HR_450) showed a high porosity development (apparent surface area around at 1500 m^2 g^{-1}), and the yield of the activation process was close to 42%, which is a significant advantage compared to conventional activation methods used to obtain activated carbon derived from lignocellulosic biomass so far. Nitrogen chemical functionalization of HTC_HR_450 was performed using a 2 M NH_4NO_3/DMF solution and pyridine following the procedure previously described [96], and this N-doped activated carbon was denoted as HTC_HR_450-N. The activated carbon heat-treated up to 900 °C and its nitrogen functionalized counterpart were also

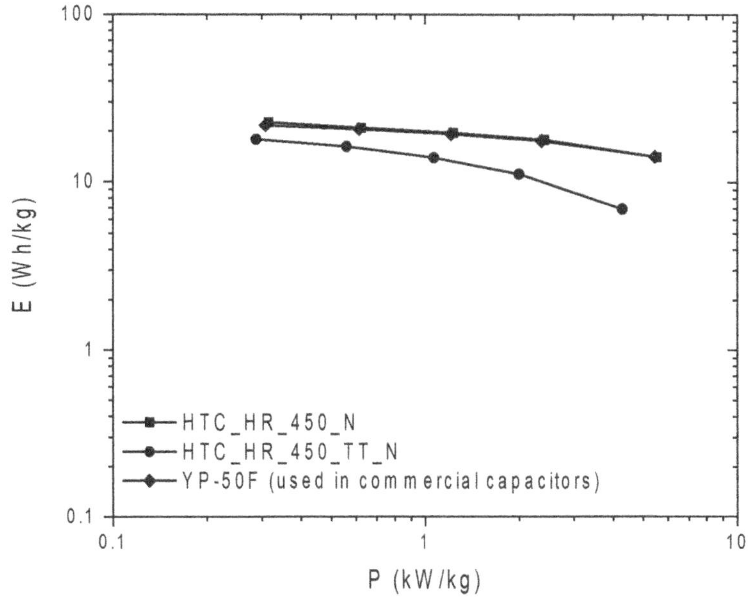

Fig. 10.6 Ragone plot of symmetric capacitors based on heat-treated N-doped activated carbons derived from hemp residue (HR) measured at a potential window of 2.5 V. Electrolyte: 1 M TEMA-BF_4/PC. Reprinted with permission from [96]. Copyright © 2020, Elsevier

prepared (HTC_HR_450_TT and HTC_HR_450_TT_N). Additionally, the same procedure was used with a commercial activated carbon as starting precursor instead of the hemp residue. The commercial activated carbon was a porous carbon prepared from wood by conventional chemical activation with H_3PO_4. The nitrogen incorporated (1.6–1.9 at.%), in the form of pyridine, amine/amide, and pyrrole groups, together with the heat treatment, increased the electrochemical stability, electrical conductivity and decreased the degradation of the activated carbon [77, 158]. The as-prepared activated carbons were used as electrodes in symmetric capacitors using an organic electrolyte (triethylmethylammonium tetrafluoroborate in propylene carbonate, TEMA-BF_4/PC). The electrochemical analysis revealed that the symmetric capacitor based on N-doped activated carbons exhibited good electrochemical performance. As Ragone plot shows, the energy density values at high power densities for N-doped HTC_HR-based capacitors were comparable to those of YP-50F-based capacitors (see Fig. 10.6), being YP-50F an activated carbon used in commercial supercapacitors. Additionally, the energy efficiency, as well as the capacitance retention obtained for such symmetric capacitors, were remarkable, being comparable to those obtained from a symmetric capacitor based on YP-50F. This was a promising result since a simple synthesis of nitrogen-containing activated carbon using a renewable precursor has led to the assembling of a competitive symmetric capacitor.

Celery-derived activated carbon (C-2) was the starting material of the activated carbon/polyaniline (PANI) composite (P/C-1), which was carbonized to obtain N-doped hierarchical porous carbon (P/C-3) [102]. Additionally, activated carbon/PANI composite was electrochemically synthesized (P/C-2) and subsequently heat treated (CP/C-2). The as-synthesized carbon materials were assessed as electrodes in supercapacitor using a three-electrode cell, but such N-doped hierarchical porous carbons were not evaluated in a capacitor configuration. However, the results of the electrochemical characterization indicated that the carbonized samples (P/C-3 and CP/C-2) displayed a better rate performance. This behavior was attributed to outstanding electrical conductivity and a higher nitrogen content compared to the non-carbonized material. It indicated that the biomass-derived activated carbons modified with PANI after carbonization are good candidates as electrodes in supercapacitors.

Wang et al. utilized sugarcane bagasse to prepare 3-dimensional N-doped porous carbon [159]. Briefly, polyaniline polymerization was carried out in a KOH-activated carbon (SBDC), and subsequent carbonization and activation processes resulted in a 3D N-doped porous carbon (named NSBDC) (Fig. 10.7). Furthermore, nitrogen-doped PANI-derived carbon was synthesized following the same process (NPDC).

From the three-electrode cell characterization in 1 M H_2SO_4, it was observed that NPDC exhibited poor capacitance retention (less than 50%) due to the poor porosity development, indicating that it might not be suitable as a supercapacitor electrode. However, SBDC/PANI was tested in a three electrode-cell, and it exhibited a noticeable rate performance. The supercapacitor performance of these N-doped carbon materials was also checked in a two-electrode cell configuration, using 1 M H_2SO_4 as electrolyte. The best-performing capacitor was the asymmetric configuration based on SBDC/PANI and NSBDC as positive and negative electrodes, respectively. The energy density of this asymmetric capacitor dropped from 32.5 Wh kg^{-1} to 12.1 Wh kg^{-1} at high power density (~7500 W kg^{-1}), while SBDC/PANI||NSBDC capacitor not only exhibited an energy density of 49.4 Wh kg^{-1} at 751 W kg^{-1}, but its energy density was 27.8 Wh kg^{-1} at 7758 W kg^{-1}. Regarding stability, this last asymmetric capacitor exhibited good capacitance retention (92.2% at 7 A g^{-1}) after 2000 cycles. Additionally, NSBDC-based symmetric capacitor was assembled and its total capacitance hardly decayed after 5000 cycles. Such an interesting capacitor performance was assigned to its unique hierarchical porous structure as well as its nitrogen functionalities, which enhanced both charge transfer and pseudocapacitance.

Polypyrrole is another N-containing polymer extensively used in the supercapacitor field. Sun et al. prepared tubular carbon aerogel with N-containing sandwich-like walls using kapok fibers and pyrrole as biomass and nitrogen precursors, respectively (See Fig. 10.8) [160]. Polymerization of pyrrole was carried out in kapok fibers using several pyrrole contents (KA-x, x = mL of pyrrole). KA-x samples were carbonized at 300 °C, 600 °C, 700 °C, 800 °C, and 900 °C under a nitrogen atmosphere, and then N_2 flow was changed to CO_2 flow. The activation process was performed for 1 h (samples denoted as KA-x-T).

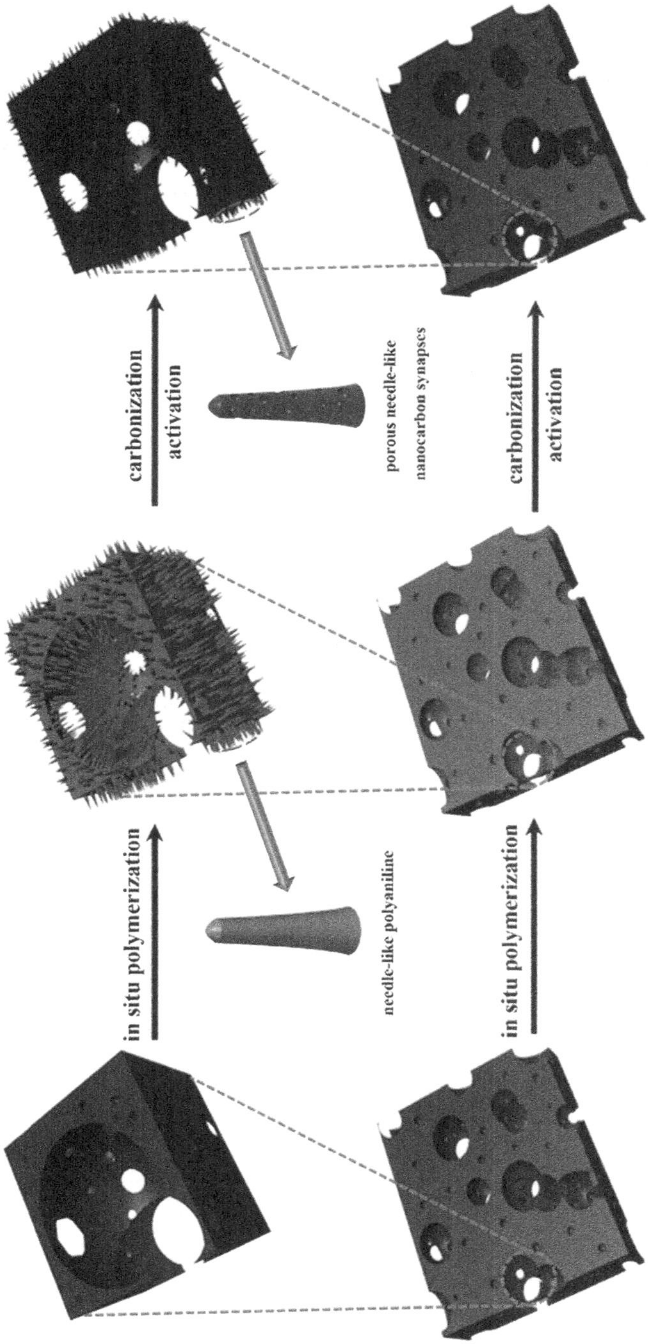

Fig. 10.7 Scheme for the synthesis of 3-dimensional N-doped porous carbon (NSBDC). Reprinted with permission from [159]. Copyright © 2018, Elsevier

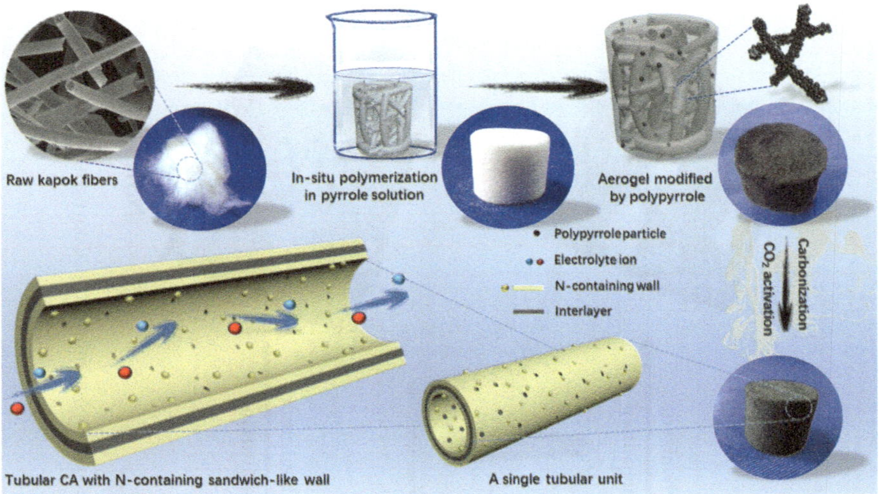

Fig. 10.8 Synthesis method of carbon aerogel with N-containing sandwich-like wall from kapok fibers. Reprinted with permission from [160]. Copyright © 2019, Elsevier

Nitrogen doping played a key role in keeping the tubular morphology of kapok fibers, as could be observed in non-doped samples, which showed an increase in the number of holes as the temperature increased. From XPS, the authors observed that kapok/polypyrrole composite only presented pyrrolic groups, while the samples that were carbonized and activated had pyridinic, pyrrolic, and quaternary groups. The amount of quaternary groups raised as temperature increased, while pyrrolic groups diminished, indicating that five-membered rings were converted into six-membered rings. CA-2-800 exhibited the best electrochemical performance in a 6 M KOH aqueous solution (three-electrode cell) even though it is not the carbon material with the largest nitrogen content. It showed noticeable stability after 4000 cycles at a high current density. The maximum energy density obtained with the symmetric capacitor was 8.51 Wh kg^{-1} at a power density of 500 W kg^{-1}. Although the stability of such a symmetric capacitor was not provided, the strategy for combining lignocellulosic biomass wastes, which can present different natural structures (tubular in this case), and the nitrogen-containing material, paves the way towards the design of unlimited storage devices.

10.4 Summary and Outlook

Biomass is a widely available renewable source that has been used since the dawn of humanity. However, even though the massive use of fossil fuels relegated biomass to second place, its great potential is again exploited. Lignocellulosic biomass has been prevalently investigated and considered as one of the most auspicious options to replace non-renewable fossil feedstock in the production of energy and products of interest in the chemical industry. The development of carbon materials from biomass sources has attracted great interest in the last years. Sophisticated materials with tunable properties have been developed by designing the synthetic route and carefully selecting the precursors. The incorporation of heteroatoms into the carbon matrix is a well-known strategy to afford carbon materials with optimized properties. N-doped biomass-derived carbon materials are excellent candidates for numerous applications. In the present chapter, several strategies for the preparation of N-doped carbon materials prepared from lignocellulosic biomass residues have been detailed. Representative examples of N-doped carbon materials in energy-related applications are discussed.

Among those applications, N-doped carbon materials have shown excellent applicability towards oxygen reduction reaction, either as catalyst or catalyst support. This outstanding electrochemical behavior is related to the formation of different N-related active sites that boost the reduction of oxygen molecules.

N-doped lignocellulosic biomass-derived activated carbons have also been successfully used as support for metal catalysts. In particular, their application in Pd-based catalysts for the dehydrogenation of formic acid is herein described. The results obtained show the importance of modulating properties of the carbon supports and emphasizing the role of nitrogen functionalities in attaining well-dispersed metal nanoparticles of small size and narrow size distribution.

N-doped activated carbons derived from lignocellulosic biomass have been profusely investigated in the field of supercapacitors. A variety of lignocellulosic precursors with a wide range of chemical compositions and structures have led to activated carbons with high quality and good performance. As a general rule, the incorporation of certain nitrogen groups endows the resulting carbon materials with improved properties, such as electrical conductivity, electron transfer, and increase in wettability. Because of this, the obtained carbon materials have suitable properties for their use as electrodes in supercapacitors.

The present chapter highlights the potential and versatility of N-doped lignocellulosic biomass-derived carbon materials and their great promise in energy-related applications. It is expected that the tremendous scope of these materials will draw further attention from the scientific community and their benefits in other environmental and energy-related applications will soon be explored.

Acknowledgments The authors thank RTI2018-095291-B-I00, PID2019-105923RB-I00 and PID2021-123079OB-I00 projects funded by MCIN/AEI/10.13039/501100011033 and "ERDF A way of making Europe". MNG gratefully acknowledges the Plan GenT project (CDEIGENT/2018/027).

References

1. Bajwa DS, Peterson T, Sharma N, Shojaeiarani J, Bajwa SG. A review of densified solid biomass for energy production. Renew Sustain Energy Rev. 2018;96:296–305. https://doi.org/10.1016/j.rser.2018.07.040.
2. Tursi A. A review on biomass: importance, chemistry, classification, and conversion. Biofuel Res J. 2019;6:962–79. https://doi.org/10.18331/BRJ2019.6.2.3.
3. Saxena RC, Adhikari DK, Goyal HB. Biomass-based energy fuel through biochemical routes: a review. Renew Sustain Energy Rev. 2009;13:167–78. https://doi.org/10.1016/j.rser.2007.07.011.
4. Shafiee S, Topal E. When will fossil fuel reserves be diminished? Energy Policy. 2009;37:181–9. https://doi.org/10.1016/j.enpol.2008.08.016.
5. Bentsen NS, Felby C. Biomass for energy in the European union - a review of bioenergy resource assessments. Biotechnol Biofuels. 2012;5:1–10. https://doi.org/10.1186/1754-6834-5-25.
6. Hussain A, Arif SM, Aslam M. Emerging renewable and sustainable energy technologies: state of the art. Renew Sustain Energy Rev. 2017;71:12–28. https://doi.org/10.1016/j.rser.2016.12.033.
7. Panwar NL, Kaushik SC, Kothari S. Role of renewable energy sources in environmental protection: a review. Renew Sustain Energy Rev. 2011;15:1513–24. https://doi.org/10.1016/j.rser.2010.11.037.
8. Kemp R, Martens P. Sustainable development: how to manage something that is subjective and never can be achieved? Sustain Sci Pract Policy. 2007;3:5–14. https://doi.org/10.1080/15487733.2007.11907997.
9. Nunes LJR, Causer TP, Ciolkosz D. Biomass for energy: a review on supply chain management models. Renew Sustain Energy Rev. 2020;120:109658. https://doi.org/10.1016/j.rser.2019.109658.
10. Doherty WOS, Mousavioun P, Fellows CM. Value-adding to cellulosic ethanol: lignin polymers. Ind Crop Prod. 2011;33:259–76. https://doi.org/10.1016/j.indcrop.2010.10.022.
11. Pasangulapati V, Ramachandriya KD, Kumar A, Wilkins MR, Jones CL, Huhnke RL. Effects of cellulose, hemicellulose and lignin on thermochemical conversion characteristics of the selected biomass. Bioresour Technol. 2012;114:663–9. https://doi.org/10.1016/j.biortech.2012.03.036.
12. Quesada-Plata F, Ruiz-Rosas R, Morallón E, Cazorla-Amorós D. Activated carbons prepared through H3PO4-assisted hydrothermal carbonisation from biomass wastes: porous texture and electrochemical performance. ChemPlusChem. 2016;81:1349–59. https://doi.org/10.1002/cplu.201600412.
13. Bonechi C, Consumi M, Donati A, Leone G, Magnani A, Tamasi G, Rossi C. Biomass: an overview. In: Dalena F, Basile A, Rossi C, editors. Bioenergy systems for the future prospects for biofuels and biohydrogen. Siena: Elsevier Ltd; 2017. p. 3–42. https://doi.org/10.1016/B978-0-08-101031-0.00001-6.
14. Kumar R, Strezov V, Weldekidan H, He J, Singh S, Kan T, Dastjerdi B. Lignocellulose biomass pyrolysis for bio-oil production: a review of biomass pre-treatment methods for production of drop-in fuels. Renew Sustain Energy Rev. 2020;123:109763. https://doi.org/10.1016/j.rser.2020.109763.
15. Ubando AT, Rivera DRT, Chen WH, Culaba AB. A comprehensive review of life cycle assessment (LCA) of microalgal and lignocellulosic bioenergy products from thermochemical processes. Bioresour Technol. 2019;291:121837. https://doi.org/10.1016/J.BIORTECH.2019.121837.
16. Caes BR, Teixeira RE, Knapp KG, Raines RT. Biomass to furanics: renewable routes to chemicals and fuels. ACS Sustain Chem Eng. 2015;3:2591–605. https://doi.org/10.1021/acssuschemeng.5b00473.

17. Kircher M. Sustainability of biofuels and renewable chemicals production from biomass. Curr Opin Chem Biol. 2015;29:26–31. https://doi.org/10.1016/j.cbpa.2015.07.010.
18. Stafford W, De Lange W, Nahman A, Chunilall V, Lekha P, Andrew J, Johakimu J, Sithole B, Trotter D. Forestry biorefineries. Renew Energy. 2020;154:461–75. https://doi.org/10.1016/j.renene.2020.02.002.
19. de Jong E, Higson A, Walsh P, Wellisch M. Task 42 biobased chemicals - value added products from biorefineries. A Rep Prep IEA Bioenergy-Task. 2011;42:1–33.
20. de Jong E, Jungmeier G. Chapter 1 - biorefinery concepts in comparison to petrochemical refineries. In: Pandey A, Höfer R, Taherzadeh M, Nampoothiri KM, Larroche C, editors. Industrial biorefineries & white biotechnology. Amsterdam: Elsevier Ltd; 2015. p. 3–33. https://doi.org/10.1016/B978-0-444-63453-5.00001-X.
21. Lin R, Man Y, Lee CKM, Ji P, Ren J. Sustainability prioritization framework of biorefinery: a novel multi-criteria decision-making model under uncertainty based on an improved interval goal programming method. J Clean Prod. 2020;251:119729. https://doi.org/10.1016/j.jclepro.2019.119729.
22. Dahiya S, Kumar AN, Shanthi Sravan J, Chatterjee S, Sarkar O, Mohan SV. Food waste biorefinery: sustainable strategy for circular bioeconomy. Bioresour Technol. 2018;248:2–12. https://doi.org/10.1016/j.biortech.2017.07.176.
23. Kamm B, Gruber PR, Kamm M. Biorefineries – industrial processes and products, Ullmann's Encyclopedia of industrial chemistry. Ullmann's Encycl Ind Chem. 2012:659–83. https://doi.org/10.1002/14356007.l04_l01.
24. Deng J, Li M, Wang Y. Biomass-derived carbon: synthesis and applications in energy storage and conversion. Green Chem. 2016;18:4824–54. https://doi.org/10.1039/C6GC01172A.
25. Liu WJ, Jiang H, Yu HQ. Development of biochar-based functional materials: toward a sustainable platform carbon material. Chem Rev. 2015;115:12251–85. https://doi.org/10.1021/acs.chemrev.5b00195.
26. Tan XF, Liu SB, Liu YG, Gu YL, Zeng GM, Hu XJ, Wang X, Liu SH, Jiang LH. Biochar as potential sustainable precursors for activated carbon production: multiple applications in environmental protection and energy storage. Bioresour Technol. 2017;227:359–72. https://doi.org/10.1016/j.biortech.2016.12.083.
27. Jain A, Ghosh M, Krajewski M, Kurungot S, Michalska M. Biomass-derived activated carbon material from native European deciduous trees as an inexpensive and sustainable energy material for supercapacitor application. J Energy Storage. 2021;34:102178. https://doi.org/10.1016/j.est.2020.102178.
28. Reza MS, Yun CS, Afroze S, Radenahmad N, Bakar MSA, Saidur R, Taweekun J, Azad AK. Preparation of activated carbon from biomass and its' applications in water and gas purification, a review. Arab J Basic Appl Sci. 2020;27:208–38. https://doi.org/10.1080/25765299.2020.1766799.
29. Tang ZE, Lim S, Pang YL, Shuit SH, Ong HC. Utilisation of biomass wastes based activated carbon supported heterogeneous acid catalyst for biodiesel production. Renew Energy. 2020;158:91–102. https://doi.org/10.1016/j.renene.2020.05.119.
30. Ogungbenro AE, Quang DV, Al-Ali KA, Vega LF, Abu-Zahra MRM. Synthesis and characterization of activated carbon from biomass date seeds for carbon dioxide adsorption. J Environ Chem Eng. 2020;8:104257. https://doi.org/10.1016/j.jece.2020.104257.
31. Titirici MM, Antonietti M. Chemistry and materials options of sustainable carbon materials made by hydrothermal carbonization. Chem Soc Rev. 2010;39:103–16. https://doi.org/10.1039/b819318p.
32. Hu B, Zhu HZ, Yu SH. Hydrothermally synthesized carbonaceous nanocomposites. In: Titirici MM, editor. Sustainable carbon materials from hydrothermal processes. Germany: John Wiley & Sons, Ltd.; 2013. p. 101–24.
33. Ipiales RP, de la Rubia MA, Diaz E, Mohedano AF, Rodriguez JJ. Integration of hydrothermal carbonization and anaerobic digestion for energy recovery of biomass waste: an overview. Energy Fuel. 2021;35:17032–50. https://doi.org/10.1021/acs.energyfuels.1c01681.

34. Kumar M, Olajire Oyedun A, Kumar A. A review on the current status of various hydrothermal technologies on biomass feedstock. Renew Sustain Energy Rev. 2018;81:1742–70. https://doi.org/10.1016/j.rser.2017.05.270.
35. Shen Y. A review on hydrothermal carbonization of biomass and plastic wastes to energy products. Biomass Bioenergy. 2020;134:105479. https://doi.org/10.1016/j.biombioe.2020.105479.
36. Cao X, Ro KS, Libra JA, Kammann CI, Lima I, Berge N, Li A, Li Y, Chen N, Yang J, Deng B, Mao J. Effects of biomass types and carbonization conditions on the chemical characteristics of hydrochars. J Agric Food Chem. 2013;61:9401–11. https://doi.org/10.1021/jf402345k.
37. Hoekman SK, Broch A, Robbins C, Zielinska B, Felix L. Hydrothermal carbonization (HTC) of selected woody and herbaceous biomass feedstocks. Biomass Convers Biorefin. 2013;3:113–26. https://doi.org/10.1007/s13399-012-0066-y.
38. Lu Y, Savage PE. Supercritical water gasification of lipid-extracted hydrochar to recover energy and nutrients. J Supercrit Fluids. 2015;99:88–94. https://doi.org/10.1016/j.supflu.2015.01.019.
39. Rosas JM, Berenguer R, Valero-Romero MJ, Rodríguez-Mirasol J, Cordero T. Preparation of different carbon materials by thermochemical conversion of lignin. Front Mater. 2014;1:1–17. https://doi.org/10.3389/fmats.2014.00029.
40. Marsh H, Rodríguez-Reinoso F. Chapter 5- activation processes (thermal or physical). In: Marsh H, Rodríguez-Reinoso F, editors. Activated carbon. Oxford: Elsevier Science Ltd; 2006. p. 243–321. https://doi.org/10.1016/B978-008044463-5/50019-4.
41. Marsh H, Rodríguez-Reinoso F. Chapter 6- activation processes (chemical). In: Marsh H, Rodríguez-Reinoso F, editors. Activated carbon. Oxford: Elsevier Science Ltd; 2006. p. 322–65. https://doi.org/10.1016/B978-008044463-5/50020-0.
42. Awasthi GP, Bhattarai DP, Maharjan B, Kim KS, Park CH, Kim CS. Synthesis and characterizations of activated carbon from wisteria sinensis seeds biomass for energy storage applications. J Ind Eng Chem. 2019;72:265–72. https://doi.org/10.1016/j.jiec.2018.12.027.
43. González-García P. Activated carbon from lignocellulosics precursors: a review of the synthesis methods, characterization techniques and applications. Renew Sustain Energy Rev. 2018;82:1393–414. https://doi.org/10.1016/j.rser.2017.04.117.
44. Sun G, Qiu L, Zhu M, Kang K, Guo X. Activated carbons prepared by hydrothermal pretreatment and chemical activation of Eucommia ulmoides wood for supercapacitors application. Ind Crop Prod. 2018;125:41–9. https://doi.org/10.1016/j.indcrop.2018.08.082.
45. Fic K, Platek A, Piwek J, Frackowiak E. Sustainable materials for electrochemical capacitors. Mater Today. 2018;21:437–54. https://doi.org/10.1016/j.mattod.2018.03.005.
46. Falco C, Marco-Lozar JP, Salinas-Torres D, Morallón E, Cazorla-Amorós D, Titirici MM, Lozano-Castelló D. Tailoring the porosity of chemically activated hydrothermal carbons: influence of the precursor and hydrothermal carbonization temperature. Carbon. 2013;62:346–55. https://doi.org/10.1016/j.carbon.2013.06.017.
47. Sun W, Lipka SM, Swartz C, Williams D, Yang F. Hemp-derived activated carbons for supercapacitors. Carbon. 2016;103:181–92. https://doi.org/10.1016/j.carbon.2016.02.090.
48. Redondo E, Carretero-González J, Goikolea E, Ségalini J, Mysyk R. Effect of pore texture on performance of activated carbon supercapacitor electrodes derived from olive pits. Electrochim Acta. 2015;160:178–84. https://doi.org/10.1016/j.electacta.2015.02.006.
49. Cheng J, Hu SC, Sun GT, Kang K, Zhu MQ, Geng ZC. Comparison of activated carbons prepared by one-step and two-step chemical activation process based on cotton stalk for supercapacitors application. Energy. 2021;215:119144. https://doi.org/10.1016/j.energy.2020.119144.
50. Vinayagam M, Suresh Babu R, Sivasamy A, Ferreira de Barros AL. Biomass-derived porous activated carbon from Syzygium cumini fruit shells and Chrysopogon zizanioides roots for high-energy density symmetric supercapacitors. Biomass Bioenergy. 2020;143:105838. https://doi.org/10.1016/j.biombioe.2020.105838.

51. Yahya MA, Al-Qodah Z, Ngah CWZ. Agricultural bio-waste materials as potential sustainable precursors used for activated carbon production: a review. Renew Sustain Energy Rev. 2015;46:218–35. https://doi.org/10.1016/j.rser.2015.02.051.
52. MacDermid-Watts K, Pradhan R, Dutta A. Catalytic hydrothermal carbonization treatment of biomass for enhanced activated carbon: a review. Waste Biomass Valorization. 2020;12: 2171–86. https://doi.org/10.1007/s12649-020-01134-x.
53. Rodríguez-Reinoso F, Molina-Sabio M, González MT. The use of steam and CO2 as activating agents in the preparation of activated carbons. Carbon. 1995;33:15–23. https://doi.org/10.1016/0008-6223(94)00100-E.
54. Lillo-Ródenas MA, Lozano-Castelló D, Cazorla-Amorós D, Linares-Solano A. Preparation of activated carbons from Spanish anthracite – II. Activation by NaOH. Carbon. 2001;39:751–9. https://doi.org/10.1016/S0008-6223(00)00186-X.
55. Sevilla M, Díez N, Fuertes AB. More sustainable chemical activation strategies for the production of porous carbons. ChemSusChem. 2021;14:94–117. https://doi.org/10.1002/cssc.202001838.
56. Linares-Solano A, Lozano-Castelló D, Lillo-Ródenas MA, Cazorla-Amorós D. Carbon activation by alkaline hydroxides preparation and reactions, porosity and performance. In: Radovic LR, editor. Chemistry and physics of carbon. 1st ed. Boca Raton: CRC Press; 2008. p. 1–62. https://doi.org/10.1201/9781420042993.
57. Maciá-Agulló JA, Moore BC, Cazorla-Amorós D, Linares-Solano A. Activation of coal tar pitch carbon fibres: physical activation vs. chemical activation. Carbon. 2004;42:1367–70. https://doi.org/10.1016/j.carbon.2004.01.013.
58. Jagtoyen M, Derbyshire F. Activated carbons from yellow poplar and white oak by H3PO4 activation. Carbon. 1998;36:1085–97. https://doi.org/10.1016/S0008-6223(98)00082-7.
59. Srinivasakannan C, Abu Bakar ZM. Production of activated carbon from rubber wood sawdust. Biomass Bioenergy. 2004;27:89–96. https://doi.org/10.1016/j.biombioe.2003.11.002.
60. Kumar A, Jena HM. Preparation and characterization of high surface area activated carbon from fox nut (Euryale ferox) shell by chemical activation with H3PO4. Results Phys. 2016;6: 651–8. https://doi.org/10.1016/j.rinp.2016.09.012.
61. Yorgun S, Yildiz D. Preparation and characterization of activated carbons from paulownia wood by chemical activation with H3PO4. J Taiwan Inst Chem Eng. 2015;53:122–31. https://doi.org/10.1016/j.jtice.2015.02.032.
62. Wiśniewski M, Pacholczyk A, Terzyk AP, Rychlicki G. New phosphorus-containing spherical carbon adsorbents as promising materials in drug adsorption and release. J Colloid Interface Sci. 2011;354:891–4. https://doi.org/10.1016/j.jcis.2010.11.072.
63. Berenguer R, Ruiz-Rosas R, Gallardo A, Cazorla-Amorós D, Morallón E, Nishihara H, Kyotani T, Rodríguez-Mirasol J, Cordero T. Enhanced electro-oxidation resistance of carbon electrodes induced by phosphorus surface groups. Carbon. 2015;95:681–9. https://doi.org/10.1016/j.carbon.2015.08.101.
64. Hulicova-Jurcakova D, Puziy AM, Poddubnaya OI, Suárez-García F, Tascón JMD, Lu GQ. Highly stable performance of Supercapacitors from phosphorus-enriched carbons. J Am Chem Soc. 2009;131:5026–7. https://doi.org/10.1021/ja809265m.
65. Pandolfo AG, Hollenkamp AF. Carbon properties and their role in supercapacitors. J Power Sources. 2006;157:11–27. https://doi.org/10.1016/j.jpowsour.2006.02.065.
66. An Y, Yang Y, Hu Z, Guo B, Wang X, Yang X, Zhang Q, Wu H. High-performance symmetric supercapacitors based on carbon nanosheets framework with graphene hydrogel architecture derived from cellulose acetate. J Power Sources. 2017;337:45–53. https://doi.org/10.1016/j.jpowsour.2016.10.112.
67. Zhao L, Baccile N, Gross S, Zhang Y, Wei W, Sun Y, Antonietti M, Titirici MM. Sustainable nitrogen-doped carbonaceous materials from biomass derivatives. Carbon. 2010;48:3778–87. https://doi.org/10.1016/j.carbon.2010.06.040.

68. Enterría M, Figueiredo JL. Nanostructured mesoporous carbons: tunin texture and surface chemistry. Carbon. 2016;108:79–102. https://doi.org/10.1016/j.carbon.2016.06.108.
69. Quílez-Bermejo J, Morallón E, Cazorla-Amorós D. Metal-free heteroatom-doped carbon-based catalysts for ORR. A critical assessment about the role of heteroatoms. Carbon. 2020;165:434–54. https://doi.org/10.1016/j.carbon.2020.04.068.
70. Inagaki M, Konno H, Tanaike O. Carbon materials for electrochemical capacitors. J Power Sources. 2010;195:7880–903. https://doi.org/10.1016/j.jpowsour.2010.06.036.
71. Salinas-Torres D, Navlani-García M, Mori K, Kuwahara Y, Yamashita H. Nitrogen-doped carbon materials as a promising platform toward the efficient catalysis for hydrogen generation. Appl Catal A Gen. 2019;571:25–41. https://doi.org/10.1016/J.APCATA.2018.11.034.
72. Matsagar BM, Yang RX, Dutta S, Ok YS, Wu KC. Recent progress in the development of biomass-derived nitrogen-doped porous carbon. J Mater Chem A. 2021;9:3703–28. https://doi.org/10.1039/d0ta09706c.
73. Kobina Sam D, Kobina Sam E, Lv X. Application of biomass-derived nitrogen-doped carbon aerogels in electrocatalysis and supercapacitors. ChemElectroChem. 2020;7:3695–712. https://doi.org/10.1002/celc.202000829.
74. Deng Y, Xie Y, Zou K, Ji X. Review on recent advances in nitrogen-doped carbons: preparations and applications in supercapacitors. J Mater Chem A. 2015;4:1144–73. https://doi.org/10.1039/C5TA08620E.
75. Inagaki M, Toyoda M, Soneda Y, Morishita T. Nitrogen-doped carbon materials. Carbon. 2018;132:104–40. https://doi.org/10.1016/j.carbon.2018.02.024.
76. Yamada Y, Tanaka H, Kubo S, Sato S. Unveiling bonding states and roles of edges in nitrogen-doped graphene nanoribbon by X-ray photoelectron spectroscopy. Carbon. 2021;185:342–67. https://doi.org/10.1016/j.carbon.2021.08.085.
77. Salinas-Torres D, Shiraishi S, Morallón E, Cazorla-Amorós D. Improvement of carbon materials performance by nitrogen functional groups in electrochemical capacitors in organic electrolyte at severe conditions. Carbon. 2015;82:205–13. https://doi.org/10.1016/j.carbon.2014.10.064.
78. Kato T, Yamada Y, Nishikawa Y, Otomo T, Sato H, Sato S. Origins of peaks of graphitic and pyrrolic nitrogen in N1s X-ray photoelectron spectra of carbon materials: quaternary nitrogen, tertiary amine, or secondary amine? J Mater Sci. 2021;56:15798–811. https://doi.org/10.1007/s10853-021-06283-5.
79. Chen Z, Jaworski A, Chen J, Budnyak TM, Szewczyk I, Rokicińska A, Dronskowski R, Hedin N, Kuśtrowski P, Slabon A. Graphitic nitrogen in carbon catalysts is important for the reduction of nitrite as revealed by naturally abundant 15N NMR spectroscopy. Dalt Trans. 2021;50:6857–66. https://doi.org/10.1039/d1dt00658d.
80. Kuroki S, Hosaka Y, Yamauchi C. A solid-state NMR study of the carbonization of polyaniline. Carbon. 2013;55:160–7. https://doi.org/10.1016/j.carbon.2012.12.022.
81. Niwa H, Horiba K, Harada Y, Oshima M, Ikeda T, Terakura K, Ozaki J-i, Miyata S. X-ray absorption analysis of nitrogen contribution to oxygen reduction reaction in carbon alloy cathode catalysts for polymer electrolyte fuel cells. J Power Sources. 2009;187:93–7. https://doi.org/10.1016/j.jpowsour.2008.10.064.
82. Kiuchi H, Kondo T, Sakurai M, Guo D, Nakamura J, Niwa H, Miyawaki J, Kawai M, Oshima M, Harada Y. Characterization of nitrogen species incorporated into graphite using low energy nitrogen ion sputtering. Phys Chem Phys. 2016;18:458–65. https://doi.org/10.1039/c5cp02305j.
83. González-Gaitán C, Ruiz-Rosas R, Morallón E, Cazorla-Amorós D. Electrochemical methods to functionalize carbon materials. In: Thakur VK, Thakur MK, editors. Chemical functionalization of carbon nanomaterials: chemistry and applications. Boca Raton: CRC Press; 2015. p. 231–62. https://doi.org/10.1201/b18724.
84. Jeon IY, Noh HJ, Baek JB. Nitrogen-doped carbon nanomaterials: synthesis, characteristics and applications. Chem - An Asian J. 2020;15:2282–93. https://doi.org/10.1002/asia.201901318.

85. Ćirić-Marjanović G, Pašti I, Mentus S. One-dimensional nitrogen-containingcarbon nanostructures. Prog Mater Sci. 2015;69:61–182. https://doi.org/10.1016/j.pmatsci.2014.08.002.
86. Shen W, Fan W. Nitrogen-containing porous carbons: synthesis and application. J Mater Chem A. 2013;1:999–1013. https://doi.org/10.1039/C2TA00028H.
87. Raymundo-Piñero E, Cazorla-Amorós D, Linares-Solano A. The role of different nitrogen functional groups on the removal of SO2 from flue gases by N-doped activated carbon powders and fibres. Carbon. 2003;41:1925–32. https://doi.org/10.1016/S0008-6223(03)00180-5.
88. Liang HW, Zhuang X, Brüller S, Feng X, Müllen K. Hierarchically porous carbons with optimized nitrogen doping as highly active electrocatalysts for oxygen reduction. Nat Commun. 2014;5:4973. https://doi.org/10.1038/ncomms5973.
89. Ma X, Fang M, Liu B, Chen R, Shi R, Wu Q, Zeng Z, Li L. Urea-assisted synthesis of biomass-based hierarchical porous carbons for the light hydrocarbons adsorption and separation. Chem Eng J. 2022;428:130985. https://doi.org/10.1016/j.cej.2021.130985.
90. Charoensook K, Huang CL, Tai HC, Lanjapalli VVK, Chiang LM, Hosseini S, Lin YT, Li YY. Preparation of porous nitrogen-doped activated carbon derived from rice straw for high-performance supercapacitor application. J Taiwan Inst Chem Eng. 2021;120:246–56. https://doi.org/10.1016/j.jtice.2021.02.021.
91. Han X, Jiang H, Zhou Y, Hong W, Zhou Y, Gao P, Ding R, Liu E. A high performance nitrogen-doped porous activated carbon for supercapacitor derived from pueraria. J Alloys Compd. 2018;744:544–51. https://doi.org/10.1016/j.jallcom.2018.02.078.
92. Zhang LS, Liang XQ, Song WG, Wu ZY. Identification of the nitrogen species on N-doped graphene layers and Pt/NG composite catalyst for direct methanol fuel cell. Phys Chem Phys. 2010;12:12055–9. https://doi.org/10.1039/c0cp00789g.
93. Yang Z, Zhang G, Guo X, Xu Y. Designing a novel N-doped adsorbent with ultrahigh selectivity for CO2: waste biomass pyrolysis and two-step activation. Biomass Convers Biorefin. 2021;11:2843–54. https://doi.org/10.1007/s13399-020-00633-0.
94. Navlani-García M, Salinas-Torres D, Vázquez-Álvarez FD, Cazorla-Amorós D. Formic acid dehydrogenation attained by Pd nanoparticles-based catalysts supported on MWCNT-C3N4 composites. Catal Today. 2021;397–399:428–35. https://doi.org/10.1016/j.cattod.2021.07.019.
95. Plavniece A, Volperts A, Dobele G, Zhurinsh A, Kaare K, Kruusenberg I, Kaprans K, Knoks A, Kleperis J. Wood and black liquor-based N-doped activated carbon for energy application. Sustainability. 2021;13:9237. https://doi.org/10.3390/su13169237.
96. Chaparro-Garnica J, Salinas-Torres D, Mostazo-López MJ, Morallón E, Cazorla-Amorós D. Biomass waste conversion into low-cost carbon-based materials for supercapacitors: a sustainable approach for the energy scenario. J Electroanal Chem. 2021;880:114899. https://doi.org/10.1016/j.jelechem.2020.114899.
97. Chaparro-Garnica J, Navlani-García M, Salinas-Torres D, Morallón E, Cazorla-Amorós D. Highly stable N-doped carbon-supported Pd-based catalysts prepared from biomass waste for H2 production from formic acid. ACS Sustain Chem Eng. 2020;8:15030–43. https://doi.org/10.1021/acssuschemeng.0c05906.
98. Quintero-Jaime AF, Cazorla-Amorós D, Morallón E. Effect of surface oxygen groups in the electrochemical modification of multi-walled carbon nanotubes by 4-amino phenyl phosphonic acid. Carbon. 2020;165:328–39. https://doi.org/10.1016/j.carbon.2020.04.062.
99. Martínez-Sánchez B, Quílez-Bermejo J, San-Fabián E, Cazorla-Amorós D, Morallón E. On the mechanism of electrochemical functionalization of carbon nanotubes with different structures with aminophenylphosphonic acid isomers: an experimental and computational approach. J Mater Chem A. 2022;10:7271–90. https://doi.org/10.1039/D1TA10322A.
100. Ramírez-Pérez AC, Quílez-Bermejo J, Sieben JM, Morallón E, Cazorla-Amorós D. Effect of nitrogen-functional groups on the ORR activity of activated carbon fiber-polypyrrole-based electrodes. Electrocatalysis. 2018;9:697–705. https://doi.org/10.1007/s12678-018-0478-y.

101. Bleda-Martínez MJ, Morallón E, Cazorla-Amorós D. Polyaniline/porous carbon electrodes by chemical polymerisation: effect of carbon surface chemistry. Electrochim Acta. 2007;52: 4962–8. https://doi.org/10.1016/j.electacta.2007.01.073.
102. Du W, Wang X, Sun X, Zhan J, Zhang H, Zhao X. Nitrogen-doped hierarchical porous carbon using biomass-derived activated carbon/carbonized polyaniline composites for supercapacitor electrodes. J Electroanal Chem. 2018;827:213–20. https://doi.org/10.1016/j.jelechem.2018.09.031.
103. Yu M, Han Y, Li Y, Li J, Wang L. Polypyrrole-anchored cattail biomass-derived carbon aerogels for high performance binder-free supercapacitors. Carbohydr Polym. 2018;199:555–62. https://doi.org/10.1016/j.carbpol.2018.04.058.
104. Ding Y, Li Y, Dai Y, Han X, Xing B, Zhu L, Qiu K, Wang S. A novel approach for preparing in-situ nitrogen doped carbon via pyrolysis of bean pulp for supercapacitors. Energy. 2021;216:119227. https://doi.org/10.1016/j.energy.2020.119227.
105. Cao X, Li Z, Chen H, Zhang C, Zhang Y, Gu C, Xu X, Li Q. Synthesis of biomass porous carbon materials from bean sprouts for hydrogen evolution reaction electrocatalysis and supercapacitor electrode. Int J Hydrog Energy. 2021;46:18887–97. https://doi.org/10.1016/j.ijhydene.2021.03.038.
106. Sekhon SS, Park JS. Biomass-derived N-doped porous carbon nanosheets for energy technologies. Chem Eng J. 2021;425:129017. https://doi.org/10.1016/j.cej.2021.129017.
107. Li Q, Wu X, Zhao Y, Miao Z, Xing L, Zhou J, Zhao J, Zhuo S. Nitrogen-doped hierarchical porous carbon through one-step activation of bean curd for high-performance supercapacitor electrode. ChemElectroChem. 2018;5:1606–14. https://doi.org/10.1002/celc.201800230.
108. Wei T, Wei X, Gao Y, Li H. Large scale production of biomass-derived nitrogen-doped porous carbon materials for supercapacitors. Electrochim Acta. 2015;169:186–94. https://doi.org/10.1016/j.electacta.2015.04.082.
109. Gao Q, Xiang H, Ni L, Hou Y, He Y, Feng Z, Yang J, Hu W, Liu Z. Nitrogen self-doped activated carbons with narrow pore size distribution from bamboo shoot shells. Colloids Surf A Physicochem Eng Asp. 2021;629:127408. https://doi.org/10.1016/j.colsurfa.2021.127408.
110. Luo W, Wang B, Heron CG, Allen MJ, Morre J, Maier CS, Stickle WF, Ji X. Pyrolysis of cellulose under ammonia leads to nitrogen-doped nanoporous carbon generated through methane formation. Nano Lett. 2014;14:2225–9. https://doi.org/10.1021/nl500859p.
111. Geng Z, Xiao Q, Lv H, Li B, Wu H, Lu Y, Zhang C. One-step synthesis of microporous carbon monoliths derived from biomass with high nitrogen doping content for highly selective CO2 capture. Sci Rep. 2016;6:30049. https://doi.org/10.1038/srep30049.
112. Tian W, Sun H, Duan X, Zhang H, Ren Y, Wang S. Biomass-derived functional porous carbons for adsorption and catalytic degradation of binary micropollutants in water. J Hazard Mater. 2020;389:121881. https://doi.org/10.1016/j.jhazmat.2019.121881.
113. Xu H, Zhang Y, Wang L, Chen Y, Gao S. Hierarchical porous biomass-derived carbon framework with ultrahigh surface area for outstanding capacitance supercapacitor. Renew Energy. 2021;179:1826–35. https://doi.org/10.1016/j.renene.2021.08.008.
114. Trogadas P, Coppens MO. Nature-inspired electrocatalysts and devices for energy conversion. Chem Soc Rev. 2020;49:3107–41. https://doi.org/10.1039/c8cs00797g.
115. Ghosh S, Basu RN. Multifunctional nanostructured electrocatalysts for energy conversion and storage: current status and perspectives. Nanoscale. 2018;10:11241–80. https://doi.org/10.1039/c8nr01032c.
116. Gao S, Geng K, Liu H, Wei X, Zhang M, Wang P, Wang J. Transforming organic-rich amaranthus waste into nitrogen-doped carbon with superior performance of the oxygen reduction reaction. Energy Environ Sci. 2015;8:221–9. https://doi.org/10.1039/c4ee02087a.
117. Liu X, Zhou Y, Zhou W, Li L, Huang S, Chen S. Biomass-derived nitrogen self-doped porous carbon as effective metal-free catalysts for oxygen reduction reaction. Nanoscale. 2015;7: 6136–42. https://doi.org/10.1039/c5nr00013k.
118. Chen P, Wang LK, Wang G, Gao MR, Ge J, Yuan WJ, Shen YH, Xie AJ, Yu SH. Nitrogen-doped nanoporous carbon nanosheets derived from plant biomass: an efficient catalyst for

oxygen reduction reaction. Energy Environ Sci. 2014;7:4095–103. https://doi.org/10.1039/c4ee02531h.
119. Quílez-Bermejo J, Strutyński K, Melle-Franco M, Morallón E, Cazorla-Amorós D. On the origin of the effect of pH in oxygen reduction reaction for nondoped and edge-type quaternary N-doped metal-free carbon-based catalysts. ACS Appl Mater Interfaces. 2020;12:54815–23. https://doi.org/10.1021/acsami.0c17249.
120. Müller-Hülstede J, Schonvogel D, Schmies H, Wagner P, Dyck A, Wark M. Incorporation of activated biomasses in Fe-N-C catalysts for oxygen reduction reaction with enhanced stability in acidic media. ACS Appl Energy Mater. 2021;4:6912–22. https://doi.org/10.1021/acsaem.1c01018.
121. Li D, Han Z, Leng K, Ma S, Wang Y, Bai J. Biomass wood-derived efficient Fe–N–C catalysts for oxygen reduction reaction. J Mater Sci. 2021;56:12764–74. https://doi.org/10.1007/s10853-021-06122-7.
122. He G, Yan G, Song Y, Wang L. Biomass juncus derived nitrogen-doped porous carbon materials for supercapacitor and oxygen reduction reaction. Front Chem. 2020;8:1–10. https://doi.org/10.3389/fchem.2020.00226.
123. Lu G, Li Z, Fan W, Wang M, Yang S, Li J, Chang Z, Sun H, Liang S, Liu Z. Sponge-like N-doped carbon materials with co-based nanoparticles derived from biomass as highly efficient electrocatalysts for the oxygen reduction reaction in alkaline media. RSC Adv. 2019;9:4843–8. https://doi.org/10.1039/c8ra10462j.
124. Huang H, Wei X, Gao S. Nitrogen-doped porous carbon derived from Malachium aquaticum biomass as a highly efficient electrocatalyst for oxygen reduction reaction. Electrochim Acta. 2016;220:427–35. https://doi.org/10.1016/j.electacta.2016.10.108.
125. He D, Zhao W, Li P, Liu Z, Wu H, Liu L, Han K, Liu L, Wan Q, Butt FK, Qu X. Bifunctional biomass-derived 3D nitrogen-doped porous carbon for oxygen reduction reaction and solid-state supercapacitor. Appl Surf Sci. 2019;465:303–12. https://doi.org/10.1016/j.apsusc.2018.09.185.
126. Luo X, Liu Z, Ma Y, Nan Y, Gu Y, Li S, Zhou Q, Mo J. Biomass derived Fe, N-doped carbon material as bifunctional electrocatalysts for rechargeable Zn-air batteries. J Alloys Compd. 2021;888:161464. https://doi.org/10.1016/j.jallcom.2021.161464.
127. Ma LL, Hu X, Liu WJ, Li HC, Lam PKS, Zeng RJ, Yu HQ. Constructing N, P-dually doped biochar materials from biomass wastes for high-performance bifunctional oxygen electrocatalysts. Chemosphere. 2021;278:130508. https://doi.org/10.1016/j.chemosphere.2021.130508.
128. Ye YY, Qian TT, Jiang H. Co-loaded N-doped biochar as a high-performance oxygen reduction reaction electrocatalyst by combined pyrolysis of biomass. Ind Eng Chem Res. 2020;59:15614–23. https://doi.org/10.1021/acs.iecr.0c03104.
129. Xia S, Guo W, Cai N, Sun L, Zhou H, Lu W, Chen X, Zhang J, Chen Y, Yang H, Sun F, Wang D, Wang X, Wang S, Chen H. Synthesis and application in oxygen reduction reaction of N-doping porous graphitic carbon from biomass waste. Fuel Process Technol. 2021;224:107028. https://doi.org/10.1016/j.fuproc.2021.107028.
130. Plavniece A, Volperts A, Dobele G, Zhurinsh A, Kaare K, Kruusenberg I, Kaprans K, Knoks A, Kleperis J. Wood and black liquor-based N-doped activated carbon for energy application. Sustain For. 2021;13:9237. https://doi.org/10.3390/su13169237.
131. Gao X, Li X, Kong Z, Xiao G, Zhu Y. Bifunctional 3D n-doped porous carbon materials derived from paper towel for oxygen reduction reaction and supercapacitor. Sci Bull. 2018;63:621–8. https://doi.org/10.1016/j.scib.2018.04.016.
132. Liu Z, Li Z, Tian S, Wang M, Sun H, Liang S, Chang Z, Lu G. Conversion of peanut biomass into electrocatalysts with vitamin B12 for oxygen reduction reaction in Zn-air battery. Int J Hydrog Energy. 2019;44:11788–96. https://doi.org/10.1016/j.ijhydene.2019.03.055.
133. Borghei M, Laocharoen N, Kibena-Põldsepp E, Johansson LS, Campbell J, Kauppinen E, Tammeveski K, Rojas OJ. Porous N,P-doped carbon from coconut shells with high

electrocatalytic activity for oxygen reduction: alternative to Pt-C for alkaline fuel cells. Appl Catal B Environ. 2017;204:394–402. https://doi.org/10.1016/j.apcatb.2016.11.029.

134. Zhong G, Meng Z, Xu M, Xie H, Xu S, Fu X, Liao W, Zheng S, Xu Y. Self-nitrogen-doped porous carbon prepared via pyrolysis of grass-blade without additive for oxygen reduction reaction. Diam Relat Mater. 2022;121:108742. https://doi.org/10.1016/j.diamond.2021.108742.
135. Kaare K, Yu E, Käämbre T, Volperts A, Dobele G, Zhurinsh A, Niaura G, TamasauskaiteTamasiunaite L, Norkus E, Kruusenberg I. Biomass-derived graphene-like catalyst material for oxygen reduction reaction. ChemNanoMat. 2021;7:307–13. https://doi.org/10.1002/cnma.202000615.
136. Li Z, Gao C, Zhao H, Meng A, Ding S, Wang X, Li S. Porous biomass-derived carbon modified by cu, N co-doping and cu nanoparticles as high-efficient electrocatalyst for oxygen reduction reaction and zinc-air battery. J Alloys Compd. 2022;897:163175. https://doi.org/10.1016/j.jallcom.2021.163175.
137. Kuo HC, Liu SH, Lin YG, Chiang CL, Tsang DCW. Synthesis of FeCo-N@N-doped carbon oxygen reduction catalysts: via microwave-assisted ammoxidation. Cat Sci Technol. 2020;10:3949–58. https://doi.org/10.1039/d0cy00376j.
138. Morales-Salas L, Salazar MR, Escobar B. Doped biochar from an invasive plant "Eichhornia crassipes" for the oxygen reduction reaction. Int J Hydrog Energy. 2022;47:30140–6. https://doi.org/10.1016/j.ijhydene.2022.06.223.
139. Hou J, Wen S, Chen J, Zhao Q, Wang L. Large-scale fabrication of biomass-derived N, S co-doped porous carbon with ultrahigh surface area for oxygen reduction. Mater Chem Phys. 2021;267:124601. https://doi.org/10.1016/j.matchemphys.2021.124601.
140. Liu F, Peng H, Qiao X, Fu Z, Huang P, Liao S. High-performance doped carbon electrocatalyst derived from soybean biomass and promoted by zinc chloride. Int J Hydrog Energy. 2014;39:10128–34. https://doi.org/10.1016/j.ijhydene.2014.04.176.
141. Zhou H, Zhang J, Amiinu IS, Zhang C, Liu X, Tu W, Pan M, Mu S. Transforming waste biomass with an intrinsically porous network structure into porous nitrogen-doped graphene for highly efficient oxygen reduction. Phys Chem Chem Phys. 2016;18:10392–9. https://doi.org/10.1039/c6cp00174b.
142. Zhao L. Natural phosphorus-doped honeycomb carbon materials as oxygen reduction catalysts derived from Pulsatilla chinensis (Bunge) regel. RSC Adv. 2017;7:13904–10. https://doi.org/10.1039/c6ra28630e.
143. Charles V, Zhang X, Yuan M, Zhang K, Cui K, Zhang J, Zhao T, Li Y, Liu Z, Li B, Zhang G. CoNi nano-alloy anchored on biomass-derived N-doped carbon frameworks for enhanced oxygen reduction and evolution reactions. Electrochim Acta. 2022;402:139555. https://doi.org/10.1016/j.electacta.2021.139555.
144. Liu A, Ma M, Zhang X, Ming J, Jiang L, Li Y, Zhang Y, Liu S. A biomass derived nitrogen doped carbon fibers as efficient catalysts for the oxygen reduction reaction. J Electroanal Chem. 2018;824:60–6. https://doi.org/10.1016/j.jelechem.2018.07.039.
145. Zhao Y, Liu X, Liu Y, Chen Y, Gao S. Favorable pore size distribution of biomass-derived N, S dual-doped carbon materials for advanced oxygen reduction reaction. Int J Hydrog Energy. 2022;47:12964–74. https://doi.org/10.1016/j.ijhydene.2022.02.064.
146. Li Q, Chen X, Yang Y. Biomass-derived nitrogen-doped porous carbon for highly efficient ambient electro-synthesis of NH3. Catalysts. 2020;10:353. https://doi.org/10.3390/catal10030353.
147. Chaparro-Garnica JA, Navlani-García M, Salinas-Torres D, Morallón E, Cazorla-Amorós D. H2 production from formic acid using highly stable carbon-supported pd-based catalysts derived from soft-biomass residues: effect of heat treatment and functionalization of the carbon support. Materials. 2021;14:6506. https://doi.org/10.3390/ma14216506.
148. Chaparro-Garnica J, Navlani-García M, Salinas-Torres D, Berenguer-Murcia Á, Morallón E, Cazorla-Amorós D. Efficient production of hydrogen from a valuable CO2-derived molecule:

formic acid dehydrogenation boosted by biomass waste-derived catalysts. Fuel. 2022;320: 123900. https://doi.org/10.1016/j.fuel.2022.123900.

149. Duan J, Xiang Z, Zhang H, Zhang B, Xiang X. Pd-Co2P nanoparticles supported on N-doped biomass-based carbon microsheet with excellent catalytic performance for hydrogen evolution from formic acid. Appl Surf Sci. 2020;530:147191. https://doi.org/10.1016/j.apsusc.2020.147191.
150. Zou L, Liu Q, Zhu D, Huang Y, Mao Y, Luo X, Liang Z. Experimental and theoretical studies of ultrafine Pd-based biochar catalyst for dehydrogenation of formic acid and application of in situ hydrogenation. ACS Appl Mater Interfaces. 2022;14:17282–95. https://doi.org/10.1021/acsami.2c00343.
151. Lee DW, Jin MH, Oh D, Lee SW, Park JS. Straightforward synthesis of hierarchically porous nitrogen-doped carbon via pyrolysis of chitosan/urea/KOH mixtures and its application as a support for formic acid dehydrogenation catalysts. ACS Sustain Chem Eng. 2017;5:9935–44. https://doi.org/10.1021/acssuschemeng.7b01888.
152. Shi Y, Xiang Z, Deng J, Nan J, Zhang B. Synthesis Pd/biomass-based carbon microsheet composite for efficient dehydrogenation from formic acid. Mater Lett. 2019;237:61–4. https://doi.org/10.1016/j.matlet.2018.11.074.
153. Cao T, Cheng J, Ma J, Yang C, Yao M, Liu F, Deng M, Wang X, Ren Y. Facile synthesis of microporous carbons from biomass waste as high performance supports for dehydrogenation of formic acid. Nanomater. 2021;11:3028. https://doi.org/10.3390/nano11113028.
154. Liu J, Deng Y, Li X, Wang L. Promising nitrogen-rich porous carbons derived from one-step calcium chloride activation of biomass-based waste for high performance supercapacitors. ACS Sustain Chem Eng. 2016;4:177–87. https://doi.org/10.1021/acssuschemeng.5b00926.
155. Yue X, Yang H, Cao Y, Jiang L, Li H, Shi F, Liu J. Nitrogen-doped cornstalk-based biomass porous carbon with uniform hierarchical pores for high-performance symmetric supercapacitors. J Mater Sci. 2022;57:3645–61. https://doi.org/10.1007/s10853-022-06891-9.
156. Yuan Y, Sun Y, Feng Z, Li X, Yi R, Sun W, Zhao C, Yang L. Nitrogen-doped hierarchical porous activated carbon derived from paddy for high-performance supercapacitors. Materials. 2021;14:1–12. https://doi.org/10.3390/ma14020318.
157. Guo N, Li M, Wang Y, Sun X, Wang F, Yang R. N-doped hierarchical porous carbon prepared by simultaneous-activation of KOH and NH3 for high performance supercapacitors. RSC Adv. 2016;6:101372–9. https://doi.org/10.1039/c6ra22426a.
158. Mostazo-López MJ, Ruiz-Rosas R, Morallón E, Cazorla-Amorós D. Generation of nitrogen functionalities on activated carbons by amidation reactions and Hofmann rearrangement: chemical and electrochemical characterization. Carbon. 2015;91:252–65. https://doi.org/10.1016/j.carbon.2015.04.089.
159. Wang B, Wang Y, Peng Y, Wang X, Wang J, Zhao J. 3-dimensional interconnected framework of N-doped porous carbon based on sugarcane bagasse for application in supercapacitors and lithium ion batteries. J Power Sources. 2018;390:186–96. https://doi.org/10.1016/j.jpowsour.2018.04.056.
160. Sun J, Li W, Lei E, Xu Z, Ma C, Wu Z, Liu S. Ultralight carbon aerogel with tubular structures and N-containing sandwich-like wall from kapok fibers for supercapacitor electrode materials. J Power Sources. 2019;438:227030. https://doi.org/10.1016/j.jpowsour.2019.227030.
161. Xu D, Su Y, Zhang S, Xiong Y. Highly porous N-doped carbons production from biomass for high-performance supercapacitors without chemical nitrogen-containing dopants. Energy Sources, Part A Recover Util Environ Eff. 2020;42:1797–807. https://doi.org/10.1080/15567036.2019.1604890.
162. Chen H, Yu F, Wang G, Chen L, Dai B, Peng S. Nitrogen and sulfur self-doped activated carbon directly derived from elm flower for high-performance supercapacitors. ACS Omega. 2018;3:4724–32. https://doi.org/10.1021/acsomega.8b00210.
163. Ma G, Zhang Z, Sun K, Peng H, Yang Q, Ran F, Lei Z. White clover based nitrogen-doped porous carbon for a high energy density supercapacitor electrode. RSC Adv. 2015;5:107707–15. https://doi.org/10.1039/c5ra20327a.

164. He J, Zhang D, Han M, Liu X, Wang Y, Li Y, Zhang X, Wang K, Feng H, Wang Y. One-step large-scale fabrication of nitrogen doped microporous carbon by self-activation of biomass for supercapacitors application. J Energy Storage. 2019;21:94–104. https://doi.org/10.1016/j.est.2018.11.015.
165. Yu D, Chen C, Zhao G, Sun L, Du B, Zhang H, Li Z, Sun Y, Besenbacher F, Yu M. Biowaste-derived hierarchical porous carbon nanosheets for ultrahigh power density supercapacitors. ChemSusChem. 2018;11:1678–85. https://doi.org/10.1002/cssc.201800202.
166. Yu J, Wang X, Peng J, Jia X, Li L, Chuan X. Porous activity of biomass-activated carbon enhanced by nitrogen-dopant towards high-performance lithium ion hybrid battery-supercapacitor. J Electrochem Soc. 2021;168:120537. https://doi.org/10.1149/1945-7111/ac42a1.
167. Yu S, Zhu X, Lou G, Wu Y, Xu K, Zhang Y, Zhang L, Zhu E, Chen H, Shen Z, Bao B, Fu S. Sustainable hierarchical porous biomass carbons enriched with pyridinic and pyrrolic nitrogen for asymmetric supercapacitor. Mater Des. 2018;149:184–93. https://doi.org/10.1016/j.matdes.2018.04.023.
168. Christina Mary AJ, Sathish CI, Murphin Kumar PS, Vinu A, Bose AC. Fabrication of hybrid supercapacitor device based on NiCo2O4@ZnCo2O4 and the biomass-derived N-doped activated carbon with a honeycomb structure. Electrochim Acta. 2020;342:136062. https://doi.org/10.1016/j.electacta.2020.136062.
169. Gandla D, Wu X, Zhang F, Wu C, Tan DQ. High-performance and high-voltage supercapacitors based on N-doped mesoporous activated carbon derived from dragon fruit peels. ACS Omega. 2021;6:7615–25. https://doi.org/10.1021/acsomega.0c06171.
170. Guan L, Pan L, Peng T, Gao C, Zhao W, Yang Z, Hu H, Wu M. Synthesis of biomass-derived nitrogen-doped porous carbon nanosheests for high-performance supercapacitors. ACS Sustain Chem Eng. 2019;7:8405–12. https://doi.org/10.1021/acssuschemeng.9b00050.
171. Ahmed S, Rafat M, Ahmed A. Nitrogen doped activated carbon derived from orange peel for supercapacitor application. Adv Nat Sci Nanosci Nanotechnol. 2018;9:035008. https://doi.org/10.1088/2043-6254/aad5d4.
172. Sahoo MK, Rao GR. A high energy flexible symmetric supercapacitor fabricated using N-doped activated carbon derived from palm flowers. Nanoscale Adv. 2021;3:5417–29. https://doi.org/10.1039/d1na00261a.
173. Shang Z, An X, Zhang H, Shen M, Baker F, Liu Y, Liu L, Yang J, Cao H, Xu Q, Liu H, Ni Y. Houttuynia-derived nitrogen-doped hierarchically porous carbon for high-performance supercapacitor. Carbon. 2020;161:62–70. https://doi.org/10.1016/j.carbon.2020.01.020.
174. Wang X, Zeng X, Cao D. Biomass-derived nitrogen-doped porous carbons (NPC) and NPC/polyaniline composites as high performance supercapacitor materials. Eng Sci. 2018;1:55–63. https://doi.org/10.30919/es.180325.
175. Ji H, Zhang C, Rao W, Guo B, Fan L, Bai Z, Bao H, Xu J. Eco-friendly Polypyrrole-coated cocozelle composites for supercapacitor application. Fibers Polym. 2020;21:1300–7. https://doi.org/10.1007/s12221-020-9375-0.

Chapter 11
Preparation of Green N-Doped Biochar Materials with Biomass Pyrolysis and Their Application to Catalytic Systems

Wei Chen, Lujiang Xu, Chengyu Dong, Huan Zhang, Shuai Gao, Yingquan Chen, and Haiping Yang

Abstract Biomass pyrolysis produces high-value chemical products and is important for developing energy applications. Green nitrogen-doped biochar, which has a porous structure, active nitrogen/oxygen-containing groups, and catalytic sites, can be prepared through activation and nitrogen doping during pyrolysis. In this chapter, preparation methods for green nitrogen-doped biochar materials via pyrolysis are comprehensively reviewed as well as one-step "pyrolysis-activation-doping". In addition, the formation mechanisms of nitrogen-doped biochars are discussed with their corresponding catalytic properties. There is a need for advanced online characterization methods and detection of the structural evolution of intermediates and radicals during reaction processes. The preparation of green N-doped biochar materials and application in biomass catalytic pyrolysis hold great promise for the economical production of chemical products from biomass.

Keywords Nitrogen-doped biochar · Catalytic pyrolysis · Biomass · Phenols · Nitrogen doping · Activation mechanism

---

W. Chen
College of Engineering, Nanjing Agricultural University, Nanjing, China

State Key Laboratory of Coal Combustion, School of Power and Energy Engineering, Huazhong University of Science and Technology, Wuhan, China

L. Xu · C. Dong · H. Zhang · S. Gao
College of Engineering, Nanjing Agricultural University, Nanjing, China

Y. Chen · H. Yang (✉)
State Key Laboratory of Coal Combustion, School of Power and Energy Engineering, Huazhong University of Science and Technology, Wuhan, China
e-mail: yanghaiping@hust.edu.cn

Z. Fang et al. (eds.), *Production of N-containing Chemicals and Materials from Biomass*, Biofuels and Biorefineries 12,
https://doi.org/10.1007/978-981-99-4580-1_11

11.1 Introduction

Biomass, as the only carbon-containing renewable resource, requires efficient conversion technologies to address energy and environmental issues [1]. The International Energy Agency predicts that renewable energy will account for 40% of the growth in global energy consumption in the next 5 years, and biomass energy will become the fastest-growing renewable resource in the world. China is a big agricultural country with abundant biomass resources. The biomass resources that can be collected for energy use reached 920 million tons of standard coal per year. Efficient and high-quality utilization of biomass is of great significance to the green and sustainable development of energy economies.

Biomass pyrolysis to produce high-value biochar products is an important development direction of biomass energy utilization [2]. Biomass pyrolysis to prepare biochar not only has good economic benefits but also can achieve negative CO_2 emissions, which has become a research hotspot in the field of biomass energy and new materials [3]. However, the biochar obtained by the direct pyrolysis of biomass has low porosity, and the specific surface area is usually less than 200 m^2/g [4]. Moreover, the active functional groups are limited, and the reactivity is not high. So the added value of biochar is not high enough to be used directly as a high-performance functional carbon material such as a catalyst, adsorption, and energy storage material [5]. Therefore, how to prepare biochar with a developed pore structure and abundant active functional groups by pyrolysis is very important for the high-quality utilization of biomass. Chemical activation and nitrogen doping are the main upgrade methods for biochar, and they are also the most widely studied methods [6, 7]. The obtained porous nitrogen-doped biochar can be used in catalysis, adsorption, energy storage, and other fields due to the developed pore structure and abundant active nitrogen-containing functional groups and oxygen-containing functional groups (shown in Fig. 11.1) [8]. Wang et al. [9] found that the activator can achieve an ideal activation effect at high temperatures, and can prepare porous biochar with developed porosity. Lin et al. [10] pointed out that nitrogen doping can convert graphite-structured carbon materials into nitrogen-doped carbons with high electrochemical activity, which can significantly improve the energy storage effect. Therefore, activation and nitrogen doping during biomass pyrolysis is a promising and green method for the preparation of porous nitrogen-doped biochar materials.

On the other hand, biomass pyrolysis to produce high-value-added chemicals is also one of the most promising biomass utilization technologies and has received extensive attention [12, 13]. The oxygen content in biomass is relatively high (~50 wt.%) [12, 14]. Therefore, rational utilization of oxygen in biomass and its conversion into oxygen-containing chemicals are crucial for efficient biomass utilization [15, 16]. Phenolic substances (such as phenol, 4-vinyl phenol, 4-ethyl phenol, etc.) are important platform compounds. Biomass pyrolysis can produce a variety of phenolic compounds [17, 18]. However, the phenolic substances obtained by the direct pyrolysis of biomass are of various types, with low selectivity [19, 20]. To

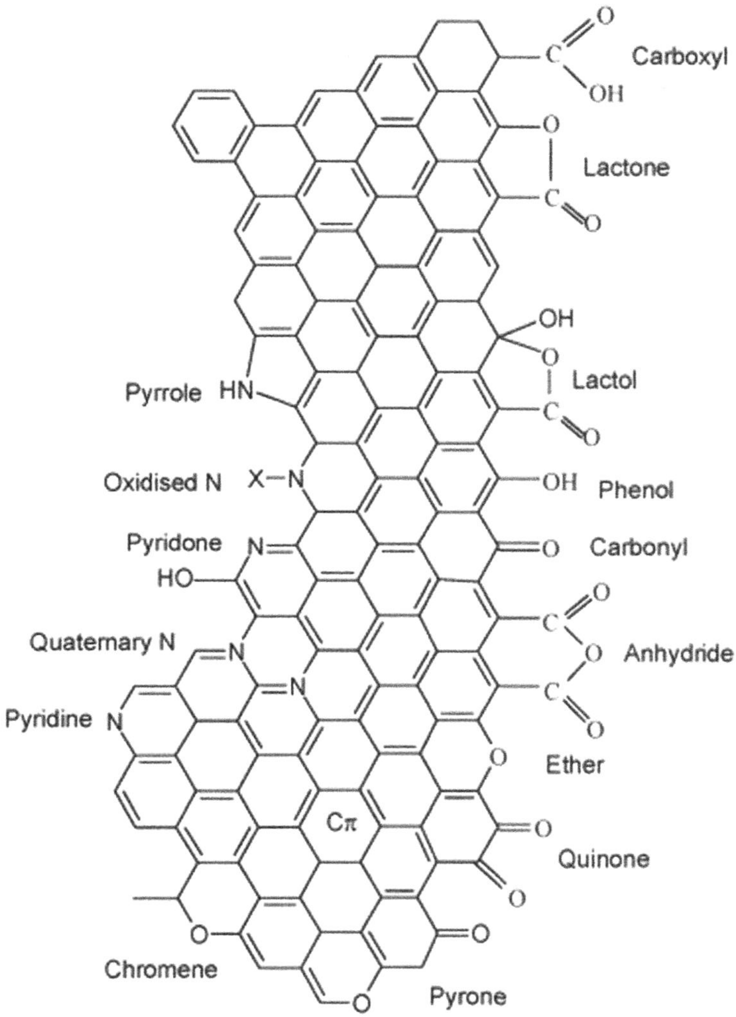

Fig. 11.1 N/O-containing functional groups on nitrogen-doped biochar. (adapted with permission from [11], Copyright © 2010 Elsevier)

improve the yield and selectivity of phenolic substances, catalysts are usually introduced in the pyrolysis process, such as basic catalysts (such as KOH, NaOH, K_2CO_3, etc.), metal oxides (such as Fe_2O_3, Al_2O_3, CaO, etc.), carbon-based catalysts (such as activated carbon, biochar, N-doped biochar, etc.) [20–22]. Among them, carbon-based catalysts have attracted much attention due to their green and cheap characteristics [23, 24]. Yang et al. [25] found that biomass-based activated carbon catalyzed the pyrolysis of wood chips, and the content of phenols could reach 90%. Chen et al. [22] found that the nitrogen-doped biochar catalyst was introduced in the pyrolysis process of bamboo, and the content of phenolic substances could reach

82%, mainly 4-vinyl phenol, 4-ethyl phenol, and phenol. The selectivity of high value-added 4-vinyl phenol reached 31%, and the yield can reach 6.65 wt.%. Therefore, it is a promising idea for the preparation of phenolic products by catalytic pyrolysis of biomass with a nitrogen-doped biochar catalyst.

Many researchers have reviewed biochar preparation methods and the use of biochar as catalysts. For example, Zou et al. [26] focused on biochar preparation from agricultural and forestry waste, and its application as carbon-based catalysts and catalyst support. Wang et al. [27] summarized the preparation of metal/biochar catalysts and their catalytic performance for biomass hydrogenation. Low et al. [28] provided the conversion techniques of lignocellulosic biomass, biochar modification, and its applications in biofuel production and pollution control. Chi et al. [29] focused on the introduction of preparation techniques of biochar and biochar-based catalyst for biofuel from algae. However, few reviews focus on the preparation of green nitrogen-doped biochar and its application in the catalytic pyrolysis of biomass for valuable chemicals.

In this chapter, the preparation of green nitrogen-doped biochar materials and their application in biomass catalytic pyrolysis for value-added products are comprehensively reviewed. Moreover, the formation mechanism of nitrogen-doped biochar materials and the catalytic pyrolysis mechanism of nitrogen-doped biochar catalysts are also discussed in depth.

11.2 Preparation Methods of Nitrogen-Doped Biochar

Researchers have successfully prepared porous nitrogen-doped biochar materials through mainly three methods: (1) nitrogen doping first and then activation using high nitrogen-containing biomass as raw material or introduction of exogenous nitrogen during biomass pyrolysis and subsequent adjustment of biochar pore structure through chemical activation; (2) activation and subsequent nitrogen doping. The nitrogen source is subjected to heat treatment or plasma treatment for nitrogen doping; (3) simultaneous activation and nitrogen doping in which the nitrogen source, activator, and biomass are mixed and pyrolyzed to prepare porous nitrogen-doped biochar materials in one step. For the method of first doping nitrogen and then activation, the severe etching reaction in the second activation process will destroy the nitrogen-containing active functional groups, so this method is unfavorable for directional nitrogen doping. Chen et al. [30] obtained nitrogen-doped biochar by bamboo pyrolysis through pyrolysis under an ammonia atmosphere, and then KOH activation of nitrogen-doped biochar at high temperature to prepare porous nitrogen-doped biochar. The results showed that although the second-step KOH activation greatly increases the specific surface area of biochar from 370 m^2/g to 2400 m^2/g, the chemical activation can also destroy the active nitrogen-containing functional groups, and its nitrogen content is reduced from 4.8 wt.% to 1.7 wt.%. For the method of first activation and then doping nitrogen, the first activation process can severely etch and remove the active functional groups inside the biomass to

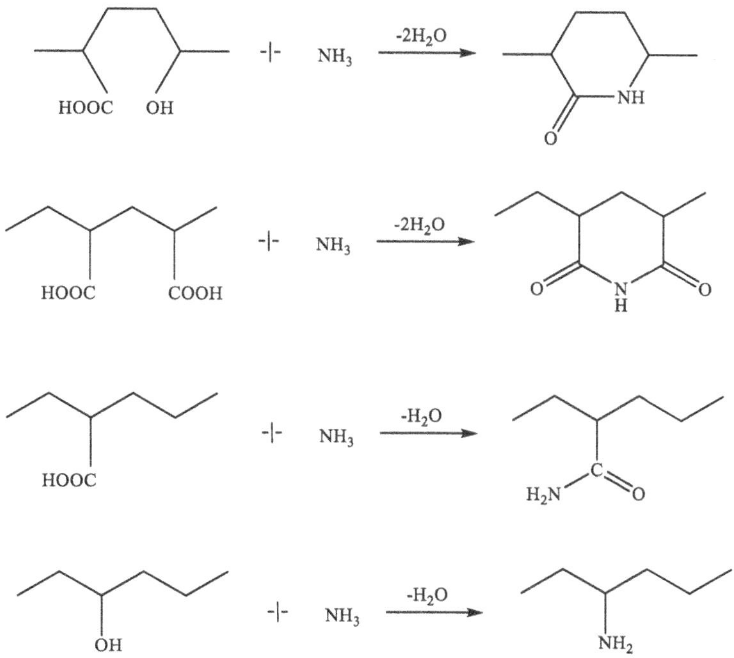

Fig. 11.2 Reaction of NH_3 with oxygen-containing functional groups for nitrogen-containing functional groups

obtain a better pore structure. However, it is difficult to introduce nitrogen for the second step of nitrogen doping, as nitrogen doping mainly depended on NH_3 reacting with surface oxygen-containing functional groups [31, 32]. So additional surface oxidation treatment of porous carbon is usually required. Hulicova-Jurcakova et al. [31] prepared the nitrogen-doped biochar through HNO_3 surface oxidation treatment first, and then heat-treated in an NH_3 atmosphere, and found that the nitrogen content increased by about 100%, while the nitrogen content only increased by about 50% for the nitrogen-doped biochar without pre-oxidation treatment. The biochar surface oxidized by HNO_3 will form a large number of carboxylic acid sites (the oxygen content increased to 15 at.%), and NH_3 can react with it to generate lactam and imide, so that the oxygen content on the biochar surface is reduced to 1.38 at.%, as shown in Fig. 11.2.

For the method of simultaneous activation and nitrogen doping, since biomass pyrolysis, chemical activation of biomass, and NH_3 reacting with active substances are carried out at the same time, the pyrolysis process will produce active small molecules, and chemical activation will help the breaking of macromolecular carbon chains in biomass, which enabled NH_3 to react with more active substances. The interaction among them can have a huge impact on the pore structure, nitrogen doping amount, and surface functional groups of the biochar. Luo et al. [33] prepared nitrogen-doped nano-porous carbon membranes by heat-treating cellulose filter in an

NH_3 atmosphere. The reaction between NH_3 and nitrogen-doped carbon materials leads to the gasification of carbon, which increases the specific surface area to 1973 m^2/g. It indicates that the NH_3 nitrogen-doped process can also produce new pores. Chen et al. [34] studied the pyrolysis of KOH-activated bamboo under an ammonia atmosphere. After adding a small proportion of activator (KOH/bamboo chips = 0.1–0.4), the nitrogen content of biochar increased to 10.4 wt.%, and the specific surface area reached 1873 m^2/g. This indicates a strong interaction among the biomass pyrolysis process, the chemical activation process, and the NH_3 nitrogen doping process. Therefore, it is necessary to further study the interaction mechanism of biomass pyrolysis, chemical activation, and nitrogen doping (NH_3 modification), and its influence on biochar pore structure, nitrogen content, and nitrogen-containing functional groups of nitrogen-doped biochar.

11.3 Chemical Activation and Nitrogen Doping During Biomass Pyrolysis for Nitrogen-Doped Biochar

Biomass pyrolysis, chemical activation, and nitrogen doping for nitrogen-doped biochar mainly include biomass pyrolysis, chemical activation to generate porous structure, and NH_3 modification to dope nitrogen. The three processes take place simultaneously and interact. By studying the interaction among the three processes, the quality of nitrogen-doped biochar products, such as pore structure, nitrogen content, and active functional groups, can be controlled directionally. At present, the biomass pyrolysis mechanism has been fully studied. Kan et al. [35] reviewed the effects of experimental conditions on the properties of pyrolysis products during lignin-based biomass pyrolysis. Many experimental parameters affect the biomass pyrolysis process, product yield, and properties, including biomass type, biomass pretreatment method, reaction atmosphere, temperature, heating rate, and residence time. Liu et al. [36] reviewed and analyzed the migration pathways of the main chemical elements (C, H, O, N, P, etc.) in the biomass pyrolysis process. After pyrolysis, C, H, and O elements mainly exist in bio-oil, syngas, biochar, tar, and polycyclic aromatic hydrocarbons, N element is mainly converted into nitrogen oxides and their precursors (NOx, NH_3, HCN, HCNO, etc.), and some N element will also enter the biochar to form nitrogen-containing functional groups. Deng et al. [37] reviewed the main methods and preparation process mechanisms of the current preparation of functional biochars. The physical structure (such as morphology, topological pore structure, pore size distribution, etc.) and surface chemical state (such as doping) can be adjusted to meet the needs of specific applications, such as energy conversion and storage. Biomass pyrolysis, activation, and modification is a mixed reaction process of biomass raw material, activator, and nitrogen doping agent under high-temperature conditions, and the interaction between them should be complicated. Therefore, it is necessary to analyze the mechanism of nitrogen doping and chemical activation.

11.3.1 Nitrogen Doping Process During Biomass Pyrolysis

Nitrogen doping during the biomass pyrolysis process can not only prepare nitrogen-doped biochar materials [33, 38], but also can produce nitrogen-containing chemicals such as pyridine, pyrrole, indole, and its derivatives [39, 40]. At present, there are two main ways to introduce nitrogen: (1) using biomass rich in nitrogen as raw material, such as algae, and nitrogen-containing wastes; (2) introducing exogenous nitrogen during biomass pyrolysis, such as NH_3, urea, melamine, aniline [41–43].

The main nitrogen source in nature is protein, which is composed of more than 20 kinds of amino acids, and the average nitrogen content reaches 16 wt.%. Therefore, the related research on self-nitrogen doping mainly focuses on the pyrolysis of protein-rich biomass, such as algae, chicken egg white, and silk products. Chen et al. [44, 45] studied microalgae pyrolysis and discussed the migration and transfer pathways of nitrogen during the pyrolysis process. The nitrogen in algae mainly exists in the protein, and the content can reach about 10 wt.%. Under high-temperature pyrolysis, amino acids in proteins undergo deamination or dehydrogenation to generate pyridinic-N and pyrrolic-N, and then part of the pyridinic-N converts into quaternary-N in biochar. Simultaneously, the amide content of the bio-oil decreased rapidly, while the content of nitriles and nitrogen-containing heterocyclic compounds (pyridine, pyrrole, and indole) increased. NH_3 reacts with fatty acids to form amides, which then undergo dehydration to form nitriles. Li et al. [38] prepared three-dimensional macroporous carbon membranes interwoven with carbon fibers by carbonizing eggshell membranes, with an oxygen content of 10 wt.% and a nitrogen content of 8 wt.%. Therefore, even though its specific surface area is small (about 220 m^2/g), the specific capacitance of the carbon electrode in the three-electrode system still reaches 297 F/g, and it has good cycling stability.

For most of the biomass with low nitrogen content in nature, NH_3 modification to doping nitrogen during biomass is an effective way to prepare nitrogen-doped biochar. Chen et al. [30, 46] prepared nitrogen-containing chemicals and nitrogen-doped biochar materials by using bamboo pyrolysis under the NH_3 atmosphere and explored the formation mechanism of nitrogen-containing products. The reactions of NH_3 and its free radicals with active oxygen-containing species generated during biomass pyrolysis produce nitrogen-containing functional groups and then form nitrogen-containing heterocyclic compounds in bio-oil and biochar. The nitrogen content in nitrogen-doped biochar can reach 4.85 wt.%, and the specific surface area can reach 370 m^2/g. Hu et al. [47] successfully prepared honeycomb-like hierarchical porous nitrogen-doped biochar with a nitrogen content of 17.72 at.% by slow pyrolysis of glycine in an ammonia atmosphere, with a specific surface area of 722 m^2/g. Benefiting from the abundant defects and active sites, the surface between the electrolyte and electrode material can be fully contacted and reacted, resulting in good electrochemical performance. Chen et al. [48] prepared nitrogen-doped nano-porous carbon nanosheets through NH_3 annealing of hydrothermal biochar, which had a high surface area (898 m^2/g) and nitrogen content (9.1 at.%).

11.3.2 Chemical Activation Process During Biomass Pyrolysis

To prepare functional nitrogen-doped porous biochar materials, in addition to nitrogen-containing groups, it is necessary to adjust their hierarchical pore structure by chemical activation [49]. Different from biomass direct pyrolysis, activators have a great influence on biomass pyrolysis and biochar properties [50], so understanding the activation mechanism is crucial for the directional preparation of ideal functional nitrogen-doped porous biochar materials.

Activators mainly include KOH, H_3PO_4, K_2CO_3, ZnCI, etc. [51]. Among them, KOH can easily produce micropore-rich and high specific surface area (up to 3000 m^2/g) of the biochar material [37]. The research on the activation mechanism of KOH is relatively mature, and it mainly focuses on the process of activating carbon-based materials, such as carbon nanotubes, coal, and graphene. The reaction mechanism of KOH activation to etch the carbon framework can be summarized into three stages [9, 52]: (1) KOH, K_2CO_3, and K_2O etch the carbon framework through a reduction reaction; (2) gaseous products from the activation process further etch the carbon fragments; (3) potassium vapor enters the carbon skeleton and further expands the pores. Lv et al. [53] used various mass ratios of KOH to raw materials to prepare mesopore-dominated porous carbons with a surface area of 1410 m^2/g. KOH-activated etching generates a large number of mesopores. At the same time, there are lots of through-holes in the mesoporous pore wall connecting the micropores and mesopores. This good topological pore structure enables the electrolyte ions to enter and exit the pores well, improving the utilization rate of the pores. Elmouwahidi et al. [54] prepared a series of activated carbons using olive waste as raw materials, and KOH and H_3PO_4 as activators, and compared the two-step method (carbonization-activation) with the one-step method (direct activation). The study showed that the pore structure and surface morphology of the activated carbon prepared by the two methods were similar when the amount of KOH added was small, while the specific surface area of the activated carbon prepared by the two-step method was larger when the amount of KOH was increased. The activated carbon prepared by H_3PO_4 activation has a small specific surface area, but it is mainly conducive to small mesopores for ion transport in the electrolyte and its surface can be doped with sufficient phosphorus-containing functional groups. Therefore, the specific capacitance of activated carbon prepared by KOH activation has a linear relationship with its specific surface area, while the specific capacitance of activated carbon prepared by H_3PO_4 activation is relatively stable and is less affected by the specific surface area.

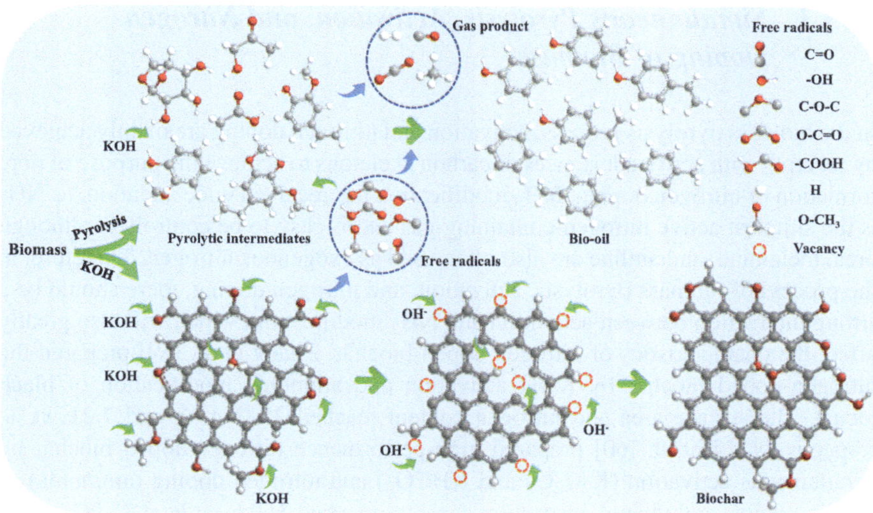

Fig. 11.3 KOH activation mechanism during biomass pyrolysis. (adapted with permission from [58], Copyright © 2020 Elsevier)

In addition to high carbon elements, biomass also contains a large number of oxygen elements and mainly exists in the form of active oxygen-containing functional groups [55]. Therefore, the reaction mechanism of the activator and carbon-based feedstock cannot be fully applied to the reaction of the activator and biomass feedstock. Lu et al. [56] used KOH to directly activate petroleum coke, and the results showed that the active functional groups on the surface of the raw material had a great influence on the KOH-activated pore formation. The oxygen-containing functional groups on the surface can first react with KOH to generate C-O-K intermediate substances, and then the intermediates react with the raw carbon skeleton to achieve the purpose of pore expansion. Qu et al. [57] prepared porous carbon with a specific surface area of 1210 m^2/g by direct one-step activation and pyrolysis of corncob residue. Chen et al. [58] explored the KOH activation mechanism during biomass pyrolysis in depth shown as in Fig. 11.3. During biomass pyrolysis, KOH first reacted with active oxygen-containing species to release free radicals and generate vacancies. At the same time, KOH also etched the carbon fragments, releasing hydrogen protons to form vacancies. Then, OH^- from KOH entered the vacancies forming lots of new oxygen-containing groups and developing porosity. Therefore, the reaction between the activator and the oxygen-containing functional groups in the biomass should be considered in the study of the simultaneous activation and pyrolysis process.

11.3.3 Simultaneous Pyrolysis, Activation, and Nitrogen Doping of Biomass

In the biomass pyrolysis process, activation and nitrogen doping are mainly achieved by reacting with active substances or carbon skeletons to achieve the purpose of pore formation or nitrogen doping. NH_3 modification has received wide attention, as NH_3 is the simplest active nitrogen-containing gas and is easy to be controlled, although urea, melamine, and aniline are also often used as exogenous nitrogen. Therefore, in the process of biomass pyrolysis, activation, and nitrogen doping, there should be a strong interaction between activation and NH_3 modification, which will also greatly affect the characteristics of nitrogen-doped biochar. Zhang et al. [59] prepared the nitrogen-doped biochar by KOH activation and ammonia modification of black locust. The surface area and nitrogen content reached 2511 m^2/g and 7.21 wt.%, respectively. Li et al. [60] prepared high-performance nitrogen-doped biochar by simultaneous activation ($K_2C_2O_4$ and $KHCO_3$) and nitrogen doping (ammonia) of bamboo chips, and phenol adsorption capacity reached 169.5 mg/g at 25 °C.

Guo et al. [61] prepared the hierarchical porous nitrogen-doped biochar by pyrolysis of hemp rods dipped and KOH activation at high temperatures in the ammonia atmosphere. The results showed that NH_3 can improve the activation degree of KOH, and KOH can also promote ammoniation and nitrogen doping by increasing the activity of carbon atoms at the edge of the carbon skeleton. The good synergy between the activation process and the nitrogen doping process can improve the specific surface area (1949 m^2/g) and nitrogen content (4.4 wt.%) of nitrogen-doped biochar. Zhu et al. [62] prepared nitrogen-doped hierarchical porous biochar through pyrolysis of the mixture of humic acid, urea, and K_2CO_3 at high temperatures. The method introduced nitrogen into the carbon precursor to provide CeN forms, which will react with the K-species and enhance the carbon etching to improve the development of pore structure. The nitrogen-doped hierarchical porous biochar had high nitrogen content (5 at.%), specific surface area (3142 m^2/g), and total pore volume (2.6 cm^3/g). Chen et al. [34] and Li et al. [63] proposed one-step chemical activation (KOH) and nitrogen doping (NH_3 modification) for nitrogen-doped biochar and found that the activator in the pyrolysis process preferentially reacted with the active oxygen-containing functional groups in the biomass, and continuously etched the carbon skeleton to form a large number of oxygen cavities. Ammonia and its free radicals can quickly occupy holes to form stable nitrogen-containing heterocyclic structures. Thus, a large amount of nitrogen was fixed in the biochar to realize the preparation of functional nitrogen-doped biochar materials (10 wt.% nitrogen content and 1800 m^2/g surface area). Moreover, this process would also generate large amounts of valuable nitrogen-containing chemicals in bio-oil products, shown as Fig. 11.4.

In general, current research uses biochar as the matrix to prepare porous nitrogen-doped biochar by a multi-step method, in which the activation process has a certain indirect effect on the nitrogen doping process by affecting the surface properties of the biochar. However, there are few studies on biomass simultaneous pyrolysis,

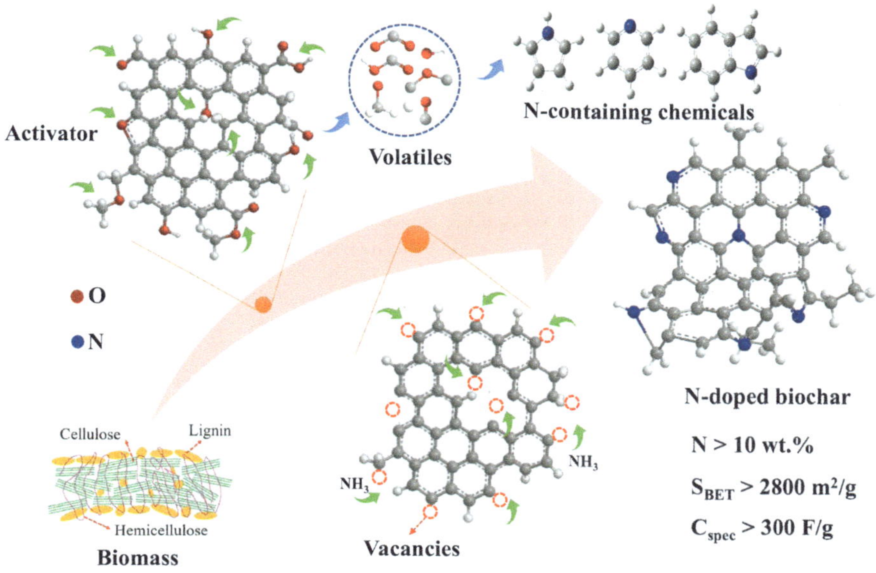

Fig. 11.4 Simultaneous pyrolysis, activation, and nitrogen doping of biomass for nitrogen-doped biochar

activation, and nitrogen doping (one-step "pyrolysis-activation-doping" method). The one-step "pyrolysis-activation-doping" method has the advantages of being simple, fast, and efficient. To realize the directionally controlled preparation of porous nitrogen-doped biochar, it is necessary to deeply study the interaction mechanism between activation and nitrogen doping during biomass pyrolysis.

11.4 Biomass Catalytic Pyrolysis with Nitrogen-Doped Biochar Catalyst

Nitrogen-doped biochar possessed a developed porous structure, and abundant active functional groups (nitrogen/oxygen-containing groups), which have the potential to be used as a catalyst for biomass pyrolysis to produce high-valued chemicals, shown as Fig. 11.5. During biomass catalytic pyrolysis process with nitrogen-doped biochar catalyst, the nitrogen content, nitrogen-containing functional groups, specific surface area of nitrogen-doped biochar have a great influence on the yield and selectivity of value chemicals (such as phenols). To understand the catalytic pyrolysis mechanism, it is necessary to study the effect of this characteristic of nitrogen-doped biochar on the biomass catalytic pyrolysis process in depth.

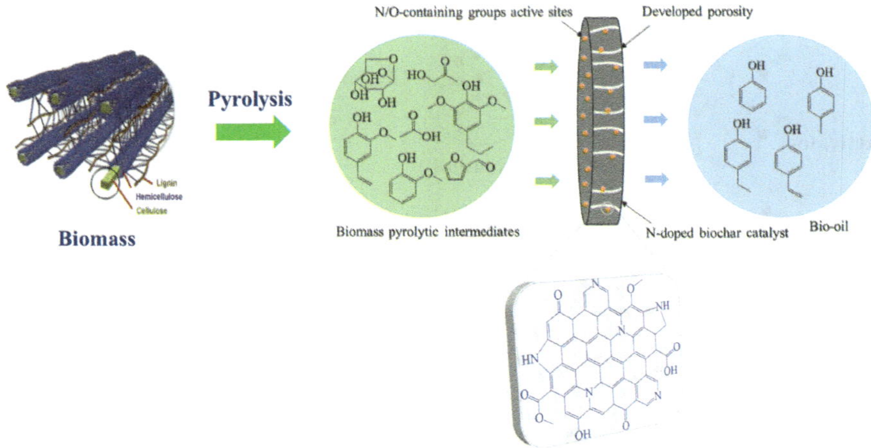

Fig. 11.5 Biomass catalytic pyrolysis for high-valued chemicals with nitrogen-doped biochar catalyst

11.4.1 Effect of the Catalytic Pyrolysis Process

The catalytic pyrolysis process of biomass has an important influence on the formation of phenolics. Yang et al. [64] found that in the catalytic pyrolysis process of wood chips, the addition of an activated biochar catalyst would reduce the yield of liquid products, and the content of phenolic substances was higher in the in situ catalytic process. Li et al. [65] pointed out that activated biochar catalyst promoted biomass pyrolysis to form more furans and phenols. They found biochar catalyst significantly promoted the reduction of aldehydes, ketones, and sugars. Higher total acidity (weak acidity and Lewis acidity) of biochar catalyst favors the formation of 2-methyl furan and phenol as shown in Fig. 11.6. Su et al. [66] found that activated carbon could promote the conversion of lignin in biomass to monophenols. Bu et al. [67] pointed out that biochar or commercial activated carbon can effectively improve the selectivity of phenols in microwave pyrolysis of biomass. Wang et al. [68] researched the catalytic pyrolysis of Douglas fir over a nanocellulose-derived biochar catalyst. Phenol concentration increased to 53.77 mg/mL from 15.76 mg/mL. After 15 reuse cycles, phenolic monomer still showed higher concentration. Ma et al. [69] found that nitrogen-doped activated carbon could promote the pyrolysis of peanut shells to generate more alkylphenols. Chen et al. [70] also pointed out that biochar shows the potential for phenols production. However, the relationship between the catalytic effect of biochar and its physiochemical properties is rarely investigated. Carbon-based catalysts can significantly promote the pyrolysis of biomass to generate phenolic substances, but few studies investigate on the effect mechanism of nitrogen-doped biochar on the biomass pyrolysis process. For the complex structure of nitrogen-doped biochar, which not only with a well-developed

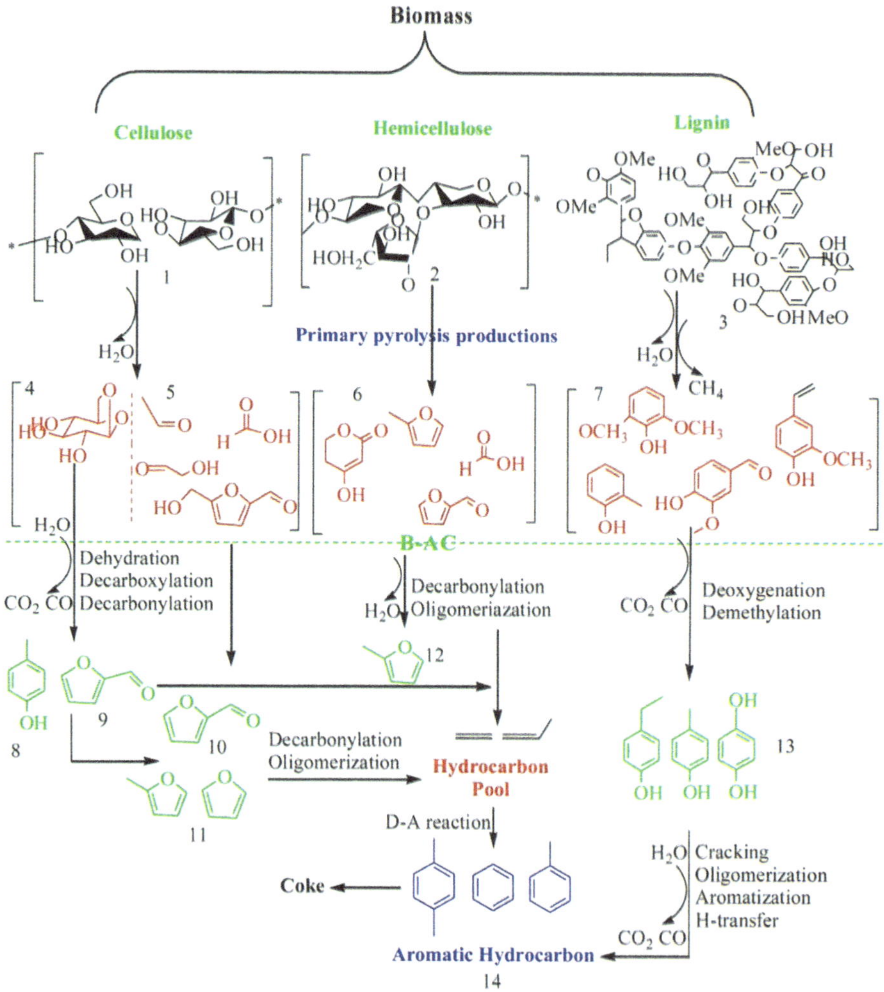

Fig. 11.6 Reaction pathways for furans and phenols through biomass catalytic pyrolysis over biochar (adapted with permission from [65], Copyright © 2020 Elsevier)

pore structure, but also abundant functional groups, its role in the biomass pyrolysis process will be more complex and diverse.

11.4.2 *Effect of Active Functional Groups in Catalyst*

The composition of active functional groups in the catalyst is an important factor affecting its catalytic activity [11, 71]. Serp et al. [72] pointed out that nitrogen-doped carbon contains abundant nitrogen-containing functional groups (pyridinic-N,

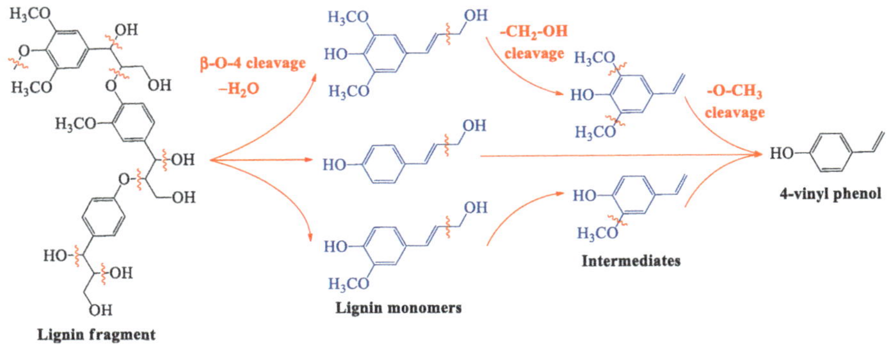

Fig. 11.7 Formation pathways of 4-vinyl phenol through biomass catalytic pyrolysis with nitrogen-doped biochar (adapted with permission from [22], Copyright © 2020 Elsevier)

pyrrolic-N, quaternary-N, pyridine-N-oxide, etc.), among which pyridinic-N and pyrrolic-N are Lewis basic sites. The nitrogen-doped biochar is alkaline and has strong catalytic activity. Guo et al. [73] found that pyridinic-N in nitrogen-doped carbon can promote the cleavage of the O-OH bond to form more -OH groups. However, Inagaki et al. [8] found that graphite-N also showed high redox reactivity during the catalytic reaction. Li et al. [71] found that nitrogen-containing functional groups in nitrogen-doped biochar can effectively combine with acidic intermediates and take place in catalytic reactions. Chen et al. [22] also found that, in the process of biomass pyrolysis, pyridinic-N and pyrrolic-N in nitrogen-doped biochar promoted the demethoxylation of phenolic substances, and the nitrogen content affected the catalytic activity largely. The formation pathway of 4-vinyl phenol was also proposed: N-doped biochar provided H radicals, which first promoted the cleavage of β-O-4 and the dehydration of lignin, resulting in the formation of pcoumaryl, coniferyl, and sinapyl alcohol subunits intermediates. These intermediates were then converted to 4-vinyl phenol with methoxy groups by removing -CH_2-OH groups, and finally they further removed -O-CH_3 groups and formed 4-vinyl phenol, shown as Fig. 11.7. The above studies show that the key to the catalytic activity of nitrogen-doped biochar is the nitrogen-containing functional group and nitrogen content, which are very important for catalytic pyrolysis, but the effect of nitrogen-containing functional groups on biomass pyrolysis is rarely reported. It is still unclear that the mechanism of action for the formation of phenols during the catalytic pyrolysis of biomass. Therefore, it is urgent to deeply explore the mechanism of nitrogen-containing functional groups in the catalytic pyrolysis process and the coupling mechanism with phenolic products.

In addition, nitrogen-doped biochar also contains abundant oxygen-containing functional groups. Su et al. [66] pointed out that the oxygen-containing functional groups in activated carbon can promote the dehydration and decarbonylation of pyrolytic volatiles, and then generate phenols. Yang et al. [74] found that the -OH, O-C=O and C-O functional groups in activated biochar exhibited high catalytic activity during biomass pyrolysis, which greatly promoted the generation of phenolic substances and aromatic hydrocarbons. Cao et al. [75] found that the hydroxyl groups existed in the seaweed-derived biochar catalyst could react with carboxylic group to form fatty acid methyl esters during catalytic pyrolysis of macroalgae. Therefore, when revealing the catalytic mechanism of nitrogen-containing functional groups, it is also necessary to further study the action mechanism of oxygen-containing functional groups in the pyrolysis process. It is very important to further explore the stability of the nitrogen-doped functional groups during catalytic pyrolysis process.

11.4.3 Effect of Pore Structure in Catalyst

Pore structure is an important channel for transferring reaction species, and is another important factor affecting the catalytic activity of catalysts [76, 77]. Zhang et al. [78] found that mesopores in activated carbon can promote the formation of phenolic species. Pena et al. [79] pointed out that activated carbon with high porosity has better catalytic effect and is more resistant to deactivation. Fuentes-Cano et al. [80] found a positive linear correlation between the catalytic performance of biochar catalysts and the mesopore area, and Ravenni et al. [81] pointed out that biochar rich in micropores could promote the conversion of macromolecular organics. Yang et al. [74] also found that the pore structure of activated biochar has a great influence on the catalytic pyrolysis products, and the content of phenolic substances has a positive linear relationship with the specific surface area of biochar, shown as Fig. 11.8. It can be seen that the pore structure of carbon-based catalysts plays an important role in the catalytic reaction process. Shen et al. [82] pointed out that for different carbon-based catalysts and catalytic conditions, micropores and mesopores showed different roles. Chen et al. [83] found that the mesopores of nitrogen-doped carbons mainly functioned as channels, and the catalytic reaction mainly occurs on the surface of the

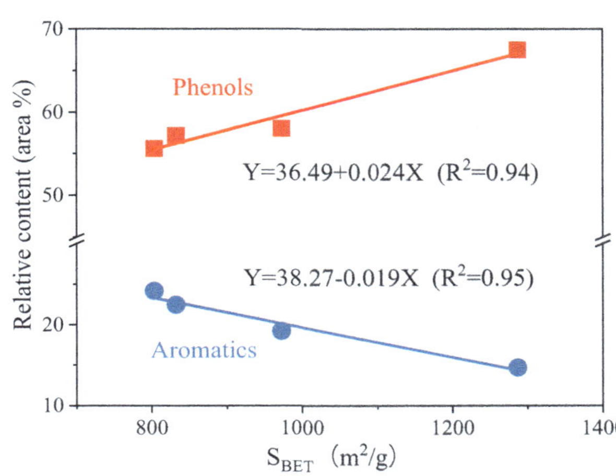

Fig. 11.8 Correlation of biochar aromatic and phenolic content with specific surface area. (adapted with permission from [74], Copyright © 2020 Elsevier)

micropores. Chen et al. [22] also found that the specific surface area and pore size of nitrogen-doped biochar had a great influence on the formation of phenolic species. However, the effect of nitrogen-doped biochar pore structure on the catalytic conversion of volatiles during biomass pyrolysis is rarely reported, and the correlation mechanism between the pore structure and the pyrolysis products is still unclear. Moreover, the renewability of the nitrogen-doped biochar catalyst is also needed to be investigated, which is critical for the catalysts.

11.4.4 Effect of Biomass Composition

There are many kinds of biomass, and its structural composition has an important influence on the catalytic pyrolysis process and the formation mechanism of phenolic products. Su et al. [66] found that activated carbon promoted the demethoxylation of lignin pyrolysis intermediates to generate monophenols (phenol and 4-ethyl phenol). Zhang et al. [78, 84] found that under the catalysis of activated carbon, cellulose pyrolysis intermediates can also generate phenols through catalytic rearrangement. Phenols were generated via two parallel reaction routes: (1) rearrangement of C6 compounds, (2) two furan oligomerization and subsequent acid catalytic reactions, shown as Fig. 11.9. Lu et al. [85] pointed out that activated carbon can promote biomass pyrolysis intermediate (4-ethyl phenol precursor) to generate more 4-ethyl phenol. However, the biomass composition is complex, and the biomass pyrolysis intermediates are numerous, and there are violent secondary reactions in the pyrolysis process. How to react with biomass pyrolysis volatiles to promote the formation of phenols (such as phenol, 4-vinyl phenol, 4-ethyl phenol, etc.) is also unclear. Therefore, it is necessary to further study the effect mechanism of nitrogen-doped biochar on the formation of phenolics, especially the regulation mechanism of nitrogen-doped biochar on typical intermediates and microstructures of biomass pyrolysis, and then clarify the regulation mechanism of phenolics. On the other hand, biomass contains inorganic components (such as K, Ca, Mg, etc.), which also have a significant impact on the biomass pyrolysis process, but its impact on the formation of phenolic substances is rarely reported.

11.5 Conclusions and Future Outlook

Biomass pyrolysis to prepare carbon materials is a negative carbon technology, and biochar materials can also be applied to catalysis, adsorption, energy storage, and soil remediation fields. Upgrading biochar is the key to realizing high-value utilization of biomass, and the one-step "pyrolysis-activation-doping" method introduced in this chapter is a promising method for preparing high-performance biochar materials, which can directly convert biomass into the functional biochar materials with developed porous structure and active functional groups. However, the

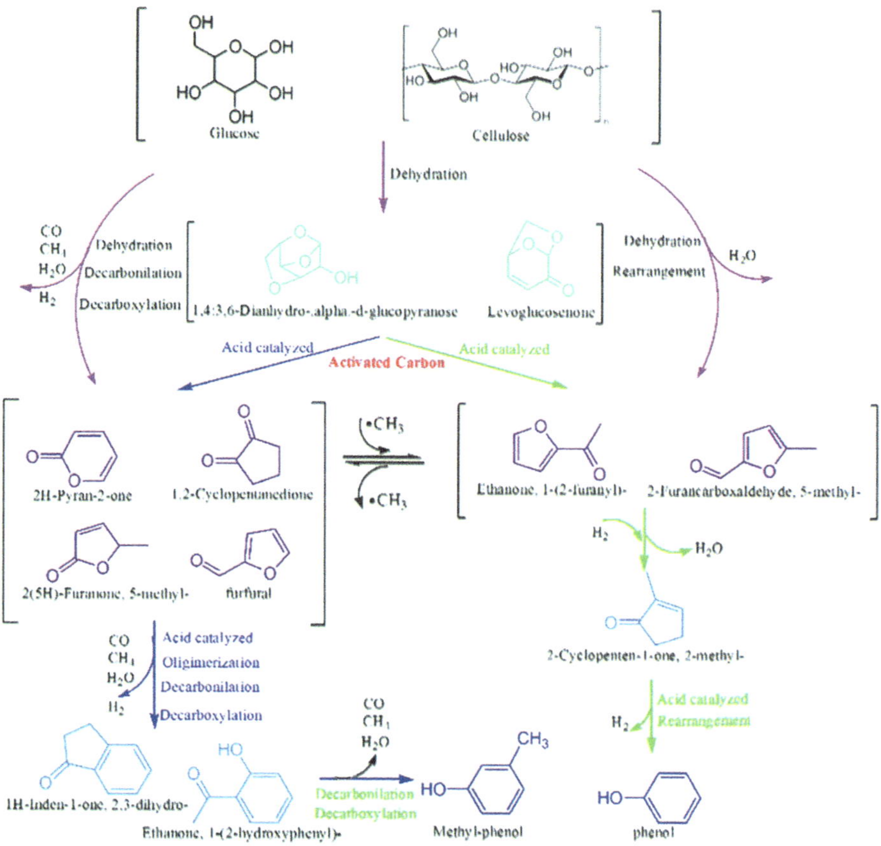

Fig. 11.9 Formation mechanism of phenols from catalytic pyrolysis of glucose and cellulose over activated carbon (adapted with permission from [84], Copyright © 2018 The Royal Society of Chemistry)

"pyrolysis-activation-doping" synergistic reaction mechanism and biochar formation mechanism need to be further studied through advanced online characterization methods, such as in situ XRD, in situ XPS, and in situ TEM, combined with theoretical methods such as quantum chemical calculations.

The mechanism of directional pyrolysis of biomass to produce valued phenols with N-doped biochar catalyst is complex, and the selectivity of target products still needs to be improved, which increases the challenge for subsequent purification of chemicals. Therefore, to realize the enrichment of biomass-directed pyrolysis products, in terms of mechanistic research, it is necessary to develop characterization techniques that can detect structural evolution of intermediates and free radicals in the reaction process in a complex atmosphere system, and improve the quality from the source, process and product. Synergistic regulation of reaction variables could be used to improve the yield of the target product.

Acknowledgements We express great appreciation of the financial support from the National Natural Science Foundation of China (52106243, 51876078, and 51861130362), National key R&D program of China (2019YFC1904003), Natural Science Foundation of Jiangsu Province (BK20221517), and China Postdoctoral Science Foundation (2018M640696 and 2019T120664).

References

1. Liao Y, Koelewijn S-F, Van den Bossche G, Van Aelst J, Van den Bosch S, Renders T, Navare K, Nicolaï T, Van Aelst K, Maesen M, Matsushima H, Thevelein JM, Van Acker K, Lagrain B, Verboekend D, Sels BF. A sustainable wood biorefinery for low–carbon footprint chemicals production. Science. 2020;367:1385–90. https://doi.org/10.1126/science.aau1567.
2. Liu W-J, Jiang H, Yu H-Q. Development of biochar-based functional materials: toward a sustainable platform carbon material. Chem Rev. 2015;115:12251–85. https://doi.org/10.1021/acs.chemrev.5b00195.
3. Yang Q, Zhou H, Bartocci P, Fantozzi F, Mašek O, Agblevor FA, Wei Z, Yang H, Chen H, Lu X, Chen G, Zheng C, Nielsen CP, McElroy MB. Prospective contributions of biomass pyrolysis to China's 2050 carbon reduction and renewable energy goals. Nat Commun. 2021;12:1698. https://doi.org/10.1038/s41467-021-21868-z.
4. Lian F, Xing B. Black carbon (biochar) inwater/soil environments: molecular structure, sorption, stability, and potential risk. Environ Sci Technol. 2017;51:13517–32. https://doi.org/10.1021/acs.est.7b02528.
5. Tian W, Zhang H, Duan X, Sun H, Shao G, Wang S. Porous carbons: structure-oriented design and versatile applications. Adv Funct Mater. 2020;30(17):1909265. https://doi.org/10.1002/adfm.201909265.
6. Li D, Chen W, Wu J, Jia CQ, Jiang X. The preparation of waste biomass-derived N-doped carbons and their application in acid gas removal: focus on N functional groups. J Mater Chem A. 2020;8:24977–95. https://doi.org/10.1039/d0ta07977d.
7. Chen Q, Tan X, Liu Y, Liu S, Li M, Gu Y, Zhang P, Ye S, Yang Z, Yang Y. Biomass-derived porous graphitic carbon materials for energy and environmental applications. J Mater Chem A. 2020;8:5773–811. https://doi.org/10.1039/C9TA11618D.
8. Inagaki M, Toyoda M, Soneda Y, Morishita T. Nitrogen-doped carbon materials. Carbon. 2018;132:104–40. https://doi.org/10.1016/j.carbon.2018.02.024.
9. Wang J, Nie P, Ding B, Dong S, Hao X, Dou H, Zhang X. Biomass derived carbon for energy storage devices. J Mater Chem A. 2017;5:2411–28. https://doi.org/10.1039/C6TA08742F.
10. Lin T, Chen I-W, Liu F, Yang C, Bi H, Xu F, Huang F. Nitrogen-doped mesoporous carbon of extraordinary capacitance for electrochemical energy storage. Science. 2015;350:1508–13. https://doi.org/10.1126/science.aab3798.
11. Figueiredo JL, Pereira MFR. The role of surface chemistry in catalysis with carbons. Catal Today. 2010;150:2–7. https://doi.org/10.1016/j.cattod.2009.04.010.
12. Wang S, Dai G, Yang H, Luo Z. Lignocellulosic biomass pyrolysis mechanism: a state-of-the-art review. Prog Energy Combust Sci. 2017;62:33–86. https://doi.org/10.1016/j.pecs.2017.05.004.
13. Che Q, Yang M, Wang X, Chen X, Chen W, Yang Q, Yang H, Chen H. Aromatics production with metal oxides and ZSM-5 as catalysts in catalytic pyrolysis of wood sawdust. Fuel Process Technol. 2019;188:146–52. https://doi.org/10.1016/j.fuproc.2019.02.016.
14. Sharifzadeh M, Sadeqzadeh M, Guo M, Borhani TN, Murthy Konda NVSN, Garcia MC, Wang L, Hallett J, Shah N. The multi-scale challenges of biomass fast pyrolysis and bio-oil upgrading: review of the state of art and future research directions. Prog Energy Combust Sci. 2019;71:1–80. https://doi.org/10.1016/j.pecs.2018.10.006.

15. Mika LT, Cséfalvay E, Németh Á. Catalytic conversion of carbohydrates to initial platform chemicals: chemistry and sustainability. Chem Rev. 2018;118:505–613. https://doi.org/10.1021/acs.chemrev.7b00395.
16. Schutyser W, Renders T, Van den Bosch S, Koelewijn SF, Beckham GT, Sels BF. Chemicals from lignin: an interplay of lignocellulose fractionation, depolymerisation, and upgrading. Chem Soc Rev. 2018;47:852–908. https://doi.org/10.1039/C7CS00566K.
17. Jeong J-Y, Lee U-D, Chang W-S, Jeong S-H. Production of bio-oil rich in acetic acid and phenol from fast pyrolysis of palm residues using a fluidized bed reactor: influence of activated carbons. Bioresour Technol. 2016;219:357–64. https://doi.org/10.1016/j.biortech.2016.07.107.
18. Cordella M, Torri C, Adamiano A, Fabbri D, Barontini F, Cozzani V. Bio-oils from biomass slow pyrolysis: a chemical and toxicological screening. J Hazard Mater. 2012;231-232:26–35. https://doi.org/10.1016/j.jhazmat.2012.06.030.
19. Amen-Chen C, Pakdel H, Roy C. Production of monomeric phenols by thermochemical conversion of biomass: a review. Bioresour Technol. 2001;79:277–99. https://doi.org/10.1016/S0960-8524(00)00180-2.
20. Kim J-S. Production, separation and applications of phenolic-rich bio-oil – a review. Bioresour Technol. 2015;178:90–8. https://doi.org/10.1016/j.biortech.2014.08.121.
21. Naron DR, Collard FX, Tyhoda L, Görgens JF. Production of phenols from pyrolysis of sugarcane bagasse lignin: catalyst screening using thermogravimetric analysis – thermal desorption – gas chromatography – mass spectroscopy. J Anal Appl Pyrolysis. 2019;138: 120–31. https://doi.org/10.1016/j.jaap.2018.12.015.
22. Chen W, Fang Y, Li K, Chen Z, Xia M, Gong M, Chen Y, Yang H, Tu X, Chen H. Bamboo wastes catalytic pyrolysis with N-doped biochar catalyst for phenols products. Appl Energy. 2020;260:114242. https://doi.org/10.1016/j.apenergy.2019.114242.
23. Fu D-q, Li X-h, Li W-y, Feng J. Catalytic upgrading of coal pyrolysis products over bio-char. Fuel Process Technol. 2018;176:240–8. https://doi.org/10.1016/j.fuproc.2018.04.001.
24. Qian K, Kumar A, Zhang H, Bellmer D, Huhnke R. Recent advances in utilization of biochar. Renew Sustain Energy Rev. 2015;42:1055–64. https://doi.org/10.1016/j.rser.2014.10.074.
25. Yang Z, Lei H, Zhang Y, Qian K, Villota E, Qian M, Yadavalli G, Sun H. Production of renewable alkyl-phenols from catalytic pyrolysis of Douglas fir sawdust over biomass-derived activated carbons. Appl Energy. 2018;220:426–36. https://doi.org/10.1016/j.apenergy.2018.03.107.
26. Zou R, Qian M, Wang C, Mateo W, Wang Y, Dai L, Lin X, Zhao Y, Huo E, Wang L, Zhang X, Kong X, Ruan R, Lei H. Biochar: from by-products of agro-industrial lignocellulosic waste to tailored carbon-based catalysts for biomass thermochemical conversions. Chem Eng J. 2022;441:135972. https://doi.org/10.1016/j.cej.2022.135972.
27. Wang S, Li H, Wu M. Advances in metal/biochar catalysts for biomass hydro-upgrading: a review. J Clean Prod. 2021;303:126825. https://doi.org/10.1016/j.jclepro.2021.126825.
28. Low YW, Yee KF. A review on lignocellulosic biomass waste into biochar-derived catalyst: current conversion techniques, sustainable applications and challenges. Biomass Bioenergy. 2021;154:106245. https://doi.org/10.1016/j.biombioe.2021.106245.
29. Chi NTL, Anto S, Ahamed TS, Kumar SS, Shanmugam S, Samuel MS, Mathimani T, Brindhadevi K, Pugazhendhi A. A review on biochar production techniques and biochar based catalyst for biofuel production from algae. Fuel. 2021;287:119411. https://doi.org/10.1016/j.fuel.2020.119411.
30. Chen W, Li K, Xia M, Chen Y, Yang H, Chen Z, Chen X, Chen H. Influence of NH_3 concentration on biomass nitrogen-enriched pyrolysis. Bioresour Technol. 2018;263:350–7. https://doi.org/10.1016/j.biortech.2018.05.025.
31. Hulicova-Jurcakova D, Kodama M, Shiraishi S, Hatori H, Zhu ZH, Lu GQ. Nitrogen-enriched nonporous carbon electrodes with extraordinary supercapacitance. Adv Funct Mater. 2009;19: 1800–9. https://doi.org/10.1002/adfm.200801100.

32. Laheaeaer A, Delpeux-Ouldriane S, Lust E, Beguin F. Ammonia treatment of activated carbon powders for supercapacitor electrode application. J Electrochem Soc. 2014;161:A568–75. https://doi.org/10.1149/2.051404jes.
33. Luo W, Wang B, Heron CG, Allen MJ, Morre J, Maier CS, Stickle WF, Ji X. Pyrolysis of cellulose under ammonia leads to nitrogen-doped nanoporous carbon generated through methane formation. Nano Lett. 2014;14:2225–9. https://doi.org/10.1021/nl500859p.
34. Chen W, Yang H, Chen Y, Chen X, Fang Y, Chen H. Biomass pyrolysis for nitrogen-containing liquid chemicals and nitrogen-doped carbon materials. J Anal Appl Pyrolysis. 2016;120:186–93. https://doi.org/10.1016/j.jaap.2016.05.004.
35. Kan T, Strezov V, Evans TJ. Lignocellulosic biomass pyrolysis: a review of product properties and effects of pyrolysis parameters. Renew Sustain Energy Rev. 2016;57:1126–40. https://doi.org/10.1016/j.rser.2015.12.185.
36. Liu WJ, Li WW, Jiang H, Yu HQ. Fates of chemical elements in biomass during its pyrolysis. Chem Rev. 2017;117:6367–98. https://doi.org/10.1021/acs.chemrev.6b00647.
37. Deng J, Li M, Wang Y. Biomass-derived carbon: synthesis and applications in energy storage and conversion. Green Chem. 2016;18:4824–54. https://doi.org/10.1039/c6gc01172a.
38. Li Z, Zhang L, Amirkhiz BS, Tan X, Xu Z, Wang H, Olsen BC, Holt CMB, Mitlin D. Carbonized chicken eggshell membranes with 3D architectures as high-performance electrode materials for supercapacitors. Adv Energy Mater. 2012;2:431–7. https://doi.org/10.1002/aenm.201100548.
39. Bae YJ, Ryu C, Jeon J-K, Park J, Suh DJ, Suh Y-W, Chang D, Park Y-K. The characteristics of bio-oil produced from the pyrolysis of three marine macroalgae. Bioresour Technol. 2011;102:3512–20. https://doi.org/10.1016/j.biortech.2010.11.023.
40. Marcilla A, Leon M, Nuria Garcia A, Banon E, Martinez P. Upgrading of tannery wastes under fast and slow pyrolysis conditions. Ind Eng Chem Res. 2012;51:3246–55. https://doi.org/10.1021/ie201635w.
41. Deng Y, Xie Y, Zou K, Ji X. Review on recent advances in nitrogen-doped carbons: preparations and applications in supercapacitors. J Mater Chem A. 2016;4:1144–73. https://doi.org/10.1039/C5TA08620E.
42. Chen Y, Zhang X, Chen W, Yang H, Chen H. The structure evolution of biochar from biomass pyrolysis and its correlation with gas pollutant adsorption performance. Bioresour Technol. 2017;246:101–9. https://doi.org/10.1016/j.biortech.2017.08.138.
43. Kasera N, Hall S, Kolar P. Effect of surface modification by nitrogen-containing chemicals on morphology and surface characteristics of N-doped pine bark biochars. J Environ Chem Eng. 2021;9:105161. https://doi.org/10.1016/j.jece.2021.105161.
44. Chen W, Yang H, Chen Y, Xia M, Yang Z, Wang X, Chen H. Algae pyrolytic poly-generation: influence of component difference and temperature on products characteristics. Energy. 2017;131:1–12. https://doi.org/10.1016/j.energy.2017.05.019.
45. Chen W, Yang H, Chen Y, Xia M, Chen X, Chen H. Transformation of nitrogen and evolution of N-containing species during algae pyrolysis. Environ Sci Technol. 2017;51:6570–9. https://doi.org/10.1021/acs.est.7b00434.
46. Chen W, Chen Y, Yang H, Li K, Chen X, Chen H. Investigation on biomass nitrogen-enriched pyrolysis: influence of temperature. Bioresour Technol. 2018;249:247–53. https://doi.org/10.1016/j.biortech.2017.10.022.
47. Hu X, Sun X, Yoo SJ, Evanko B, Fan F, Cai S, Zheng C, Hu W, Stucky GD. Nitrogen-rich hierarchically porous carbon as a high-rate anode material with ultra-stable cyclability and high capacity for capacitive sodium-ion batteries. Nano Energy. 2019;56:828–39. https://doi.org/10.1016/j.nanoen.2018.11.081.
48. Chen P, Wang L-K, Wang G, Gao M-R, Ge J, Yuan W-J, Shen Y-H, Xie A-J, Yu S-H. Nitrogen-doped nanoporous carbon nanosheets derived from plant biomass: an efficient catalyst for oxygen reduction reaction. Energy Environ Sci. 2014;7:4095–103. https://doi.org/10.1039/C4EE02531H.

49. Matsagar BM, Yang R-X, Dutta S, Ok YS, Wu KCW. Recent progress in the development of biomass-derived nitrogen-doped porous carbon. J Mater Chem A. 2021;9:3703–28. https://doi.org/10.1039/D0TA09706C.
50. Zhang B, Jiang Y, Balasubramanian R. Synthesis, formation mechanisms and applications of biomass-derived carbonaceous materials: a critical review. J Mater Chem A. 2021;9:24759–802. https://doi.org/10.1039/D1TA06874A.
51. Jin C, Nai J, Sheng O, Yuan H, Zhang W, Tao X, Lou XW. Biomass-based materials for green lithium secondary batteries. Energy Environ Sci. 2021;14:1326–79. https://doi.org/10.1039/D0EE02848G.
52. Wang J, Kaskel S. KOH activation of carbon-based materials for energy storage. J Mater Chem. 2012;22:23710–25. https://doi.org/10.1039/C2JM34066F.
53. Lv Y, Zhang F, Dou Y, Zhai Y, Wang J, Liu H, Xia Y, Tu B, Zhao D. A comprehensive study on KOH activation of ordered mesoporous carbons and their supercapacitor application. J Mater Chem. 2012;22:93–9. https://doi.org/10.1039/C1JM12742J.
54. Elmouwahidi A, Bailón-García E, Pérez-Cadenas AF, Maldonado-Hódar FJ, Carrasco-Marín F. Activated carbons from KOH and H_3PO_4-activation of olive residues and its application as supercapacitor electrodes. Electrochim Acta. 2017;229:219–28. https://doi.org/10.1016/j.electacta.2017.01.152.
55. Huber GW, Iborra S, Corma A. Synthesis of transportation fuels from biomass: chemistry, catalysts, and engineering. Chem Rev. 2006;106:4044–98. https://doi.org/10.1021/cr068360d.
56. Chunlan L, Shaoping X, Yixiong G, Shuqin L, Changhou L. Effect of pre-carbonization of petroleum cokes on chemical activation process with KOH. Carbon. 2005;43:2295–301. https://doi.org/10.1016/j.carbon.2005.04.009.
57. Qu W-H, Xu Y-Y, Lu A-H, Zhang X-Q, Li W-C. Converting biowaste corncob residue into high value added porous carbon for supercapacitor electrodes. Bioresour Technol. 2015;189:285–91. https://doi.org/10.1016/j.biortech.2015.04.005.
58. Chen W, Gong M, Li K, Xia M, Chen Z, Xiao H, Fang Y, Chen Y, Yang H, Chen H. Insight into KOH activation mechanism during biomass pyrolysis: chemical reactions between O-containing groups and KOH. Appl Energy. 2020;278:115730. https://doi.org/10.1016/j.apenergy.2020.115730.
59. Zhang C, Song W, Ma Q, Xie L, Zhang X, Guo H. Enhancement of CO_2 capture on biomass-based carbon from black locust by KOH activation and ammonia modification. Energy Fuels. 2016;30:4181–90. https://doi.org/10.1021/acs.energyfuels.5b02764.
60. Li Y, Li Z, Xing B, Li H, Ma Z, Zhang W, Reubroycharoen P, Wang S. Green conversion of bamboo chips into high-performance phenol adsorbent and supercapacitor electrodes by simultaneous activation and nitrogen doping. J Anal Appl Pyrolysis. 2021;155:105072. https://doi.org/10.1016/j.jaap.2021.105072.
61. Guo N, Li M, Wang Y, Sun X, Wang F, Yang R. N-doped hierarchical porous carbon prepared by simultaneous-activation of KOH and NH_3 for high performance supercapacitors. RSC Adv. 2016;6:101372–9. https://doi.org/10.1039/C6RA22426A.
62. Zhu Y, Chen M, Zhang Y, Zhao W, Wang C. A biomass-derived nitrogen-doped porous carbon for high-energy supercapacitor. Carbon. 2018;140:404–12. https://doi.org/10.1016/j.carbon.2018.09.009.
63. Li K, Chen W, Yang H, Chen Y, Xia S, Xia M, Tu X, Chen H. Mechanism of biomass activation and ammonia modification for nitrogen-doped porous carbon materials. Bioresour Technol. 2019;280:260–8. https://doi.org/10.1016/j.biortech.2019.02.039.
64. Yang Z, Lei H, Qian K, Zhang Y, Villota E. Renewable bio-phenols from in situ and ex situ catalytic pyrolysis of Douglas fir pellet over biobased activated carbons. Sustain Energy Fuels. 2018;2:894–904. https://doi.org/10.1039/c7se00607a.
65. Li W, Wang D, Zhu Y, Chen J, Lu Y, Li S, Zheng Y, Zheng Z. Efficient ex-situ catalytic upgrading of biomass pyrolysis vapors to produce methylfurans and phenol over bio-based activated carbon. Biomass Bioenergy. 2020;142:105794. https://doi.org/10.1016/j.biombioe.2020.105794.

66. Su Y, Liu L, Xu D, Du H, Xie Y, Xiong Y, Zhang S. Syngas production at low temperature via the combination of hydrothermal pretreatment and activated carbon catalyst along with value-added utilization of tar and bio-char. Energy Convers Manag. 2020;205:112382. https://doi.org/10.1016/j.enconman.2019.112382.
67. Bu Q, Lei H, Wang L, Wei Y, Zhu L, Zhang X, Liu Y, Yadavalli G, Tang J. Bio-based phenols and fuel production from catalytic microwave pyrolysis of lignin by activated carbons. Bioresour Technol. 2014;162:142–7. https://doi.org/10.1016/j.biortech.2014.03.103.
68. Wang C, Lei H, Zhao Y, Qian M, Kong X, Mateo W, Zou R, Ruan R. Integrated harvest of phenolic monomers and hydrogen through catalytic pyrolysis of biomass over nanocellulose derived biochar catalyst. Bioresour Technol. 2021;320:124352. https://doi.org/10.1016/j.biortech.2020.124352.
69. Ma S, Li H, Zhang G, Iqbal T, Li K, Lu Q. Catalytic fast pyrolysis of walnut shell for alkylphenols production with nitrogen-doped activated carbon catalyst. Front Environ Sci Eng. 2021;15:25. https://doi.org/10.1007/s11783-020-1317-y.
70. Chen X, Che Q, Li S, Liu Z, Yang H, Chen Y, Wang X, Shao J, Chen H. Recent developments in lignocellulosic biomass catalytic fast pyrolysis: strategies for the optimization of bio-oil quality and yield. Fuel Process Technol. 2019;196:106180. https://doi.org/10.1016/j.fuproc.2019.106180.
71. Li M, Xu F, Li H, Wang Y. Nitrogen-doped porous carbon materials: promising catalysts or catalyst supports for heterogeneous hydrogenation and oxidation. Cat Sci Technol. 2016;6:3670–93. https://doi.org/10.1039/C6CY00544F.
72. Serp P, Machado B. Nanostructured carbon materials for catalysis. Royal Society of Chemistry; 2015.
73. Guo D, Shibuya R, Akiba C, Saji S, Kondo T, Nakamura J. Active sites of nitrogen-doped carbon materials for oxygen reduction reaction clarified using model catalysts. Science. 2016;351:361–5. https://doi.org/10.1126/science.aad0832.
74. Yang H, Chen Z, Chen W, Chen Y, Wang X, Chen H. Role of porous structure and active O-containing groups of activated biochar catalyst during biomass catalytic pyrolysis. Energy. 2020;210:118646. https://doi.org/10.1016/j.energy.2020.118646.
75. Cao B, Yuan J, Jiang D, Wang S, Barati B, Hu Y, Yuan C, Gong X, Wang Q. Seaweed-derived biochar with multiple active sites as a heterogeneous catalyst for converting macroalgae into acid-free biooil containing abundant ester and sugar substances. Fuel. 2021;285:119164. https://doi.org/10.1016/j.fuel.2020.119164.
76. Krerkkaiwan S, Mueangta S, Thammarat P, Jaisat L, Kuchonthara P. Catalytic biomass-derived tar decomposition using char from the co-pyrolysis of coal and giant leucaena wood biomass. Energy Fuels. 2015;29:3119–26. https://doi.org/10.1021/ef502792x.
77. Wang M, Jin L, Li Y, Lv J, Wei B, Hu H. In-situ catalytic upgrading of coal pyrolysis tar coupled with CO_2 reforming of methane over Ni-based catalysts. Fuel Process Technol. 2018;177:119–28. https://doi.org/10.1016/j.fuproc.2018.04.022.
78. Zhang Y, Lei H, Yang Z, Qian K, Villota E. Renewable high-purity mono-phenol production from catalytic microwave-induced pyrolysis of cellulose over biomass-derived activated carbon catalyst. ACS Sustain Chem Eng. 2018;6:5349–57. https://doi.org/10.1021/acssuschemeng.8b00129.
79. Pena J, Villot A, Gerente C. Pyrolysis chars and physically activated carbons prepared from buckwheat husks for catalytic purification of syngas. Biomass Bioenergy. 2020;132:105435. https://doi.org/10.1016/j.biombioe.2019.105435.
80. Fuentes-Cano D, Parrillo F, Ruoppolo G, Gómez-Barea A, Arena U. The influence of the char internal structure and composition on heterogeneous conversion of naphthalene. Fuel Process Technol. 2018;172:125–32. https://doi.org/10.1016/j.fuproc.2017.12.015.
81. Ravenni G, Sárossy Z, Ahrenfeldt J, Henriksen UB. Activity of chars and activated carbons for removal and decomposition of tar model compounds – a review. Renew Sustain Energy Rev. 2018;94:1044–56. https://doi.org/10.1016/j.rser.2018.07.001.

82. Shen Y. Chars as carbonaceous adsorbents/catalysts for tar elimination during biomass pyrolysis or gasification. Renew Sustain Energy Rev. 2015;43:281–95. https://doi.org/10.1016/j.rser.2014.11.061.
83. Chen XY, Chen C, Zhang ZJ, Xie DH, Deng X, Liu JW. Nitrogen-doped porous carbon for supercapacitor with long-term electrochemical stability. J Power Sources. 2013;230:50–8. https://doi.org/10.1016/j.jpowsour.2012.12.054.
84. Zhang Y, Lei H, Yang Z, Duan D, Villota E, Ruan R. From glucose-based carbohydrates to phenol-rich bio-oils integrated with syngas production via catalytic pyrolysis over an activated carbon catalyst. Green Chem. 2018;20:3346–58. https://doi.org/10.1039/C8GC00593A.
85. Lu Q, Ye X-n, Zhang Z-b, Cui M-S, Guo H-q, Qi W, Dong C-q, Yang Y-p. Catalytic fast pyrolysis of bagasse using activated carbon catalyst to selectively produce 4-ethyl phenol. Energy Fuels. 2016;30:10618–26. https://doi.org/10.1021/acs.energyfuels.6b02628.

Part IV
N Transformations During Thermal Processes

Chapter 12
Evaluating the Role of Gasification Stages on Evolution of Fuel-N to Deepen in Sustainable Production of NH_3

Fernando Léo, Noemí Gil-Lalaguna, Zainab Afailal, Rubenildo Andrade, Electo Lora, and Isabel Fonts

Abstract In this chapter, pyrolysis of meat and bone meal (MBM) with two types of reactors and gasification of produced pyrolysis char under different atmospheres are shown for estimating contributions to fuel-N distribution from MBM gasification. Formation of the main nitrogen-containing (N-containing) compounds was studied: char-N, tar-N, NH_3-N, HCN-N and N_2-N. Argon was used in pyrolysis experiments and argon+O_2 and argon+O_2 + steam mixtures were used for char gasification that allowed quantification of N_2 generated. The experimental results showed that NH_3 formation was hardly affected by pyrolysis temperature in the studied range from 600 °C to 800 °C or by the reactor type (fixed or fluidized). However, on the matter of tar-N yield, the range and type of reactor greatly affected its production from 21% at 600 °C for fixed beds to 3% at 800 °C in the fluidized bed. Pyrolysis char gasification using steam as one of the gasifying agents greatly affects NH_3 production, from 0.1% with O_2 + Ar to 13.8% with O_2 + Ar + steam. The most abundant N-containing compound obtained in the pyrolysis step was NH_3 (33–35%). For the char gasification step, the most abundant N-containing compounds were N_2 (35–45%) and NH_3 (13.8%). An extensive analysis of tar-N was performed that allowed point relevant differences to be determined between the tar-N composition of pyrolysis at 600 °C in the fixed bed and at 800 °C in the fluidized bed. Neither

F. Léo
Thermochemical Processes Group, Aragon Institute for Engineering Research (I3A), Chemical and Environmental Engineering Department, University of Zaragoza, Zaragoza, Spain

Center for Excellence in Thermal and Distributed Generation (NEST), Institute of Mechanical Engineering, Federal University of Itajubá, Itajubá, Brazil

N. Gil-Lalaguna · Z. Afailal · I. Fonts (✉)
Thermochemical Processes Group, Aragon Institute for Engineering Research (I3A), Chemical and Environmental Engineering Department, University of Zaragoza, Zaragoza, Spain
e-mail: isabelfo@unizar.es

R. Andrade · E. Lora
Center for Excellence in Thermal and Distributed Generation (NEST), Institute of Mechanical Engineering, Federal University of Itajubá, Itajubá, Brazil

Z. Fang et al. (eds.), *Production of N-containing Chemicals and Materials from Biomass*, Biofuels and Biorefineries 12,
https://doi.org/10.1007/978-981-99-4580-1_12

HCN nor NO were obtained in appreciable quantities in this work. The results obtained throughout the experiments explains 50% of the NH_3 and N_2 yields on the overall MBM gasification process. Among the stages studied, the pyrolysis stage and char gasification in the presence of steam contribute significantly to NH_3 production, although the tar-N cracking and reforming reactions could also have a noteworthy effect on NH_3 production.

Keywords Biological · Nitrogen · Rich residues · Fuel-N · Pyrolysis · Step · Gasification · Tar-N · Ammonia

Abbreviations

%	used when the percentage is the same referred to moles or mass, as for example un fuel-N distribution
ATR-FTIR	attenuated total reflection—fourier transform infrared
BAT	Best available techniques
DKP	Diketopiperazines
$GasArO_2$	Char gasification at 800 °C with Ar and O_2
$GasArO_2H_2O$	Char gasification at 800 °C with Ar, O_2 and H_2O
GC-MS	Gas chromatography-mass spectrometry
GJ	Giga Joule
HACA	Hydrogen abstraction acetylene addition
MBM	Meat and bone meal
Mt.	Million tonnes
PAH	Polycyclic aromatic hydrocarbons
Pyr600Fix	Pyrolysis at 600 °C in fixed bed
Pyr600Flu	Pyrolysis at 600 °C in fluidized bed
Pyr800Flu	Pyrolysis at 800 °C in fluidized bed
STP	Standard temperature and pressure (0 °C and 1 atm)
vol%	vol percentage
wt%	Weight percentage

12.1 Introduction

Due to its intensive use for the formulation of synthetic fertilizers, ammonia (NH_3) is currently one of the most produced chemicals in the world, with 176 Mt. per year [1] and it is estimated that its production may multiply until more than 900 Mt. per year in 2050 [2]. NH_3 is synthesized industrially by the Haber-Bosch process, which is based on a reversible chemical reaction in which the stable atmospheric N_2 reacts with H_2, mainly produced from steam reforming of natural gas, in a catalyst bed at elevated pressures (150–300 bar) and temperatures (350–500 °C) to give the aforementioned ammonia ($N_2(g) + 3H_2(g) \leftrightarrows 2NH_3(g) + heat$). Even though this is an exothermic reaction, the global process is one of the most energy-intensive industrial chemical processes, requiring 28 GJ·per tonne NH_3 when using the best available techniques (BAT) [3], which is around 2% ($8.6 \cdot 10^9$ GJ) of total final energy consumption in industrial processes [2]. When using natural gas as a H_2 source without CO_2 capture, this high energy consumption also translates into emissions ranging between 1.56 (with BAT) and 2.1 tonne of CO_2 for each tonne of NH_3 produced [1], which means a total of about 500 Mt. of CO_2 (about 1.8% of global CO_2 emissions) [1, 4]. In addition to these environmental drawbacks, natural gas accounts for up to 80% of the production costs of NH_3-based fertilizers, and the current exorbitant natural gas prices have led to the temporary shutdown of some fertilizer plants, which may have an impact on agricultural yields and the food supply chain. Consequently, there is great interest in developing sustainable processes that do not require the use of fossil fuels for NH_3 synthesis.

To obtain H_2 by a more sustainable process than steam reforming of natural gas, biomass gasification, steam reforming of biomass derived liquids or water electrolysis using renewable energies is being studied. Novel methods in which NH_3 synthesis does not take place by the Haber-Bosch process are still operating on a basic research level, e.g. NH_3 produced by bacteria in a biological process, by a catalytic electrochemical process from N_2 and H_2O and by chemical looping involving a series of chemical and electrochemical reactions [1]. However, the Haber-Bosch process as well as these last newfangled processes also present drawbacks in relation to the excessive fixation of stable atmospheric N_2, which is one of the anthropogenic causes of the imbalance in the biogeochemical N cycle. According to researchers at the Stockholm Environment Institute, a drastic reduction in the amount of N_2 fixation from 121 Mt. $year^{-1}$ to 35 Mt. $year^{-1}$ is necessary to restore the balance in the biogeochemical N cycle. Therefore, it is of great interest to establish environmentally sustainable alternatives for NH_3 production involving the use of hydrogen (H) from renewable sources, but also the use of a reactive nitrogen (N) source instead of the stable atmospheric N_2. Nitrogen-rich (N-rich) biological wastes are a possible source of both elements, H and N.

These N-rich biowastes include animal by-products (ABPs), which are materials obtained from animals and not intended for human consumption (slaughterhouse waste, fallen stock or dead pets), as for example meat and bone meal (MBM). Meat and bone meal (MBM) were selected as raw material in this work due to their high nitrogen content ($\approx$10 wt%), high-level of annual production (1.2 Mt. in Spain and 20 Mt. in EU) and their challenging management way [5]. If not processed properly, animal by-products bear the risk of transmitting diseases to humans or animals. They are sorted into three categories, being Category 1 the highest-risk material and Category 3 the one with the lowest associated risk. Over the years, European Food Security Agency (EFSA) has evaluated several alternative processing methods to produce biofuels or for other purposes such as compost, biogas or animal feeding. As a result, some EU countries have approved establishments to treat, above all, residues belonging to Category 3 [6]. Alternatives for Categories 1 and 2 are still scarce [6]. Same trends in the approved establishments are also followed by other countries as Great Britain [7]. That is why are mainly landfilled or incinerated. Incineration of these residues achieves the destruction of pathogens and prions, but it presents the drawback of the high generation of NOx, due to their elevated N content. Moreover, because the existing treatments do not allow the joint recovery of both nutrients (NH_3) and energy, there is a need for more sustainable treatments for these N-rich wastes.

Recently, our research group proposed a strategy in which NH_3 could be produced by gasification of a N-rich biological waste, meat and bone meal (MBM) (9.9 wt% of N), while the waste was energetically valorized [8] and achieved a fuel nitrogen (fuel-N) conversion to NH_3 of 67%. Once NH_3 was condensed and recovered, the combustion of the gasification gas could achieve a potential energy production up to 102 GJ ton^{-1} NH_3, turning the NH_3 synthesis from an energy demanding process (28 GJ ton^{-1} NH_3 when using the best existing industrial plants) into a power producing one. The maximum conversion obtained in that work would allow 10% of the annual NH_3 production in Europe to be produced from the proposed process [8]. Based on these results, the joint production of NH_3 and combustible gas by the gasification of wastes with high N content could have environmental benefits that would help to mitigate the problems associated with the alteration of the biogeochemical cycles of C and N. This positive effect could be even enhanced by increasing the conversion of fuel-N to NH_3. However, optimizing the gasification operating conditions for which NH_3 conversion can be increased or maximized is complicated due to the large number of reactions involving NH_3 either as a reactive or a product in the gasification process.

There are few references in the literature on the gasification of MBM [9–11], most of them focusing on the energetic valorization of the waste. In the air-gasification of MBM, Marculescu et al. achieved a gas with a low heating value of about 4.3 MJ m^{-3}(STP), which is enough to be combusted in an engine [10]. The LHV of the syngas may be improved by modifying the operating conditions. For example, Soni et al. obtained a gas with a very high H_2 content (36.2–47.1 vol%) when gasified MBM in the presence of steam, with mass ratios of steam/MBM between 0.4 and 0.8 [11]. Regarding fuel-N, its final distribution after a gasification process can be understood as the joint contribution of the N-involving reactions taking place in the following five stages: (1) pyrolysis, (2) char-N gasification, (3) tar-N cracking, (4) tar-N reforming and (5) gas phase reactions. The main N-containing products obtained in gasification are: char-N, tar-N, HCN-N, NH_3-N and N_2-N, the latter being determined by difference in the majority of cases. Previously, fuel-N distribution and NH_3 evolved during gasification were only studied from the point of view of its minimization as a pollutant. However, since then, several studies have evaluated the production of NH_3 following a similar route [12–16]. In these works, the promotional effect of Ca on perovskites for NH_3 production via steam catalytic gasification of microalgae was demonstrated, obtaining a maximum conversion yield of fuel-N into NH_3-N of 62.2%. To either reduce the production of N-containing pollutants in thermochemical processes or maximize the production of a specific added-value N-containing compound, such as NH_3, it is of utmost importance to know the role of each one of the aforementioned stages on the fuel-N distribution obtained from the whole process. The novelty of this work lies in the following aspects: (1) to produce ammonia via gasification of biological N-rich residues, (2) the experimental quantification of the N_2 generated and (3) to assess the contribution of each stage of the gasification process on the fuel-N distribution.

In this context, this work's main objective is to determine the importance of some of the aforementioned stages on the final fuel-N distribution obtained in the gasification of MBM with O_2-steam as gasifying agents. For this purpose, it has been assumed that the final fuel-N distribution can be broken down into the sum of individual contributions of each stage. Specifically, the fuel-N distribution obtained in the stages of pyrolysis, char gasification with O_2 and char gasification with O_2-steam will be individually and experimentally studied. As the gas residence time in both processes, pyrolysis and gasification, is usually not negligible, the final fuel-N distribution obtained experimentally will be influenced not only by gas-solid reactions, but also by gas-phase reactions involving the tar and gas species. Therefore, depending on the temperature and the type of reactor, the contribution of tar-N cracking, tar-N reforming and gas-phase reactions may have a significant impact on the fuel-N distribution. The contribution of tar-N cracking and secondary reactions will be assessed in this work by comparison of tar composition obtained from pyrolysis and gasification processes conducted at different temperatures and in different types of reactors, fixed and fluidized beds.

12.2 Materials and Methods

12.2.1 Materials: MBM Characterization

Meat and bone meal (MBM) used in this work was supplied by the Spanish company Residuos Aragón S.A., in which MBM are stabilized in an autoclave reactor (steam at 120 °C). The proximate and ultimate composition of this waste, as well as its concentration in metals and higher heating value, can be found elsewhere in other previous work of our research group [8]. The content of the main macro-components present in MBM were analyzed by different techniques, protein fraction by the quantification of the total (after hydrolysis) and free amino-acid content by ionic chromatography in a Biochrom 30 (sensitivity 10 pmol, further explanations of the procedure can be found elsewhere [8]), fats by Soxhlet extraction with n-hexane (further explanations elsewhere [17]), ash by combustion at 550 °C to avoid the decomposition of the carbonates following ISO-21656:2021 and moisture according to ISO-18134:2021. Attenuated Total Reflection-Fourier Transform Infrared (ATR-FTIR) spectra of MBM (and of chars obtained in the pyrolysis and gasification experiments) were obtained with an Agilent Cary 630 ATR-FTIR spectrometer. The range of wavenumber used was 4000–400 cm^{-1} with a resolution of 4 cm^{-1} over 24 scans. The results of these analysis with an in-depth discussion about the amino-acids present in the sample and about the FTIR spectrum is provided in Sect. 12.3.1.

12.2.2 Pyrolysis and Gasification Experiments

Two different types of reactors, a fixed bed and a fluidized bed, were utilized in the pyrolysis experiments, while the gasification experiments were carried out exclusively in a fluidized bed reactor.

The fixed bed reactor (90 mm diameter and 320 mm length) was a laboratory-scale batch reactor that works discontinuously for the solid (i.e. semi-batch) and continuously for the gas (see Fig. 12.1). In each fixed bed pyrolysis experiment, the reactor was filled with 300 g of MBM and placed inside an electric furnace that allowed control of the heating rate (approximately 8 °C min^{-1}); final temperature in the pyrolysis experiments was maintained during 1 h. This temperature was set at 600 °C in the fixed bed reactor, while it was increased up to 800 °C in the fluidized bed reactor. The gas residence time in the fixed bed pyrolysis experiments varies from around 15 s at the beginning of the heating ramp, to 9 s during the highest gas stream generation period.

MBM was used as raw material in all the pyrolysis experiments, while the char obtained in the fixed bed pyrolysis experiments (carried out at 600 °C) was the feedstock for the gasification experiments (evaluation of char-N gasification stage). Gasification experiments were carried out at 800 °C under different reaction media: (1) a mixture of O_2 and Ar and (2) a mixture of O_2, Ar and steam as

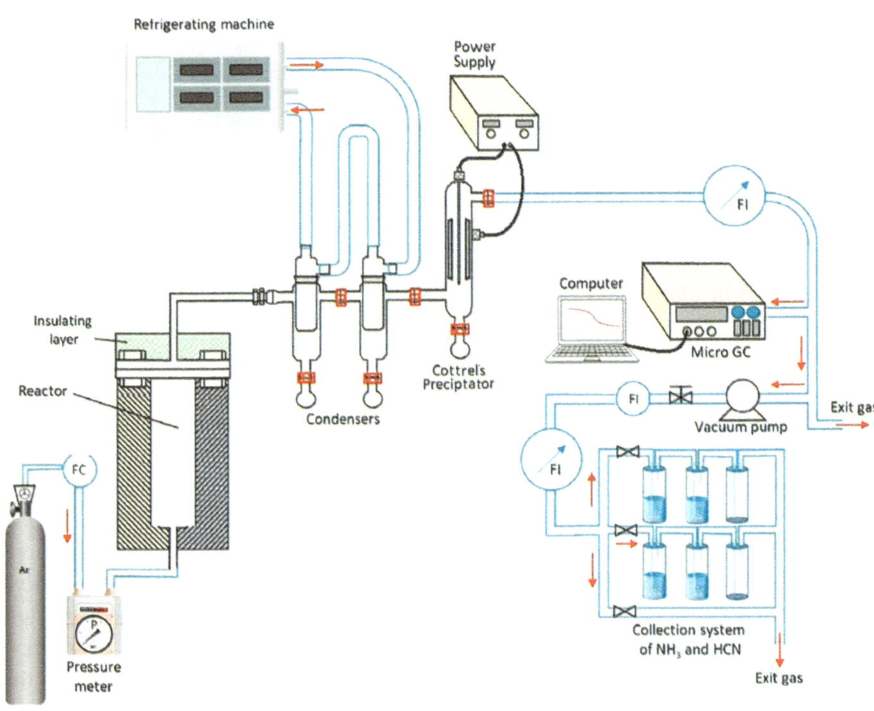

Fig. 12.1 Lab-scale fixed bed plant used in pyrolysis experiments

Table 12.1 Peak wavenumber assignments to functional groups and biomolecules

Peak number	Wavenumber on MBM spectrum (cm^{-1})	Wavenumber literature range (cm^{-1})	Functional group assignment	Biomolecule
1	3277	3288[a]; 3285[b]	ν(O-H) (broad band) and ν(N-H) of peptide bond (secondary amide)	Carbohydrates and proteins
2	3007	3010[c]	ν(C-H) unsaturated bonds (=C-H) typical of fatty acids	Lipids
3	2954	2954[c]; 2958[d]	ν_{as}(C-H) CH_3 saturated bonds of fatty acids	Lipids
4	2919	2925[c]; 2920[d]	ν_{as}(C-H) CH_2 saturated bonds of fatty acids	Lipids
5	2852	2855[c]; 2852[d]	ν_s(C-H) CH_2 saturated bonds of fatty acids	Lipids
6	1738	1738[d]; 1745[e]	ν(C=O) fatty acid esters in triglycerides and cholesterol esters	Lipids
7	1640	1645[a]; 1650[f];	ν(C=O) peptide bond in amide I band (1700–1600 cm^{-1})	Proteins
8	1520	1542[a]	δ_s(N-H) and ν(C-N) peptide bond in amide II band (1600–1500 cm^{-1})	Proteins
9	1458	1460[c]; 1465[d]	δ_s(C-H) CH_2 saturated bonds of fatty acids	Lipids
10	1380	1377[d]; 1380[e]	δ_s(C-H) CH_3 saturated bonds of fatty acids	Lipids
11	1240	1243[g]	ν(C-N) and δ_s(N-H) of peptide bond and ν_{as}(P=O) of >PO_2 in phosphodiesters	Nucleic acids, polyphosphates, hydroxyapatite and phospholipids
12	1163	1059–1065[a,h]	ν(C-O) and ν(C-C)	Carbohydrates
13	1080	1079[g]; 1080[e]; 1083[i]	ν_s(P=O) of >PO_2 in phosphodiesters and ν(C-O)	Nucleic acids, phospholipids, hydroxyapatite and carbohydrates
14	1030	1024[i]	ν(C-C)	Carbohydrates
15	876	875[e]	ν(P-O-P)	Nucleic acids and polyphosphates

as asymmetric, *s* symmetric, δ_s bending scissoring, ν stretching
[a][20]
[b][24]
[c][25]
[d][26]
[e][27]
[f][28]
[g][29]
[h][30]
[i][31]

gasifying-fluidizing agents. A more detailed description and a schematic diagram of the fluidized bed reactor plant can be found elsewhere [8, 18].

In the fluidized bed experiments, the material, either MBM for pyrolysis or char for gasification experiments, was manually fed through a double valve system by pulses (1.5 g min^{-1}). The equivalence ratio was 20% in both gasification conditions and the steam/carbon utilized in the air-steam gasification was 0.5 g g^{-1}. The duration of the pyrolysis or gasification experiments carried out in the fluidized bed was 1 h. To prevent N_2 from entering the fluidized bed reactor during the introduction of the raw material, the volume between the two valves was purged with Ar. The utilization of Ar as carrier or fluidizing agent allowed the quantification of the N_2 generated in the reactions, which represents an improvement over our previous work, where the N_2 generated was only roughly determined by difference. The gas residence time at high temperature in the pyrolysis and gasification experiments in fluidized bed was around 12 s (pyrolysis), 19 s (gasification with Ar + O_2) and 16 s (gasification with Ar + O_2 + steam), respectively. The residence times were calculated considering the total flow of gas (gasifying/fluidizing agent and gas generated) going through the volume of the reactor.

12.2.3 Characterization of N-Containing Products

The fuel-N distribution in the pyrolysis and the gasification experiments was assessed by quantifying the yield of N in the main N-containing products, which were char-N, tar-N, NH_3-N, HCN-N and N_2-N. Moreover, in the gasification experiments, the yields of NO-N and N_2O-N were also quantified. NH_3 and HCN were determined by two wet methods in the tar condensates and also in the gas stream exiting from the condensation system. The detailed procedure can be found elsewhere [8, 19]. N_2 was analyzed on-line using a micro gas chromatograph (Agilent 3000-A), which determines, every 2.5 min, the volume percentage of the following compounds in the gas stream exiting the condensation system: H_2, N_2, CO_2, CO, CH_4, C_2H_6, C_2H_4, C_2H_2, and H_2S. NO and N_2O concentration in the gasification gas were determined by infrared spectroscopy on a gas sample taken at the outlet of the condensation zone. Char-N and tar-N were determined considering the mass yields of char and tar, respectively, and their nitrogen contents (elemental composition). Moreover, the tar-N composition was analyzed by GC-MS (Gas Chromatography–Mass spectrometry) GC-MS in an Agilent 7890A gas chromatograph coupled to an Agilent 5975C mass selective detector using a DB-17 ms column (60 m × 250 μm × 0.25 μm). The compounds identified were classified in chemical families according to their main functional groups.

12.2.4 Stage Contribution to Final Fuel-N Distribution

The comparison between the fuel-N distribution determined in the MBM pyrolysis experiments and in the char gasification experiments with our previous results found for the fuel-N distribution in the single gasification of MBM (with a mixture of air and steam) [4] was used to show the contribution of the different stages. In the pyrolysis experiments, where MBM was used as feedstock, the aforementioned fuel-N distribution was calculated over the MBM-N fed to the experiment. In the gasification experiments, carried out with pyrolysis char as feedstock, the fuel-N distribution was calculated over the char-N fed to the gasification stage, but also globally over the MBM-N fed to the pyrolysis stage conducted previously to the gasification one. The closures of the N mass balance were between 70 wt% and 77 wt% for all experiments conducted, except for pyrolysis in the fluidized bed at 800 °C, whose closure was lower (60 wt%). The non-quantified N could be, for example, in the form of HNCO, which is known to be present in this type of thermochemical processes. The pyrolysis experiments conducted in the fixed and the fluidized bed reactors were useful to evaluate the contribution of pyrolysis to NH_3-N, tar-N, HCN-N and N_2-N yields.

The difference between the tar-N yield obtained from the pyrolysis of MBM at 600 °C with the one determined in the single gasification of MBM may be used to highlight the contribution of the tar-N cracking and reforming during the overall gasification process. Specifically, the effect of tar-N reforming reactions could be assessed by the comparison of the tar-N yield obtained from the pyrolysis in fluidized bed at 800 °C with the one obtained from the direct gasification of MBM. In the same way, the impact of tar-N cracking and gas phase secondary reactions in the tar-N yield have been assessed by the comparison of the values obtained in the pyrolysis experiments at 600 °C and 800 °C in the fixed and the fluidized bed reactors. Moreover, to know more in depth the effect of tar-N cracking and secondary gas phase reactions, the composition of the tar-N obtained from these pyrolysis experiments at 600 °C and 800 °C has been determined quantitatively (% area) by GC-MS.

In the overall gasification process, the char gasification stage can be expected to add more NH_3-N and HCN-N to the gas phase from nitrogen that still remains in the char. Therefore, the results obtained from the char gasification experiments conducted using Ar + O_2 and Ar + O_2 + H_2O allows estimating their share in the final yield to these N-containing products, although they could be also affected by gas-phase reactions due to the relatively high gas residence time.

12.3 Results

12.3.1 MBM Characterization

MBM used in this work presents a fraction of protein origin (58.8 wt%), ash (16.5 wt % measured at 550 °C), fats (11.3 wt%) [17] and moisture (4.5 wt%); the remaining fraction is mainly composed of carbohydrates and nucleic acids. The proteinaceous fraction and the nucleic acids are the two macro-components that provide most of the N to MBM.

The total amino-acid content after total hydrolysis of MBM and the free amino-acid content after extraction from MBM are 58.8 wt% and 49.8 wt%, determined by the quantification of the 17 most abundant amino-acids in living things. As a result, the content of polypeptides and proteins calculated as the difference between the aforementioned amino-acid experimental quantifications is 9 wt%. The significantly higher content of free amino acids than polypeptides and proteins could be due to the stabilization treatment applied to MBM (autoclaved with steam at 120 °C). N in proteins and polypeptides is taking part of the peptide bond, that is to say, in the form of a secondary amide. The peptide bond is formed between a primary amine group (-NH_2) of an amino-acid and the carboxylic group (–COOH) of another amino-acid. As an example, Fig. 12.2 shows the formation of a peptide bond between two amino-acids, glutamic acid and lysine. Besides proteins, MBM also contains free amino acids, whose N functional group is a primary amine (and not the peptide group as in proteins). Amino-acids, can be classified as non-polar, polar, acidic and basic, depending on their side chain character. Glutamic acid is an example of an acidic amino-acid due to its acid group at the end of the side chain. Figure 12.3 shows other examples of amino-acids with non-polar and basic side chains. Figure 12.4 shows

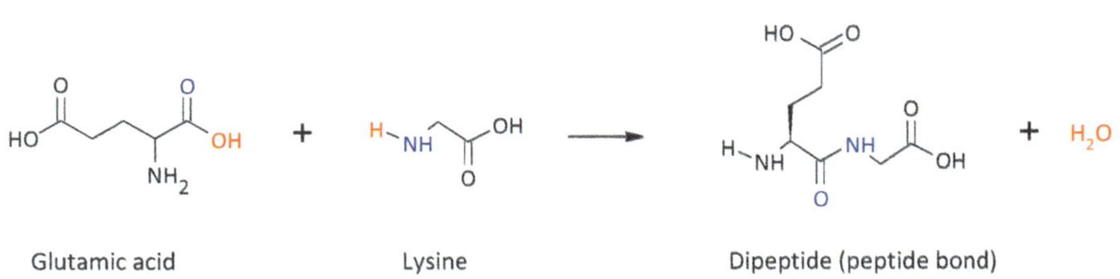

Fig. 12.2 Peptide bond formation between glutamic acid and lysine

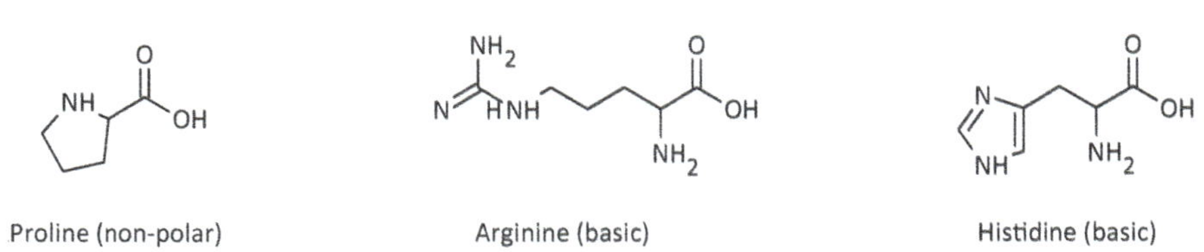

Fig. 12.3 Examples of amino-acids

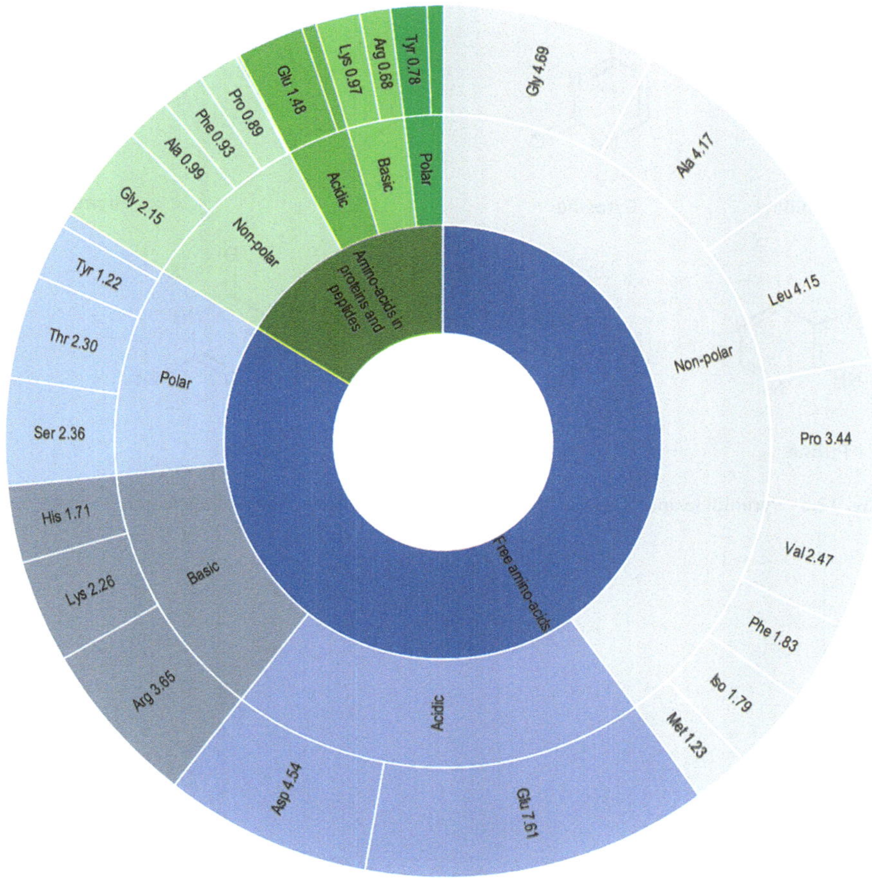

Fig. 12.4 Quantification of different types of amino-acids present in free amino-acid fraction and protein and peptide fraction of meat and bone meal

the content of each amino-acid and its classification in the free amino-acid fraction and the protein and peptide fraction of MBM. Considering the N content of MBM (9.9 wt%), it can be said that N in the form of free amino-acids represents about 72.1% of the total fuel-N in MBM, while 12.9% of the fuel-N is in the form of proteins and peptides, that is to say 85% of the fuel-N has a proteinaceous origin. Apart from proteins, N in living beings is also present in nucleic acids, which are the fourth largest group of biomolecules and therefore would accumulate most of the remaining N.

A nucleic acid is formed by a chain of nucleotides, containing each of them a pentose, a phosphate group and a nitrogenous base. This nitrogenous base may be based on a pyrimidine structure like cytosine, uracil and thymine or on a purine structure (imidazole ring fused to a pyrimidine) like adenine and guanine (Fig. 12.5).

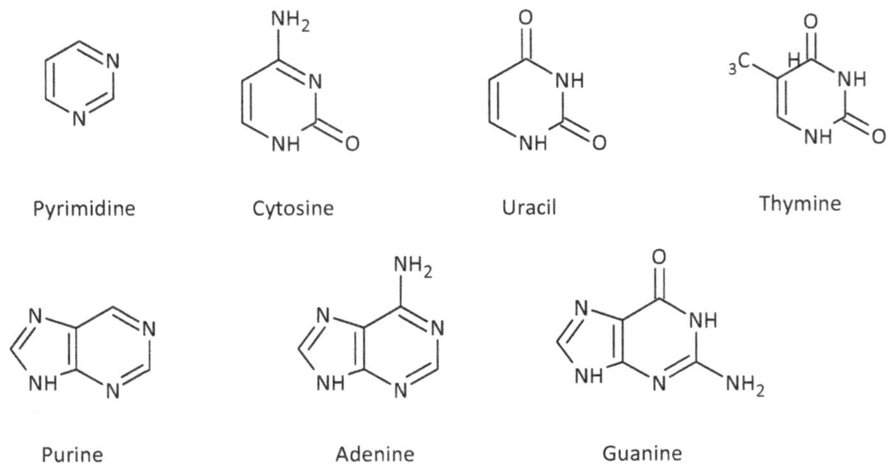

Fig. 12.5 Pyrimidine and purine-based structures of nitrogenous bases of nucleotides

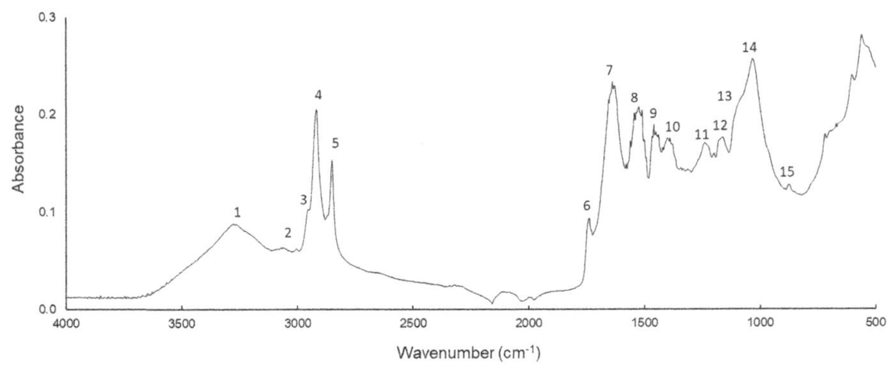

Fig. 12.6 ATR-FTIR spectrum of meat and bone meal and characteristic peaks (see Table 12.1 for peak identification)

Apart from the N contained in the pyrimidine and purine structures, there are nitrogenous bases presenting other N functionalities, such as primary amine in cytosine, adenine and guanine, and secondary amide (cyclic) in guanine, thymine, cytosine and uracil.

ATR-FTIR spectrum of MBM is shown in Fig. 12.6; the char spectra will not be shown since no distinguishable functional groups are present.

ATR-FTIR spectrum of MBM shows peaks that denote clearly the presence of lipids and proteins if it is compared with literature spectra of meats containing both biomolecules [20]. The existence of nucleic acids, phospholipids, polyphosphates and hydroxyapatite from bones in MBM has also been confirmed by some of the peak wavenumbers. The polypeptide and protein repeat units give rise to nine characteristic absorption bands, namely, amides A, B, and I–VII, being the amide I (1700–1600 cm^{-1}) and II (1600–1500 cm^{-1}) bands the two most prominent vibrational bands [21]. Specifically, the broad unique peak around 3260 cm^{-1} can be assigned to the hydrogen bond due to the presence of O–H and N–H bonds. The N–H stretching of the spectrum is caused by secondary amides (proteins), since primary amides should show two peaks around this wavenumber corresponding to the 2 N–H bonds of the NH_2 group. The weak peak of about 3040 cm^{-1} could be attributed to C–H stretching in unsaturated bonds (=C–H), typical in some of the fatty acids that form triglycerides. The prominent peaks at 2950 cm^{-1}, 2920 cm^{-1} and 2850 cm^{-1} correspond to C–H saturated bonds, which are mainly present in the fatty acid chains of triglycerides and phospholipids, but also in the structure of proteins. Two stretching vibrations of the carbonyl group (C=O) can be distinguished, one belonging to the C=O bond in fatty acid esters and cholesterol esters (1738 cm^{-1}) and the other one in amides (1640 cm^{-1}). The amide II band in secondary amides (1600–1500 cm^{-1}), with a peak wavenumber of 1520 cm^{-1}, is the result of the scissoring of the N–H bond (40–60% of the potential energy) and the stretching of the C–N bond (18–40%) [22]. The presence of biomolecules, such as nucleic acids, phospholipids, hydroxyapatite and other polyphosphates, containing P=O bonds, have been evidenced by two peak wavenumbers (1240 cm^{-1} and 1080 cm^{-1}), being the one at 1080 cm^{-1} characteristic in the spectrum of hydroxyapatite [23], which is the major component of bones.

12.3.2 Fuel-N Distribution Obtained in Pyrolysis Stage

The product distribution (solid, liquid and gas) and the fuel-N distribution obtained in the pyrolysis experiments are shown in Table 12.2.

Table 12.2 Product distribution and fuel-N distribution obtained in the pyrolysis experiments

	Pyrolysis at 600 °C in fixed bed	Pyrolysis at 800 °C in fluidized bed
	Pyr600Fix	Pyr800Flu
Product distribution (wt%)		
Char	31 ± 2	35
Condensate (bio-oil and water)	50 ± 3	22
Gas	9 ± 1	29
Fuel-N distribution (%)		
Char-N	19 ± 1	14
Tar-N	21 ± 2	3
NH_3-N	33 ± 2	35
HCN-N	0 ± 0	0.13
N_2-N	3 ± 2	8

The yields to condensate, usually named bio-oil, and to gas in the Pyr600Fix are in concordance with those shown in literature for the same type of reactor and pyrolysis temperature [32]. As expected, the temperature (600 °C and 800 °C) and the type of reactor (fixed and fluidized) had a strong effect on the yields of both condensate and gas. Because of the effect of temperature, the gas yield obtained in Pyr800Flu seems to increase at the expense of decreasing the condensate yield if compared with those obtained in Pyr600Fix [32]. This trend could be justified by the thermal cracking of some of the tar compounds and the formation of light gaseous compounds. In the same way, in the fluidized bed experiments, volatile compounds are released can faster and all together in the same reaction atmosphere (and not gradually as in the fixed bed occurs), which could favour the occurrence of secondary gas phase reactions, leading to the formation of light gaseous compounds. The yields to the solid product (char) obtained in Pyr600Fix and Pyr800Flu are similar in spite of the different temperature and type of reactor used. The high values of char yields (31–35 wt%) are in the upper range of those obtained in the pyrolysis of lignocellulosic biomass even at lower temperatures) and independently of the type of reactor [32]. These high solid yields may be related with the high ash content of MBM and also with the high fraction of proteins in MBM; other characteristic fractions of this type of waste, such as the animal fats, have been proved to produce exiguous char yields (<2 wt%) even at lower pyrolysis temperatures than those studied in this work [33].

Regarding the fuel-N distribution, the greatest impact of the pyrolysis conditions is found on the tar-N yield, which decreases from 21% to 3%. This sharp reduction may be justified by the significant impact of the secondary gas phase reactions on the release of heteroatoms from tar, as it is well-known for lignocellulosic pyrolysis condensates [34] and also for other rich N-containing raw materials such as sewage sludge [35]. The presence in the protein fraction of amino-acids with reactive side chains could justify the high char-N yields, since these reactive side chains enhance the occurrence of cross-linking reactions producing char-N following two types of cross-linking reaction [36]. One of them has been found to occur between the amine side group of lysine residue and that of glutamine or aspargine residues, yielding an isopeptide and NH_3. The other crosslinking reaction happens between the acidic side chain of glutamic acid or aspartic acid and the amine side groups in either free amino-acids or protein's backbone. This reaction allows obtaining water and a chemical structure containing a N atom joint to two carbonyl groups. ATR-FTIR analyses of the char do not allow determination of the functional groups in char. However, in the tar obtained from Pyr600Fix a compound named 3-ethyl-4-methyl-1H-pyrrole-2,5-dione, which has one N joint to two carbonyl groups (similar structure), could be identified (see Sect. 12.3.2). The high content in glutamic acid and aspartic acid in MBM suggests that this last cross-linking reaction may have a significant importance on char-N formation. As expected, the yield of char-N decreases with the pyrolysis temperature.

The yield of N_2-N is higher for Pyr800Flu (8%) than for Pyr600Flu (3%). In reducing atmospheres, the decomposition reaction of NH_3 into N_2 and H_2 $\left(2NH_3(g) \leftrightarrows N_2(g) + 3H_2(g) \Delta H_r^\circ > 0\right)$ is of particular relevance and a known source of N_2 [37]. This decomposition reaction is shifted towards the products (H_2 and N_2) at temperatures above 180 °C and its endothermic character could justify the increase of the N_2-N yield with the pyrolysis temperature.

In both experiments, most of the fuel-N ends up in the form of NH_3, reaching yields of about 35%. There is not a clear effect of temperature or residence time since the NH_3-N yield is similar for both pyrolysis experiments (Pyr600Fix and Pyr800Flu). Therefore, although it could be expected to obtain some extra NH_3 at 800 °C because of further decomposition of char-N and thermal cracking of tar-N, this extra NH_3 could be subsequently decomposed to N_2 and H_2 due to the higher temperature. The yield of fuel-N to HCN-N was negligible in both experiments.

12.3.3 Characterization of Tar Obtained in Pyrolysis Experiments

The composition of the tar obtained from the experiments Pyr600Fix and Pyr800Flu is shown (Fig. 12.7) grouped in chemical families with the aim of evaluating the transformation of tar compounds due to thermal cracking.

The composition obtained in both experiments is significantly different. In the experiment Pyr600Fix, a high proportion of aliphatic amides, fatty acids and fatty nitriles is present in the liquid, while the composition of the liquid from the experiment Pyr800Flu only includes cyclic and aromatic compounds, but not aliphatic ones. At 300 °C and above, fatty acids are directly devolatilized as free acids from the triglycerides present in MBM and may lead to the formation of alkanes and alkenes, which are also present in the condensate [38]. During the pyrolysis, these fatty acids react with ammonia, coming from the thermal decomposition of proteins, to yield fatty amides, which in turn can be converted to fatty nitriles [39]. Some of the aliphatic amides, such as hexanamide or phenylpropanamide, do not have a long aliphatic chain, being the pyrolysis of the protein fraction their most likely origin [40]. Numerous cyclic saturated and unsaturated amides, coming also from the pyrolysis of the protein fraction [40], were detected in the condensate of the experiment Pyr600Fix, but not in the one from the experiment Pyr800Flu. Many research works have pointed out that 6-member cyclic dipeptide-based compounds, named diketopiperazines (DKPs), are the main products from the pathway of protein pyrolysis [41]. However, in the Pyr600Fix condensate the proportion of 5-member cyclic dipeptides, most of them are 5-substituted-2,4 imidazolidinedione derivates and lactams, and are significantly higher in concentration than DKPs. Same results were obtained in a study about human hair pyrolysis, where different 5-substituted 2,4-imidazolidinediones compounds were identified, but not the expected DKPs [42]. The formation of imidazolines in the pyrolysis of dipeptides has also been suggested as an alternative route to the formation of DKPs [40]. The proportion of lactams (cyclic monopeptides), such as pyrrolidone, caprolactam, succinimide or piperidinone, in the condensate from Pyr600Fix is also significantly higher than the

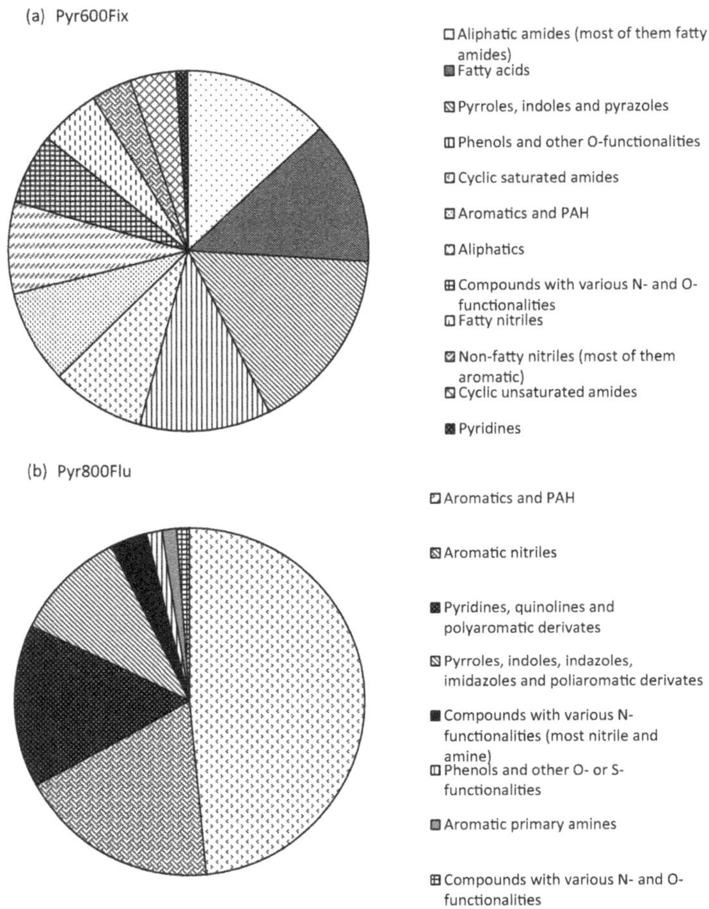

Fig. 12.7 Composition (area %) of tar obtained in pyrolysis experiments (Pyr600Fix and Pyr800Flu)

one of DKPs derivates. Except for the acid group, pyrrolidone has a structure similar to pyro-L-glutamic acid, which was obtained by Douda and Basiuk (2000) in the pyrolysis of glutamic acid rather than DKP [43]. A significant percentage of alkyl phenols were determined in the Pyr600Fix; which are known to be formed not only in the pyrolysis of lignin, but also in the pyrolysis of amino-acids, such as tyrosine, tryptophan, phenylalanine and proteins [44, 45]. In the condensate Pyr600Fix, a compound named [1, 2, 4]triazolo[1,5-a]pyrimidin-7-ol has been identified. The structure of this compound, similar to guanine, could indicate that this compound could come from the pyrolysis of nucleic acids instead of proteins.

Aromatic and polycyclic aromatic hydrocarbons (PAH) are the most abundant tar compounds from the experiment Pyr800Flu. Most of these compounds are benzene alkyl derivates, such as toluene, styrene, benzene, 2-propenyl- or benzene,

1,3-dimethyl-, although some polyaromatic compounds, such as indene and naphthalene, 2-methyl- were also found. Secondary reactions of the alkyl phenols observed in the condensate of the experiment Pyr600Fix could be the origin of the benzene alkyl derivates found in the pyrolysis experiment carried out at higher temperature (Pyr800Flu).

Lastly, both liquids present a significant proportion of N-heterocyclic aromatic compounds such as, pyrroles, pyrazoles, indoles, pyridines, quinolines, indazoles and imidazoles. However, the proportion of these compounds is much greater in the Pyr800Flu. Moreover, the tar from Pyr600Fix contains a greater proportion of 5-member ring compounds with nitrogen, such as pyrroles, indoles or pyrazoles, than 6-member ring ones, such as pyridines, whose proportion is exiguous. The proportion of 5-member ring compounds with nitrogen (pyrroles, indoles, indazoles, imidazoles) is lower in the condensate from Pyr800Flu than in the one from Pyr600Fix, but the proportion of 6-member ring compounds (pyridines and quinolines) is substantially greater in Pyr800Flu. This increase in the proportion of 6-member ring could be related with a mechanism of formation similar to the Hydrogen Abstraction Acetylene Addition (HACA) mechanism, since it accelerates the formation of benzenoid PAHs and inhibits the formation of cyclopentaring-fused products [46].

12.3.4 Fuel-N Distribution Obtained from Char Gasification Stage

The product distribution and the fuel-N distribution obtained from the char gasification experiments are shown in Table 12.3.

As the total amount of gasifying agent utilized in these gasification experiments (O_2 or O_2 + H_2O) is different, the product distribution in these tests has been analyzed only over the mass of MBM pyrolysis char fed to make them comparable. A greater conversion of the pyrolysis char is achieved during the gasification step when using only O_2 as gasifying agent, thus indicating that the presence of steam makes slower the char gasification reactions with O_2. This could be explained by the fact that jointly introducing steam leads to an increased total flow of gasifying agent and, therefore, a reduced contact time with the solid in the reactor. Moreover, the concentration of oxygen (more reactive gas) is diluted when adding steam (less reactive gas), thus negatively affecting combustion kinetics. This finding was also observed in other works studying char gasification in which, for example, the fraction of the charcoal carbon converted into solid remained almost constant (about 50%) when the steam to carbon ratio was increased from 0 g g^{-1} to 0.5 g g^{-1}, even though the ER was kept constant (30%) [47]. Gil-Lalaguna et al. [18] also reported that, during the air-steam gasification of char derived from sewage sludge pyrolysis, carbon combustion occurred faster than steam gasification reactions, with values that ranged 35–45% of carbon fraction remaining as solid. As

Table 12.3 Product distribution and characterization of products obtained in char gasification experiments

	Char gasification at 800 °C with O_2 ($GasArO_2$)	Char gasification at 800 °C with O_2 and H_2O ($GasArO_2H_2O$)
Product distribution over char fed (wt%)[a]		
Solid residue	58	70
Condensate	4	21
Gas	40	55
Elemental analysis of the solid residue from gasification (concentration, wt%)		
C	21.7	36.8
H	0.47	0.67
N	3.1	4.2
Condensate characterization (species concentration in the condensate, wt%)		
Water (by Karl-Fisher)	100	~95
NH_3 (by titration)	0	5.0
Tar (by difference)	0	0
Gas species yield over char fed (wt%)		
H_2	0.20	1.7
CO	9.1	25.8
CH_4	0.09	0.54
CO_2	28.4	22.8
NH_3	0.009	1.10
HCN	0.035	0.04
N_2	2.2	3.0
N_2O	0.11	0.14
$C_2Hx + H_2S$	0.18	0.18
Fuel-N distribution (%) over char-N fed		
Solid residue-N	28	44
Tar-N	0	0
NH_3-N	0.1	13.8
HCN-N	0.3	0.4
N_2-N	35	45
N_2O-N	1.1	1.3
Fuel-N distribution (%) over MBM-N fed to the previous pyrolysis stage		
Solid residue-N	5.5	9.6
Tar-N	0	0
NH_3-N	0.02	3.0
HCN-N	0.06	0.08
N_2-N	6.9	6.8
N_2O-N	0.2	0.2

[a]Product yields obtained from $GasArO_2H_2O$ experiment are greater than 100% due to the condensation of steam introduced as gasifying agent

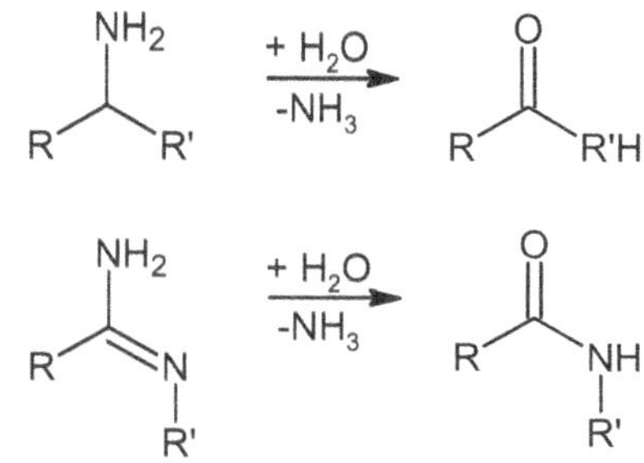

Fig. 12.8 Examples of possible deamination routes of char-amine group in presence of steam

expected, the yield of condensate is significantly greater when steam is used as gasifying agent, since the unreacted steam condenses in the tar recovery system, thus increasing the yield of this product. The condensates from both gasification experiments mainly contain water (see Table 12.3) and tar is not detected gravimetrically (by mass difference with water quantified by Karl-Fisher titration), indicating that the gasification of pyrolysis char instead of the raw MBM is a good strategy to drastically diminish tar concentration in the gasification gas. The same result was found in our research group when comparing the production of tar during the gasification of sewage sludge (about 11–45 g m^{-3}(STP)) and the gasification of char derived from sewage sludge pyrolysis, in which, tar production was almost negligible and the results of Karl Fischer titration of the condensate were 100 wt% water [48]. Apart from water, 5.0 wt% of NH_3 was quantified by titration in the condensate from the experiment carried out with the mixture of O_2 and steam, while this compound was not detected in the condensate from the experiment carried out with only O_2 as gasifying agent. This is a very interesting result implying that the presence of steam promotes formation of NH_3 in char gasification reactions. In this respect, Aljbour and Kawamoto (2013), who investigated the gasification of cedarwood, quantified an increase in the fuel-N conversion to NH_3-N from 6% at S/C = 0 to 55% at S/C = 2 g·g^{-1} [49]. As NH_3 is formed only when steam is present, it is thought that the origin could be the deamination of amine functional groups present in the char and the subsequent formation of a carbonyl group in the char surface (see Fig. 12.8). Deamination route is also suggested by other authors studying gasification of microalgae [13] and hydrothermal degradation of amino-acids [50]. Negligible conversions of fuel-N to tar-N and HCN-N were found when using either O_2 or a mixture of O_2 and steam as gasifying agent in char gasification experiments.

The gas yield increases significantly when the gasifying agent includes steam apart from O_2. As can be seen in Table 12.3, this increase is mainly due to the significant augmentation in CO production. A priori, this CO augmentation in the presence of steam is not an expected result since the presence of steam usually promotes the formation of H_2 and CO_2 and the disappearance of CO via the water gas shift reaction ($CO(g) + H_2O(g) \rightarrow H_2(g) + CO_2(g)$). Although other reactions, such as the Boudouard reaction ($C(s) + CO_2(g) \leftrightarrow 2CO(g)$) or the water-gas reaction ($C(s) + H_2O(g) \leftrightarrow CO(g) + H_2(g)$), could favor the formation of CO, it seems that such a pronounced increase should be due to another source. In this respect, Schafer et al. [51] found that HCN undergoes hydrolysis into NH_3 and CO at temperatures over 700 °C ($HCN(g) + H_2O(g) \rightarrow NH_3(g) + CO(g)$), that matches with both the increase on the CO and NH_3 yields observed in this work when H_2O is used in the gasification atmosphere. In the same way, maybe the decarbonylation of the carbonyl group (C=O) formed in the deamination reaction (see Fig. 12.8) could be another additional source.

Regarding the fuel-N distribution (see Table 12.3), the N atomic balance of char gasification with the mixture of O_2 and steam closes near 100%. However, a significant amount of fuel-N has not been quantified in the N-containing products of the experiment $GasArO_2$. The lack of precision on the N atomic balance could be due to some N-containing products not being quantified, such as HNCO, or to the error in the quantification of any of the products, as for example, the solid residue-N due its heterogeneity. After gasification under both atmospheres of reaction, char-N is mainly recovered in the form of solid residue-N and N_2. The yield of N remaining in the solid residue of these char gasification experiments is significantly higher than that obtained from the direct gasification of MBM [8], which could be explained due to the lower reactivity of MBM pyrolysis char in comparison with the raw MBM. Under both atmospheres of reaction, a similar amount of the fuel-N is transformed to N_2-N (~7%), so the use of steam does not have any effect on the production of N_2. According to calculations, if the thermodynamic equilibrium was reached at the temperature studied in this work, most of the fuel-N would end in the form of N_2 either in the gasification with Ar + O_2 or in the gasification with Ar + O_2 + steam. As no NH_3 is observed in the experiment $GasArO_2$, the origin of N_2 could not be the decomposition of NH_3 in N_2 and H_2, but it could be related to the reaction of HCN with O_2 or NO [51].

12.3.5 Contribution of Each Stage to Final Fuel-N Distribution

In previous work carried out by our research group [4], gasification of MBM with a mixture of air and steam at 800 °C led to the following distribution of fuel-N: NH_3 (67%), HCN (0.83%), solid residue-N (2.6%), tar-N (0.37%), NO-N (0.49%) and N_2-N (28%).

The results of the pyrolysis experiments conducted in this work, both in the fluidized and fixed bed reactors, have shown that the yield of NH_3-N (calculated over the N present in the MBM) was around 33–35% (see Table 12.3) which means that around 50% (33/67) of the NH_3 generated in the whole gasification process is due to the contribution of the pyrolysis stage and that the stages of char-N gasification and secondary gas phase reactions do not entail a net conversion of NH_3. Neither the increase of the temperature from 600 °C to 800 °C nor secondary reactions that may happen in the fluidized bed due to the joint presence of all the primary pyrolysis products in the reaction atmosphere can imply an increase in the NH_3-N yield.

Moreover, it has been found that the gasification of the pyrolysis char with only O_2 as gasifying does not incorporate more NH_3 to the final production. However, the gasification of the pyrolysis char with the mixture of O_2 and steam produced additional NH_3, yielding around 3.0% of NH_3-N with respect to the original MBM-N. Therefore, around 4.5% (3.0/67) of the NH_3 produced in the overall gasification process would come from this stage. With these data, the sum of the NH_3 produced in MBM pyrolysis and char gasification (with O_2 + steam) only explains the origin of 54.5% of the NH_3 generated in the overall gasification process.

The generation of N_2 has been confirmed by conducting experiments of MBM pyrolysis and char gasification, as it has been clearly identified and quantified by gas chromatography thanks to the use of Ar instead of N_2 in pyrolysis or Ar and O_2 instead of air in gasification. The yield of N_2-N obtained in the pyrolysis experiments ranges from 3% (at 600 °C) to 8% (at 800 °C), while it is around 6.9% (calculated over the content of N in the original MBM) in the char gasification stage, either with O_2 or with a mixture of O_2 and steam as gasification agents. The greater conversion to N_2 obtained at higher temperatures and with higher gas residence times (fluidized bed) points to its formation in gas-phase reactions favored at high temperatures. In the same way as happened for NH_3, the sum of the N_2 produced in the stages of MBM pyrolysis (8%) and char gasification (6.9%) accounts only for half (14.9/28) of the total N_2 generated in the overall gasification process (28%). In this respect, the exiguous tar-N yield in the direct gasification of MBM (0.37%) in comparison with the one obtained from the Pyr600Fix (21%) points to a significant effect of tar-N cracking and reforming reaction on the final fuel-N distribution and probably on the further production of NH_3 and N_2. The solid residue-N yield obtained from the MBM char gasification stage (between 5% and 10% depending on the gasifying mixture) is higher than the one obtained from the direct gasification of MBM (solid residue-N: 2.6%) due to the char is more refractory than the raw MBM. In the direct gasification of MBM, this extra solid residue-N also will end up in the form of other N-containing products. Comparing the tar-N yield obtained in experiment Pyr600Fix (21%) with that obtained in experiment Pyr800Flu (3%), it can be said that the thermal cracking reactions have a strong effect on the distribution of the fuel-N. However, the impact of the tar-N reforming reaction is less important since the tar-N yield in the Pyr800Flu experiment (3%) is just slightly higher to the one obtained in the direct gasification of MBM (0.37%).

None of the minority N-containing products (HCN-N, NO-N) previously obtained in the direct air-steam gasification of MBM were obtained in appreciable quantities in the individual stages studied in this work, confirming that these compounds are barely produced in the gasification of this waste.

12.4 Conclusions and Future Outlook

12.4.1 Conclusions

In this work, the contribution of each stage of meat and bone meal (MBM) gasification to the final fuel-N distribution into different gasification products, including NH_3 as the main interest compound, has been studied.

Regarding the first stage involving pyrolysis, any clear connection between temperature (600 °C and 800 °C) nor residence time with the production of NH_3 was found, while a significant decrease on tar-N yield was observed when increasing the pyrolysis temperature.

In the subsequent conversion of pyrolysis char, the generation of NH_3 was considerably increased by the use of steam in the char gasification step (up to 13.8% of char-N yielded NH_3-N). Using only O_2 as gasifying agent did not produce additional NH_3 in this stage. Negligible conversion of char-N to tar-N and HCN-N was found at this gasification stage.

By comparing the results from these individual stages with global data of MBM gasification, it can be stated that the pyrolysis stage produced around 50% of the total NH_3 generated during the direct gasification of MBM, while steam-gasification of char-N was responsible for less than 5% of total NH_3. The amount of N_2 produced during the stages of MBM pyrolysis (8% of fuel-N) and char gasification (7% of fuel-N) accounted for only 50% of the total N_2 produced during the entire gasification process (28% of fuel-N). The low generation of HCN and NO in any of the stages studied confirms the low conversion of fuel-N to these products in the MBM gasification process.

Final fuel-N distribution from MBM gasification is explained just in part considering the individual stages of MBM pyrolysis and char gasification (with air and/or steam), indicating that gas phase reactions involving tar-N cracking and reforming process (out of the scope of this work) have a significant impact on the final fuel-N distribution and NH_3 generation. To better comprehend the NH_3 production, and what followed with other N-containing compounds, additional research work attempting to reproduce tar-N cracking and reforming reactions is required.

12.4.2 Future Outlook

Human population depends on synthetic ammonia-based fertilizers, recent estimates indicate that without ammonia only half of the world's population could be fed [52]. Nitrogen is an essential plant nutrient that has no substitute. Apart from its use for the production of synthetic fertilizers, a future boost in ammonia production from 185 Mt. to 900 Mt. is expected due to its promising use as a fuel in marine transport and power production. Table 12.4 shows the data reported in the U.S. Geological

Table 12.4 Annual production of ammonia in 2021 country-by-country breakdown [55]

Country	Annual Production (Mt)	Country	Annual Production (Mt)
China	42,000	Algeria	2600
Russia	16,300	Germany	2290
United States	12,700	Ukraine	2170
India	12,100	Poland	2100
Indonesia	6000	Netherlands	2000
Saudi Arabia	4300	Oman	1730
Trinidad and Tobago	4050	Australia	1700
Egypt	4000	Malaysia	1400
Iran	4000	Uzbekistan	1200
Canada	3760	Nigeria	1100
Pakistan	3400	Vietnam	1050
Qatar	3270	Other countries	14,500
		World total (rounded)	150,000

Table 12.5 Annual generation of N-rich residues in EU, nitrogen content and yields of NH_3-N

	Annual Generation in EU-27 (Mt)	Estimated nitrogen content (wt%)	Yields of NH_3-N over fuel-N (%) [5]
Sewage sludge	10 [54]	2.5–7.9 [55]	30
MBM	20 [5]	7.8–10.4 [56, 57]	67
Manure (any excrement and/or urine of farmed animals)	1400 (EU-27 and UK) [58] 135 (Spain) [58]	It depends on many factors type of livestock, excrement (faeces or urine), sex and age of the animals, feed, ...	

Survey about the world annual production of ammonia in 2021 country-by-country breakdown [53].

The high GCH emissions, the perspectives about fossil fuels depletion, the strong pressure on the nitrogen biogeochemical cycle, as well as the future boost on ammonia demand make obligatory the development of new technologies, such as gasification, to be able to produce ammonia from renewable sources of both hydrogen and nitrogen. Nitrogen-rich biological residues such as, sewage sludge, MBM and livestock manure, either raw or for example the residue obtained from their anaerobic digestion, can be considered for this purpose. In the same way, as some plastic residues, that cannot be easily recycled, such as wood panels containing urea-formaldehyde resins, polyurethane or nylon. Table 12.5 shows generation of N-rich biological residues with an estimated nitrogen content for 27 European countries in the EU.

If the conversions of fuel-N to ammonia from sewage sludge and MBM reported in a previous work [5] are considered, approximately 10% of the total EU production (10 Mt. of NH_3) could be produced via gasification (67%). Manure could be another interesting feedstock for this process. In the European Union (EU-27) and UK, animal farming (pigs, cattle and chickens) generated annually more than 1.4 billion tons of manure during the period 2016–2019 [58]. The management of this manure and its application to soil causes ammonia emissions that in 2010 amounted to 2.086 Mt. and 1.66 Mt. [59]. A correct treatment of manure could contribute to the recovery of at least the N emitted in the form of NH_3 during manure management, which is not a negligible amount when compared to the amount of industrial NH_3 produced (11.2 Mt) [60] and to the amount of mineral nitrogen fertilizer consumed by agriculture (10.0 Mt) [61] in EU.

On the other hand, livestock farms are enormous energy consumers. Paris et al. reported energy consumptions discerning between direct consumptions of energy (such as animal housing, manure management or milking processes) and indirect consumptions (such as the production of fertilizers and pesticides, and energy used to produce animal feed). These energy consumptions per kg of product ranged between 43.73 MJ kg^{-1} and 59.2 MJ kg^{-1} for beef, 15.9 MJ kg^{-1} and 22.7 MJ kg^{-1} for pork, 9.6 MJ kg^{-1} and 19.1 MJ kg^{-1} for broiler and 3.5 MJ kg^{-1} and 20.5 MJ kg^{-1} for chicken egg in regular livestock installations in the EU-27 [62]. The authors concluded that energy use is concentrated in feed, housing, and manure management. Specifically, animal feed (including feed and fertilizers) requires around three-quarters of all energy inputs in most livestock (cow milk, beef, pork, broiler and chicken egg). In this context, the manure management via a treatment able to recover not only the nitrogen but also the energy would reduce the impact of the sector from an energetic and an environmental point of view applying the circular economy and biorefinery concepts.

Finally, the contribution of this strategy to the circular economy would boost the competitiveness by protecting businesses against scarcity of resources and volatile prices. Moreover, it will create local jobs at all skills levels and opportunities for social integration and cohesion.

References

1. The Royal Society (2020) Green ammonia. Ammonia: zero-carbon fertiliser, fuel and energy store. (2020) policy briefing. ISBN: 978–1–78252-448-9.
2. Meng F, Wagner A, Kremer AB, Kanazawa D, Leung JJ, Goult P, Guan M, Herrman S, Spellman E, Sauter P, Lingeswaran S, Stuchtey MM, Hansen K, Masanet E, Serrenho AC, Ishii N, Kikuchi Y, Cullen JM. Planet compatible pathways for transitioning the chemical industry. Cambridge Open Engage-ChemRxiv. 2022;120(8):e2218294120. https://doi.org/10.26434/chemrxiv-2022-hx17h.
3. Rafiqul I, Weber C, Lehmann B, Voss A. Energy efficiency improvements in ammonia production—perspectives and uncertainties. Energy. 2005;30:2487–504. https://doi.org/10.1016/j.energy.2004.12.004.

4. International Energy Agency. (2021) Ammonia technology roadmap. https://www.iea.org/reports/ammonia-technology-roadmap. Accessed 27 Jan 2023.
5. European Food Safety Authority (2022) Animal by-products. https://www.efsa.europa.eu/en/topics/animal-by-products. Accessed 27 Jan 2023.
6. Food Safety–European Commission (2022) Approved establishments - ABP. https://food.ec.europa.eu/safety/animal-products/approved-establishments-abp_en. Accessed 26 Jan 2023.
7. United Kingdom Government (2023) Animal by-product operating plants: approved premises - GOV.UK. https://www.gov.uk/government/publications/animal-by-product-operating-plants-approved-premises. Accessed 27 Jan 2023.
8. Gil-Lalaguna N, Afailal Z, Aznar M, Fonts I. Exploring the sustainable production of ammonia by recycling N and H in biological residues: evolution of fuel-N during glutamic acid gasification. J Clean Prod. 2020;282:124417. https://doi.org/10.1016/j.jclepro.2020.124417.
9. Campoy M, Gómez-Barea A, Ollero P, Nilsson S. Gasification of wastes in a pilot fluidized bed gasifier. Fuel Process Technol. 2014;121:63–9. https://doi.org/10.1016/j.fuproc.2013.12.019.
10. Marculescu C, Alexe F. Assessing the power generation solution by thermal-chemical conversion of meat processing industry waste. Energy Procedia. 2014;50:738–43. https://doi.org/10.1016/J.EGYPRO.2014.06.091.
11. Soni CG, Dalai AK, Pugsley T, Fonstad T. Steam gasification of meat and bone meal in a two-stage fixed-bed reactor system. Asia-Pacific J Chem Eng. 2011;6:71–7. https://doi.org/10.1002/apj.454.
12. Wang P, Shen C, Wang B, Xu P, Shen L. Ammonia production from nitrogen-rich biomass gasification: nitrogen transformation from model amino acids. Fuel. 2022;326:125071. https://doi.org/10.1016/J.FUEL.2022.125071.
13. Wang P, Xu P, Shen C, Wang B, Shen L, Song T. Green ammonia production via microalgae steam catalytic gasification process over $LaFeO_3$ perovskite. SSRN Electron J. 2022;318:123322. https://doi.org/10.2139/SSRN.3978731.
14. Niu X, Shen L. Evolution of carbon and nitrogen during chemical looping gasification of rapeseed cake with Ca-Fe oxygen carrier. Chem Eng J. 2022;431:134232. https://doi.org/10.1016/J.CEJ.2021.134232.
15. Wang P, Xu P, Shen L. Promotional effect of Ca on La1-xCaxFeO3 perovskites for ammonia production via steam catalytic gasification of microalgae. Chem Eng J. 2023;455:140483. https://doi.org/10.1016/J.CEJ.2022.140483.
16. Wang P, Yan J, Wang S, Xu P, Laihong S, Song T. Synergistic effects of lanthanum ferrite perovskite and hydrogen to promote ammonia production during microalgae catalytic pyrolysis process. Bioresour Technol. 2021;340:125641. https://doi.org/10.1016/J.BIORTECH.2021.125641.
17. Cascarosa E, Fonts I, Mesa JM, Sánchez JL, Arauzo J. Characterization of the liquid and solid products obtained from the oxidative pyrolysis of meat and bone meal in a pilot-scale fluidised bed plant. Fuel Process Technol. 2011;92:1954–62. https://doi.org/10.1016/J.FUPROC.2011.05.015.
18. Gil-Lalaguna N, Sánchez JL, Murillo MB, Ruiz V, Gea G. Air-steam gasification of char derived from sewage sludge pyrolysis. Comparison with the gasification of sewage sludge. Fuel. 2014;129:147–55. https://doi.org/10.1016/j.fuel.2014.03.059.
19. Aznar M, San Anselmo M, Manya JJ, Benita Murillo M. Experimental study examining the evolution of nitrogen compounds during the gasification of dried sewage sludge. Energy Fuel. 2009;23:3236–45. https://doi.org/10.1021/ef801108s.
20. Lozano M, Rodríguez-Ulibarri P, Echeverría JC, Beruete M, Sorolla M, Beriain MJ. Mid-infrared spectroscopy (MIR) for simultaneous determination of fat and protein content in meat of several animal species. Food Anal Methods. 2017;10:3462–70. https://doi.org/10.1007/S12161-017-0879-1.
21. Kong J, Yu S. Fourier transform infrared spectroscopic analysis of protein secondary structures. Acta Biochim Biophys Sin Shanghai. 2007;39:549–59. https://doi.org/10.1111/J.1745-7270.2007.00320.X.

22. Krimm S, Bandekar J. Vibrational spectroscopy and conformation of peptides, polypeptides, and proteins. Adv Protein Chem. 1986;38:181–364. https://doi.org/10.1016/S0065-3233(08)60528-8.
23. Mujahid M, Sarfraz S, Amin S. On the formation of hydroxyapatite nano crystals prepared using cationic surfactant. Mater Res. 2015;18:468–72. https://doi.org/10.1590/1516-1439.298014.
24. Wang P, Wang S, Wang B, Shen L, Song T. Green production of ammonia from nitrogen-rich biomass pyrolysis: evolution of fuel-N under H_2-rich atmosphere. Fuel Process Technol. 2022;227:107126. https://doi.org/10.1016/j.fuproc.2021.107126.25.
25. Forfang K, Zimmermann B, Kosa G, Kohler A, Shapaval V. FTIR spectroscopy for evaluation and monitoring of lipid extraction efficiency for oleaginous fungi. PLoS One. 2017;12: e0170611. https://doi.org/10.1371/JOURNAL.PONE.0170611.
26. Oleszko A, Olsztyńska-Janus S, Walski T, Grzeszczuk-Kuć K, Bujok J, Gałecka K, Czerski A, Witkiewicz W, Komorowska M. Application of FTIR-ATR spectroscopy to determine the extent of lipid peroxidation in plasma during haemodialysis. Biomed Res Int. 2015;2015: 245607. https://doi.org/10.1155/2015/245607.
27. Shapaval V, Brandenburg J, Blomqvist J, Tafintseva V, Passoth V, Sandgren M, Kohler A. Biochemical profiling, prediction of total lipid content and fatty acid profile in oleaginous yeasts by FTIR spectroscopy. Biotechnol Biofuels. 2019;12:1–12. https://doi.org/10.1186/S13068-019-1481-0.
28. Cestelli Guidi M, Mirri C, Fratini E, Licursi V, Negri R, Marcelli A, Amendola R. In vivo skin leptin modulation after 14 MeV neutron irradiation: a molecular and FT-IR spectroscopic study. Anal Bioanal Chem. 2012;404:1317–26. https://doi.org/10.1007/S00216-012-6018-3.
29. Mateus T, Almeida I, Costa A, Viegas D, Magalhães S, Martins F, Herdeiro MT, da Cruz E, Silva OAB, Fraga C, Alves I, Nunes A, Rebelo S. Fourier-transform infrared spectroscopy as a discriminatory tool for myotonic dystrophy type 1 metabolism: a pilot study. Int J Environ Res Public Health. 2021;18(7):3800. https://doi.org/10.3390/IJERPH18073800.
30. Gómez-Ordóñez E, Rupérez P. FTIR-ATR spectroscopy as a tool for polysaccharide identification in edible brown and red seaweeds. Food Hydrocoll. 2011;25:1514–20. https://doi.org/10.1016/J.FOODHYD.2011.02.009.
31. Wiercigroch E, Szafraniec E, Czamara K, Pacia MZ, Majzner K, Kochan K, Kaczor A, Baranska M, Malek K. Raman and infrared spectroscopy of carbohydrates: a review. Spectrochim Acta A Mol Biomol Spectrosc. 2017;185:317–35. https://doi.org/10.1016/J.SAA.2017.05.045.
32. Garcia-Nunez JA, Pelaez-Samaniego MR, Garcia-Perez ME, Fonts I, Abrego J, Westerhof RJM, Garcia-Perez M. Historical developments of pyrolysis reactors: a review. Energy Fuel. 2017;31:5751–75. https://doi.org/10.1021/acs.energyfuels.7b00641.
33. Ben Hassen-Trabelsi A, Kraiem T, Naoui S, Belayouni H. Pyrolysis of waste animal fats in a fixed-bed reactor: production and characterization of bio-oil and bio-char. Waste Manag. 2014;34:210–8. https://doi.org/10.1016/J.WASMAN.2013.09.019.
34. Elliott DC. Relation of reaction time and temperature to chemical composition of pyrolysis oils. ACS Symp Ser. 1988;376:55–65. https://doi.org/10.1021/BK-1988-0376.CH006.
35. Fonts I, Azuara M, Lázaro L, Gea G, Murillo MB. Gas chromatography study of sewage sludge pyrolysis liquids obtained at different operational conditions in a fluidized bed. Ind Eng Chem Res. 2009;48:5907–15. https://doi.org/10.1021/ie900421a.
36. Hansson KM. Principles of biomass pyrolysis with emphasis on the formation of the nitrogen-containing gases HNCO, HCN and NH_3. In: Doktorsavhandlingar vid Chalmers Tekniska Hogskola. Gothenburg: Chalmers University of Technology; 2003. p. 1–80. ISBN: 91-7291-349-5.
37. Benés M, Pozo G, Abián M, Millera A, Bilbao R, Alzueta MU. Experimental study of the pyrolysis of NH_3 under flow reactor conditions. Energy Fuel. 2021;35:7193–200. https://doi.org/10.1021/acs.energyfuels.0c03387.

38. Maher KD, Bressler DC. Pyrolysis of triglyceride materials for the production of renewable fuels and chemicals. Bioresour Technol. 2007;98:2351–68. https://doi.org/10.1016/J.BIORTECH.2006.10.025.
39. Simoneit BRT, Rushdi AI, Abas MRB, Didyk BM. Alkyl Amides and Nitriles as Novel Tracers for Biomass Burning. Environ Sci Technol. 2002;37:16–21. https://doi.org/10.1021/ES020811Y.
40. Moldoveanu SC. Pyrolysis of Organic Molecules with Applications to Health and Environmental Issues. In: Moldoveanu SC, editor. Techniques and Instrumentation in Analytical Chemistry. Winston-Salem: Elsevier; 2009. p. 724. ISBN: 9780444531131.
41. Ren Q, Zhao C. Evolution of fuel-N in gas phase during biomass pyrolysis. Renew Sust Energ Rev. 2015;50:408–18. https://doi.org/10.1016/j.rser.2015.05.043.
42. Munson TO, Vick J. Comparison of human hair by pyrolysis-capillary column gas chromatography and gas chromatography-mass spectrometry. J Anal Appl Pyrolysis. 1985;8:493–501. https://doi.org/10.1016/0165-2370(85)80047-4.
43. Douda J, Basiuk VA. Pyrolysis of amino acids: recovery of starting materials and yields of condensation products. J Anal Appl Pyrolysis. 2000;56:113–21. https://doi.org/10.1016/S0165-2370(00)00078-4.
44. Chiavari G, Galletti GC. Pyrolysis—gas chromatography/mass spectrometry of amino acids. J Anal Appl Pyrolysis. 1992;24:123–37. https://doi.org/10.1016/0165-2370(92)85024-F.
45. Saiz-Jimenez C, Martin-Sanchez PM, Gonzalez-Perez JA, Hermosin B. Analytical pyrolysis of the fungal melanins from ochroconis spp. isolated from Lascaux cave, France. Appl Sci. 2021;11:1198. https://doi.org/10.3390/APP11031198.
46. Shukla B, Koshi M. A novel route for PAH growth in HACA based mechanisms. Combust Flame. 2012;159:3589–96. https://doi.org/10.1016/J.COMBUSTFLAME.2012.08.007.
47. Moreira R, Vaz R, Portugal A, Gil-Lalaguna N, Sánchez JL, Bimbela F. Gasification of Charcoal in Air, Oxygen, and Steam Mixtures over a γ-Al_2O_3 Fluidized Bed. Energy Fuel. 2018;32:406–15. https://doi.org/10.1021/ACS.ENERGYFUELS.7B02257.
48. Gil-Lalaguna N, Sánchez JL, Murillo MB, Rodríguez E, Gea G. Air–steam gasification of sewage sludge in a fluidized bed. Influence of some operating conditions. Chem Eng J. 2014;248:373–82. https://doi.org/10.1016/J.CEJ.2014.03.055.
49. Aljbour SH, Kawamoto K. Bench-scale gasification of cedar wood – Part II: Effect of operational conditions on contaminant release. Chemosphere. 2013;90:1501–7. https://doi.org/10.1016/j.chemosphere.2012.08.030.
50. Körner P. Hydrothermal degradation of amino acids. ChemSusChem. 2021;14:4947. https://doi.org/10.1002/CSSC.202101487.
51. Schäfer S, Bonn B. Hydrolysis of HCN as an important step in nitrogen oxide formation in fluidised combustion. Part 1. Fuel. 2000;79(10):1239–46. https://doi.org/10.1016/S0016-2361(99)00254-9.
52. Our World in Data (2017) How many people does synthetic fertilizer feed?. https://ourworldindata.org/how-many-people-does-synthetic-fertilizer-feed. Accessed 16 Sep 2022.
53. U.S. Geological Survey. Mineral commodity summaries 2023. U.S. Geological Survey; 2023. p. 210. https://doi.org/10.3133/mcs2023.
54. Eurostat (2022) Sewage sludge production and disposal from urban wastewater in EU. https://ec.europa.eu/eurostat/databrowser/view/ten00030/default/table?lang=en. Accessed 5 Jan 2023.
55. Fonts I, Gea G, Azuara M, Ábrego J, Arauzo J. Sewage sludge pyrolysis for liquid production: a review. Renew Sust Energ Rev. 2012;16:2781–805. https://doi.org/10.1016/j.rser.2012.02.070.
56. Fedorowicz EM, Miller SF, Miller BG. Biomass gasification as a means of carcass and specified risk materials disposal and energy production in the beef rendering and meatpacking industries. Energy Fuel. 2007;21:3225–32. https://doi.org/10.1021/ef7003128.
57. Conesa JA, Fullana A, Font R. Thermal decomposition of meat and bone meal. J Anal Appl Pyrolysis. 2003;70:619–30. https://doi.org/10.1016/S0165-2370(03)00044-5.

58. Köninger J, Lugato E, Panagos P, Kochupillai M, Orgiazzi A, Briones MJI. Manure management and soil biodiversity: towards more sustainable food systems in the EU. Agric Syst. 2021;194:103251. https://doi.org/10.1016/J.AGSY.2021.103251.
59. Eurostat (2010) Share of NH_3 emissions from livestock manure and agricultural soils in 2010, EU-27. https://ec.europa.eu/eurostat/statistics-explained/index.php?title=File:Share_of_NH3_emissions_from_livestock_manure_and_agricultural_soils_(kilotonnes),_2010,_EU-27.png#file. Accessed 6 Feb 2023.
60. Eurostat (2021) Annual total production of anhydrous ammonia in 2021 in EU. https://ec.europa.eu/eurostat/databrowser/view/DS-056121__custom_4844612/default/table?lang=en. Accessed 6 Feb 2023.
61. Eurostat (2022) Mineral nitrogen fertiliser consumption by agriculture in EU. https://ec.europa.eu/eurostat/statistics-explained/index.php?title=Agri-environmental_indicator_-_mineral_fertiliser_consumption. Accessed 6 Feb 2023.
62. Paris B, Vandorou F, Tyris D, Balafoutis AT, Vaiopoulos K, Kyriakarakos G, Manolakos D, Papadakis G. Energy use in the eu livestock sector: a review recommending energy efficiency measures and renewable energy sources adoption. Appl Sci. 2022;12(4):2142. https://doi.org/10.3390/app12042142.

Index

Z. Fang et al. (eds.), *Production of N-containing Chemicals and Materials from Biomass*, Biofuels and Biorefineries 12,
https://doi.org/10.1007/978-981-99-4580-1

B

F

I

N

Q

R

T

The manufacturer's authorised representative in the EU is Springer Nature Customer Service Centre GmbH, Europaplatz 3, 69115 Heidelberg, Germany. If you have any concerns regarding our products, please contact ProductSafety@springernature.com

Printed and bound by CPI Group (UK) Ltd, Croydon, CR0 4YY
30/09/2026
02225319-0016